I0502982

FEDERAL EXECUTIVE TEAM

Acting Director, Climate Change Science Program: ....William J. Brennan
Director, Climate Change Science Program Office:...Peter A. Schultz
Lead Agency Principal Representative to CCSP;
NOAA Acting Deputy Under Secretary of Commerce
for Oceans and Atmosphere: ....Mary M. Glackin
Chair, Synthesis and Assessment Product Advisory Group,
Associate Director, EPA National Center for Environmental Assessment: ....Michael W. Slimak
Synthesis and Assessment Product Coordinator,
Climate Change Science Program Office: ....Fabien J.G. Laurier

AGENCY EXECUTIVE COMMITTEE (AEC) AND CARBON CYCLE INTERAGENCY WORKING
GROUP (CCIWG) MEMBERS WHO FACILITATED THE DEVELOPMENT OF THIS REPORT:

Lead Agency Coordinator for SAP 2.2; member AEC.......................................Krisa M. Arzayus, NOAA
Chair, AEC; member CCIWG...Diane E. Wickland, NASA
Member AEC; Co-Chair, CCIWG ....Roger C. Dahlman, DOE
Member AEC; Co-Chair, CCIWG ....Edwin J. Sheffner, NASA
Member AEC and CCIWG...James H. Butler, NOAA
Member AEC and CCIWG...David Hofmann, NOAA
Member AEC and CCIWG ....Patricia Jellison, USGS
Member AEC and CCIWG...Fredric Lipschultz, NSF
Member AEC and CCIWG...Allen M. Solomon, USDA
Member CCIWG...Paula Bontempi, NASA
Member CCIWG...Nancy Cavallaro, USDA
Member CCIWG...William Emanuel, NASA
Member CCIWG...Roger Hanson, CCSPO
Member CCIWG...Carolyn G. Olson, USDA
Member CCIWG...Kathy Tedesco, NOAA
Member CCIWG...Luis Tupas, USDA
Member CCIWG...Charlie Walthall, USDA

PRODUCTION TEAM

Technical Advisor: David J. Dokken
Graphic Design Lead   Sara W. Veasey, NOAA
Graphic Design Co-Lead   Deborah B. Riddle, NOAA
Graphic Design   Jamie P. Payne, ORNL
Graphic Design   Brandon Farrar, STG, Inc.
Graphic Design   Glenn M. Hyatt, NOAA
Graphic Design   Deborah Misch, STG, Inc.
Copy Editor Lead   Anne Markel, STG, Inc.
Copy Editor   Walter Koncinski, ORNL
Copy Editor   Deborah Counce, ORNL
Scientific Editor   Anne Waple, STG, Inc.
Logistical and Data Management Support   Sherry B. Wright, ORNL
Other Technical Support   Mieke van der Wansem, Consensus
                                                      Building Institute, Inc.
                                                      Ona Ferguson, Consensus Building
                                                      Institute, Inc.
                                                      Dan Wei, The Pennsylvania State
                                                      University

The First State of the Carbon Cycle Report (SOCCR)

The North American Carbon Budget and Implications for the Global Carbon Cycle

Synthesis and Assessment Product 2.2
Report by the U.S. Climate Change Science Program
and the Subcommittee on Global Change Research

EDITED BY THE SCIENTIFIC COORDINATION TEAM:

Anthony W. King (Lead), Lisa Dilling (Co-Lead),

Gregory P. Zimmerman (Project Coordinator), David M. Fairman,

Richard A. Houghton, Gregg Marland, Adam Z. Rose, and Thomas J. Wilbanks

November 2007

Members of Congress:

On behalf of the National Science and Technology Council, the U.S. Climate Change Science Program (CCSP) is pleased to transmit to the President and the Congress this report, *North American Carbon Budget and Implications for the Global Carbon Cycle*, as part of a series of Synthesis and Assessment Products produced by the CCSP. This series of 21 reports is aimed at providing current evaluations of climate change science to inform public debate, policy, and operational decisions. These reports are also intended to inform CCSP's consideration of future program priorities.

CCSP's guiding vision is to provide the Nation and the global community with the science-based knowledge to manage the risks and opportunities of change in the climate and related environmental systems. The Synthesis and Assessment Products are important steps toward that vision, helping translate CCSP's extensive observational and research base into informational tools directly addressing key questions that are being asked of the research community.

This product will contribute to and enhance the understanding of the North American carbon budget and the implications for the global carbon cycle. It was developed with broad scientific input and in accordance with the Guidelines for Producing CCSP Synthesis and Assessment Products, Section 515 of the Treasury and General Government Appropriations Act for Fiscal Year 2001 (Public Law 106-554), and the Information Quality Act guidelines issued by the Department of Commerce and the National Oceanic and Atmospheric Administration pursuant to Section 515.

We commend the report's authors for both the thorough nature of their work and their adherence to an inclusive review process.

Samuel W. Bodman
Secretary of Energy
Vice Chair, Committee on Climate
Change Science and Technology
Integration

Carlos M. Gutierrez
Secretary of Commerce
Chair, Committee on Climate Change
Science and Technology Integration

John H. Marburger III
Director, Office of Science and
Technology Policy
Executive Director, Committee
on Climate Change Science and
Technology Integration

TABLE OF CONTENTS

Preface   Anthony W. King, ORNL; Lisa Dilling, Univ. Colo./NCAR; Gregory P. Zimmerman, ORNL; David M. Fairman, Consensus Building Inst., Inc.; Richard A. Houghton, Woods Hole Research Center; Gregg Marland, ORNL and Mid Sweden Univ. (Östersund); Adam Z. Rose, The Pa. State Univ. and Univ. Southern Calif.; Thomas J. Wilbanks, ORNL

Executive Summary Anthony W. King, ORNL; Lisa Dilling, Univ. Colo./NCAR; Gregory P. Zimmerman, ORNL; David M. Fairman, Consensus Building Inst., Inc.; Richard A. Houghton, Woods Hole Research Center; Gregg Marland, ORNL and Mid Sweden Univ. (Östersund); Adam Z. Rose, The Pa. State Univ. and Univ. Southern Calif.; Thomas J. Wilbanks, ORNL

Chapter 1   Anthony W. King, ORNL; Lisa Dilling, Univ. Colo./NCAR; Gregory P. Zimmerman, ORNL; David M. Fairman, Consensus Building Inst., Inc.; Richard A. Houghton, Woods Hole Research Center; Gregg Marland, ORNL and Mid Sweden Univ. (Östersund); Adam Z. Rose, The Pa. State Univ. and Univ. Southern Calif.; Thomas J. Wilbanks, ORNL

Chapter 2   **Coordinating Lead Author:** Christopher B. Field, Carnegie Inst.
Lead Authors: Jorge Sarmiento, Princeton Univ.; Burke Hales, Oreg. State Univ.

Chapter 3   **Coordinating Lead Author:** Stephen Pacala, Princeton Univ.
Lead Authors: Richard A. Birdsey, USDA Forest Service; Scott D. Bridgham, Univ. Oreg.; Richard T. Conant, Colo. State Univ.; Kenneth Davis, The Pa. State Univ.; Burke Hales, Oreg. State Univ.; Richard A. Houghton, Woods Hole Research Center; Jennifer C. Jenkins, Univ. Vt.; Mark Johnston, Saskatchewan Research Council; Gregg Marland, ORNL and Mid Sweden Univ. (Östersund); Keith Paustian, Colo. State Univ;
Contributing Authors: John Caspersen, Univ. Toronto; Robert Socolow, Princeton Univ.; Richard S. J. Tol, Hamburg Univ.

Chapter 4   **Coordinating Lead Author:** Erik Haites, Margaree Consultants, Inc.
Lead Authors: Ken Caldeira, Carnegie Inst.; Patricia Romero Lankao, NCAR and UAM-Xochimilco; Adam Z. Rose, The Pa. State Univ. and Univ. Southern Calif.; Thomas J. Wilbanks, ORNL
Contributing Authors: Skip Laitner, U.S. EPA; Richard Ready, The Pa. State Univ.; Roger Sedjo, Resources for the Future

Chapter 5   **Coordinating Lead Authors:** Lisa Dilling, Univ. Colo./NCAR; Ronald Mitchell, Univ. Oreg.
Lead Author: David M. Fairman, Consensus Building Inst., Inc.
Contributing Authors: Myanna Lahsen, IGBP (Brazil) and Univ. Colo.; Susanne Moser, NCAR; Anthony Patt, Boston Univ./IIASA; Chris Potter, NASA; Charles Rice, Kans. State Univ.; Stacy VanDeveer, Univ. N.H.

Part II Overview   **Coordinating Lead Author:** Gregg Marland, ORNL and Mid Sweden Univ. (Östersund)
Contributing Authors: Robert J. Andres, Univ. N. Dak.; T.J. Blasing, ORNL; Thomas A. Boden, ORNL; Christine T. Broniak, Oreg. State Univ.; Jay S. Gregg, Univ. Md.; London M. Losey, Univ. N. Dak.; Karen Treanton, IEA (Paris)

Chapter 6   **Lead Author:** Thomas J. Wilbanks, ORNL
Contributing Authors: Marilyn Brown, Ga. Inst. Tech.; Ken Caldeira, Carnegie Inst.; William Fulkerson, Univ. Tenn.; Erik Haites, Margaree Consultants, Inc; Stephen Pacala, Princeton Univ.; David M. Fairman, Consensus Building Inst., Inc.

Chapter 7   **Lead Author:** David L. Greene, ORNL

Chapter 8   **Lead Author:** John Nyboer, Simon Fraser Univ.
Contributing Authors: Mark Jaccard, Simon Fraser Univ.; Ernst Worrell, LBNL

Chapter 9   **Lead Author:** James E. McMahon, LBNL
    Contributing Authors: Michael A. McNeil, LBNL; Itha Sánchez Ramos, Instituto de Investigaciones Eléctricas (Mexico)

Part III Overview   **Lead Author:** Richard A. Houghton, Woods Hole Research Center

Chapter 10   **Lead Authors:** Richard T. Conant, Colo. State Univ.; Keith Paustian, Colo. State Univ.
    Contributing Authors: Felipe García-Oliva, UNAM; H. Henry Janzen, Agriculture and Agri-Food Canada; Victor J. Jaramillo, UNAM; Donald E. Johnson, Colo. State Univ. (deceased); Suren N. Kulshreshtha, Univ. Saskatchewan

Chapter 11   **Lead Authors:** Richard A. Birdsey, USDA Forest Service; Jennifer C. Jenkins, Univ. Vt.; Mark Johnston, Saskatchewan Research Council; Elisabeth Huber-Sannwald, Instituto Potosino de Investigación Científica y Tecnológica
    Contributing Authors: Brian Amiro, Univ. Manitoba; Ben de Jong, ECOSUR; Jorge D. Etchevers Barra, Colegio de Postgraduado; Nancy French, Altarum Inst.; Felipe García-Oliva, UNAM; Mark Harmon, Oreg. State Univ.; Linda S. Heath, USDA Forest Service; Victor J. Jaramillo, UNAM; Kurt Johnsen, USDA Forest Service; Beverly E. Law, Oreg. State Univ.; Erika Marín-Spiotta, Univ. Calif. Berkeley; Omar Masera, UNAM; Ronald Neilson, USDA Forest Service; Yude Pan, USDA Forest Service; Kurt S. Pregitzer, Mich. Tech. Univ.

Chapter 12   **Lead Author:** Charles Tarnocai, Agriculture and Agri-Food Canada
    Contributing Authors: Chien-Lu Ping, Univ. Alaska; John Kimble, USDA NRCS (retired)

Chapter 13   **Lead Author:** Scott D. Bridgham, Univ. Oreg.
    Contributing Authors: J. Patrick Megonigal, Smithsonian Environmental Research Center; Jason K. Keller, Smithsonian Environmental Research Center; Norman B. Bliss, SAIC, USGS Center for Earth Resources Observation and Science; Carl Trettin, USDA Forest Service

Chapter 14   **Lead Author:** Diane E. Pataki, Univ. Calif., Irvine
    Contributing Authors: Alan S. Fung, Dalhousie Univ.; David J. Nowak, USDA Forest Service; E. Gregory McPherson, USDA Forest Service; Richard V. Pouyat, USDA Forest Service; Nancy Golubiewski, Landcare Research; Christopher Kennedy, Univ. Toronto; Patricia Romero Lankao, NCAR and UAM-Xochimilco; Ralph Alig, USDA Forest Service

Chapter 15   **Lead Authors:** Francisco P. Chavez, MBARI; Taro Takahashi, Columbia Univ.
    Contributing Authors: Wei-Jun Cai, Univ. Ga.; Gernot Friederich, MBARI; Burke Hales, Oreg. State Univ.; Rik Wanninkhof, NOAA; Richard A. Feely, NOAA

Appendix A   See Chapter 3 Author List
Appendix B   See Chapter 3 Author List
Appendix C   See Chapter 8 Author List
Appendix D   See Chapter 11 Author List
Appendix E   See Chapter 11 Author List
Appendix F   See Chapter 13 Author List
Appendix G   See Chapter 15 Author List

ACKNOWLEDGEMENT

The idea for a State of the Carbon Cycle Report (SOCCR) was first developed by the Carbon Cycle Interagency Working Group (CCIWG) of the U.S. Climate Change Science Program in consultation with its Carbon Cycle Science Steering Group. A subcommittee of the CCIWG, the Agency Executive Committee (AEC) facilitated the development of this report. The AEC included representatives of the lead and supporting agencies assigned to Synthesis and Assessment Product 2.2 (SAP 2.2) and the assigned Lead Agency Coordinator for SAP 2.2. Funding for the  preparation and production of SAP 2.2 was provided by NASA, NOAA, DOE, and NSF. The peer review was led by NOAA, in collaboration with the Agency Executive Committee. Additionally, USDA and USGS contributed by supporting several of their scientists' participation on the Scientific Coordination Team and as chapter authors.

This report has been peer reviewed in draft form by individuals chosen for their diverse perspectives and technical expertise. The expert review and selection of reviewers followed the OMB's Information Quality Bulletin for Peer Review. The purpose of this independent review is to provide candid and critical comments that will assist the Climate Change Science Program in manuscript, and responses to the peer review comments are publicly available at: www.climatescience.gov/Library/sap/sap2-2/default.php.

The AEC and the Scientific Coordination Team thank the following individuals for their peer review of this report: **Dr. Dominique Blain**, Environment Canada; **Dr. James G. Bockheim**, Professor, University of Wisconsin; **Dr. Richard A. Bourbonniere**, Environment Canada; **Dr. Josep Canadell**, CSIRO Division of Marine and Atmospheric Research; **Dr. Robert Dickinson**, Georgia Institute of Technology; **Dr. Phillip M. Dougherty**, MeadWestvaco; **Dr. George C. Eads**, CRI International; **William L. Fang**, Edison Electric Institute; **Dr. Christoph Gerbig**, Max-Planck-Institute for Biogeochemistry; **Dr. Patrick Gonzalez**, The Nature Conservancy; **Dr. Kevin Gurney**, Purdue University; **Dr. Richard A. Jahnke**, Skidaway Institute of Oceanography; **Dr. Dale W. Johnson**, University of Nevada; **John Kinsman**, Edison Electric Institute; **Dr. Christopher J. Kucharik**, University of Wisconsin-Madison; **Dr. Corinne Le Quere**, University of East Anglia; **Dr. Ingeborg Levin**, University of Heidelberg; **Dr. Alan A. Lucier**, National Council for Air and Stream Improvement, Inc.; **Dr. Loren Lutzenhiser**, Portland State University; **Susann Nordrum**, Chevron Energy Technology Company; **Naomi Pena**, Pew Center on Global Climate Change; **Dr. Michael Raupach**, CSIRO Marine and Atmospheric Research; **Dr. Jeffrey Richey**, University of Washington; **Dr. Jonathan Rubin**, University of Maine; **Dr. David Schimel**, National Center for Atmospheric Research; **Dr. Joshua Schimel**, University of California Santa Barbara; **Dr. Lee Schipper**, World Resources Institute; **Jeffrey B. Tschirley**, Food and Agriculture Organization of the United Nations; **Dr. John R. Trabalka**, SENES Oak Ridge Inc., Center for Risk Analysis; **Dr. Susan M. Wachter**, University of Pennsylvania; and **Dr. Douglas W.R. Wallace**, Leibniz-Institut für Meereswissenschaften.

The Scientific Coordination Team would also like to thank all of the many individuals from the public, private, and non-profit sectors who participated in the development of this report by providing feedback, attending workshops, being interviewed about the initial outline, and providing comments during the public comment period. Their time and thoughtful participation was invaluable to the editors and authors in crafting a document that aims to be broadly useful for decision making. The public review comments, draft manuscript, and response to public comments are publicly available at: www.climatescience.gov/Library/sap/sap2-2/default.php.

ABSTRACT

North America is currently a net source of carbon dioxide to the atmosphere, contributing to the global buildup of greenhouse gases in the atmosphere and associated changes in the Earth's climate. In 2003, North America emitted nearly two billion metric tons of carbon to the atmosphere as carbon dioxide. North America's fossil-fuel emissions in 2003 (1856 million metric tons of carbon ± 10% with 95% certainty) were 27% of global emissions. Approximately 85% of those emissions were from the United States, 9% from Canada, and 6% from Mexico. The combustion of fossil fuels for commercial energy (primarily electricity) is the single largest contributor, accounting for approximately 42% of North American fossil emissions in 2003. Transportation is the second largest, accounting for 31% of total emissions.

There are also globally important carbon sinks in North America. In 2003, growing vegetation in North America removed approximately 500 million tons of carbon per year (± 50%) from the atmosphere and stored it as plant material and soil organic matter. This land sink is equivalent to approximately 30% of the fossil-fuel emissions from North America. The imbalance between the fossil-fuel source and the sink on land is a net release to the atmosphere of 1350 million metric tons of carbon per year (± 25%).

Approximately 50% of North America's terrestrial sink is due to the regrowth of forests in the United States on former agricultural land that was last cultivated decades ago, and on timberland recovering from harvest. Other sinks are relatively small and not well quantified with uncertainties of 100% or more. The future of the North American terrestrial sink is also highly uncertain. The contribution of forest regrowth is expected to decline as the maturing forests grow more slowly and take up less carbon dioxide from the atmosphere. But, how regrowing forests and other sinks will respond to changes in climate and carbon dioxide concentration in the atmosphere is highly uncertain.

The large difference between current sources and sinks and the expectation that the difference could become larger if the growth of fossil-fuel emissions continues and land sinks decline suggest that addressing imbalances in the North American carbon budget will likely require actions focused on reducing fossil-fuel emissions. Options to enhance sinks (growing forests or sequestering carbon in agricultural soils) can contribute, but enhancing sinks alone is likely insufficient to deal with either the current or future imbalance. Options to reduce emissions include efficiency improvement, fuel switching, and technologies such as carbon capture and geological storage. Implementing these options will likely require an array of policy instruments at local, regional, national, and international levels, ranging from the encouragement of voluntary actions to economic incentives, tradable emissions permits, and regulations. Meeting the demand for information by decision makers will likely require new modes of research characterized by close collaboration between scientists and carbon management stakeholders.

RECOMMENDED CITATIONS

For the Report as a whole:

CCSP, 2007. *The First State of the Carbon Cycle Report (SOCCR): The North American Carbon Budget and Implications for the Global Carbon Cycle.* A Report by the U.S. Climate Change Science Program and the Subcommittee on Global Change Research [King, A.W., L. Dilling, G.P. Zimmerman, D.M. Fairman, R.A. Houghton, G. Marland, A.Z. Rose, and T.J. Wilbanks (eds.)]. National Oceanic and Atmospheric Administration, National Climatic Data Center, Asheville, NC, USA, 242 pp.

For the Preface:

King, A.W., L. Dilling, G.P. Zimmerman, D.M. Fairman, R.A. Houghton, G. Marland, A.Z. Rose, and T.J. Wilbanks, 2007: Preface. In: *The First State of the Carbon Cycle Report (SOCCR): The North American Carbon Budget and Implications for the Global Carbon Cycle.* A Report by the U.S. Climate Change Science Program and the Subcommittee on Global Change Research [King, A.W., L. Dilling, G.P. Zimmerman, D.M. Fairman, R.A. Houghton, G. Marland, A.Z. Rose, and T.J. Wilbanks (eds.)]. National Oceanic and Atmospheric Administration, National Climatic Data Center, Asheville, NC, USA, pp. XIII-XVIII.

For the Executive Summary:

King, A.W., L. Dilling, G.P. Zimmerman, D.M. Fairman, R.A. Houghton, G. Marland, A.Z. Rose, and T.J. Wilbanks, 2007: Executive Summary. In: *The First State of the Carbon Cycle Report (SOCCR): The North American Carbon Budget and Implications for the Global Carbon Cycle.* A Report by the U.S. Climate Change Science Program and the Subcommittee on Global Change Research [King, A.W., L. Dilling, G.P. Zimmerman, D.M. Fairman, R.A. Houghton, G. Marland, A.Z. Rose, and T.J. Wilbanks (eds.)]. National Oceanic and Atmospheric Administration, National Climatic Data Center, Asheville, NC, USA, pp. 1-14.

For Chapter 1:

King, A.W., L. Dilling, G.P. Zimmerman, D.M. Fairman, R.A. Houghton, G. Marland, A.Z. Rose, and T.J. Wilbanks, 2007: What Is the Carbon Cycle and Why Care? In: *The First State of the Carbon Cycle Report (SOCCR): The North American Carbon Budget and Implications for the Global Carbon Cycle.* A Report by the U.S. Climate Change Science Program and the Subcommittee on Global Change Research [King, A.W., L. Dilling, G.P. Zimmerman, D.M. Fairman, R.A. Houghton, G. Marland, A.Z. Rose, and T.J. Wilbanks (eds.)]. National Oceanic and Atmospheric Administration, National Climatic Data Center, Asheville, NC, USA, pp. 15-20.

For Chapter 2:

Field, C.B., J. Sarmiento, and B. Hales, 2007: The Carbon Cycle of North America in a Global Context. In: *The First State of the Carbon Cycle Report (SOCCR): The North American Carbon Budget and Implications for the Global Carbon Cycle.* A Report by the U.S. Climate Change Science Program and the Subcommittee on Global Change Research [King, A.W., L. Dilling, G.P. Zimmerman, D.M. Fairman, R.A. Houghton, G. Marland, A.Z. Rose, and T.J. Wilbanks (eds.)]. National Oceanic and Atmospheric Administration, National Climatic Data Center, Asheville, NC, USA, pp. 21-28.

For Chapter 3:

Pacala, S., R.A. Birdsey, S.D. Bridgham, R.T. Conant, K. Davis, B. Hales, R.A. Houghton, J.C. Jenkins, M. Johnston, G. Marland, and K. Paustian, 2007: The North American Carbon Budget Past and Present. In: *The First State of the Carbon Cycle Report (SOCCR): The North American Carbon Budget and Implications for the Global Carbon Cycle.* A Report by the U.S. Climate Change Science Program and the Subcommittee on Global Change Research [King, A.W., L. Dilling, G.P. Zimmerman, D.M. Fairman, R.A. Houghton, G. Marland, A.Z. Rose, and T.J. Wilbanks (eds.)]. National Oceanic and Atmospheric Administration, National Climatic Data Center, Asheville, NC, USA, pp. 29-36.

For Chapter 4:

Haites, E., K. Caldeira, P. Romero Lankao, A.Z. Rose, T.J. Wilbanks, S. Laitner, R. Ready, and R. Sedjo, 2007: What Are the Options That Could Significantly Affect the North American Carbon Cycle? In: *The First State of the Carbon Cycle Report (SOCCR): The North American Carbon Budget and Implications for the Global Carbon Cycle.* A Report by the U.S. Climate Change Science Program and the Subcommittee on Global Change Research [King, A.W., L. Dilling, G.P.

Zimmerman, D.M. Fairman, R.A. Houghton, G. Marland, A.Z. Rose, and T.J. Wilbanks (eds.)]. National Oceanic and Atmospheric Administration, National Climatic Data Center, Asheville, NC, USA, pp. 37-48.

For Chapter 5:

Dilling, L., R. Mitchell, D.M. Fairman, M. Lahsen, S. Moser, A. Patt, C. Potter, C. Rice, and S. VanDeveer, 2007: How Can We Improve the Usefulness of Carbon Science for Decision Making? In: *The First State of the Carbon Cycle Report (SOCCR): The North American Carbon Budget and Implications for the Global Carbon Cycle*. A Report by the U.S. Climate Change Science Program and the Subcommittee on Global Change Research [King, A.W., L. Dilling, G.P. Zimmerman, D.M. Fairman, R.A. Houghton, G. Marland, A.Z. Rose, and T.J. Wilbanks (eds.)]. National Oceanic and Atmospheric Administration, National Climatic Data Center, Asheville, NC, USA, pp. 49-56.

For Part II Overview:

Marland, G., R.J. Andres, T.J. Blasing, T.A. Boden, C.T. Broniak, J.S. Gregg, L.M. Losey, and K. Treanton, 2007: Energy, Industry, and Waste Management Activities: An Introduction to CO_2 Emissions From Fossil Fuels. In: *The First State of the Carbon Cycle Report (SOCCR): The North American Carbon Budget and Implications for the Global Carbon Cycle*. A Report by the U.S. Climate Change Science Program and the Subcommittee on Global Change Research [King, A.W., L. Dilling, G.P. Zimmerman, D.M. Fairman, R.A. Houghton, G. Marland, A.Z. Rose, and T.J. Wilbanks (eds.)]. National Oceanic and Atmospheric Administration, National Climatic Data Center, Asheville, NC, USA, pp. 57-64.

For Chapter 6:

Wilbanks, T.J., M. Brown, K. Caldeira, W. Fulkerson, E. Haites, S. Pacala, and D.M. Fairman, 2007: Energy Extraction and Conversion. In: *The First State of the Carbon Cycle Report (SOCCR): The North American Carbon Budget and Implications for the Global Carbon Cycle*. A Report by the U.S. Climate Change Science Program and the Subcommittee on Global Change Research [King, A.W., L. Dilling, G.P. Zimmerman, D.M. Fairman, R.A. Houghton, G. Marland, A.Z. Rose, and T.J. Wilbanks (eds.)]. National Oceanic and Atmospheric Administration, National Climatic Data Center, Asheville, NC, USA, pp. 65-72.

For Chapter 7:

Greene, D.L., 2007: Transportation. In: *The First State of the Carbon Cycle Report (SOCCR): The North American Carbon Budget and Implications for the Global Carbon Cycle*. A Report by the U.S. Climate Change Science Program and the Subcommittee on Global Change Research [King, A.W., L. Dilling, G.P. Zimmerman, D.M. Fairman, R.A. Houghton, G. Marland, A.Z. Rose, and T.J. Wilbanks (eds.)]. National Oceanic and Atmospheric Administration, National Climatic Data Center, Asheville, NC, USA, pp. 73-84.

For Chapter 8:

Nyboer, J., M. Jaccard, and E. Worrell, 2007: Industry and Waste Management. In: *The First State of the Carbon Cycle Report (SOCCR): The North American Carbon Budget and Implications for the Global Carbon Cycle*. A Report by the U.S. Climate Change Science Program and the Subcommittee on Global Change Research [King, A.W., L. Dilling, G.P. Zimmerman, D.M. Fairman, R.A. Houghton, G. Marland, A.Z. Rose, and T.J. Wilbanks (eds.)]. National Oceanic and Atmospheric Administration, National Climatic Data Center, Asheville, NC, USA, pp. 85-94.

For Chapter 9:

McMahon, J.E., M.A. McNeil, and I.S. Ramos, 2007: Buildings. In: *The First State of the Carbon Cycle Report (SOCCR): The North American Carbon Budget and Implications for the Global Carbon Cycle*. A Report by the U.S. Climate Change Science Program and the Subcommittee on Global Change Research [King, A.W., L. Dilling, G.P. Zimmerman, D.M. Fairman, R.A. Houghton, G. Marland, A.Z. Rose, and T.J. Wilbanks (eds.)]. National Oceanic and Atmospheric Administration, National Climatic Data Center, Asheville, NC, USA, pp. 95-102.

For Part III Overview:

Houghton, R.A., 2007: The Carbon Cycle in Land and Water Systems. In: *The First State of the Carbon Cycle Report (SOCCR): The North American Carbon Budget and Implications for the Global Carbon Cycle*. A Report by the U.S. Climate Change Science Program and the Subcommittee on Global Change Research [King, A.W., L. Dilling, G.P. Zimmerman, D.M. Fairman, R.A. Houghton, G. Marland, A.Z. Rose, and T.J. Wilbanks (eds.)]. National Oceanic and Atmospheric Administration, National Climatic Data Center, Asheville, NC, USA, pp. 103-106.

For Chapter 10:

Conant, R.T., K. Paustian, F. Garcia-Oliva, H.H. Janzen, V.J. Jaramillo, D.E. Johnson, and S.N. Kulshreshtha, 2007: Agricultural and Grazing Lands. In: *The First State of the Carbon Cycle Report (SOCCR): The North American Carbon Budget and Implications for the Global Carbon Cycle*. A Report by the U.S. Climate Change Science Program and the Subcommittee on Global Change Research [King, A.W., L. Dilling, G.P. Zimmerman, D.M. Fairman, R.A. Houghton, G. Marland, A.Z. Rose, and T.J. Wilbanks (eds.)]. National Oceanic and Atmospheric Administration, National Climatic Data Center, Asheville, NC, USA, pp. 107-116.

For Chapter 11:

Birdsey, R.A., J.C. Jenkins, M. Johnston, E. Huber-Sannwald, B. Amero, B. de Jong, J.D.E. Barra, N. French, F. Garcia-Oliva, M. Harmon, L.S. Heath, V.J. Jaramillo, K. Johnsen, B.E. Law, E. Marín-Spiotta, O. Masera, R. Neilson, Y. Pan, and K.S. Pregitzer, 2007: North American Forests. In: *The First State of the Carbon Cycle Report (SOCCR): The North American Carbon Budget and Implications for the Global Carbon Cycle*. A Report by the U.S. Climate Change Science Program and the Subcommittee on Global Change Research [King, A.W., L. Dilling, G.P. Zimmerman, D.M. Fairman, R.A. Houghton, G. Marland, A.Z. Rose, and T.J. Wilbanks (eds.)]. National Oceanic and Atmospheric Administration, National Climatic Data Center, Asheville, NC, USA, pp. 117-126.

For Chapter 12:

Tarnocai, C., C.-L. Ping, and J. Kimble, 2007: Carbon Cycles in the Permafrost Region of North America. In: *The First State of the Carbon Cycle Report (SOCCR): The North American Carbon Budget and Implications for the Global Carbon Cycle*. A Report by the U.S. Climate Change Science Program and the Subcommittee on Global Change Research [King, A.W., L. Dilling, G.P. Zimmerman, D.M. Fairman, R.A. Houghton, G. Marland, A.Z. Rose, and T.J. Wilbanks (eds.)]. National Oceanic and Atmospheric Administration, National Climatic Data Center, Asheville, NC, USA, pp. 127-138.

For Chapter 13:

Bridgham, S.D., J.P. Megonigal, J.K. Keller, N.B. Bliss, and C. Trettin, 2007: Wetlands. In: *The First State of the Carbon Cycle Report (SOCCR): The North American Carbon Budget and Implications for the Global Carbon Cycle*. A Report by the U.S. Climate Change Science Program and the Subcommittee on Global Change Research [King, A.W., L. Dilling, G.P. Zimmerman, D.M. Fairman, R.A. Houghton, G. Marland, A.Z. Rose, and T.J. Wilbanks (eds.)]. National Oceanic and Atmospheric Administration, National Climatic Data Center, Asheville, NC, USA, pp. 139-148.

For Chapter 14:

Pataki, D.E., A.S. Fung, D.J. Nowak, E.G. McPherson, R.V. Pouyat, N. Golubiewski, C. Kennedy, P. Romero Lankao, and R. Alig, 2007: Human Settlements and the North American Carbon Cycle. In: *The First State of the Carbon Cycle Report (SOCCR): The North American Carbon Budget and Implications for the Global Carbon Cycle*. A Report by the U.S. Climate Change Science Program and the Subcommittee on Global Change Research [King, A.W., L. Dilling, G.P. Zimmerman, D.M. Fairman, R.A. Houghton, G. Marland, A.Z. Rose, and T.J. Wilbanks (eds.)]. National Oceanic and Atmospheric Administration, National Climatic Data Center, Asheville, NC, USA, pp. 149-156.

For Chapter 15:

Chavez, F.P., T. Takahashi, W.-J. Cai, G. Friederich, B. Hales, R. Wanninkhof, and R.A. Feely, 2007: Coastal Oceans. In: *The First State of the Carbon Cycle Report (SOCCR): The North American Carbon Budget and Implications for the Global Carbon Cycle*. A Report by the U.S. Climate Change Science Program and the Subcommittee on Global Change Research [King, A.W., L. Dilling, G.P. Zimmerman, D.M. Fairman, R.A. Houghton, G. Marland, A.Z. Rose, and T.J. Wilbanks (eds.)]. National Oceanic and Atmospheric Administration, National Climatic Data Center, Asheville, NC, USA, pp. 157-166.

For Appendix A

Pacala, S., R.A. Birdsey, S.D. Bridgham, R.T. Conant, K. Davis, B. Hales, R.A. Houghton, J.C. Jenkins, M. Johnston, G. Marland, and K. Paustian, 2007: Historical Overview of the Development of United States, Canadian, and Mexican Ecosystem Sources and Sinks for Atmosperic Carbon. In: *The First State of the Carbon Cycle Report (SOCCR): The North American Carbon Budget and Implications for the Global Carbon Cycle*. A Report by the U.S. Climate Change Science Program and the Subcommittee on Global Change Research [King, A.W., L. Dilling, G.P. Zimmerman, D.M. Fairman, R.A. Houghton, G. Marland, A.Z. Rose, and T.J. Wilbanks (eds.)]. National Oceanic and Atmospheric Administration, National Climatic Data Center, Asheville, NC, USA, pp. 167-168.

For Appendix B

Pacala, S., R.A. Birdsey, S.D. Bridgham, R.T. Conant, K. Davis, B. Hales, R.A. Houghton, J.C. Jenkins, M. Johnston, G. Marland, and K. Paustian, 2007: Eddy-Covariance Measurements Now Confirm Estimates of Carbon Sinks From Forest Inventories. In: *The First State of the Carbon Cycle Report (SOCCR): The North American Carbon Budget and Implications for the Global Carbon Cycle*. A Report by the U.S. Climate Change Science Program and the Subcommittee on Global Change Research [King, A.W., L. Dilling, G.P. Zimmerman, D.M. Fairman, R.A. Houghton, G. Marland, A.Z. Rose, and T.J. Wilbanks (eds.)]. National Oceanic and Atmospheric Administration, National Climatic Data Center, Asheville, NC, USA, pp. 169-170.

For Appendix C

Nyboer, J., M. Jaccard, and E. Worrell, 2007: Industry and Waste Management - Supplemental Material. In: *The First State of the Carbon Cycle Report (SOCCR): The North American Carbon Budget and Implications for the Global Carbon Cycle*. A Report by the U.S. Climate Change Science Program and the Subcommittee on Global Change Research [King, A.W., L. Dilling, G.P. Zimmerman, D.M. Fairman, R.A. Houghton, G. Marland, A.Z. Rose, and T.J. Wilbanks (eds.)]. National Oceanic and Atmospheric Administration, National Climatic Data Center, Asheville, NC, USA, pp. 171-172.

For Appendix D

Birdsey, R.A., J.C. Jenkins, M. Johnston, E. Huber-Sannwald, B. Amero, B. de Jong, J.D.E. Barra, N. French, F. Garcia-Oliva, M. Harmon, L.S. Heath, V.J. Jaramillo, K. Johnsen, B.E. Law, E. Marín-Spiotta, O. Masera, R. Neilson, Y. Pan, and K.S. Pregitzer, 2007: Ecosystem Carbon Fluxes. In: *The First State of the Carbon Cycle Report (SOCCR): The North American Carbon Budget and Implications for the Global Carbon Cycle*. A Report by the U.S. Climate Change Science Program and the Subcommittee on Global Change Research [King, A.W., L. Dilling, G.P. Zimmerman, D.M. Fairman, R.A. Houghton, G. Marland, A.Z. Rose, and T.J. Wilbanks (eds.)]. National Oceanic and Atmospheric Administration, National Climatic Data Center, Asheville, NC, USA, pp. 173-174.

For Appendix E

Birdsey, R.A., J.C. Jenkins, M. Johnston, E. Huber-Sannwald, B. Amero, B. de Jong, J.D.E. Barra, N. French, F. Garcia-Oliva, M. Harmon, L.S. Heath, V.J. Jaramillo, K. Johnsen, B.E. Law, E. Marín-Spiotta, O. Masera, R. Neilson, Y. Pan, and K.S. Pregitzer, 2007: Principles of Forest Management for Enhancing Carbon Sequestration. In: *The First State of the Carbon Cycle Report (SOCCR): The North American Carbon Budget and Implications for the Global Carbon Cycle*. A Report by the U.S. Climate Change Science Program and the Subcommittee on Global Change Research [King, A.W., L. Dilling, G.P. Zimmerman, D.M. Fairman, R.A. Houghton, G. Marland, A.Z. Rose, and T.J. Wilbanks (eds.)]. National Oceanic and Atmospheric Administration, National Climatic Data Center, Asheville, NC, USA, pp. 175-176.

For Appendix F

Bridgham, S.D., J.P. Megonigal, J.K. Keller, N.B. Bliss, and C. Trettin, 2007: Wetlands - Supplemental Materials. In: *The First State of the Carbon Cycle Report (SOCCR): The North American Carbon Budget and Implications for the Global Carbon Cycle*. A Report by the U.S. Climate Change Science Program and the Subcommittee on Global Change Research [King, A.W., L. Dilling, G.P. Zimmerman, D.M. Fairman, R.A. Houghton, G. Marland, A.Z. Rose, and T.J. Wilbanks (eds.)]. National Oceanic and Atmospheric Administration, National Climatic Data Center, Asheville, NC, USA, pp. 177-192.

For Appendix G

Chavez, F.P., T. Takahashi, W.-J. Cai, G. Friederich, B. Hales, R. Wanninkhof, and R.A. Feely, 2007: New pCO_2 Database for Coastal Ocean Waters Surrounding North America. In: *The First State of the Carbon Cycle Report (SOCCR): The North American Carbon Budget and Implications for the Global Carbon Cycle*. A Report by the U.S. Climate Change Science Program and the Subcommittee on Global Change Research [King, A.W., L. Dilling, G.P. Zimmerman, D.M. Fairman, R.A. Houghton, G. Marland, A.Z. Rose, and T.J. Wilbanks (eds.)]. National Oceanic and Atmospheric Administration, National Climatic Data Center, Asheville, NC, USA, pp. 193-194.

PREFACE

Report Motivation and Guidance for Using This Synthesis/Assessment Report

Authors: Anthony W. King, ORNL; Lisa Dilling, Univ. Colo./NCAR; Gregory P. Zimmerman, ORNL; David M. Fairman, Consensus Building Inst., Inc.; Richard A. Houghton, Woods Hole Research Center; Gregg Marland, ORNL and Mid Sweden Univ. (Östersund); Adam Z. Rose, The Pa. State Univ. and Univ. Southern Calif.; Thomas J. Wilbanks, ORNL

A primary objective of the U.S. Climate Change Science Program (CCSP) is to provide the best possible scientific information to support public discussion, as well as government and private sector decision making, on key climate-related issues. To help meet this objective, the CCSP has identified an initial set of 21 Synthesis and Assessment Products (SAPs) that address its highest priority research, observation, and decision support needs.

This report—CCSP SAP 2.2—addresses Goal 2 of the CCSP Strategic Plan: Improve quantification of the forces bringing about changes in the Earth's climate and related systems. The report provides a synthesis and integration of the current knowledge of the North American carbon budget and its context within the global carbon cycle. In a format useful to decision makers, it (1) summarizes our knowledge of carbon cycle properties and changes relevant to the contributions of and impacts[1] upon North America and the rest of the world, and (2) provides scientific information for decision support focused on key issues for carbon management and policy. Consequently, this report is aimed at both the decision-maker audience and to the expert scientific and stakeholder communities.

Background

This report addresses carbon emissions; natural reservoirs and sequestration (absorption and storage); rates of transfer; the consequences of changes in carbon cycling on land and the ocean; effects of

purposeful carbon management; effects of agriculture, forestry, and natural resource management on the carbon cycle; and the socio-economic drivers and consequences of changes in the carbon cycle. It covers North America's land, atmosphere, inland waters, and coastal oceans, where "North America" is defined as Canada, the United States of America (excluding Hawaii), and Mexico. Coastal oceans are defined as coastal waters less than 100 km from the North American coastline, where surface water concentrations of carbon dioxide (CO_2) are influenced by coastal processes. The report focuses on the current carbon budget for North America defined by the availability of most recent published data circa 2003. Historical trends and processes from 1750 (beginning of the Industrial Revolution) and 1850 (expanding use of fossil fuels in the Industrial Revolution) to present are included where appropriate and needed to explain the current carbon budget. Near term (to 2020), mid term (2020-2040), and long-term (2040-2100) projections of current trends are considered where available (published) and appropriate. The report includes an analysis of North America's carbon budget that documents the state of knowledge and quantifies the best estimates (*i.e.*, consensus, accepted, official) and uncertainties. This analysis provides a baseline against which future results from the North American Carbon Program (NACP) www.nacarbon.org/nacp/about.html can be compared.

The focus of this report follows the *Prospectus* developed by the Climate Change Science Program and posted on its website at www.climatescience.gov. The audience for SAP 2.2 includes scientists, decision makers in the public sector (*e.g.*, national, provincial, state, and local governments), the private sector (carbon-related industry, including energy, transportation, agriculture, and forestry sectors; and

---

[1] The term "impacts" as used in this report refers to specific effects of changes in the carbon cycle, such as acidification of the ocean, the effect of increased CO_2 on plant growth and survival, and changes in concentrations of carbon in the atmosphere. The term is not used as a shortened version of "climate impacts," as was adopted for the *Strategic Plan for the U.S. Climate Change Science Program*.

climate policy and carbon management interest groups), the international community, and the general public. This broad audience is indicative of the diversity of stakeholder groups interested in knowledge of carbon cycling in North America and of how such knowledge might be used to influence or make decisions. Not all the scientific information needs of this broad audience can be met in this first SAP, but the scientific information provided herein is designed to be understandable by all. The primary users of SAP 2.2 are likely to be officials involved in formulating climate policy, individuals responsible for managing carbon in the environment, and scientists involved in assessing the state of knowledge concerning carbon cycling and the carbon budget of North America.

It is envisioned that SAP 2.2 will be used (1) as a state-of-the-art assessment of our knowledge of carbon cycle properties and changes relevant to the contributions of and carbon-specific impacts upon North America in the context of the rest of the world; (2) as a contribution to relevant national and international assessments; (3) to provide the scientific basis for decision support that will guide management and policy decisions that affect carbon fluxes, emissions, and sequestration; (4) as a means of informing policymakers and the public concerning the general state of our knowledge of the global carbon cycle with respect to the contributions of and impacts on North America; and (5) to inform future efforts for carbon science to support decision making. For example, well-quantified regional and continental-scale carbon source and sink estimates, error terms, and associated uncertainties will be available for use in climate policy formulation and by resource managers interested in quantifying carbon emissions reductions or carbon uptake and storage. This report is also intended for senior managers and members of the general public who desire to improve their overall understanding of North America's role in the global carbon budget and to gain perspective on what is and is not known.

The questions addressed by this report include:
- What is the carbon cycle and why should we care?
- How do North American carbon sources and sinks relate to the global carbon cycle?
- What are the primary carbon sources and sinks in North America, and how are they changing and why?
- What are the direct, non-climatic effects of increasing atmospheric CO_2 or other changes in the carbon cycle on the land and oceans of North America?
- What options can be implemented in North America that could significantly affect the North American and global carbon cycles (e.g., North American sinks and global atmospheric concentrations of CO_2)?
- How can we improve the usefulness of carbon science

for decision making?
- What additional knowledge is needed for effective carbon management?

Suggestions for Reading, Using, and Navigating This Report

The above questions provide the basis for the five chapters in Part I of this SAP. These five chapters focus on integrating and synthesizing information presented in Parts II and III of this report in combination with additional peer-reviewed published information from outside the report. The report's assessment of the North American carbon budget is, for example, presented in Chapter 3. The *Executive Summary* further distills and synthesizes information from across the report to address the questions above, which structure the report.

Part II of the report focuses on the energy- and industrial-related components of the North American carbon cycle and discusses the carbon emissions and other aspects of (a) energy extraction and conversion, (b) the transportation sector, (c) industry and waste management, and (d) the buildings sector. Part III provides information about land and water systems, including human settlements, and their roles in the carbon cycle. Both Parts II and III are introduced by an *Overview* of the subject matter and information in the chapters of the respective sections.

A reader interested in cross-sector integration and synthesis at the national and continental scale might, therefore, first read the *Executive Summary* followed by reading Chapters 1 through 5, referring to Chapters 6-15 and the *Overviews* of Parts II and III for more expanded discussion of information specific to individual sectors or ecosystems. Conversely, if a reader is more interested in sectoral-specific information, he or she might want to peruse the appropriate chapters in Part II as a first step. Chapter 1 is intended as a background "primer" for those less familiar with concepts of carbon cycling and its importance in considerations of climate change. Those familiar with those issues might choose to skip that chapter or use it for a quick review.

Definitions and Conventions

Throughout this report, quantification of carbon sources and sinks follows the following convention. *Sources*, such as fossil-fuel emissions, that add carbon to the atmosphere are indicated with positive numbers. *Sinks*, such as forest growth, that remove carbon from the atmosphere are indicated with negative numbers. The difference between a source and a sink is *net* exchange with the atmosphere, and may be either positive or negative (*i.e.*, a source or sink), depending on which is larger. Sources and sinks, unless otherwise indicated, are given in units of million metric tons of carbon per year (Mt C per year).

Additional definitions of terms, acronyms, and units are provided in the *Glossary and Acronyms* section of this report.

The Treatment of Uncertainty in This Report

Communicating confidence in the findings of scientific syntheses and assessments, including the characterization of certainty in numbers reported by those assessments, is an important part of making scientific assessments useful to decision makers and other stakeholders. That communication is sometimes challenged by nuanced differences among participants in their understanding of terms such as uncertainty or confidence. The challenge is heightened when attempting to integrate and synthesize analyses from a broad spectrum of sectors and disciplines, each with its own methods, conventions, and sometimes language for addressing and communicating "uncertainty."

Variability in physical processes (*e.g.*, carbon sequestration by woody vegetation) in time and space, measurement error, and sampling error (itself intimately linked to temporal and spatial variability) all contribute to uncertainty in quantifying elements of the North American carbon budget. Uncertainties may be compounded by the use of "expansion factors"—the analytical models used to interpolate and extrapolate local measurements to represent larger areas. Methods for translating from the readily measurable to quantities that are difficult or costly to measure (such as the use of allometric relationships to estimate whole tree biomass from measurements of stem diameter and tree height) can also compound uncertainty. The magnitudes of these and other sources of uncertainty vary across sectors and elements of the carbon cycle. Consequently, so do the emphases and methods for dealing with uncertainty vary across the different disciplines that study these elements. There is no single applicable quantitative method for integrating these variable sources and methods. There exist, of course, statistical techniques, such as the meta-analysis widely used in epidemiology and biomedical clinical trials to combine results from previous separate but related studies. But only

rarely, even within a sector or discipline, are the statistical pre-requisites of meta-analysis met by the diverse studies of carbon cycle elements.

To address this challenge, and to provide for synthesis across and comparability among carbon cycle elements, a convention has been adopted for characterizing uncertainty in the report's synthetic findings and results (for example, in the synthesized carbon budget for North America of Chapter 3 and in the Executive Summary). Uncertainty is characterized using asterisks to represent the five categories described in the accompanying text box.

Unless otherwise noted, values presented as "y ± x%" should be interpreted to mean that the authors are 95% certain the actual value is between y − x% and y + x%. Where appropriate, the absolute range is sometimes reported rather than the relative range: y ± z, where z = y × x% ÷ 100. The system of asterisks is used as shorthand for the categories in tables and text.

These are informed categorizations. They reflect expert judgment, using all known published descriptions of uncertainty surrounding the "best available" or "most likely" estimate. There is always a chance, something like 1 in 20, that the actual value lies outside the range surrounding the best/most likely estimate, but it is much more likely that the actual value is in that range. Some things are known well, and one can be highly (95%) certain that the actual value is within ± 10% of the estimate. Some things are known less well, perhaps there are fewer studies, a broader, more variable range of estimates from different studies, or more variability or measurement and sampling error reported by individual studies, and one can only be highly certain that the actual value is captured by the estimate by increasing the relative range around the estimate to say ± 25 or 50%. With very few and variable or conflicting studies, there is very little certainty and confidence in the estimate, the relative range of likely values is large and uncertainty is characterized as being greater than 100%.

The 95% boundary was chosen to communicate the extremely high certainty or confidence that the actual value was in the reported range, and the low likelihood that it was outside that range. However, this characterization is not a statistical property of the estimate, and should not be confused with 95% confidence intervals based

---

CCSP SAP 2.2 Uncertainty Conventions

***** = 95% certain that the actual value is within 10% of the estimate reported,
**** = 95% certain that the estimate is within 25%,
*** = 95% certain that the estimate is within 50%,
** = 95% certain that the estimate is within 100%, and
* = uncertainty greater than 100%.
† = The magnitude and/or range of uncertainty for the given numerical value(s) is not provided in the references cited.

on parametric statistical estimation of the standard error of the mean.

The authors have used this system for categorizing uncertainty only where they have synthesized diverse published information and compared across this diversity. When citing an existing published estimate, authors were encouraged to include the reported characterizations of uncertainty, whether quantitative or qualitative. Chapters in this report, especially those of Parts II and III, therefore, include several different ways of characterizing uncertainty: simple ranges, standard deviations, standard error, and confidence intervals.

In all cases, the form and character of the uncertainty being expressed should be clear either from the context of the text or as described in a footnote. There are circumstances in which no characterization of the uncertainty of data or information is shown, such as when a number is taken from a published source that itself did not include a characterization of uncertainty. In these cases, the authors have not provided a characterization of uncertainty, and the reader should assume that no characterization of uncertainty was available to the authors.

The Treatment of Greenhouse Gases in This Report

Atmospheric CO_2 is recognized as the largest single human-mediated agent of climate change. While CO_2's importance as a greenhouse gas is a primary motivator for understanding how carbon cycles through the atmosphere and other parts of the Earth system, this report is about the carbon cycle and carbon budgets, and not about greenhouse gases. Accordingly, this report focuses on the North American carbon budget as it influences, and is influenced by, concentrations of atmospheric CO_2. Methane (CH_4) is also an important greenhouse gas and a potential contributor to human-caused climate change. However, CH_4 and other non-CO_2 carbon gases are not typically included in global carbon budgets because their sources and sinks are not well understood. For this reason, and to manage scope and focus, we too follow that convention, and this report is limited primarily to carbon and CO_2. There is significant discussion of CH_4 in individual chapters where appropriate (*e.g.*, Chapter 8 on industry and waste management, Chapter 10 on agricultural and grazing lands, and Chapter 13 on wetlands), but the report's coverage of CH_4 is not comprehensive. We made no effort towards an across-sector, continental-scale synthesis and assessment of CH_4 as part of the North American carbon budget. Similarly, we provide no comprehensive treatment of black carbon, isoprene, or other volatile organic carbon compounds that represent a small fraction of global or continental carbon budgets. We make no consideration of nitrous oxide (N_2O) or other non-carbon greenhouse gases.

The Treatment of Emissions Data Sources in This Report

Part II of this report (Chapters 6 through 9) discusses patterns and trends of CO_2 emissions by sector (the transportation sector, for example). Estimating emissions by sector brings special challenges in defining sectors and assembling the requisite data. Readers will find that there is consistency and coherence within each of the report's chapters but will encounter differences across chapters. Different experts and different disciplines with different perspectives on the carbon cycle use different sector boundaries, different data sources, different conversion factors, *etc.* Different analysts and literature sources will use data for different base years and may treat, for example, electricity and biomass fuels differently. The national reports of the United States, Canada, and Mexico do not cover the same time periods nor do they present data in the same way. In this report, the chapter authors have chosen the system boundaries and data they find most useful for their sectors and perspectives, even though it makes for some differences across chapters. However, the database of the International Energy Agency (IEA; www.iea.org) allows for summary of CO_2 emissions for the three countries defined as North America in this report according to sectors that closely correspond to the sectoral division of Chapters 6 through 9 (See the Part II Overview). Similarly, the database of the Energy Information Administration (EIA; www.eia.doe.gov) provides total global and North American fossil-fuel emissions (by country) as a reference against which the relative size and contribution of sector emissions and carbon sinks can be compared (Chapters 2 and 3).

The Synthesis and Assessment Product Team

A full list of the Authorship Team (in addition to the list of lead authors provided at the beginning of each chapter) is provided on page iv of this report. The Scientific Coordination Team, as described below, reviewed the scientific/technical input and managed the formatting, editing, assembly, and preparation of the report.

The SAP 2.2 *Prospectus* identified a Scientific Coordination Team responsible for organizing and outlining this SAP and for its final content and submission. The Coordination Team was also responsible for identifying chapter authors, coordinating all of the inputs to this report, and leading the overall synthesis and integration of this report. The Coordination Team provided oversight and editorial review of individual chapters and, with the assistance of the respective chapter authors, prepared the *Part II Overview* and *Part III Overview*, as well as the *Abstract* and the *Executive Summary* for this report. The "Key Findings" accompanying Chapters 2–15 were developed in collaboration between the Scientific Coordination Team and the respective chapter authors. These findings were compiled and edited for length, style, and consistency by the Coordination Team as part of the synthesis and integration across the report. Therefore, any error or misrepresentation in the "Key Findings" is the responsibility of the Scientific Coordination Team, and not of the chapter authors.

The members of the Coordination Team and their roles are:
- Dr. Anthony W. King, Overall Lead
- Dr. Lisa Dilling, Co-Lead, Stakeholder Interaction Lead
- Dr. David M. Fairman, Stakeholder Interaction
- Dr. Richard A. Houghton, Scientific Content (Land Use)
- Dr. Gregg Marland, Scientific Content (Emissions)
- Dr. Adam Z. Rose, Scientific Content (Economics)
- Dr. Thomas J. Wilbanks, Scientific Content (Human Dimensions)

The activities of the Scientific Coordination Team were managed by:
- Mr. Gregory P. Zimmerman, Project Coordinator

The Scientific Coordination Team recruited one or more scientific experts to be responsible for writing each individual chapter of SAP 2.2. This person (or persons) was designated as either the Coordinating Lead author or the Lead Chapter author. For the individual chapters in Part I, the respective Coordinating Lead author had responsibility for orchestrating the preparation of the chapter. For each chapter in Parts II and III, the respective Lead Author had that responsibility. These Coordinating Lead authors and Lead Chapter authors are recognized leaders in their fields, drawn from the wide and diverse scientific community of North America and the world, as well as other qualified stakeholder groups. Their qualifications include the quality and relevance of current publications in the peer-reviewed literature pertaining to

their chapter topics, past or present positions of leadership in the topic fields, and other documented experience and knowledge of high relevance. Each Coordinating Lead author and Lead Chapter author was responsible for the review and synthesis of current knowledge and production of text for his/her respective chapter. The Coordinating Lead authors and Lead Chapter Authors were responsible for recruiting well-qualified contributing authors in their areas of expertise and responsibility. The Coordinating Lead authors and Lead Chapter Authors, along with the Scientific Coordination Team, were also responsible for ensuring that scientific expert, stakeholder, and public review comments on their chapters are reflected in this report.

Stakeholder Involvement Process

Research suggests that in order for an assessment to be useful for decision making, it must be not only scientifically accurate and rigorous, but also relevant to the near-term concerns of decision makers and their constituencies ("stakeholders"). It must also be created in a way that stakeholders perceive as fair and unbiased; this last point is especially important when the assessment deals with a controversial public issue.

To make the SAP 2.2 as useful for decision making as possible, we dedicated significant effort and resources to developing a stakeholder engagement process. Because the North American carbon cycle involves a vast array of interactions between human activities and the environment, and because changes in the carbon cycle may have far-reaching economic, social, and political implications, the stakeholders for this report arguably include the entire population of the continent.

To focus the stakeholder engagement process, the Coordination Team sought to identify and involve representatives of government (national and subnational) with current or potential responsibility for carbon management, businesses with a substantial interest in carbon management, and environmental groups active in carbon cycle issues, along with academic and consulting experts in carbon cycle issues. We were partially successful in our efforts to involve a broad and representative group of stakeholders. Our extensive outreach efforts generated public comments from only a limited number of individuals, and attendance at our individual workshops was not equally balanced across all stakeholder groups. We did, however, succeed in generating participation and public comment from all the major stakeholder groups. What the process lacked in numbers, it arguably made up for in the quality of interaction and feedback received.

The stakeholder engagement process involved a combination of interviews, workshops, and online communication tools such as a website and email. Stakeholders' interests were considered and represented at all stages. However, the responsibility for content of the report rested with the authors themselves.

We began involving stakeholders early in the process, at a point where they might have significant opportunity to provide input into the shape and overall structure of the report. Our first activity was to conduct a "rapid stakeholder assessment" which consisted of approximately 30 phone interviews with stakeholders from government, academia, business, and environmental groups. During this assessment, we asked stakeholders about their impressions of our tentative outline for the report, and for suggestions on chapter authors.

We then conducted the first of our stakeholder workshops, also focusing on the draft outline and asking how we might make the report as useful as possible to a wide range of stakeholders. At this workshop, we significantly changed the structure of the report based on valuable input from the group assembled. After the workshop, we then posted our draft outline online, and provided an open comment period for anyone to send in comments, which were also considered in constructing the next draft and formal SAP 2.2 *Prospectus* outline. We also created an online email listserv early in the process, which now has over 350 members subscribed. Our second workshop occurred mid-way through the process, when the authors had created an early draft of their chapters. At the workshop, stakeholders and authors met together, so that input and feedback could be direct and interactive. Through the Climate Change Program Office, we then received feedback on a peer-reviewed draft through a formal public comment process. Finally, we conducted a third stakeholder workshop during the public comment process, in order to have one more opportunity for direct dialogue on the document. We also maintained a public website from the start of the process with our names and contact information, and communicated via email and phone with stakeholders. The website can be accessed at http://cdiac.ornl.gov/SOCCR.

EXECUTIVE SUMMARY

Lead Author: Scientific Coordination Team
Scientific Coordination Team Members: Anthony W. King (Lead), ORNL; Lisa Dilling (Co-Lead), Univ. Colo./NCAR; Gregory P. Zimmerman (Project Coordinator), ORNL; David M. Fairman, Consensus Building Inst., Inc.; Richard A. Houghton, Woods Hole Research Center; Gregg Marland, ORNL; Adam Z. Rose, The Pa. State Univ. and Univ. Southern Calif.; Thomas J. Wilbanks, ORNL

Abstract

North America is currently a net source of CO_2 to the atmosphere, contributing to the global buildup of greenhouse gases in the atmosphere and associated changes in the Earth's climate. In 2003, North America emitted nearly two billion metric tons of carbon to the atmosphere as CO_2. North America's fossil-fuel emissions in 2003 (1856 million metric tons of carbon ±10% with 95% certainty) were 27% of global emissions. Approximately 85% of those emissions were from the United States, 9% from Canada, and 6% from Mexico. The combustion of fossil fuels for commercial energy (primarily electricity) is the single largest contributor, accounting for approximately 42% of North American fossil emissions in 2003. Transportation is the second largest, accounting for 31% of total emissions.

There are also globally important carbon sinks in North America. In 2003, growing vegetation in North America removed approximately 500 million tons of carbon per year (±50%) from the atmosphere and stored it as plant material and soil organic matter. This land sink is equivalent to approximately 30% of the fossil-fuel emissions from North America. The imbalance between the fossil-fuel source and the sink on land is a net release to the atmosphere of 1350 million metric tons of carbon per year (± 25%).

Approximately 50% of North America's terrestrial sink is due to the regrowth of forests in the United States on former agricultural land that was last cultivated decades ago, and on timberland recovering from harvest. Other sinks are relatively small and not well quantified with uncertainties of 100% or more. The future of the North American terrestrial sink is also highly uncertain. The contribution of forest regrowth is expected to decline as the maturing forests grow more slowly and take up less CO_2 from the atmosphere. But this expectation is surrounded by uncertainty because how regrowing forests and other sinks will respond to changes in climate and CO_2 concentration in the atmosphere is highly uncertain.

The large difference between current sources and sinks and the expectation that the difference could become larger if the growth of fossil-fuel emissions continues and land sinks decline suggest that addressing imbalances in the North American carbon budget will likely require actions focused on reducing fossil-fuel emissions. Options to enhance sinks (growing forests or sequestering carbon in agricultural soils) can contribute, but enhancing sinks alone is likely insufficient to deal with either the current or future imbalance. Options to reduce emissions include efficiency improvement, fuel switching, and technologies such as carbon capture and geological storage. Implementing these options will likely require an array of policy instruments at local, regional, national, and international levels, ranging from the encouragement of voluntary actions to economic incentives, tradable emissions permits, and regulations. Meeting the demand for information by decision makers will likely require new modes of research characterized by close collaboration between scientists and carbon management stakeholders.

ES.1 SYNTHESIS AND ASSESSMENT OF THE NORTH AMERICAN CARBON BUDGET

Understanding the North American carbon budget, both sources and sinks, is critical to the United States Climate Change Science Program goal of providing the best possible scientific information to support public discussion, as well as government and private sector decision making, on key climate-related issues. In response, this report provides a synthesis, integration, and assessment of the current knowledge of the North American carbon budget and its context within the global carbon cycle. The report focuses on the carbon cycle as it influences the concentration of carbon dioxide (CO_2) in the atmosphere. Methane (CH_4), nitrous oxide, and other greenhouse gases are also relevant to climate issues, but their consideration is beyond the scope and mandate of this report.

> The rate of CO_2 released to the atmosphere is far larger than can be balanced by the biological and geological processes that naturally remove CO_2 from the atmosphere and store it in terrestrial and marine environments.

The report is organized as a response to questions relevant to carbon management and to a broad range of stakeholders charged with understanding and managing energy and land use. The questions were identified through early and continuing dialogue with these stakeholders, including scientists; decision makers in the public and private sectors, including national and sub-national government; carbon-related industries, such as energy, transportation, agriculture, and forestry; and climate policy and carbon management interest groups.

The questions and the answers provided by this report are summarized below. The reader is referred to the indicated chapters for further, more detailed, discussion. Unless otherwise referenced, all values, statements of findings and conclusions are taken from the chapters of this report where the attribution and citation of the primary sources can be found.

> Trends in fossil-fuel use and tropical deforestation are accelerating.

ES.2 WHAT IS THE CARBON CYCLE AND WHY SHOULD WE CARE?

The carbon cycle, described in Chapters 1 and 2, is the combination of many different physical, chemical, and biological processes that transfer carbon between the major storage pools (known as reservoirs): the atmosphere, plants, soils, freshwater systems, oceans, and geological sediments. Hundreds of millions of years ago, and over millions of years, this carbon cycle was responsible for the formation of coal, petroleum, and natural gas, the fossil fuels that are the primary sources of energy for our modern societies.

Humans have altered the Earth's carbon budget. Today, the cycling of carbon among atmosphere, land, and freshwater and marine environments is in rapid transition and out of balance. Over tens of years, the combustion of fossil fuels is releasing into the atmosphere quantities of carbon that were accumulated in the Earth system over millions of years. Furthermore, tropical forests that once held large quantities of carbon are being converted to agricultural lands, releasing additional carbon to the atmosphere as a result. Both the fossil-fuel and land-use related releases are sources of carbon to the atmosphere. The combined rate of release is far larger than can be balanced by the biological and geological processes that naturally remove CO_2 from the atmosphere and store it in terrestrial and marine environments as part of the Earth's carbon cycle. These processes are known as sinks. Therefore, much of the CO_2 released through human activity has "piled up" in the atmosphere, resulting in a dramatic increase in the atmospheric concentration of CO_2. The concentration increased by 31% between 1850 and 2003, and the present concentration is higher than at any time in the past 420,000 years. Because CO_2 is an important greenhouse gas, the imbalance between sources and sinks and the subsequent increase in concentration in the atmosphere is very likely causing changes in Earth's climate (IPCC, 2007).

Furthermore, these trends in fossil-fuel use and tropical deforestation are accelerating. The magnitude of the changes raises concerns about the future behavior of the carbon cycle. Will the carbon cycle continue to function as it has in recent history, or will a CO_2-caused warming result in a weakening of the ability of sinks to take up CO_2, leading to further warming? Drought, for example, may reduce forest growth. Warming can release carbon stored in soil, and warming and drought may increase forest fires. Conversely, will elevated concentrations of CO_2 in the atmosphere stimulate plant growth as it is known to do in laboratory and field experiments and thus strengthen global or regional sinks?

The question is complicated because CO_2 is not the only substance in the atmosphere that affects the Earth's surface temperature and climate. Other greenhouse gases include CH_4, nitrous oxide, the halocarbons, and ozone, and all of these gases, together with water vapor, aerosols, solar radiation, and properties of the Earth's surface, are involved in the evolution of climate change. Carbon dioxide, alone, is responsible for approximately 55-60% of the change in the Earth's radiation balance due to increases in well-mixed atmospheric greenhouse gases and CH_4 for about another 20% (values are for the late 1990s; with a relative uncertainty of

10%; IPCC, 2001). These two gases are the primary gases of the carbon cycle, with CO_2 being particularly important. Furthermore, the consequences of increasing atmospheric CO_2 extend beyond climate change alone. The accumulation of carbon in the oceans as a result of more than a century of fossil-fuel use and deforestation has increased the acidity of the surface waters, with serious consequences for corals and other marine organisms that build their skeletons and shells from calcium carbonate.

Inevitably, the decision to influence or control atmospheric concentrations of CO_2 as a means to prevent, minimize, or forestall future climate change, or to avoid damage to marine ecosystems from ocean acidification, will require management of the carbon cycle. That management involves both reducing sources of CO_2 to the atmosphere and enhancing sinks for carbon on land or in the oceans. Strategies may involve both short- and long-term solutions. Short-term solutions may help to slow the rate at which carbon accumulates in the atmosphere while longer-term solutions are developed. In any case, formulation of options by decision makers and successful management of the Earth's carbon budget as part of a portfolio of climate-change mitigation and adaptation strategies will require solid scientific understanding of the carbon cycle.

Understanding the current carbon cycle may not be enough, however. The concept of managing the carbon cycle carries with it the assumption that the carbon cycle will continue to operate as it has in recent centuries. A major concern is that the carbon cycle, itself, is vulnerable to land-use or climate change that could bring about additional releases of carbon to the atmosphere from either land or the oceans.

Over recent decades both terrestrial ecosystems and the oceans have been natural sinks for carbon. If either, or both, of those sinks were to become sources, slowing or reversing the accumulation of carbon in the atmosphere could become much more difficult. Thus, understanding the current global carbon cycle is necessary for managing carbon, but is not sufficient. Projections of the future behavior of the carbon cycle in response to human activity and to climate and other environmental change are also important to understanding system vulnerabilities.

Perhaps even more importantly, effective management of the carbon cycle requires more than basic understanding of the current or future carbon cycle. It also requires cost-effective, feasible, and politically palatable options for carbon management. Just as carbon cycle knowledge must be assessed and evaluated, so must management options and tradeoffs. See Chapter 1 for further discussion of why the general public, as well as individuals and institutions interested in carbon management, should care about the carbon cycle.

> A major concern is that the carbon cycle, itself, is vulnerable to land-use or climate change that could bring about additional releases of carbon to the atmosphere from either land or the oceans.

ES.3 HOW DO NORTH AMERICAN CARBON SOURCES AND SINKS RELATE TO THE GLOBAL CARBON CYCLE?

In 2004, North America was responsible for approximately 25% of the CO_2 emissions produced globally by fossil-fuel combustion (Chapter 2 this report). The United States, the world's largest emitter of CO_2, accounted for 86% of the North American total in 2004 (85% in 2003). In 2003, Canada accounted for 9% and Mexico for 6%, of the total. Among all countries, the United States, Canada, and Mexico ranked, respectively, as the first, seventh, and eleventh largest emitters of CO_2 from fossil fuels in 2003 (Marland et al., 2006). The United States ranked eleventh in *per capita* emissions (5.43 tons carbon per year) in 2003; Canada ranked thirteenth (4.88 tons carbon per year); and Mexico eighty-ninth (1.10 tons carbon per year). *Per capita* emissions of the United States and Canada were, respectively, 4.8 and 4.3 times the global *per capita* emissions of 1.14 tons carbon per year. Mexico's *per capita* emissions were slightly below the global value. Combined, these three countries contributed almost one third (32%) of the cumulative global fossil-fuel CO_2 emissions between

> In 2004, North America was responsible for approximately 25% of the CO_2 emissions produced globally by fossil-fuel combustion.

BOX ES.1: Treatment of Uncertainty

Sources of uncertainty vary widely across the many sectors and elements of the North American carbon cycle. The attention to uncertainty and the methods for dealing with uncertainty also vary across the disciplines that study these elements and across individual studies and publications. There is no single applicable quantitative method for integrating these variable sources, methods, and characterizations.

To provide for synthesis across and comparability among carbon cycle elements, the following convention has been adopted for characterizing uncertainty in the report's synthetic findings and results (for example, in the synthesized carbon budget for North America of Chapter 3 and in the Executive Summary). Uncertainty is characterized using five categories:

(1) ***** = 95% certain that the actual value is within 10% of the estimate reported,
(2) **** = 95% certain that the estimate is within 25%,
(3) ***   = 95% certain that the estimate is within 50%,
(4) **   = 95% certain that the estimate is within 100%, and
(5) *   = uncertainty greater than 100%.

Unless otherwise noted, values presented as "y ± x%" should be interpreted to mean that the authors are 95% certain the actual value is between y − x% and y + x%. Where appropriate, the absolute range is sometimes reported rather than the relative range: y ± z, where z = y × x% ÷ 100. The system of asterisks is used as short-hand for the categories in tables and text.

These are informed categorizations. They reflect expert judgment, using all known published descriptions of un-certainty surrounding the "best available" or "most likely" estimate. The 95% boundary was chosen to commu-nicate the high degree of certainty that the actual value was in the reported range and the low likelihood (1/20) that it was outside that range. This characterization is not, however, a statistical property of the estimate, and should not be confused with statistically defined 95% confidence intervals.

The authors of this report have used this system for categorizing uncertainty only where they have synthesized diverse published information and compared across this diversity. When citing an existing published estimate, authors were encouraged to include the characterizations of uncertainty reported by those publications (e.g., ranges, standard error, or confidence intervals). There are circumstances in which no characterization of the uncertainty of data or information is shown, such as when a number is taken from a published source that itself did not include a characterization of uncertainty. In these cases, the authors have not provided a characteriza-tion of uncertainty, and the reader should assume that no characterization of uncertainty was available to the authors. Additional discussion of sources of uncertainty and their treatment in this report can be found in the Preface under "The Treatment of Uncertainty in this Report."

1751 and 2002. Emissions from parts of Asia are increasing at a growing rate and may surpass those of North America in the near future, but North America is incontrovertibly a major source of atmospheric CO_2, historically, at present, and in the immediate future.

The contribution of North American carbon sinks to the global carbon budget is less clear. The global terrestrial sink is quite uncertain, averaging somewhere in the range of 0 to 3800 million tons of carbon per year during the 1980s, and in the range of 1000 to 3600 million tons of carbon per year

in the 1990s (IPCC, 2000). This report estimates a North American sink of approximately 500 million tons of carbon per year for 2003, with 95% certainty that the actual value is within plus or minus 50% of that estimate, or between 250 and 750 million tons carbon per year (Chapter 3 this report) (see the Text Box on Treatment of Uncertainty). Assuming a global terrestrial sink of approximately two bil-lion tons of carbon per year (as inferred by the atmospheric analyses for the 1990s), the North American terrestrial sink reported here of approximately 500 million tons of carbon per year suggests that the North American sink is perhaps

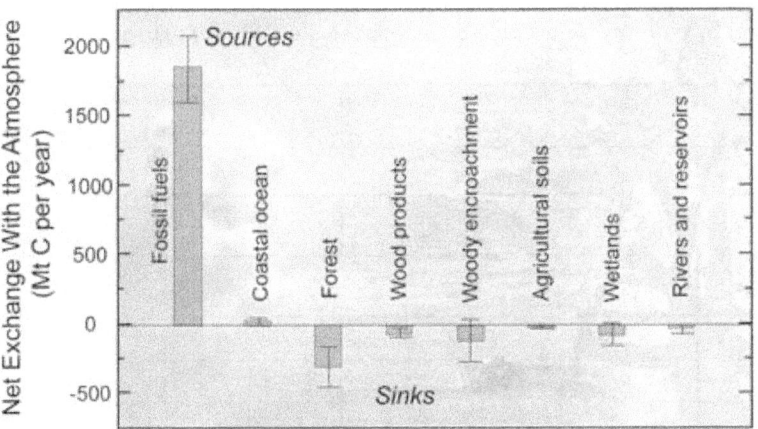

Figure ES.1 North American carbon sources and sinks (million tons of carbon per year) in 2003. Height of a bar indicates a best estimate for net carbon exchange between the atmosphere and the indicated element of the North American carbon budget. Sources add CO_2 to the atmosphere; sinks remove it. Error bars indicate the uncertainty in that estimate, and define the range of values that include the actual value with 95% certainty. See Chapter 3 and Chapters 6-15 of this report for details and discussion of these sources and sinks.

25% of the global sink. In contrast, previous analyses using global models of CO_2 transport in the atmosphere estimate a North American sink for 1991-2000 of approximately one billion tons of carbon per year, or approximately 50% of a global sink of roughly two billion tons of carbon per year (Chapter 2 this report). The North American sink estimate of this report is derived from studies using ground-based inventories, and the difference between estimates is likely influenced by the methodology employed and the period of the analysis (Chapters 2 and 3 this report). Developments in the use of atmospheric models to estimate terrestrial sinks concurrent with the production and publication of this report will continue to refine and improve those estimates.

The global terrestrial sink is predominantly in northern lands, most likely as a consequence of forest regrowing on abandoned agricultural land in northern temperate regions (*e.g.*, the eastern United States) and patterns of forest fire and recovery in the boreal forests of Canada and Eurasia. The sink north of 30° N alone is estimated to be 600 to 2300 million tons of carbon per year for the 1980s (IPCC, 2001). Thus, the sink of approximately 500 million tons of carbon per year in North America is consistent with the fraction of northern land area in North America (37%), as opposed to Eurasia (63%). Rates of forest clearing in the tropics, including those of Mexico, currently exceed rates of recovery, and thus tropical regions dominated by rainforests or other forest types are currently a source of carbon to the atmosphere.

It is clear that the global carbon cycle of the 21st century will continue to be influenced by large fossil-fuel emissions from North America, and that the North American carbon budget will continue to be dominated by the fossil-fuel

sources. The future trajectory of carbon sinks in North America and their contribution to the global terrestrial sink is less certain, in part because the role of regrowing forests is likely to decline as the forests mature, and in part because the response of forests and other ecosystems to future climate change and increases in atmospheric CO_2 concentrations is uncertain. The variation among model projections and scenarios of where and how future climate will change contribute to that uncertainty. Additionally, response to a particular future change will likely vary among ecosystems and the response will depend on a variety of incompletely understood environmental factors.

ES.4 WHAT ARE THE PRIMARY CARBON SOURCES AND SINKS IN NORTH AMERICA, AND HOW AND WHY ARE THEY CHANGING?

ES.4.1 The Sources

The primary source of human-caused carbon emissions in North America that contributes to the increase of CO_2 in the atmosphere is the release of CO_2 during the combustion of fossil fuels (Figure ES.1) (Chapter 3 this report). Fossil-fuel carbon emissions in the United States, Canada, and Mexico totaled approximately 1900 million tons of carbon in 2003 (with 95% confidence that the actual value lies within 10% of that estimate[1]) and have increased at an average rate of approximately 1% per year for the last 30 years. The United States was responsible for approximately 85% of North America's fossil-fuel emissions in 2003, Canada for 9%, and Mexico 6% (Table ES.1). The overall 1% growth in United States' emissions masks faster than 1% growth in some sectors (*e.g.*, transportation) and slower growth in others (*e.g.*, increased manufacturing energy efficiency).

Total United States' emissions have grown at close to the North American average rate of about 1.0% per year over the past 30 years, but United States' *per capita* emissions have been roughly constant, while the carbon intensity (carbon emitted/dollar of real [inflation adjusted] GDP) of the United States' economy has de-

> Fossil-fuel carbon emissions in the United States, Canada, and Mexico have increased at an average rate of approximately 1% per year for the last 30 years.

---
[1] See Text Box ES.1 for a discussion of numerical data and estimates.

Table ES.1 North American annual net carbon emissions (source = positive) or uptake (land sink = negative) (million tons carbon per year) by country. See Table 3.1, Chapter 3 of this report for references to sources of data.

Source (positive) or Sink (negative)	United States	Canada	Mexico	North America
Fossil source (positive)				
Fossil fuel (oil, gas,coal)	1582***** (681, 328, 573)	164***** (75, 48, 40)	110***** (71, 29, 11)	1856***** (828,405,624)
Non-fossil carbon sink (negative) or source (positive)				
Forest	-256***	-28**	+52**	-233***
Wood products	-57***	-11***	ND	-68***
Woody encroachment	-120*	ND	ND	-120*
Agricultural soils	-8***	-2***	ND	-10***
Wetlands	-23*	-23*	-4*	-49*
Rivers and lakes	-25**	ND	ND	-25*
Coastal oceans [a]				
Total carbon source or sink	-489***	-64**	48*	-505***
Net carbon source (positive)	1093****	100***	158***	1351****

Uncertainty:
*****(95% confidence within 10%)
****(95% confidence within 25%)
***(95% confidence within 50%)
**(95% confidence within 100%)
*(95% confidence bounds >100%)
ND = No data available
[a] Coastal waters within 100 km of the North American coastline, defined by the region in which the surface water concentration of CO_2 is inflluenced by coastal processes, may be a source of 19 million tons of carbon per year but with 95% confidence bounds greater than 100% (*i.e.*, they may be a small sink). See discussion of coastal ocean sources and sinks in Chapters 3 and 15 of this report, and their distribution by ocean region rather than country in Chapter 15 of this report.

creased at a rate of about 2% per year (Chapter 3 this report). The decline in the carbon intensity of the United States' economy was caused both by increased energy efficiency, particularly in the manufacturing sector, and structural changes in the economy with growing contributions from sectors such as services with lower energy consumption and carbon intensity. The service sector is likely to continue to grow. Accordingly, carbon emissions will likely continue to grow more slowly than GDP (Chapter 3 this report).

The extraction of fossil-fuels and other primary energy sources and their conversion to energy commodities and services, including electricity generation, is the single largest contributor to the North American fossil-fuel source, accounting for approximately 42% of North American fossil

> The extraction of fossil-fuels and their conversion to energy commodities and services, including electricity generation, is the single largest contributor to the North American fossil-fuel source.

emissions in 2003 (Chapter 6 this report). Electricity generation is responsible for the largest share of those emissions: approximately 94% in the United States in 2004, 65% in Canada in 2003, and 67% in Mexico in 1998. Again, United States' emissions dominate. United States' emissions from electricity generation are approximately 17 times larger than those of Canada and 23 times those of Mexico, reflecting in part the relatively greater population of the United States in both cases and its much higher level of development than Mexico. On a *per capita* basis, the emissions from electricity generation are 2.14 tons of carbon for the United States in 2004, 1.15 tons of carbon for Canada in 2003, and 0.28 tons of carbon for Mexico in 1998 (note these are the latest years for which data are available).

More than half of electricity produced in North America (67% in the United States) is consumed in buildings, making that single use one of the largest factors in North American emissions (Chapter 9 this report). In fact, in 2003 the CO_2 emissions from United States' buildings alone were greater than total CO_2 emissions of any country in the world, except China. Energy use in buildings in the United States and Canada (including the use of natural gas, wood, and other fuels as well as electricity) has increased by 30% since 1990, corresponding to an annual growth rate of 2.1%. In the United States, the major drivers of energy consumption in the buildings sector are growth in commercial floor space and increase in the size of the average home. Carbon emissions from buildings are expected to grow with population and income. Furthermore, the shift from family to single-occupant households means that the number of households will increase faster than population growth—each household with its own heating and cooling systems and electrical appliances. Certain electrical appliances (such as air-conditioning equipment) once considered a luxury are now becoming commonplace. Technology- and market-driven improvements in the efficiency of appliances are expected to continue, but the improvements will probably not be sufficient to curtail emissions growth in the buildings sector without government intervention.

The transportation sector of North America accounted for 31% of total North American emissions in 2003, most (87%) of it from the United States (Chapter 7 this report). The growth in transportation and associated CO_2 emissions has been steady during the past forty years and has been most rapid in Mexico, the country most dependent upon road transport. The growth of transportation is driven by population, *per capita* income, and economic output, and energy use in transportation is expected to increase by 46% in North America between 2003 and 2025. If the mix of fuels is assumed to remain the same, CO_2 emissions would increase from 587 million tons of carbon in 2003 to 859 million tons of carbon in 2025.

Emissions from North American industry (not including fossil-fuel mining and processing or electricity generation) are a relatively small (12%) and declining component of North America's emissions (Chapter 8 this report). Emissions decreased nearly 11% between 1990 and 2002, while energy consumption in the United States and Canada increased by 8-10% during that period. In both countries, a shift in production toward less energy-intensive industries and dissemination of more energy efficient equipment has kept the rate of growth in energy demand lower than the rate of growth of industrial GDP. Emission reductions in industry have also resulted from the voluntary, proactive initiatives of both individual corporations and trade associations in response to climate change issues (Chapter 4 this report).

The remaining portion (approximately 15%) of North American fossil-fuel emissions includes those from other sectors. This includes natural gas and other non-electrical

fossil energy used in residential and commercial buildings and fuels used in agriculture.

ES.4.2 The Sinks

Approximately 30% of North American fossil-fuel emissions are offset by a sink of approximately 500±250 million tons of carbon per year. The uncertainty in the North American sink of ±50% is substantially larger than the ±10% uncertainty in the emissions source. The total sink is a combination of many factors, including forest regrowth, fire suppression, and agricultural soil conservation (Figure ES.1, Chapter 3, Part III: Chapters 10-15 this report). The sink is currently about 490 million tons of carbon per year in the United States and approximately 60 million tons of carbon per year in Canada. Mexican ecosystems are a net source of about 50 million tons of carbon per year, mostly as a consequence of ongoing deforestation. The coastal ocean surrounding North America is perhaps an additional small net source of carbon to the atmosphere of approximately 20 million tons of carbon per year. The coastal ocean is, however, highly variable, and that number is highly uncertain with variability (standard deviation) of greater than 100%. North America's coastal waters could be a small sink and in some places are. How much the coastal carbon exchange with the atmosphere is influenced by humans is also unknown.

The primary carbon sink in North America (approximately 50% of the total) is in the forests of the United States and Canada (Table ES.1). These forests are still growing (accumulating carbon) after their re-colonization of farmland 100 or more years ago. Forest regrowth takes carbon out of the atmosphere and stores most of it in above-ground vegetation

been reduced by more than 95% from the pre-settlement levels, and this reduction favors shrubs and trees in competition with grasses. The sink may be as large as 20% of the North American sink, but it may also be negligible. The uncertainty of this estimate is greater than 100%. If that highly uncertain sink is excluded (see Overview of Part III this report), the estimate of the North American sink falls to 385 million tons of carbon per year or approximately 20% of fossil-fuel emissions in 2003. Woody encroachment might actually be a source, maybe even a relatively large one. The state of the science is such that we simply don't know (Chapter 3 and Part III Overview this report).

Wood products are thought to account for about 13% of the total North American sink. The uncertainty in this sink is ±50%. Wood products are a sink because they are increasing, both in use (*e.g.*, furniture, house frames, etc.) and in landfills. The wetland sink, about 9% of the North American sink but with an uncertainty of greater than 100%, is in both the peats of Canada's extensive frozen (permafrost) and unfrozen wetlands and the mineral soils of Canadian and United States' wetlands. Drainage of peatlands in the United States has released carbon to the atmosphere, and the very large volume of carbon in North American wetlands (the single largest carbon reservoir of any North American ecosystem) is vulnerable to release in response to both climate change and the further drainage of wetlands for development. Either change might shift the current modest sink to a potentially large source, although many aspects of wetlands and their future behavior are poorly known.

The primary carbon sink in North America (approximately 50% of the total) is in the forests of the United States and Canada.

(wood), with as much as a third of it in soils. The suppression of forest fires also increases net accumulation of carbon in forests. As the recovering forests mature, however, the rate of net carbon uptake (the sink) declines. In Canada, the estimated forest sink declined by nearly a third between 1990 and 2004, but with high year-to year variability. Over that period, the annual changes in above-ground carbon stored in managed Canadian forests varied from between a sink of approximately 50 million tons of carbon per year to a source of approximately 40 million tons of carbon per year. Years when the forests were a source were generally years with high forest fire activity.

The very large volume of carbon in North American wetlands (the single largest carbon reservoir of any North American ecosystem) is vulnerable to release in response to both climate change and the further drainage of wetlands for development.

Woody encroachment, the invasion of woody plants into grasslands or of trees into shrublands, is a potentially large, but highly uncertain carbon sink. It is caused by a combination of fire suppression and grazing. Fire inside the United States has

Two processes determine the carbon balance of agricultural lands: management and changes in environmental factors. The effects of management (*e.g.*, cultivation, conservation tillage) are reasonably well known and have been responsible for historic losses of carbon in Canada and the United States (and current losses in Mexico), albeit with some increased carbon uptake and storage in recent years. Agricultural lands in North America are nearly neutral with respect to carbon, with mineral soils absorbing carbon and organic soils releasing it. The balance of these sinks and sources is a net sink of 10±5 million tons of carbon per year (Table ES.1). The effects of climate on this balance are not well known.

Soil erosion leads to the accumulation of carbon containing sediments in streams, rivers, and lakes (both natural and man-made). This represents a carbon sink, estimated at approximately 25 million tons of carbon per year for the United States. We know of no similar analysis for Canada or Mexico. The result is a highly uncertain estimate for North America known to no better than the estimate for the United States alone, plus or minus more than 100%.

The density and development patterns of human settlements are drivers of fossil-fuel emissions, especially in the important residential and transportation sectors. Conversion of agricultural and wildlands to cities and other human settlements reduces carbon stocks, while the growth of urban and suburban trees increases them. The growth of urban trees in North America produces a sink of approximately 16 to 49 million tons of carbon per year, which is 1 to 3% of North American fossil-fuel emissions in 2003. Settlements in North America are thus almost certainly a net source of atmospheric CO_2.

ES.5 WHAT ARE THE DIRECT, NON-CLIMATIC EFFECTS OF INCREASING ATMOSPHERIC CARBON DIOXIDE OR OTHER CHANGES IN THE CARBON CYCLE ON THE LAND AND OCEANS OF NORTH AMERICA?

The potential impacts of increasing concentrations of atmospheric CO_2 (and other greenhouse gases) on the Earth's climate are well documented (IPCC, 2007) and are the dominant reason for societal interest in the carbon cycle. However, the consequences of a carbon cycle imbalance and the buildup of CO_2 in the atmosphere extend beyond climate change alone. Ocean acidification and "CO_2 fertilization" of land plants are foremost among these direct, non-climatic effects.

The uptake of carbon by the world's oceans as a result of human activity over the last century has made them more acidic (Chapters 1 and 2 this report). This acidification negatively impacts corals and other marine organisms that build their skeletons and shells from calcium carbonate. Future changes could dramatically alter the composition of ocean ecosystems of North America and elsewhere, possibly eliminating coral reefs by 2100.

Rates of photosynthesis of many plant species often increase in response to elevated concentrations of CO_2, thus potentially increasing plant growth and even agricultural crop yields in the future (Chapters 2, 3, 10-13 this report). There is, however, continuing scientific debate about whether such "CO_2 fertilization" will continue into the future with prolonged exposure to elevated CO_2, and whether the fertilization of photosynthesis will translate into increased plant growth and net uptake and storage of carbon by terrestrial ecosystems. Recent studies provide many conflicting results. Experimental treatment with elevated CO_2 can lead to consistent increases in plant growth. On the other hand, it can also have little effect on plant growth, with an initial stimulation of photosynthesis but limited long-term effects on carbon accumulation in the plants. Moreover, it is unclear how plants and ecosystem might respond simultaneously to both "CO_2 fertilization" and climate change. While there is some experimental evidence that plants may use less water when exposed to elevated CO_2, extended deep drought or other unfavorable climatic conditions could reduce the positive effects of elevated CO_2 on plant growth. Thus, it is far from clear that elevated concentrations of atmospheric CO_2 have led to terrestrial carbon uptake and storage or will do so over large areas in the future. Moreover, elevated carbon dioxide is known to increase CH_4 emissions from wetlands, further increasing the uncertainty in how plant response to elevated CO_2 will affect the global atmosphere and climate.

The carbon cycle also intersects with a number of critical Earth system processes, including the cycling of both water and nitrogen. Virtually any change in the lands or waters of North America as part of purposeful carbon management will consequently affect these other processes and cycles. Some interactions may be beneficial. For example, an increase in organic carbon in soils is likely to increase the availability of nitrogen for plant growth

> The growth of urban trees in North America produces a sink of approximately 1 to 3 percent of North American fossil-fuel emissions in 2003.

> The carbon cycle also intersects with a number of critical Earth system processes, including the cycling of both water and nitrogen.

and enhance the water-holding capacity of the soil. Other interactions, such as nutrient limitation, fire, insect attack, increased respiration from warming, may be detrimental. However, very little is known about the complex web of interactions between carbon and other systems at continental scales, or the effect of management on these interactions.

ES.6 WHAT POTENTIAL MANAGEMENT OPTIONS IN NORTH AMERICA COULD SIGNIFICANTLY AFFECT THE NORTH AMERICAN AND GLOBAL CARBON CYCLES (E.G., NORTH AMERICAN SINKS AND GLOBAL ATMOSPHERIC CARBON DIOXIDE CONCENTRATIONS)?

Addressing imbalances in the North American and global carbon cycles requires a mix of options, no single option being sufficient, focused on reducing carbon emissions (Chapter 4 this report). Options focused on enhancing carbon sinks in soils and vegetation in North America can contribute as well, but the potential of these options alone is insufficient to deal with the magnitude of current imbalances in the North American carbon budget and their contributions to the global imbalance.

> Addressing imbalances in the North American and global carbon cycles requires a mix of options focused on reducing carbon emissions.

Currently, options for reducing carbon emissions include:
- Reducing emissions from the transportation sector through efficiency improvement, higher prices for carbon-based fuels, liquid fuels derived from vegetation (ethanol from corn or other biomass feedstock, for example), and in the longer run (after 2025), hydrogen generated from non-fossil sources of energy;
- Reducing the carbon emissions associated with energy use in buildings through efficiency improvements and energy-saving passive design measures;
- Reducing emissions from the industrial sector through efficiency improvement, fuel-switching, and innovative process designs;
- Reducing emissions from energy extraction and conversion through efficiency improvement, fuel-switching, technological change (including carbon sequestration and capture and storage), and reduced demands due to increased end-use efficiency; and
- Capturing the CO_2 emitted from fossil-fired generating units and injecting it into a suitable geological formation or deep in the sea for long-term storage (carbon capture and storage).

Options for managing terrestrial carbon stocks include:
- Maintaining existing terrestrial carbon stocks in vegeta-

tion and soils and in wood products;
- Reducing carbon loss associated with land management practices, including those of agriculture (e.g., reduced tillage in expanding croplands) and forest harvest (e.g., minimizing soil disturbance); and
- Increasing terrestrial carbon sequestration through afforestation, reforestation, planting of urban "forests," reduced tillage in established crop lands, and similar practices.

In many cases, significant progress with such options would require a combination of technology research and development, policy interventions, and information and education programs.

Opinions differ about the relative mitigation impact of emission reduction versus carbon sequestration. Assumptions about the cost of mitigation and the policy instruments used to promote mitigation significantly affect assessments of mitigation potential. For example, appropriately designed carbon emission cap and trading policies could achieve a given level of carbon emissions reduction at lower cost than some other policy instruments by providing incentives to use the least-cost combination of mitigation/sequestration alternatives.

However, the evaluation of any policy instrument should consider technical, institutional, and socioeconomic constraints that would affect its implementation, such as the ability of sources to monitor their actual emissions and the constitutional authority of national and/or provincial/state governments to impose emissions taxes, regulate emissions, and/or regulate efficiency standards. Also, practically every policy (except cost-saving energy conservation options), no matter what instrument is used to implement it, has a cost in terms of utilization of resources and ensuing price increases that leads to reductions in output, income, employment, or other measures of economic well-being. These costs must be weighed against the benefits (or avoided costs) of reducing carbon emissions. In addition to the standard reduction in damages noted above, many options and measures that reduce emissions and increase sequestration also have significant co-benefits in terms of economic efficiency (where market failures are being corrected, as in many cases of energy conservation), environmental management, and energy security.

The design of carbon management systems must also consider unintended consequences involving other greenhouse gases. For instance, carbon sequestration strategies such as reduced tillage can increase emissions of CH_4 and nitrous oxide, which are also greenhouse gases. Strategies for dealing with climate change will have to consider these other gases as well as other components of the climate systems,

such as small airborne particles and the physical aspects of plant communities.

Direct reductions of carbon emissions from fossil-fuel use are considered "permanent" reductions, while carbon sequestration in plants or soils is a "non-permanent" reduction, in that carbon stored through conservation practices could potentially be re-emitted if management practices revert back to the previous state or otherwise change. This *permanence* issue applies to all forms of carbon sinks. For example, the carbon sink associated with forest regrowth could be slowed or reversed from sink to source if the forests are burnt in wildfires or forest harvest and management practices change.

Changes in land management (*e.g.*, tillage reduction, pasture improvement, afforestation) will stimulate the uptake and sequestration of carbon for only a finite period. Over time, the processes of carbon gain and loss from vegetation and soil come into a new balance with the change in land use and land management. The amount of carbon stored in the plants and soil will tend to level off at a new maximum with the altered processes of uptake balanced by altered processes of release, after which there is no further accumulation (sequestration) of carbon. For example, following changes in tillage to promote carbon absorption in agricultural soils (Chapter 10 this report) the amount of carbon in the soil will tend to reach a new constant level after 15–30 years. The sink declines, then disappears, or nearly so, as the amount of carbon being added to the soil is balanced by losses. The same pattern is observed as forests are planted, as they re-grow on abandoned farmland or as they recover from fire, harvest, or other disturbance. It takes significantly longer for forests to reach a new balance of uptake and release with many forests sequestering significant amounts of carbon 125 years after establishment, but as forests mature, the rate of sequestration declines and in old growth forests processes of carbon uptake are very nearly balanced by processes of release (Chapters 3 and 11 this report).

Mitigation actions in one area (*e.g.*, geographic region, production system) can inadvertently result in additional emissions elsewhere. This phenomenon, commonly referred to as *leakage*, can occur when a policy of emission reduction by one country shifts emission-intensive industry or energy production toward other countries, increasing their emissions and thus reducing the overall benefit. Similarly, leakage can be a concern for sequestration and storage of carbon in forests. Reducing harvest rates in one area, for example, can stimulate increased cutting and reduction in stored carbon in other areas. Leakage may be of minor concern for agricultural carbon storage, since most practices would have little or no effect on the supply and demand of agricultural commodities. Chapter 4 further compares measures taken to reduce emissions with those taken to sequester carbon.

Options and measures can be implemented in a variety of ways at a variety of scales, not only at international or national levels. For example, a number of municipalities, state governments, and private firms in North America have made commitments to voluntary greenhouse gas emission reductions. For cities, one focus has been the Cities for Climate Protection program of International Governments for Local Sustainability (formerly ICLEI). For some states and provinces, the Regional Greenhouse Gas (Cap and Trade) Initiative is nearing implementation. For industry, one focus has been membership in the Pew Center and in the Environmental Protection Agency (EPA) Climate Leaders Program.

> Many options and measures that reduce emissions and increase sequestration also have significant co-benefits in terms of economic efficiency, environmental management, and energy security.

ES. 7 HOW CAN WE IMPROVE THE USEFULNESS OF CARBON SCIENCE FOR DECISION MAKING?

Effective carbon management requires that relevant, appropriate science be communicated to the wide variety of people whose decisions affect carbon cycling (Chapter 5 this report). Because the field is relatively new and the demand for policy-relevant information has been limited, carbon cycle science has rarely been organized or conducted to inform carbon management. To generate information that can systematically inform carbon management decisions, scientists and decision makers should clarify what information would be most relevant in specific sectors and arenas for carbon management, adjust research priorities as necessary, and develop mechanisms that enhance the credibility and legitimacy of the information being generated.

> A number of municipalities, state governments, and private firms in North America have made commitments to voluntary greenhouse gas emission reductions.

In the United States, the federal carbon science enterprise does not yet have many mechanisms to assess emerging demands for carbon information across scales and sectors. Federally funded carbon science has focused predominantly on basic research to reduce uncertainties about the carbon cycle. Initiatives are now underway to promote coordinated, interdisciplinary research that is strategically prioritized to address societal needs. The need for this type of research is increasing. Interest in carbon management across sectors suggests that there may be substantial demand for information in the energy, transportation, agriculture, forestry, and industrial sectors, at scales ranging from local to global.

To ensure that carbon science is as useful as possible for decision making, carbon scientists and carbon managers need to create new forums and institutions for communication and coordination. Research suggests that in order to make a significant contribution to management, scientific and technical information intended for decision making must be perceived not only as credible (worth believing), but also as salient (relevant to decision making on high priority issues) and legitimate (conducted in a way that stakeholders believe is fair, unbiased, and respectful of divergent views and interests). To generate information that meets these tests, carbon stakeholders and scientists need to collaborate to develop research questions, design research strategies, and review, interpret, and disseminate results. Transparency and balanced participation are important for guarding against politicization and enhancing usability.

Initiatives are now underway to promote coordinated, interdisciplinary research that is strategically prioritized to address societal needs.

To make carbon cycle science more useful to decision makers in the United States and elsewhere in North America, leaders in the carbon science community might consider the following steps:

- Identify specific categories of decision makers for whom carbon cycle science is likely to be salient, focusing on policy makers and private sector managers in carbon-intensive sectors (energy, transport, manufacturing, agriculture, and forestry);
- Identify and evaluate existing information about carbon impacts of decisions and actions in these arenas, and assess the need and demand for additional information. In some cases, demand may need to be nurtured and fostered through a two-way interactive process;
- Encourage scientists and research programs to experiment with new and different ways of making carbon cycle science more salient, credible, and legitimate to carbon managers;
- Involve not just physical or biological disciplines in scientific efforts to produce useable science, but also social scientists, economists, and communication experts; and
- Consider initiating participatory pilot research projects and identifying existing "boundary organizations" (or establishing new ones) to bridge carbon management and carbon science.

ES.8 WHAT ADDITIONAL KNOWLEDGE IS NEEDED FOR EFFECTIVE CARBON MANAGEMENT?

Scientists and carbon managers need to improve their joint understanding of the top priority questions facing carbon-related decision-making. Priority needs specific to individual ecosystem or sectors are described in Chapters 6-15 of this report. To further prioritize those needs across disciplines and sectors, scientists need to collaborate more effectively with decision makers in undertaking research and interpreting results in order to answer those questions. More deliberative processes of consultation with potential carbon managers at all scales can be initiated at various stages of the research process. This might include workshops, focus groups, working panels, and citizen advisory groups. Research on the effective production of science that can be used for decision making suggests that ongoing, iterative processes that involve decision makers are more effective than those that do not (Chapter 5 this report).

In the light of changing views on the impacts of CO_2 released to the atmosphere, research and development will likely focus on the extraction of energy while preventing CO_2 release. Fossil fuels might well remain economically competitive and socially desirable as a source of energy in some circumstances, even when one includes the extra cost of capturing the CO_2 and preventing its atmospheric release

be a useful investment. Quantitative estimates of land-use change and the impact of various management practices are also highly uncertain, as are the interactions among CO_2, CH_4, and nitrous oxide as greenhouse gas emissions. If carbon accounting becomes a critical feature of carbon management, improved data are needed on the relationship of forest management practices to carbon storage, as well as inexpensive tools and techniques for monitoring. An assessment of agroforestry practices in Mexico as well as in temperate landscapes would also be helpful. Importantly, there is a need for multi-criteria analysis of various uses of landscapes—tradeoffs between carbon storage and other uses of the land must be considered. If markets emerge more fully for trading carbon credits, the development of such decision support tools will likely be encouraged.

when converting these fuels into non-carbon secondary forms of energy like electricity, hydrogen, or heat. Research and development needs in the energy and conversion arena include clarifying potentials for carbon capture and storage, exploring how to make renewable energy affordable at large scales of deployment, examining societal concerns about nuclear energy, and learning more about policy options for distributed energy and energy transitions. There is also need for better understanding of the public acceptability of policy incentives for reducing dependence on carbon intensive energy sources.

In the transportation sector, improved data on Mexican greenhouse gas emissions and trends is needed, as well as on the potential for mitigating transportation-related emissions in North America. Advances in transportation mitigation technologies and policies are also needed. In the industry and waste management sectors, work on materials substitution and energy efficient technologies in production processes holds promise for greater emissions reductions. Needs for the building sector include: further understanding the total societal costs of CO_2 as an externality of buildings costs, economic and market analyses of various reduced emission features at various time scales of availability, and construction of cost curves for emission reduction options.

Turning to the ecosystem arena, the synthesis and assessment of this report provides a baseline against which future results from the North American Carbon Program (NACP) can be compared. The report also highlights key uncertainties in North American sources and sinks. For example, in the agricultural and grazing land sectors, inventories still carry a great deal of uncertainty, especially in the arena of woody encroachment. If such inventories are to be the basis for future decision making, reducing such uncertainties may

Soils in the permafrost region store vast amounts of carbon and are currently a small sink. There is, however, little certainty about how these soils will respond to changes brought about by climate. While these regions are likely not subject to management options, improved information on carbon storage and the trajectory of these reservoirs may provide additional insight into the likelihood of release of large amounts of carbon to the atmosphere that may affect global decision making. Similarly, there is great uncertainty in the response of the carbon pools of wetlands to climate changes, and very little data on freshwater mineral soils and estuarine carbon both in Canada and Mexico.

With respect to human settlements, additional studies of the carbon balance of settlements of varying densities, geographical location, and patterns of development are needed to quantify the potential impacts of various policy and planning alternatives on net greenhouse gas emissions. In coastal regions, additional information on carbon fluxes will help to constrain continental carbon balance estimates should information on that scale become useful for decision making. Research on ocean carbon uptake and storage is also needed in order to fully inform decision making on options for carbon management.

With respect to carbon management, there is a need for more insight into how incentives to reduce emissions affect the behavior of households and businesses, the influence of reducing uncertainty on the willingness of decision makers to make commitments, the affect of increased R&D spending

on technological innovation, the socioeconomic distribution of mitigation/sequestration costs and benefits, and the manner in which mitigation costs and policy instrument design affect the macroeconomy. Improvements in decision analysis in the face of irreducible uncertainty would be helpful as well.

Finally, CH_4 is second only to CO_2 as an important human-caused greenhouse gas. Methane sources and sinks are, however, not nearly as well understood as those for CO_2, and the consideration of CH_4 as part of the North American carbon budget is consequently well beyond the scope of this report. Research to better understand CH_4 sources and sinks and better integrate CH_4 into understanding of the carbon cycle could improve knowledge of how carbon management might influence both CO_2 and CH_4 in the atmosphere.

What Is the Carbon Cycle and Why Care?

Author Team: Anthony W. King, ORNL; Lisa Dilling, Univ. Colo./ NCAR; Gregory P. Zimmerman, ORNL; David M. Fairman, Consensus Building Inst., Inc.; Richard A. Houghton, Woods Hole Research Center; Gregg Marland, ORNL and Mid Sweden Univ. (Östersund); Adam Z. Rose, The Pa. State Univ. and Univ. Southern Calif.; Thomas J. Wilbanks, ORNL

1.1 WHY A REPORT ON THE CARBON CYCLE?

The concept of a carbon cycle is probably unfamiliar to most people other than scientists and some decision makers in the public and private sectors. More familiar is the water cycle, where precipitation falls on the earth to supply water bodies and evaporation returns water vapor to the clouds, which then renew the cycle through precipitation. In an analogous way, carbon—a fundamental requirement for life on Earth—cycles through exchanges among stores (or reservoirs) of carbon on and near the Earth's surface (mainly in plants and soils), in the atmosphere (mainly as gases), and in water and sediments in the ocean. Stated in oversimplified terms, plants take up carbon dioxide (CO_2) from the atmosphere through photosynthesis and create sugars and other carbohydrates, which animals and humans use for food, shelter, and energy to sustain life. Emissions from plants, other natural systems, and human activities return carbon to the atmosphere, which renews the cycle (Figure 1.1).

All of the components of this cycle—the atmosphere, the terrestrial vegetation, soils, freshwater lakes and rivers, the ocean, and geological sediments—are reservoirs (stores) of carbon. As carbon cycles through the system, it is exchanged between reservoirs, transferred from one to the next, with exchanges often in both directions. The carbon budget is an accounting of the balance of exchanges of carbon among the reservoirs: how much carbon is stored in a reservoir at a particular time, how much is coming in from other reservoirs, and how much is going out. When the inputs to a reservoir (the sources) exceed the outputs (the sinks), the amount of carbon in the reservoir is increased. The myriad physical, chemical, and biological processes that transfer carbon among reservoirs, and transform carbon among its various

molecular forms during those transfers, are responsible for the cycling of carbon through reservoirs. That cycling determines the balance of the carbon budget observed at any particular time. Quantifying the carbon budget over time can reveal whether the budget is or is not in balance (carbon accumulating in a reservoir would indicate an imbalance). If found to be out of balance, this quantification can provide understanding about why such a condition exists (for example, which sources exceed which sinks over what periods) (Sabine *et al.*, 2004, Chapter 2 this report). If the imbalance is deemed undesirable, the understanding of source and sinks can provide clues into how it might be managed (for example, which sinks are large relative to sources and might, if managed, provide leverage on changes in a reservoir) (Caldeira *et al.*, 2004; Chapter 4 this report). The global carbon budget is currently out of balance, with carbon accumulating in the form of CO_2 and methane (CH_4) in the atmosphere since the preindustrial era (*circa* 1750). Human use of coal, petroleum, and natural gas, combined with agriculture and other land-use change is primarily responsible. Documented by the Intergovernmental Panel on Climate Change for the 1990s (IPCC, 2001, p. 4), these trends continue in the early twenty-first century (Keeling and Whorf, 2005; Marland *et al.*, 2006).

> The global carbon budget is currently out of balance, with carbon accumulating in the form of CO_2 and methane (CH_4) in the atmosphere since the preindustrial era (circa 1750).

The history of the Earth's carbon balance as reflected in changes in atmospheric CO_2 concentration can be reconstructed from geological records, geochemical reconstructions, measurements on air bubbles trapped in glacial ice, and in recent decades, direct measurements of the atmosphere. Over the millennia, tens and hundreds of millions of

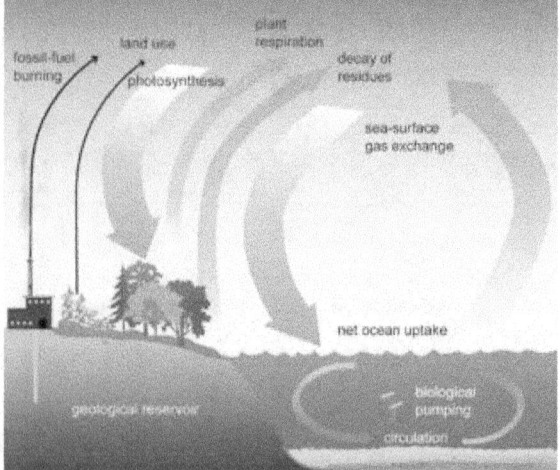

Figure 1.1 The Earth's carbon cycle. Carbon cycles through pools or reservoirs of carbon on land, in the ocean, and in sedimentary rock formations over daily, seasonal, annual, millennial, and geological time scales. See the accompanying text box. Figure adapted from http://www.esd.ornl.gov/iab/iab2-2.htm.

BOX 1.1: The Earth's Carbon Cycle

The burning of fossil fuels transfers carbon from geological reservoirs of coal, oil, and gas and releases carbon dioxide (CO_2) into the atmosphere. Tropical deforestation and other changes in land use also release carbon to the atmosphere as vegetation is burned and dead material decays. Photosynthesis transfers CO_2 from the atmosphere and the carbon is stored in wood and other plant tissues. The respiration that accompanies plant metabolism transfers some of the carbon back to the atmosphere as CO_2. When plants die, their decay also releases CO_2 to the atmosphere. A fraction of the dead organic material is resistant to decay and that carbon accumulates in the soil. Chemical and physical processes are responsible for the exchange of CO_2 across the sea surface. The small difference between the flux into and out of the surface ocean is responsible for net uptake of CO_2 by the ocean. Phytoplankton, small plants floating in the surface ocean, use carbon dissolved in the water to build tissue and calcium carbonate shells. When they die, they begin to sink and decay. As they decay, most of the carbon is redissolved into the surface water, but a fraction sinks into the deeper ocean, the so-called "biological pump", eventually reaching the ocean sediments. Currents within the ocean also circulate carbon from surface waters to the deep ocean and back. Carbon accumulated in soils and ocean sediments millions of years of ago was slowly transformed to produce the geological reservoirs of today's fossil fuels. For a more detailed, quantitative description, see Prentice et al. (2001), Houghton (2003), Sundquist and Visser (2003), Sabine et al. (2004), and Chapter 2 of this report.

years ago, vast quantities of carbon were stored in residues from dead plant and animal life that sank into the earth and became fossilized. On these time scales, small imbalances in the carbon cycle and geological processes, acting over millions of years, produced large but slow changes in atmospheric CO_2 concentrations of greater than 3000 parts per million (ppm) over periods of 150-200 million years (Prentice et al., 2001). By perhaps 20 million year ago, atmospheric CO_2 concentrations were less than 300 ppm (Prentice et al., 2001). Subsequently, imbalances in the carbon cycle linked with climate variations, especially the large glacial-interglacial cycles of the last 420,000 years, resulted in changes of approximately 100 ppm over periods of 50-75 thousand years (Prentice et al., 2001; Sabine et al., 2004). During the current interglacial climate, for at least the last 11,000 years, variations in atmospheric CO_2, also likely climate driven, were less than 20 ppm (Joos and Prentice, 2004). For 800-1000 years prior to the Industrial Revolution of the 1700s and 1800s, atmospheric CO_2 concentrations varied by less than 10 ppm (Prentice et al., 2001).

With the advent of the steam engine, the internal combustion engine, and other technological and economic elements of the Industrial Revolution, human societies found that the fossilized carbon formed hundreds of millions of years ago had great value as energy sources for economic growth. The 1800s and 1900s saw a dramatic rise in the combustion of these "fossil fuels" (e.g., coal, petroleum, and natural gas), releasing into the atmosphere, over decades, quantities of carbon that had been stored in the Earth system over millennia. These fossil-fuel emissions combined with and soon exceeded (circa 1910) the CO_2 emissions from burning and decomposition of dead plant material that accompanied clearing of forests for agricultural land use (Houghton, 2003).

It is not surprising, then, that measurements of CO_2 in the Earth's atmosphere have shown a steady increase in concentration over the twentieth century (Keeling and Whorf, 2005). The global CO_2 concentration has increased by approximately 100 ppm over the past 200 years, from a preindustrial concentration of 280 ± 10 ppm (Prentice et al., 2001) to a concentration (measured at Mauna Loa, Hawaii) of 369 ppm in 2000 and 377 ppm in 2004 (Keeling and Whorf, 2005). Methane shows a similar pattern, with relatively stable concentrations prior to about 1800 followed by a rapid increase (Ehhalt et al., 2001). Roughly, 20% of CH_4 emissions are from gas released in the extraction and transportation of fossil fuels; the rest is from biological sources including expanding rice and cattle production (Prinn, 2004). Such large increases in atmospheric carbon over such a short period of time relative to historical variations, together with patterns of human activity that will likely continue into the twenty-first century, such as trends in fossil-fuel use and tropical deforestation, raises concerns about imbalances in the carbon cycle and their implications.

1.2 THE CARBON CYCLE AND CLIMATE CHANGE

Most of the carbon in the Earth's atmosphere is in the form of CO_2 and CH_4. Both CO_2 and CH_4 are important "greenhouse gases." Along with water vapor and other "radiatively active" gases in the atmosphere, they absorb heat radiated from the Earth's surface, heat that would otherwise be lost into space. As a result, these gases help to warm the Earth's atmosphere. Rising concentrations of atmospheric CO_2 and other greenhouse gases can alter the Earth's radiant energy balance. The Earth's energy budget determines the global circulation of heat and water through the atmosphere and the patterns of temperature and precipitation we experience as weather and climate. Thus the human disturbance of the Earth's global carbon cycle during the industrial era and the resulting imbalance in the Earth's carbon budget and buildup of atmospheric CO_2 have consequences for climate and climate change. According to the IPCC, CO_2 is the largest single forcing agent of climate change (IPCC, 2001)[1].

In addition to the relationship between climate change and atmospheric CO_2 as a greenhouse gas, research is beginning to reveal the feedbacks between a changing carbon cycle and changing climate, and the associated implications for future climate change. Simulations with climate models that include an interactive global carbon cycle indicate a positive feedback between climate change and atmospheric CO_2 concentrations. The magnitude of the feedback varies considerably among models; but in all cases, future atmospheric CO_2 concentrations are higher and temperature increases are larger in the coupled climate-carbon cycle simulations than in simulations without the coupling and feedback between climate change and changes in the carbon cycle (Friedlingstein et al., 2006). The research is in its early stages, but 8 of the 11 models, in a recent comparison among models (Friedlingstein et al., 2006), attributed most of the feedback to changes in land carbon, with the majority locating those changes in the tropics. Differences among models in almost every aspect of plant and soil response to climate were responsible for the differences in model results, including

---

[1] Methane is also an important contributor (IPCC, 2001). However, CH4 and other non-CO2 carbon gases are not typically included in global carbon budgets because their sources and sinks are not well understood (Sabine et al., 2004). For this reason, and to manage scope and focus, we too follow that convention and this report is limited primarily to the carbon cycle and carbon budget of North America as it influences and is influenced by atmospheric CO2. Methane is discussed in individual chapters where appropriate, but the report makes no effort to provide a comprehensive synthesis and assessment of CH4 as part of the North American carbon budget. Similarly we provide no comprehensive treatment of black carbon, isoprene, or other volatile organic carbon compounds that represent a small fraction of global or continental carbon budgets.

plant growth in response to atmospheric CO_2 concentrations and climate and accelerated decomposition of dead organic matter in response to warmer temperatures.

Changes in temperature, precipitation, and other climate variables also contribute to year-to-year changes in carbon cycling. Nearly all of the biological, chemical, and physical processes responsible for exchange of carbon between atmosphere, land, and ocean are influenced to some degree by climate variables, and both ocean-atmosphere and land-atmosphere exchanges (sources and sinks) show year-to-year variation attributable to variability in climate (Prentice et al., 2001; Schaefer et al., 2002; Houghton, 2003; Sabine et al., 2004; Greenblatt and Sarmiento, 2004; Chapter 2 this report). This variability is believed to be responsible for the large year-to-year differences in the accumulation of CO_2 in the atmosphere; annual changes differ by as much as 3000 to 4000 million metric tons of carbon (Mt C) per year (Prentice et al., 2001; Houghton, 2003). Both land and ocean show changes, for example, in apparent response to climate conditions linked to El Niño events, although the variability in the net

> The human disturbance of the Earth's global carbon cycle during the industrial era and the resulting imbalance in the Earth's carbon budget and buildup of atmospheric CO_2 have consequences for climate and climate change.

land-atmosphere exchange is larger (Prentice et al., 2001; Houghton, 2003; Sabine et al., 2004). Figure 1.2 illustrates this variability, showing for North America year-to-year variation in satellite observations of the annual net transfer of carbon from the atmosphere to plants. Variability of this sort, in both land and ocean, contributes uncertainty to carbon budgeting and may appear as "noise" when attempting to detect "signals" of longer-term climate relevant trends (Sabine et al., 2004) or, eventually, signals of effective carbon management.

Many of the currently proposed options to prevent, minimize, or forestall future climate change will likely require management of the carbon cycle and concentrations of CO_2 in the atmosphere. That management includes both reducing sources, such as the combustion of fossil fuels, and enhancing sinks, such as uptake and storage (sequestration) in vegetation and soils. In either case, the formulation of options by decision makers and successful management of the Earth's carbon budget requires solid scientific understanding of the carbon cycle and the "ability to account for all carbon stocks, fluxes, and changes and to distinguish the effects of human actions from those of natural system variability" (CCSP, 2003).

So, why care about the carbon cycle? In short, because people care about the potential consequences of global climate change, they also, necessarily, care about the carbon cycle and the balance between carbon sources and sinks, natural and human, which determine the budget imbalance and accumulation of carbon in the atmosphere as CO_2.

1.3 OTHER IMPLICATIONS OF AN IMBALANCE IN THE CARBON BUDGET

The consequences of an unbalanced carbon budget with carbon accumulating in the atmosphere as CO_2 and CH_4 are not completely understood, but it is known that they extend beyond climate change alone. Experimental studies, for example, show that for many plant species, rates of photosynthesis often increase in response to elevated concentrations of CO_2 thus potentially increasing plant growth and even agricultural crop yields in the future. There is, however, considerable uncertainty about whether such "CO_2 fertilization" will continue into the future with prolonged exposure to elevated CO_2; and, of course, its potential beneficial effects on plants presume climatic conditions that are also favorable to plant and crop growth.

It is also increasingly evident that atmospheric CO_2 concentrations are responsible for increased acidity of the surface

ocean (Caldeira and Wickett, 2003), with potentially dire future consequences for corals and other marine organisms that build their skeletons and shells from calcium carbonate. Ocean acidification is a powerful reason, in addition to climate change, to care about the carbon cycle and the accumulation of CO_2 in the atmosphere (Orr *et al.*, 2005).

1.4 WHY THE CARBON BUDGET OF NORTH AMERICA?

The continent of North America has been identified as both a significant source and a significant sink of atmospheric CO_2 (IPCC, 2001; Pacala *et al.*, 2001; Goodale *et al.*, 2002; Gurney *et al.*, 2002; EIA, 2005). More than a quarter (27%) of global carbon emissions, from the combination of fossil-fuel burning and cement manufacturing, are attributable to North America (United States, Canada, and Mexico) (Marland *et al.*, 2003). North American plants remove CO_2 from the atmosphere and store it as carbon in plant biomass and soil organic matter, mitigating to some degree the human-caused (anthropogenic) sources. The magnitude of the "North American sink" has been previously estimated at anywhere from less than 100 Mt C per year to slightly more than 2000 Mt C per year (Turner *et al.*, 1995; Fan *et al.*, 1998), with a value near 350 to 750 Mt C per year most likely (Houghton *et al.*, 1999; Goodale *et al.*, 2002; Gurney *et al.*, 2002). The North American sink is thus, a substantial, if highly uncertain, fraction, from 15% to essentially 100%, of the extra-tropical Northern Hemisphere terrestrial sink estimated to be in the range of 600 to 2300 Mt C per year during the 1980s (Prentice *et al.*, 2001). It is also a reasonably large fraction (perhaps near 30%) of the global terrestrial sink estimated at 1900 Mt C per year for the 1980s (but with a range of uncertainty from a large sink of 3800 Mt C per year to a small source of 300 Mt C per year (Prentice *et al.*, 2001). The global terrestrial sink absorbs approximately one quarter of the carbon added to the atmosphere by human activities, but with uncertainties linked to the uncertainties in the size of that sink. Global atmospheric carbon concentrations would be substantially higher than they are without the partially mitigating influence of the sink in North America. However, estimates of that sink vary widely, and it needs to be better quantified.

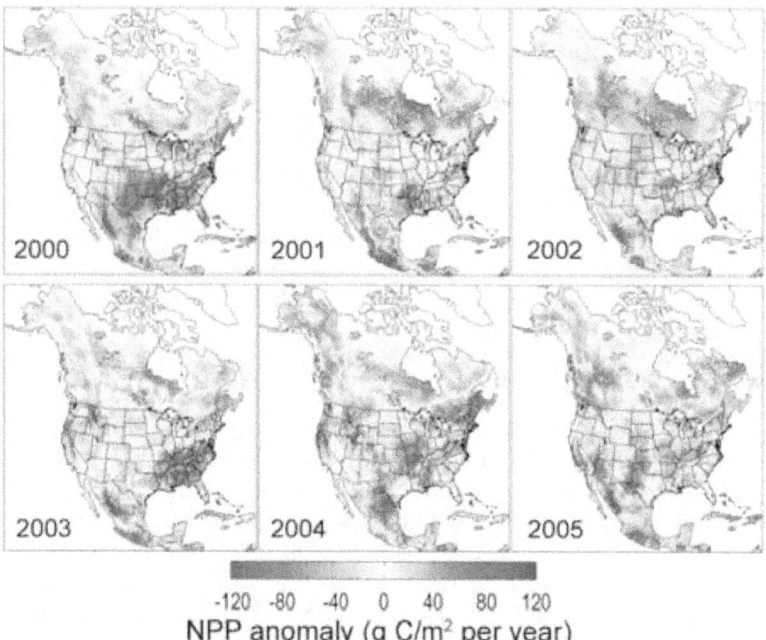

-120 -80 -40 0 40 80 120
NPP anomaly (g C/m² per year)

Figure 1.2 Variability in net primary production (NPP) for North America from 2000-2005. Values are the deviation from 6-year average annual NPP estimated by the MOD17 1-km resolution data product from the Moderate Resolution Imaging Spectroradiometer (MODIS) aboard the National Aeronautics and Space Administration (NASA) Terra and Aqua satellites. Blue indicates regions where that year's NPP, the net carbon fixed by vegetation from the atmosphere, was greater than average; red indicates where annual NPP was less than the average. See Running *et al.* (2004) for further information on the MODIS NPP product. Figure courtesy of Dr. Steven W. Running, University of Montana.

Some mechanisms that might be responsible for the North American terrestrial sink are reasonably well known. These mechanisms include, but are not limited to, the regrowth of forests following abandonment of agriculture, changes in fire and other disturbance regimes, historical climate change, and fertilization of ecosystem production by nitrogen deposition and elevated atmospheric CO_2 (Dilling *et al.*, 2003; Foley *et al.*, 2004). Recent studies have indicated that some of these processes are likely more important than others for the current North American carbon sink, with regrowth of forests on former agricultural land generally considered to be a major contributor, and with, perhaps, a significant contribution from enhanced plant growth in response to higher concentrations of atmospheric CO_2 (CO_2 fertilization) (Caspersen *et al.*, 2000; Schimel *et al.*, 2000; Houghton, 2002). But significant uncertainties remain (Caspersen *et al.*, 2000; Schimel *et al.*, 2000; Houghton, 2002), with some arguing that even the experimental evidence for CO_2 fertilization is equivocal at the larger spatial scales necessary for a significant terrestrial sink (*e.g.*, Nowak *et al.*, 2004; Friedlingstein *et al.*, 2006). The future of the current North American terrestrial sink is highly uncertain, and it depends on which mechanisms are the dominant drivers now and in the future.

Estimates of coastal carbon cycling and input of carbon from the land are equally uncertain (Liu *et al.*, 2000). Coastal processes are also difficult to parameterize in global carbon cycle models, which are often used to derive best-guess estimates for regional carbon budgets (Liu *et al.*, 2000). It is very important to quantify carbon fluxes in coastal margins of the area adjacent to the North American continent, lest regional budgets of carbon on land be misattributed.

North America is a major player in the global carbon cycle, in terms of both sources and sinks. Accordingly, understanding the carbon budget of North America is a necessary part of understanding the global carbon cycle. Such understanding is helpful for successful carbon management strategies to mitigate fossil-fuel emissions or stabilize concentrations of greenhouse gases in the atmosphere. Moreover, a large North American terrestrial sink generated by "natural" processes is an ecosystem service that would be valued at billions of dollars if purchased or realized through direct human economic and technological intervention. Its existence will likely influence carbon-management decision making, and it is important that its magnitude and its dynamics be well understood (Kirschbaum and Cowie, 2004; Canadell *et al.*, 2007).

It is particularly important to understand the likely future behavior of carbon in North America, including terrestrial and oceanic sources and sinks. Decisions made about future carbon management with expectations of the future behavior of the carbon cycle that proved to be significantly in error, could be costly. For example, future climate-carbon feedbacks could change the strength of terrestrial sinks and put further pressure on emission reductions to achieve atmospheric stabilization targets (Jones *et al.*, 2006; Canadell *et al.*, 2007). The future cannot be known, but understanding the current and historical carbon cycle will increase confidence in projections for appropriate consideration by decision makers.

> More than a quarter (27%) of global carbon emissions are attributable to North America.

> North America is a major player in the global carbon cycle, in terms of both sources and sinks.

1.5 CARBON CYCLE SCIENCE IN SUPPORT OF CARBON MANAGEMENT DECISIONS

Beyond understanding the science of the North American carbon budget and its drivers, increasing attention is now being given to deliberate management strategies for carbon (DOE, 1997, Hoffert *et al.*, 2002; Dilling *et al.*, 2003). Carbon management is now being considered at a variety of scales in North America. There are tremendous opportunities for carbon cycle science to improve decision making in this arena, whether in reducing carbon emissions from the use of fossil fuels, or in managing terrestrial carbon sinks. Many decisions in government, business, and everyday life are connected with the carbon cycle. They can relate to driving forces behind changes in the carbon cycle (such as consumption of fossil fuels) and strategies for managing them, and/or impacts of changes in the carbon cycle (such as climate change or ocean acidification) and responses to reduce their severity. Carbon cycle science can help to inform these decisions by providing timely and reliable information about facts, processes, relationships, and levels of confidence.

In seeking ways to use scientific information more effectively in decision making, we must pay particular attention to the importance of developing constructive scientist–stakeholder interactions. Studies of these interactions all indicate that neither scientific research nor assessments can be assumed to be relevant to the needs of decision makers if conducted in isolation from the context of those users' needs (Cash and Clark, 2001; Cash et al., 2003; Dilling et al., 2003; Parson, 2003). Carbon cycle science's support of decision making is more likely to be effective if the science connected with communication structures is considered by both scientists and users to be legitimate and credible. Well-designed scientific assessments can be one of these effective communication media.

The climate and carbon research community of North America, and a diverse range of stakeholders, recognize the need for an integrated synthesis and assessment focused on North America to (a) summarize what is known and what is known to be unknown, documenting the maturity as well as the uncertainty of this knowledge; (b) convey this information to scientists and to the larger community; and (c) ensure that our studies are addressing the questions of concern to society and decision-making communities. As the most comprehensive synthesis to date of carbon cycle knowledge and trends for North America, incorporating stakeholder interactions throughout its production[2], this report, the *First State of the Carbon Cycle Report (SOCCR)*, focused on *The North American Carbon Budget and Implications for the Global Carbon Cycle* is intended as a step in that direction.

---

[2] A discussion of stakeholder participation in the production of this report can be found in the *Preface* of this report.

2

The Carbon Cycle of North America in a Global Context

Coordinating Lead Author: Christopher B. Field, Carnegie Inst.

Lead Authors: Jorge Sarmiento, Princeton Univ.; Burke Hales, Oreg. State Univ.

KEY FINDINGS

- Human activity over the last two centuries, including combustion of fossil fuel and clearing of forests, has led to a dramatic increase in the concentration of atmospheric carbon dioxide. Global atmospheric carbon dioxide concentrations have risen by 31% since 1850 and are now higher than they have been for at least 420,000 years.
- North America is responsible for approximately 25% of the emissions produced globally in 2004 by fossil-fuel combustion, with the United States accounting for 86% of the North American total.
- Human-caused emissions (a carbon source) dominate the carbon budget of North America. Largely unmanaged, unintentional processes capture a fraction of this carbon in plants, soils, and other sinks. Currently, these sinks (970 ± 360 million metric tons of carbon (Mt C) per year, based on atmospheric inversion studies, or 530 ± 265 Mt C per year, based on the inventories used in this report) capture approximately 30-50% of the North American emissions, 7-13% of global fossil-fuel emissions, and 30-50% of the global terrestrial sink inferred from global budget analyses and atmospheric inversions.

- While the future trajectory of carbon sinks in North America is uncertain (substantial climate change could convert current sinks into sources), it is clear that the carbon cycle of the next few decades will be dominated by the large sources from fossil-fuel emissions.
- Because North American carbon emissions are at least a quarter of global emissions, a reduction in North American emissions would have global consequences.

2.1 THE GLOBAL CARBON CYCLE

The modern global carbon cycle is a collection of many different kinds of processes, with diverse drivers and dynamics, that transfer carbon among major pools in rocks, fossil fuels, the atmosphere, the oceans, and plants and soils on land (Sabine *et al.*, 2004b) (Figure 2.1). During the last two centuries, human actions, especially the combustion of fossil fuel and the clearing of forests, have altered the global carbon cycle in important ways. Specifically, these actions have led to a rapid, dramatic increase in the concentration of carbon dioxide (CO_2) in the atmosphere (Figure 2.2), changing the radiation balance of the Earth (Hansen *et al.*, 2005), and very likely causing much of the warming observed over the last 50 years (Hegerl et al., 2007). The cause of the recent increase in atmospheric CO_2 is confirmed beyond a reasonable doubt (Prentice, 2001). This does not imply, however, that the other components of the carbon cycle have remained unchanged during this period. In fact, the background, or unmanaged parts, of the carbon cycle have changed dramatically over the past two centuries. The consequence of these changes

is that only about 40% ± 15%[1] of the CO_2 emitted to the atmosphere from fossil-fuel combustion and forest clearing has remained there (Sabine *et al.*, 2004b). In essence, human actions have received a large subsidy from the unmanaged parts of the carbon cycle. This subsidy has sequestered, or hidden from the atmosphere, approximately 279 ± 160 billion tons (gigatons [Gt]) of carbon[2].

---

[1] Most of the uncertainty in this number is due to the approximately 100% uncertainty in carbon lost from forest clearing. This includes uncertainties in areas deforested, in conditions at the time of deforestation, and in the fate following deforestation (Houghton, 1999). Except where otherwise noted, the uncertainty bounds on the numbers in this chapter are expert assessments by the authors of the cited literature, based on synthesizing a wide range of empirical and modeling studies. The details of the approaches to assessing uncertainty are discussed in the literature cited.

[2] Unless specified otherwise, throughout this chapter, the pools and fluxes in the carbon cycle are presented in Gt C [1 Gt = 1 billion tons or 1×10^{15} g]. The mass of CO_2 is greater than the mass of carbon by the ratio of their molecular weights, 44/12 or 3.67 times; 1 km³ of coal contains approximately 1 Gt C.

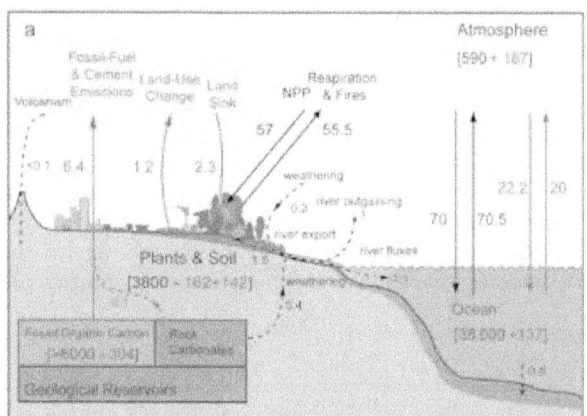
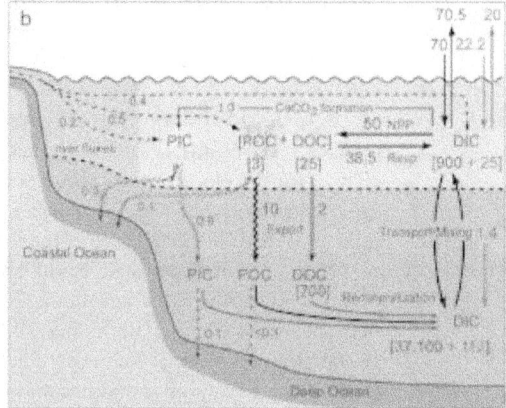
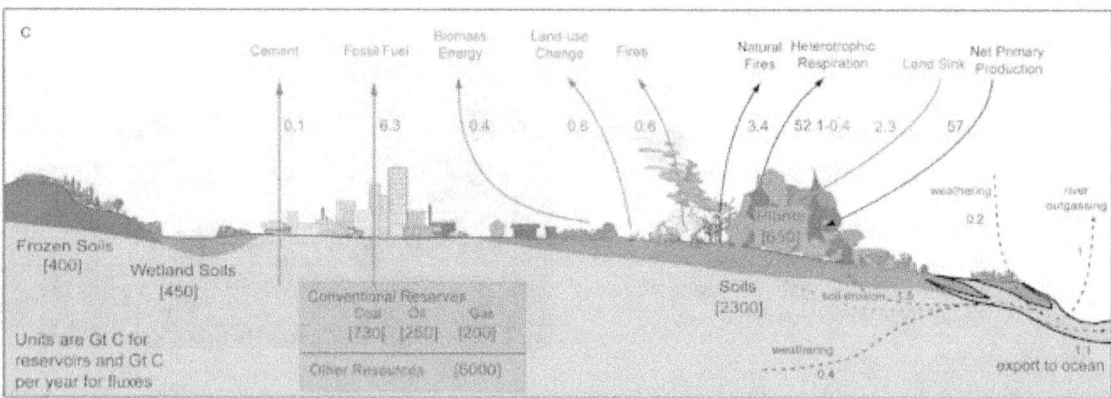

Figure 2.1 Schematic representation of the components of the global carbon cycle. The three panels show (a) the overall cycle, (b) the details of the ocean cycle, and (c) the details of the land cycle. For all panels, carbon stocks are in brackets, and fluxes have no brackets. Stocks and fluxes prior to human-influence are in black. Human-induced perturbations are in red. For stocks, the human-induced perturbations are the cumulative total through 2003. H uman-caused fluxes are means for the 1990s (the most recent available data for some fluxes). Redrawn from Sabine *et al.* (2004b) with updates through 2003 as discussed in the text.

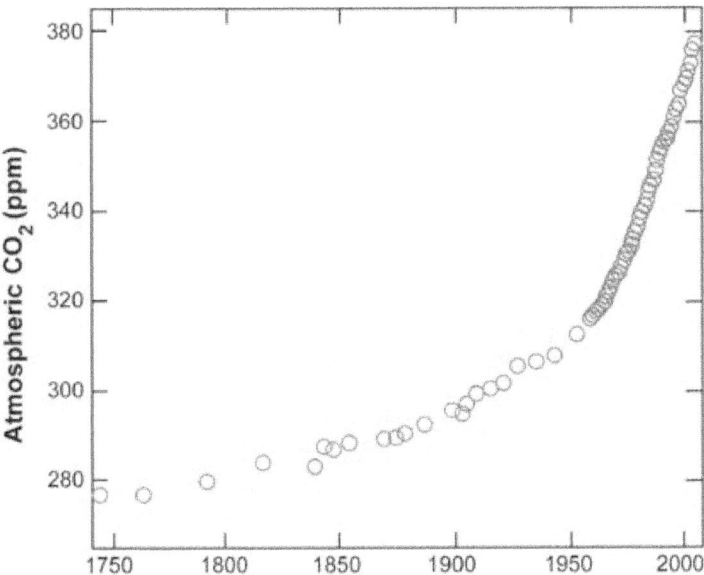

Figure 2.2 Atmospheric CO_2 concentration from 1750 to 2005. The data prior to 1957 (red circles) are from the Siple ice core (Friedli et al., 1986). The data since 1957 (blue circles) are from continuous atmospheric sampling at the Mauna Loa Observatory (Hawaii) (Keeling et al., 1976; Thoning et al., 1989) (with updates available at http://cdiac.ornl.gov/trends/co2/sio-mlo.htm).

terrestrial carbon cycle: plant growth on land annually fixes about 57 ± 9 Gt of atmospheric carbon, approximately ten times the annual emission from fossil-fuel combustion, into carbohydrates. Respiration by land plants, animals, and microorganisms, which provides the energy for growth, activity, and reproduction, returns a slightly smaller amount to the atmosphere. Part of the difference between photosynthesis and respiration is burned in wildfires, and part is stored as plant material or soil organic carbon. The second comprises the ocean carbon cycle: about 92 Gt of atmospheric carbon dissolves annually in the oceans, and about 90 Gt per year moves from the oceans to the atmosphere (While the gross fluxes have a substantial uncertainty, the difference is known to within ± 0.2 Gt)[3]. These air-sea fluxes are driven by cycling within the oceans that governs exchanges between pools of dissolved CO_2, bicarbonate (HCO_3), carbonate (CO_3), organic matter, and calcium carbonate ($CaCO_3$).

The recent subsidy, or sequestration, of carbon by the unmanaged parts of the carbon cycle, makes them critical for an accurate understanding of climate change. Future increases in carbon uptake in the unmanaged parts of the cycle could moderate the risks from climate change, while decreases or transitions from uptake to release could amplify the risks, perhaps dramatically.

In addition to its role in the climate, the carbon cycle intersects with a number of critical Earth system processes. Because plant growth is essentially the removal of CO_2 from the air through photosynthesis, agriculture and forestry contribute important fluxes. Wildfire is a major release of carbon from plants and soils to the atmosphere (Sabine *et al.*, 2004b). The increasing concentration of CO_2 in the atmosphere has already made the world's oceans more acid (Caldeira and Wickett, 2003). Future changes could dramatically alter the composition of ocean ecosystems (Feely *et al.*, 2004; Orr *et al.*, 2005).

2.1.1 The Unmanaged Global Carbon Cycle

The modern background, or unmanaged, carbon cycle includes the processes that occur in the absence of human actions. However, these processes are currently so altered by human influences on the carbon cycle that it is not appropriate to label them natural. This background part of the carbon cycle is dominated by two pairs of gigantic fluxes with annual uptake and release that are close to balanced (Sabine *et al.*, 2004b) (Figure 2.1). The first of these comprises the

Before the beginning of the industrial revolution, carbon uptake and release through these two pairs of large fluxes were almost balanced, with carbon uptake on land of approximately 0.45 ± 0.18 Gt C per year transferred to the oceans by rivers and released from the oceans to the atmosphere (Jacobson *et al.*, 2007). As a consequence, the level of CO_2 in the atmosphere varied by less than 25 parts per million (ppm) in the 10,000 years prior to 1850 (Joos and Prentice, 2004). However, atmospheric CO_2 was not always so stable. During the preceding 420,000 years, atmospheric CO_2 was 180-200 ppm during the ice ages and approximately 275 ppm during interglacial periods (Petit *et al.*, 1999). The lower ice-age concentrations in the atmosphere most likely reflect a transfer of carbon from the atmosphere to the oceans, possibly driven by changes in ocean circulation and sea-ice cover (Sigman and Boyle, 2000; Keeling and Stephens, 2001). Enhanced biological activity in the oceans, stimulated by increased delivery of iron-rich terrestrial dust, may have also contributed to this increased uptake (Martin, 1990).

> The increasing concentration of CO_2 in the atmosphere has already made the world's oceans more acid. Future changes could dramatically alter the composition of ocean ecosystems.

---

[3] This uncertainty is one-half the range among the subset of the 19 Ocean Carbon-Cycle Model Intercomparison Project (OCMIP) models that are consistent with the available ^{14}C and CFC-11 data (Matsumoto *et al.*, 2004).

In the distant past, the global carbon cycle was out of balance in a different way. Fossil fuels are the product of prehistorically stored plant growth, especially 354 to 290 million years ago in the Carboniferous period. During this time, luxuriant plant growth and geological activity combined to bury a small fraction of each year's growth. Over millions of years, this gradual burial led to the accumulation of vast stocks of fossil fuel. The total accumulation of fossil fuels is uncertain, but probably in the range of 6000 ± 3000 Gt (Sabine *et al.*, 2004b). This burial of carbon also led to a near doubling of atmospheric oxygen (Falkowski *et al.*, 2005).

2.1.2 Human-induced Perturbations to the Carbon Cycle

Since the beginning of the industrial revolution, there has been a massive release of carbon from fossil-fuel combustion and deforestation. Cumulative carbon emissions from fossil-fuel combustion, natural gas flaring, and cement manufacturing from 1751 through 2003 are 304 ± 30 Gt (Marland and Rotty, 1984; Andres *et al.*, 1999)[4]. Land-use change from 1850 to 2003, mostly from forest clearing, added another 162±160 Gt (DeFries *et al.*, 1999; Houghton, 1999)[5]. The rate of fossil-fuel consumption in any recent year would have required, for its production, more than 400 times the current global primary production (total plant growth) of the land and oceans combined (Dukes, 2003). This has led to a rapid increase in the concentration of CO_2 in the atmosphere since the mid-1800s, with atmospheric CO_2 rising by 31% (*i.e.*, from 287 ppm to 375 ppm in 2003; the increase from the mid-1700s was 35%).

In 2004, the three major countries of North America (Canada, Mexico, and the United States) together accounted for carbon emissions from fossil-fuel combustion of approximately 1.88 ± 0.2 Gt C, (about 25%) of the global total[6]. The United States, the world's largest emitter of CO_2, was responsible for 86% of the North American total. *Per capita* emissions in 2004 were 5.5 ± 0.5 metric tons in the United States, 4.9 ± 0.5 metric tons in Canada, and 1.0 ± 0.1 metric tons in Mexico. *Per capita* emissions in the United States were nearly 5 times the world average, 2.5 times the *per capita* emissions for Western Europe, and more than 8 times the average for Asia and Oceania (DOE EIA, 2006). The world's largest

countries, China and India, have total carbon emissions from fossil-fuel combustion and the flaring of natural gas that are growing rapidly. The 2004 total for China was 80% of that in the United States, and the total for India was 18% of that in the United States. *Per capita* emissions for China and India in 2004 were 18% and 5%, respectively, of the United States rate (DOE EIA, 2006).

2.2 ASSESSING GLOBAL AND REGIONAL CARBON BUDGETS

Changes in the carbon content of the oceans and plants and soils on land can be evaluated with at least five different approaches—flux measurements, inventories, inverse estimates based on atmospheric CO_2, process models, and calculation as a residual. The first method, direct measurement of carbon flux, is well developed over land for measurements over the spatial scale of up to 1 km[2], using the eddy flux technique (Wofsy *et al.*, 1993; Baldocchi and Valentini, 2004). Although eddy flux measurements are now collected at more than 100 networked sites, spatial scaling presents formidable challenges due to spatial heterogeneity. To date, estimates of continental-scale fluxes based on eddy flux must be regarded as preliminary. Over the oceans, eddy flux is possible (McGillis, 2001), but estimates based on air-sea CO_2 concentration difference are more widely used (Takahashi *et al.*, 1997).

Inventories, based on measuring trees on land (Birdsey and Heath, 1995) or carbon in ocean-water samples (Takahashi *et al.*, 2002; Sabine *et al.*, 2004a) can provide useful constraints on changes in the size of carbon pools, though their utility for quantifying short-term changes is limited. Inventories were the foundation of the recent conclusion that 118 ± 19 Gt of human-caused carbon entered the oceans through 1994 (Sabine *et al.*, 2004a) and that forests in the mid latitudes of the Northern Hemisphere absorbed and stored 0.6 to 0.7 Gt C per year in the 1990s (Goodale *et al.*, 2002). Changes in the atmospheric inventory of oxygen (O_2) (Keeling *et al.*, 1996) and carbon-13 (^{13}C) in CO_2 (Siegenthaler and Oeschger, 1987) provide a basis for partitioning CO_2 flux into land and ocean components.

Process models and inverse estimates based on atmospheric CO_2 (or CO_2 in combination with ^{13}C or O_2) also provide use-

---

[4] Updates through 2003 available at http://cdiac.ornl.gov/trends/mis/tre_glob.html.

[5] Updates through 2000 online at http://cdiac.ornl.gov/trends/landuse/houghton/houghton.html. The total through 2003 was extrapolated based on the assumption that the annual fluxes in 2001-2003 were the same as in 2000.

[6] Uncertainties in national and *per capita* emissions are based on data reported by individual countries.

ful constraints on carbon stocks and fluxes. Process models build from understanding the underlying principles of atmosphere/ocean or atmosphere/ecosystem carbon exchange to make estimates over scales of space and time that are relevant to the global carbon cycle. For the oceans, calibration against observations with tracers (*e.g.*, carbon-14 [^{14}C] and chlorofluorocarbons) (Broecker *et al.*, 1980) tends to nudge a wide range of models toward similar results. Sophisticated models with detailed treatment of the ocean circulation, chemistry, and biology all reach about the same estimate for the current ocean carbon sink, 1.5 to 1.8 Gt C per year (Greenblatt and Sarmiento, 2004) and are in quantitative agreement with data-inventory approaches. Models of the land carbon cycle take a variety of approaches. They differ substantially in the data used as constraints, in the processes simulated, and in the level of detail (Cramer *et al.*, 1999; Cramer *et al.*, 2001). Models that take advantage of satellite data have the potential for comprehensive coverage at high spatial resolution (Running *et al.*, 2004), but only over the time domain with available satellite data. Flux components related to human activities, deforestation, for example, have been modeled based on historical land use (Houghton *et al.*, 1999). At present, model estimates are uncertain enough that they are often used most effectively in concert with other kinds of estimates (*e.g.*, Peylin *et al.*, 2005).

Inverse estimates based on atmospheric gases (CO_2, ^{13}C in CO_2, or O_2) infer surface fluxes based on the spatial and temporal pattern of atmospheric gas concentration, coupled with information on atmospheric transport (Newsam and Enting, 1988). The atmospheric concentration of CO_2 is now measured with high precision at approximately 100 sites worldwide, with many of the stations added in the last decade (Masarie and Tans, 1995). The ^{13}C in CO_2 and high-precision O_2 are measured at far fewer sites. The basic approach is a linear Bayesian inversion (Tarantola, 1987; Enting, 2002), with many variations in the time scale of the analysis, the number of regions used, and the transport model. Inversions have more power to resolve year-to-year differences than mean fluxes (Rodenbeck *et al.*, 2003; Baker *et al.*, 2006). Limitations in the accuracy of atmospheric inversions come from the limited density of concentration measurements (especially in the tropics), uncertainty in the transport, and errors in the inversion process (Baker *et al.*, 2006). Recent studies that use a number of sets of CO_2 monitoring stations (Rodenbeck *et al.*, 2003), models (Gurney *et al.*, 2003; Law *et al.*, 2003; Gurney *et al.*, 2004; Baker *et al.*, 2006), temporal scales, and spatial regions (Pacala *et al.*, 2001), highlight the sources of the uncertainties and appropriate steps for managing them.

A final approach to assessing large-scale CO_2 fluxes is solving as a residual. At the global scale, the net flux to or from the land is often calculated as the residual left after

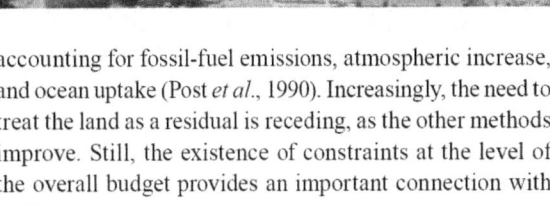

accounting for fossil-fuel emissions, atmospheric increase, and ocean uptake (Post *et al.*, 1990). Increasingly, the need to treat the land as a residual is receding, as the other methods improve. Still, the existence of constraints at the level of the overall budget provides an important connection with reality.

2.3 RECENT DYNAMICS OF THE UNMANAGED CARBON CYCLE

Of the approximately 466 ± 160 Gt C added to the atmosphere by human actions through 2003, only about 187 ± 5 Gt remain. The "missing carbon" must be stored, at least temporarily, in the oceans and in ecosystems on land. Based on a recent ocean inventory, 118 ± 19 Gt of the missing carbon was in the oceans, as of 1994 (Sabine *et al.*, 2004a). Extending this calculation, based on recent sinks (Takahashi *et al.*, 2002; Gloor *et al.*, 2003; Gurney *et al.*, 2003; Matear and McNeil, 2003; Matsumoto *et al.*, 2004), leads to an estimate of 137 ± 24 Gt C through 2003. This leaves about 142 ± 160 Gt that must be stored on land (with most of the uncertainty due to the uncertainty in emissions from land use). Identifying the processes responsible for the uptake on land, their spatial distribution, and their likely future trajectory has been one of the major goals of carbon cycle science over the last decade.

Much of the recent research on the global carbon cycle has focused on annual fluxes and their spatial and temporal variation. The temporal and spatial patterns of carbon flux provide a pathway to understanding the underlying mechanisms. Based on several different approaches, carbon

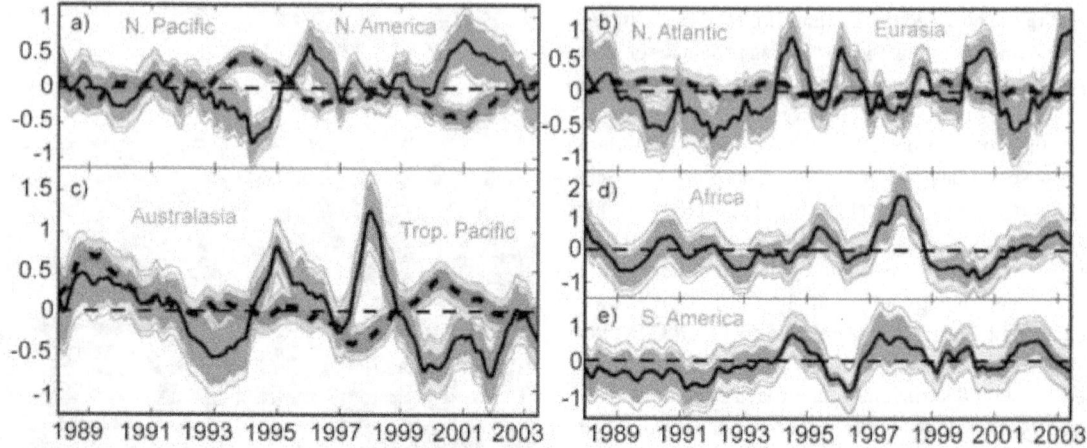

Figure 2.3 The 13-model mean CO_2 flux interannual variability (Gt C per year) for several continents (solid lines) and ocean basins (dashed lines). In each panel, the dark inner band is the 1σ intermodel spread, the lighter adjacent band is the 1σ estimation uncertainty on interannual variability, and the outer band (visible only for the land) is the root sum of squares of the two uncertainty components. (a) North Pacific and North America, (b) Atlantic north of 15°N and Eurasia, (c) Australasia and Tropical Pacific, (d) Africa, and (e) South America (note the different scales for Africa and South America) (Baker et al., 2006).

uptake by the oceans averaged 1.7 ± 0.2 Gt C per year[7] for the period from 1992-1996 (Takahashi et al., 2002; Gloor et al., 2003; Gurney et al., 2003; Matear and McNeil, 2003; Matsumoto et al., 2004). The total human-caused flux is this amount, plus 0.45 Gt per year of preindustrial outgasing, for a total of 2.2 ± 0.4 Gt per year. This rate represents an integral over high-latitude areas, which are gaining carbon, and the tropics, which are losing carbon (Takahashi et al., 2002; Gurney et al., 2003; Gurney et al., 2004; Jacobson et al., 2007). Interannual variability in the ocean sink for CO_2, though substantial (Greenblatt and Sarmiento, 2004), is much smaller than interannual variability on the land (Baker et al., 2006).

In the 1990s, carbon releases from land-use change were more than balanced by ecosystem uptake, leading to a net sink on land (without accounting for fossil-fuel emissions) of 1.1 ± 1.5 Gt C per year (Schimel et al., 2001; Sabine et al., 2004b). The dominant sources of recent interannual variation in the net land flux were El Niño and the eruption of Mount Pinatubo in 1991 (Bousquet et al., 2000; Rodenbeck et al., 2003; Baker et al., 2006), with most of the year-to-year variation in the tropics (Figure 2.3). Fire likely plays a large role in this variability (van der Werf et al., 2004).

On a time scale of thousands of years, the ocean will be the sink for more than 90% of the carbon released to the atmosphere by human activities (Archer et al., 1998). The rate of CO_2 uptake by the oceans is, however, limited. Carbon dioxide enters the oceans by dissolving in seawater. The rate of this process is determined by the concentration difference between the atmosphere and the surface waters and by an air-sea exchange coefficient related to wave action, wind, and turbulence (Le Quéré and Metzl, 2004). Because the surface waters represent a small volume with limited capacity to store CO_2, the major control on ocean uptake is at the level of moving carbon from the surface to intermediate and deep waters. Important contributions to this transport come from the large-scale circulation of the oceans, especially the sinking of cold water in the Southern Ocean and, to a lesser extent, the North Atlantic.

On land, numerous processes contribute to carbon storage and carbon loss. Some of these are directly influenced through human actions (e.g., the planting of forests, conversion to no-till agriculture, or the burying of organic wastes in landfills). The human imprint on others is indirect. This category includes ecosystem responses to climate change (e.g., warming and changes in precipitation), changes in the composition of the atmosphere (e.g., increased CO_2 and increased tropospheric ozone), and delayed consequences of past actions (e.g., regrowth of forests after earlier harvesting). Early analyses of the global carbon budget (e.g., Bacastow and Keeling, 1973) typically assigned all of the net flux on land to a single mechanism, fertilization of plant growth by increased atmospheric CO_2. Recent evidence emphasizes the diversity of mechanisms.

In the 1990s, carbon releases from land-use change were more than balanced by ecosystem uptake, leading to a net sink on land (without accounting for fossil-fuel emissions).

[7] This uncertainty is one-half the range among the subset of the 19 Ocean Carbon-Cycle Model Intercomparison Project (OCMIP) models that are consistent with the available ^{14}C and CFC-11 data (Matsumoto et al., 2004).

2.3.1 The Carbon Cycle of North America

The land area of North America is a large source of carbon, but the residual (without emissions from fossil-fuel combustion) is, by most estimates, currently a sink for carbon. This conclusion for the continental scale is based mainly on the results of atmospheric inversions. Several studies address the carbon balance of particular ecosystem types (*e.g.*, forests [Kurz and Apps, 1999; Goodale *et al.*, 2002; Chen *et al.*, 2003]). Pacala and colleagues (2001) used a combination of atmospheric and land-based techniques to estimate that the 48 contiguous United States are currently a carbon sink of 0.3 to 0.6 Gt C per year. This estimate and a discussion of the processes responsible for recent sinks in North America are updated in Chapter 3 of this report. Based on inversions using 13 atmospheric transport models, North America was a carbon sink of 0.97 ± 0.36 Gt C per year from 1991-2000 (Baker *et al.*, 2006)[8]. Over the area of North America, this amounts to an annual carbon sink of 39.6 g C per square meter per year, similar to the sink inferred for all northern lands (North America, Europe, Boreal Asia, and Temperate Asia) of 32.5 g C per square meter per year (Baker *et al.*, 2006).

Very little of the current carbon sink in North America is a consequence of deliberate action to absorb and store (sequester) carbon. Some is a collateral benefit of steps to improve land management, for increasing soil fertility, im-

proving wildlife habitat, *etc*. Much of the current sink is unintentional, a consequence of historical changes in technologies and preferences in agriculture, transportation, and urban design.

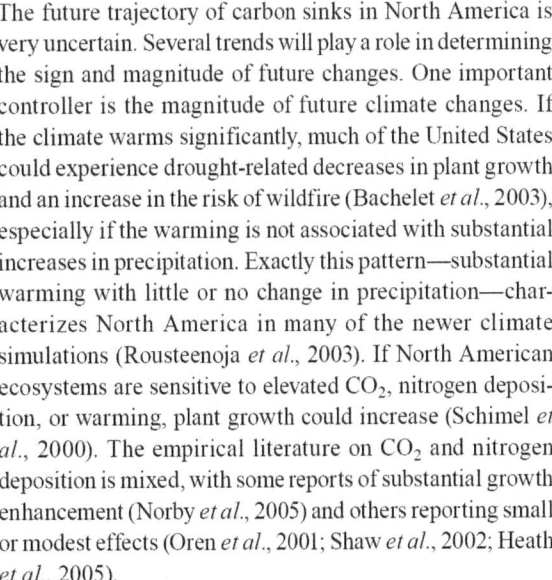

2.4 CARBON CYCLE OF THE FUTURE

The future trajectory of carbon sinks in North America is very uncertain. Several trends will play a role in determining the sign and magnitude of future changes. One important controller is the magnitude of future climate changes. If the climate warms significantly, much of the United States could experience drought-related decreases in plant growth and an increase in the risk of wildfire (Bachelet *et al.*, 2003), especially if the warming is not associated with substantial increases in precipitation. Exactly this pattern—substantial warming with little or no change in precipitation—characterizes North America in many of the newer climate simulations (Rousteenoja *et al.*, 2003). If North American ecosystems are sensitive to elevated CO_2, nitrogen deposition, or warming, plant growth could increase (Schimel *et al.*, 2000). The empirical literature on CO_2 and nitrogen deposition is mixed, with some reports of substantial growth enhancement (Norby *et al.*, 2005) and others reporting small or modest effects (Oren *et al.*, 2001; Shaw *et al.*, 2002; Heath *et al.*, 2005).

Overall, the carbon budget of North America is dominated by carbon releases from the combustion of fossil fuels. Recent sinks, largely from carbon uptake in plants and soils, may approach 50% of the recent fossil-fuel source (Baker *et al.*, 2006). Most of this uptake appears to be a rebound, as natural and managed ecosystems recover from past disturbances. Little evidence supports the idea that these ecosystem sinks will increase in the future. Substantial climate change could convert current sinks into sources (Gruber *et al.*, 2004).

In the future, trends in the North American energy economy may intersect with trends in the natural carbon cycle. A large-scale investment in afforestation could offset substantial future emissions (Graham, 2003). However, costs of this kind of effort would include loss of the new-forested area from its previous uses (including grazing or agriculture), the energy costs of managing the new

---

[8] This uncertainty is a sample standard deviation across monthly output from 13 models.

forests, and any increases in emissions of non-CO_2 greenhouse gases from the new forests. Large-scale investments in biomass energy (energy produced from vegetative matter) would have similar costs but would result in offsetting emissions from fossil-fuel combustion, rather than sequestration (Giampietro *et al.*, 1997). The relative costs and benefits of investments in afforestation and biomass energy will require careful analysis (Kirschbaum, 2003). Investments in other energy technologies, including wind and solar, will require some land area, but the impacts on the natural carbon cycle are unlikely to be significant or widespread (Hoffert *et al.*, 2002; Pacala and Socolow, 2004).

Like the present, the carbon cycle of North America during the next several decades will be dominated by fossil-fuel emissions. Deliberate geological sequestration may become an increasingly important component of the budget sheet. Still, progress in controlling the net release to the atmosphere must be centered on the production and consumption of energy rather than the processes of the unmanaged carbon cycle. North America has many opportunities to decrease emissions (Chapter 4 this report). Nothing about the status of the unmanaged carbon cycle provides a justification for assuming that it can compensate for emissions from fossil-fuel combustion.

Nothing about the status of the unmanaged carbon cycle provides a justification for assuming that it can compensate for emissions from fossil-fuel combustion.

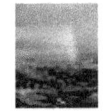

The North American Carbon Budget Past and Present

Coordinating Lead Author: Stephen Pacala, Princeton Univ.

Lead Authors: Richard A. Birdsey, USDA Forest Service; Scott D. Bridgham, Univ. Oreg.; Richard T. Conant, Colo. State Univ.; Kenneth Davis, The Pa. State Univ.; Burke Hales, Oreg. State Univ.; Richard A. Houghton, Woods Hole Research Center; Jennifer C. Jenkins, Univ. Vt.; Mark Johnston, Saskatchewan Research Council; Gregg Marland, ORNL and Mid Sweden Univ. (Östersund); Keith Paustian, Colo. State Univ.

Contributing Authors: John Caspersen, Univ. Toronto; Robert Socolow, Princeton Univ.; Richard S. J. Tol, Hamburg Univ.

KEY FINDINGS

- Fossil-fuel carbon emissions in the United States, Canada, and Mexico totaled 1856 million tons per year in 2003 (plus or minus 10%). This represents 27% of global fossil-fuel emissions.
- Approximately 30% of North American fossil-fuel emissions are offset by a natural sink estimated at 505 million tons of carbon per year (plus or minus 50%) for the period including 2003 caused by a variety of factors, including forest regrowth, wildfire suppression, and agricultural soil conservation.
- In 2003, North America emitted a net of 1351 million tons of carbon per year (plus or minus 25%) to the atmosphere.
- North American carbon dioxide emissions from fossil fuel have increased at an average rate of approximately 1% per year for the last 30 years.
- Growth in emissions accompanies the historical growth in the industrial economy and Gross Domestic Product (GDP) of North America. However, at least in the United States and Canada, the rate of emissions growth is less than the growth in GDP, reflecting a decrease in the carbon intensity of these economies.
- Fossil-fuel emissions from North America are expected to continue to grow, but more slowly than GDP.
- Historically, the plants and soils of the United States and Canada were sources for atmospheric carbon dioxide, primarily as a consequence of the expansion of croplands into forests and grasslands. In recent decades these regions have shifted from source to sink as forests recover from agricultural abandonment, fire suppression is practiced, and logging is reduced, and as a result, these regions are now accumulating carbon. In Mexico, emissions of carbon continue to increase due to net deforestation.
- The future of the North American carbon sink is highly uncertain. The contribution of recovering forests to this sink is likely to decline as these forests mature, but we do not know how much of the sink is due to fertilization of the ecosystems by nitrogen in air pollution and by increasing carbon dioxide concentrations in the atmosphere, nor do we understand the impact of ozone in the lower atmosphere or how the sink will change as the climate changes. Increases in decomposition and wildfire caused by climate change could, in principle, convert the sink into a source.
- The current magnitude of the North American sink offers the possibility that significant mitigation of fossil-fuel emissions could be accomplished by managing forests, rangelands, and croplands to increase the carbon stored in them. However, the range of uncertainty in these estimates is at least as large as the estimated values themselves.
- Current trends towards lower carbon intensity of United States' and Canadian economies increase the likelihood that a portfolio of carbon management technologies will be able to reduce the 1% annual growth in fossil-fuel emissions. This same portfolio might be insufficient if carbon emissions were to begin rising at the approximately 3% growth rate of GDP.

3.1 FOSSIL FUEL

Fossil-fuel carbon emissions in the United States, Canada, and Mexico totaled 1856 million metric tons of carbon (Mt C) per year in 2003 and have increased at an average rate of approximately 1% per year for the last 30 years (United States = 1582, Canada = 164, Mexico = 110 Mt C per year, see Figure 3.1)[1]. This represents 27% of global emissions, from a continent with 7% of the global population and 25% of global GDP (EIA, 2005).

The United States is the world's largest emitter in absolute terms (EIA, 2005). The United States' *per capita* emissions are also among the largest in the world (5.4 t C per year), but the carbon intensity of its economy (emissions per unit GDP)

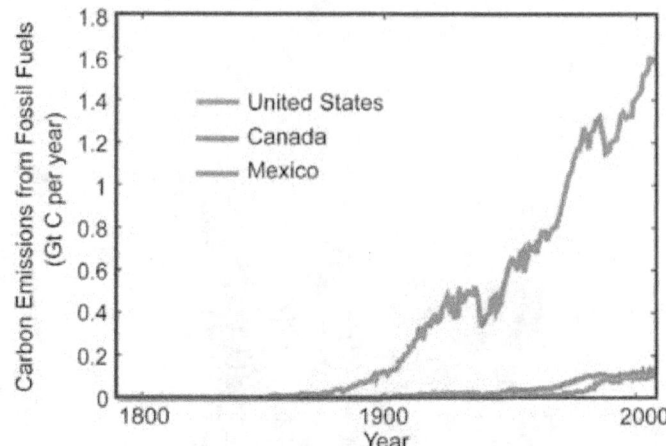

Figure 3.1 Historical carbon emissions from fossil fuel in the United States, Canada, and Mexico. Data from the U.S. Energy Information Administration (EIA, 2005).

at 0.15 metric tons of emitted carbon per dollar of GDP is close to the world's average of 0.14 t C/$ (EIA, 2005). Total United States' emissions have grown at close to the North American average rate of about 1.0% per year over the past 30 years, but the United States' *per capita* emissions have been roughly constant, while the carbon intensity of the United States' economy has decreased at a rate of about 2% per year (see Figures 3.1 to 3.4).

Absolute emissions grew at 1% per year even though *per capita* emissions were roughly constant simply because of population growth at an average rate of 1%. The constancy of United States' *per capita* values masks faster than 1% growth in some sectors (*e.g.*, transportation) that was balanced by slower growth in others (*e.g.*, increased manufacturing energy efficiency) (Figures 3.2, 3.3, and 3.4).

Historical decreases in United States' carbon intensity began early in the twentieth century and continue despite the approximate stabilization of *per capita* emissions (Figure 3.2). Why has the United States' carbon intensity declined? This question is the subject of extensive literature on the so-called structural decomposition of the energy system and on the relationship between GDP and the environment (*i.e.*, Environmental Kuznets Curves; Grossman and Krueger, 1995; Selden and Song, 1994). See, for example, Greening *et al.* (1997, 1998), Casler and Rose (1998), Golove and Schipper (1998), Rothman (1998), Suri and

Chapman (1998), Greening *et al.* (1999), Ang and Zhang (2000), Greening *et al.* (2001), Davis *et al.* (2002), Kahn (2003), Greening (2004), Lindmark (2004), Aldy (2005), and Lenzen *et al.* (2006).

Possible causes of the decline in United States' carbon intensity include: structural changes in the economy, technological improvements in energy efficiency, behavioral changes by consumers and producers, the growth of renewable and nuclear energy, and the displacement of oil consumption

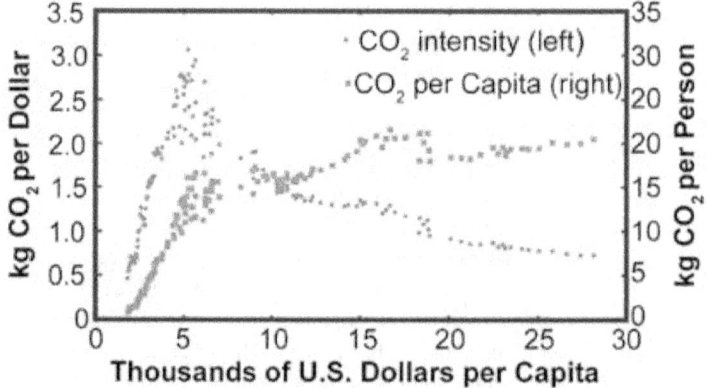

Figure 3.2 The historical relationship between United States' *per capita* GDP and United States' carbon intensity (green symbols, kg CO_2 emitted per 1995 dollar of GDP) and *per capita* carbon emissions (blue symbols, kg CO_2 per person). Each symbol shows a different year and each of the two time series progresses roughly chronologically from left (early) to right (late) and ends in 2002. *Source:* Maddison (2003), Marland *et al.* (2005). Thus, the blue square farthest to the right shows United States' *per capita* CO_2 emissions in 2002. The square second farthest to the right shows *per capita* emissions in 2001. The third farthest to the right shows 2000, and so on. Note that *per capita* emissions have been roughly constant over the last 30 years (squares corresponding to *per capita* GDP greater than approximately $16,000).

---

[1] Uncertainty estimates for the numerical data presented in this chapter can be found in Tables 3.1 through 3.3.

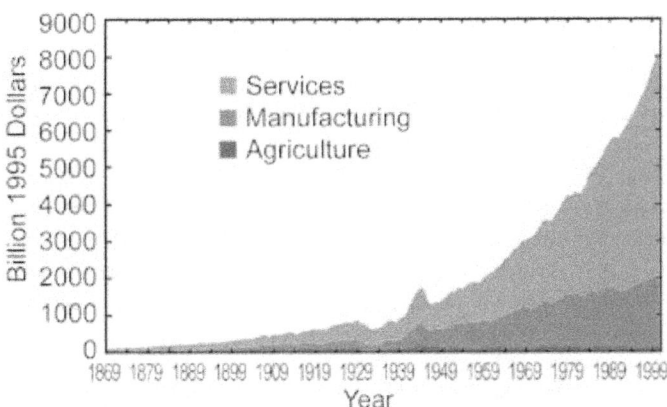

Figure 3.3 Historical United States' GDP divided among the manufacturing, services, and agricultural sectors. *Source:* Mitchell (1998), WRI (2005).

by gas and/or of coal consumption by oil and gas (if we produce the same amount of energy from coal, oil, and gas, then the emissions from oil are only 80% of those from coal, and from gas only 75% of those from oil) (Casler and Rose, 1998; Ang and Zhang, 2000). The last two items on this list are not dominant causes because we observe that both primary energy consumption and carbon emissions grew at close to 1% per year over the past 30 years (EIA, 2005). At least in the United States, there has been no significant decarbonization of the energy system during this period. However, all of the other items on the list play a significant role. The economy has grown at an annual rate of 2.8% over the last three decades because of 3.6% growth in the service sector; manufacturing grew at only 1.5% per year (Figure 3.3). Because the service sector has much lower carbon intensity than manufacturing, this faster growth of services reduces the country's carbon intensity. If all of the growth in the service sector had been in manufacturing from 1971 to 2001, then the emissions would have grown at 2% per year instead of 1% (here we equate the manufacturing sector in Figure 3.3 with the industrial sector in Figure 3.4). So, structural change is at least one-half of the answer. Because the service sector is likely to continue to grow more rapidly than other sectors of the economy, we expect that carbon emissions will continue to grow more slowly than GDP. This is important because it implies considerable elasticity in the relationship between emissions growth and economic growth. It also widens the range of policy options that are now technologically possible. For example, a portfolio of current technologies able to convert the 1% annual growth in emissions into a 1% annual decline, might be insufficient if carbon emissions were to begin rising at the ~3% growth rate of GDP (Pacala and Socolow, 2004).

However, note that industrial emissions are approximately constant (Figure 3.4) despite 1.5% economic growth in manufacturing (Figure 3.3). This decrease in carbon intensity is caused both by within-sector structural shifts (*i.e.*, from

heavy to light manufacturing) and by technological improvements (See Part II of this report). Emissions from the residential sector are growing at roughly the same rate as the population (Figure 3.4; 30-year average of 1.0% per year), while emissions from transportation are growing faster than the population, but slower than GDP (Figure 3.4; 30-year average of 1.4% per year). The difference between the 3% growth rate of GDP and the 1.6% growth in emissions from transportation is not primarily due to technological improvement because carbon emissions per mile traveled have been level or increasing over the period (Chapter 7 this report).

3.2 CARBON SINKS[2]

Approximately 30% of North American fossil-fuel emissions are offset by a natural sink estimated at 505 Mt C per year caused by a variety of factors, including forest regrowth, fire suppression, and agricultural soil conservation. The sink absorbs 489 Mt C per year in the United States and 64 Mt C per year in Canada. Mexican ecosystems create a net source of 48 Mt C per year. Rivers and international trade also export a net of 161 Mt C per year that was captured from the atmosphere by the continent's ecosystems, and so North America absorbs 666 Mt C per year of atmospheric CO_2 (666 = 505 + 161). Because most of these net exports will return

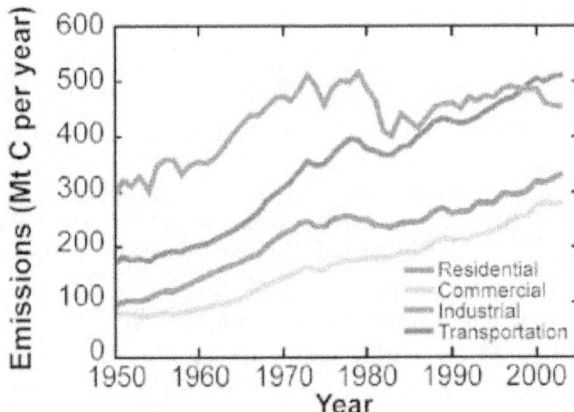

Figure 3.4 Historical United States' carbon emissions divided among the residential, services, manufacturing, and transportation sectors. *Source:* EIA (2005).

[2] See Tables 3.1 and 3.2 for estimates, citations, and uncertainty of estimates

to the atmosphere elsewhere within 1 year (*e.g.* carbon in exported grain will be eaten, metabolized, and exhaled as CO_2), the net North American sink is rightly thought of as 505 Mt C per year even though the continent absorbs a net of 666 Mt C per year. Moreover, coastal waters may be small net emitters to the atmosphere at the continental scale (19 Mt C per year), but this flux is highly uncertain (Chapter 15 this report). The portion of the coastal flux caused by human activity is thought to be close to zero, so coastal sea-air exchanges should be excluded from the continental carbon sink.

As reported in Chapter 2, the sink in the United States is approximately 40% (plus or minus 20%) the size of the global carbon sink, while the sink in Canada is about 7% (plus or minus 7%) the size of the global sink. The source in Mexico reduces the global sink by ~4% (plus or minus more than 4%). The reason for the disproportionate importance of United States' sinks is probably the unique land-use history of the country (summary in Appendix A). During European settlement, large amounts of carbon were released from the harvest of virgin forests and the plowing of virgin soils to create agricultural lands. The abandonment of many of the formerly agricultural lands in the east and the regrowth of forest is a unique event globally and is responsible for about one-half of the United States' sink (Houghton *et al.*, 2000). Most of the United States' sink thus represents a one-time recapture of some of the carbon that was released to the atmosphere during settlement. In contrast, Mexican ecosystems, like those of many tropical nations, are still a net carbon source because of ongoing deforestation (Masera *et al.*, 1997).

The non-fossil fluxes in Tables 3.1 and 3.2 are derived exclusively from inventory methods in which the total amount of carbon in a pool (*i.e.*,

living forest trees plus forest soils) is measured on two occasions. The difference between the two measurements shows if the pool is gaining (sink) or losing (source) carbon. Carbon inventories are straightforward in principle, but of uneven quality in practice. For example, we know the carbon in living trees in the United States relatively accurately because the U.S. Forest Service Forest Inventory program measures trees systematically in more than 200,000 locations. However, we must extrapolate from a few measurements of forest soils with models because there is no national inventory of carbon in forest soils.

Although the fluxes in Tables 3.1 and 3.2 represent the most recent published estimates, with most less than five years old, a few are older than ten years (see the citations at the bottom of each table). Also, the time interval between inventories varies among the elements of the tables, with most covering a five to ten year period. In these tables and throughout this document we report uncertainties using the six categories outlined in Box 3.1.

Table 3.1 Annual net emissions (source = positive) or uptake (land sink = negative) of carbon in millions of tons circa 2003 (see Box 3.1 for uncertainty conventions).

Source (positive) or Sink (negative)	United States	Canada	Mexico	N. America
Fossil source (positive)				
Fossil fuel (oil, gas, coal)	1582[a]***** (681, 328, 573)	164[a]***** (75, 48, 40)	110[a]***** (71, 29, 11)	1856***** (828, 405, 624)
Non-fossil carbon sink (negative) or source (positive)				
Forest	−256[b]***	−28[c]***	+52[d]**	−233***
Wood products	−57[e]***	−11[f]***	ND	−68***
Woody encroachment	−120[g]*	ND	ND	−120*
Agricultural soils	−8[h]***	−2[h]***	ND	−10[h]***
Wetlands	−23[i]*	−23[i]*	−4[i]*	−49*
Rivers and reservoirs	−25[j]**	ND	ND	−25*
Total carbon source or sink	−489***	−64**	48*	−505***
Net carbon source (positive)	1093****	100***	158***	1351****

[a] http://www.eia.doe.gov/env/inlenv.htm
[b] Smith and Heath (2005) for above-ground carbon, but including 20 Mt C per year for United States' urban and suburban forests from Chapter 14, and Pacala *et al.* (2001) for below-ground carbon.
[c] Environment Canada (2006), Chapter 11, plus 11 Mt C per year for Canadian urban and suburban forests, Chapter 14.
[d] Masera *et al.* (1997)
[e] Skog *et al.* (2004), Skog and Nicholson (1998)
[f] Goodale *et al.* (2002)
[g] Houghton *et al.* (1999), Hurtt *et al.* (2002), Houghton and Hackler (1999).
[h] Chapter 10; Uncertain; Could range from -7 Mt C per year to -14 Mt C per year for North America.
[i] Chapter 13
[j] Stallard (1998); Pacala *et al.* (2001)
ND indicates that no data are available.

BOX 3.1: CCSP SAP 2.2 Uncertainty Conventions

*****	=	95% certain that the actual value is within 10% of the estimate reported,
****	=	95% certain that the estimate is within 25%,
***	=	95% certain that the estimate is within 50%,
**	=	95% certain that the estimate is within 100%, and
*	=	uncertainty greater than 100%.
†	=	The magnitude and/or range of uncertainty for the given numerical value(s) is not provided in the references cited.

In addition to inventory methods, it is also possible to estimate carbon sources and sinks by measuring carbon dioxide (CO_2) in the atmosphere. For example, if air exits the border of a continent with more CO_2 than it contained when it entered, then there must be a net source of CO_2 somewhere inside the continent. We do not include estimates obtained in this way because they are still highly uncertain at continental scales. Pacala *et al.* (2001) found that atmosphere- and inventory-based methods gave consistent estimates of United States' ecosystem sources and sinks but that the range of uncertainty from the former was considerably larger than the range from the latter. For example, by far the largest published estimate for the North American carbon sink was produced by an analysis of atmospheric data by Fan *et al.* (1998) (-1700 Mt C per year). The appropriate inventory-based estimate to compare this to is our -666 Mt C per year of net absorption (atmospheric estimates include net horizontal exports by rivers and trade), and this number is well within the wide uncertainty limits in Fan *et al.* (1998). The allure of estimates from atmospheric data is that they do not risk missing critical uninventoried carbon pools. But in practice, they are still far less accurate at continental scales than a careful inventory (Pacala *et al.*, 2001). Using today's technology, it should be possible to complete a comprehensive inventory of the sink at national scales with the same accuracy as the United States' forest inventory currently achieves for above-ground carbon in forests (25%, Smith and Heath, 2005). Moreover, this inventory would provide disaggregated information about the sink's causes and geographic distribution. In contrast, estimates

from atmospheric methods rely on the accuracy of atmospheric models, and estimates obtained from different models vary by 100% or more at the scale of the United States, Canada, or Mexico (Gurney *et al.*, 2004). Nonetheless, extensions of the atmospheric sampling network should improve the accuracy of atmospheric methods and might allow them to achieve the accuracy of inventories at regional and whole-country scales. In addition, atmospheric methods will continue to provide an independent check on inventories to make sure that no large flux is missed, and atmospheric methods will remain the only viable method to assess interannual variation in the continental flux of carbon.

The current magnitude of the North American sink (documented in Tables 3.1 and 3.2) offers the possibility that significant carbon mitigation could be accomplished by managing forests, rangelands, and croplands to increase the carbon stored in them. However, many of the estimates in Tables 3.1 and 3.2 are highly uncertain; for some, the range of uncertainty is larger than the value reported. The largest contributors to the uncertainty in the United States' sink are the amount of carbon stored on rangelands because of the encroachment of woody vegetation and the lack of comprehensive and continuous inventory of Alaskan lands. A carbon inventory of these lands would do more to constrain the size of the United States' sink than would any other measurement program of similar cost. Also, we still lack

Table 3.2 Annual net horizontal transfers of carbon in millions of tons (see Box 3.1 for uncertainty conventions).

Net horizontal transfer: imports exceed exports = positive; exports exceed imports = negative	United States	Canada	Mexico	North America
Wood products	14[c],*****	−74[a],****	−1[b],*	−61****
Agriculture products	−65[d],***	ND	ND	−65***
Rivers to ocean	−35[d],**	ND	ND	−35*
Total net absorption (Total carbon source or sink in Table 3.1 plus exports)	−575***	−138**	47*	−666**
Net absorption (negative) or emission (positive) by coastal waters	ND	ND	ND	19[e],*

[a] Environment Canada (2005), World Forest Institute (2006)

[b] Masera *et al.* (1997)

[c] Skog *et al.* (2004), Skog and Nicholson (1998)

[d] Pacala *et al.* (2001)

[e] Chapter 15

ND indicates that no data are available.

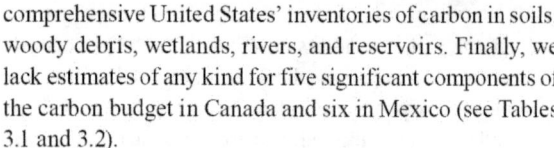

comprehensive United States' inventories of carbon in soils, woody debris, wetlands, rivers, and reservoirs. Finally, we lack estimates of any kind for five significant components of the carbon budget in Canada and six in Mexico (see Tables 3.1 and 3.2).

The cause and future of the North American carbon sink is also highly uncertain. Although we can document the accumulation of carbon in ecosystems and wood products, we do not know how much of the sink is due to fertilization of the ecosystems by the nitrogen in air pollution and by the added CO_2 in the atmosphere. We do not fully understand the impact of tropospheric ozone, nor do we understand precisely how the sink will change as the climate changes. Research is mixed about the importance of nitrogen and CO_2 fertilization (Casperson *et al.*, 2000; Oren *et al.*, 2001; Hungate *et al.*, 2003; Luo, 2006; Körner *et al.*, 2005). If these factors are weak, then, all else being equal, we expect the North American sink to decline over time as ecosystems complete their recovery from past exploitation (Hurtt *et al.*, 2002). However, if these factors are strong, then the sink could grow in the future. Similarly, global warming is expected to lengthen the growing season in most parts of North America, which should increase the sink (but see Goetz *et al.*, 2005). But warming is also expected to increase forest fire and the rate of decomposition of dead organic matter, which should decrease the sink and might convert it into a source (Gillett *et al.*, 2004; Flannigan *et al.*, 2005; Schaphoff *et al.*, 2006; Westerling *et al.*, 2006). The relative strength of the various opposing factors is still difficult to predict. Experimental manipulations of climate, atmospheric CO_2, tropospheric ozone, and nitrogen, at the largest possible scale, will be required to reduce uncertainty about the future of the carbon sink.

In what follows, we provide additional detail about the elements in Tables 3.1 and 3.2.

3.2.1 Forests

Based on U.S. Forest Service inventories, forest ecosystem carbon stocks in the United States, excluding soil carbon, have increased since 1953. The rate of increase has recently slowed because of increasing harvest and declining growth in some areas with maturing forests. The current average annual increase in carbon in trees is 146 Mt C per year (Smith and Heath, 2005, uncertainty ****) plus 20 Mt C per year from urban and suburban trees (the midpoint of the range in Chapter 14, uncertainty ***). The total estimate of the carbon sink in forested ecosystems is -256 Mt C per year and includes a sink of 90 Mt C per year (uncertainty **) from the accumulation of nonliving carbon in the soil (-90-146-20 = -256) (Pacala *et al.*, 2001; Goodale *et al.*, 2002). Although the magnitude of the forest soil sink has always been uncertain, it is now possible to measure the total above-and below-ground sink in a few square kilometers by monitoring the atmospheric CO_2 that flows into and out of the site over the course of a year. Note that these spatially intensive methods, appropriate for monitoring the sink over a few square kilometers, are unrelated to the spatially extensive methods described above, which attempt to constrain the sink at continental scales. As described in Appendix B, these studies are producing data that, so far, confirm the estimates of inventories and show that most of the forest sink is above-ground.

According to Canada's Greenhouse Gas Inventory (Environment Canada 2006, Chapter 11 this report), managed forests in Canada (comprising 83% of the total forest area) sequestered an average of 17 Mt C per year in trees and soils between 1990 and 2004 (uncertainty **). In addition, Chapter 14 estimates a sink of 11 (2-20) Mt C per year in urban and suburban trees of Canada (uncertainty ***) that were not included in the Environment Canada (2006) accounting. The total estimate for the Canadian forest sink is thus 28 Mt C per year (Table 3.1).

The two published carbon inventories for Mexican forests (Masera *et al.*, 1997 and Cairns *et al.*, 2000) both report substantial losses of forest carbon, primarily because of deforestation in the tropical south. However, both of these studies rely on calculations of carbon loss from remote imagery, rather than direct measurements, and both report results for a period that ended more than 10 years ago. Thus, in addition to being highly uncertain, the estimates for Mexican forests in Table 3.1 are not recent. Chapter 14 estimates a small urban forest sink of 2 (0-3) Mt C per year in Mexico. Whether the small urban forest sink would have been detected in changes in remote imagery and included in the Mexican inventories is uncertain, and accordingly is not included in Table 3.1.

3.2.2 Wood Products

Wood products create a carbon sink because they accumulate both in use (*e.g.*, furniture, house frames, *etc.*) and in landfills. The wood products sink is estimated at -57 Mt C per year in the United States (Skog and Nicholson, 1998) and

Table 3.3 Carbon stocks in North America in billions of tons, (see Box 3.1 for uncertainty conventions).

	United States	Canada	Mexico	North America
Forest	67[a],***	86[a],***	19[d],**	171***
Cropland	14[b],***'	4[b],****	1[b],*	19****
Grazing lands	33[b],***	12[b],***	10[b],***	55***
Wetlands	64[c],***	157[c],***	2[c],*	223***
Total	178***	259***	33**	468***

[a] Goodale *et al.* (2002)

[b] Chapter 10

[c] Chapter 13

[d] Masera *et al.* (1997)

-11 Mt C per year in Canada (Goodale *et al.*, 2002, Chapter 11 this report). We know of no estimates for Mexico.

3.2.3 Woody Encroachment

Woody encroachment is the invasion of woody plants into grasslands or the invasion of trees into shrublands. It is caused by a combination of fire suppression and grazing. Fire inside the United States has been reduced by more than 95% from the pre-settlement level of approximately 80 million hectares burned per year, and this favors shrubs and trees in competition with grasses (Houghton *et al.*, 2000). Field studies show that woody encroachment both increases the amount of living plant carbon and decreases the amount of dead carbon in the soil (Guo and Gifford, 2002; Jackson *et al.*, 2002). Although the total gains and losses are ultimately of similar magnitude (Jackson *et al.*, 2002), the losses occur within approximately a decade after the woody plants invade (Guo and Gifford, 2002), while the gains occur over a period of up to a century or more. Thus, the net source or sink depends on the distribution of times since woody plants invaded, and this is not known. Estimates for the size of the current United States' woody encroachment sink (Houghton *et al.*, 1999, Houghton and Hackler, 2000; and Hurtt *et al.*, 2002) all rely on methods that do not account for the initial rapid loss of carbon from soil when grasslands were converted to shrublands or forest. The estimate of -120 Mt C per year in Table 3.1 is from Houghton *et al.* (1999), but is similar to the estimates from the other two studies (-120 and -130 Mt C per year). No estimates are currently available for Canada or Mexico. Note the error estimate of more than 100% in Table 3.1. A comprehensive set of measurements of woody encroachment would reduce the error in the national and continental carbon budgets more than any other inventory.

3.2.4 Agricultural Lands

Soils in croplands and grazing lands have been historically depleted of carbon by humans and their animals, especially if the land was converted from forest to non-forest use. Harvest or consumption by animals reduces the input of organic matter to the soil, while tillage and manure inputs increase the rate of decomposition. Changes in cropland management, such as the adoption of no-till agriculture (Chapter 10 this report), have reversed the losses of carbon on some croplands, but the losses continue on the remaining lands. The net is a small sink of -2 Mt C per year for agricultural soils in Canada and, for the United States, is a sink of between -5 and -12 Mt C per year.

3.2.5 Wetlands

Peatlands are wetlands that have accumulated deep soil carbon deposits because plant productivity has exceeded decomposition over thousands of years. Thus, wetlands form the largest carbon pool of any North American ecosystem (Table 3.3). If drained for development, this soil carbon pool is rapidly lost. Canada's extensive frozen and unfrozen wetlands create a net sink of -23 Mt C per year, with from -6 to -11 Mt C per year of that sink in areas underlain by permafrost (Chapters 12 and 13, this report). Drainage of peatlands in the conterminous United States has created a source of 6 Mt C per year, but other wetlands, including those in Alaska, are a sink of -29 Mt C per year for a net United States wetland sink of -23 Mt C per year (Chapter 13, this report). The very large pool of peat in northern wetlands is vulnerable to climate change and could add more than 100 ppm to the atmosphere (1 ppm ≈ 2.1 billion tons of carbon [Gt C]) during this century, if released, because of global warming (see the model result in Cox *et al.*, 2000 for an example).

> Wetlands form the largest carbon pool of any North American ecosystem (Table 3.3). If drained for development, this soil carbon pool is rapidly lost.

The carbon sink due to sedimentation in wetlands is estimated to be 4 Mt C per year in Canada and 22 Mt C per year in the United States,

but this estimate is highly uncertain (Chapter 13 this report). Another important priority for research is to better constrain carbon sequestration due to sedimentation in wetlands, lakes, reservoirs, and rivers.

The focus on this chapter is on CO_2; we do not include estimates for other greenhouse gases. However, wetlands are naturally an important source of methane (CH_4). Methane emissions effectively cancel out the positive benefits of any carbon storage, such as peat in Canada, and make United States' wetlands a source of warming on a decadal time scale (Chapter 13 this report). Moreover, if wetlands become warmer and remain wet with future climate change, they have the potential to emit large amounts of CH_4. This is probably the single most important consideration, and unknown, in the role of wetlands and future climate change.

3.2.6 Rivers and Reservoirs
Organic sediments accumulate in artificial lakes and in alluvium (deposited by streams and rivers) and colluvium (deposited by wind or gravity) and represent a carbon sink. Pacala *et al.* (2001) extended an analysis of reservoir sedimentation (Stallard, 1998) to an inventory of the 68,000 reservoirs in the United States and also estimated net carbon burial in alluvium and colluvium. Table 3.1 includes the midpoint of their estimated range of 10 to 40 Mt C per year in the coterminous United States. This analysis has also recently been repeated and produced an estimate of 17 Mt C per year (E. Sundquist, personal communication; unreferenced). We know of no similar analysis for Canada or Mexico.

3.2.7 Exports Minus Imports of Wood and Agricultural Products

---

Fossil-fuel emissions currently dominate the net carbon balance in the United States, Canada, and Mexico.

The United States imports more wood products (14 Mt C per year) than it exports and exports more agricultural products (35 Mt C per year) than it imports (Pacala *et al.*, 2001). The large imbalance in agricultural products is primarily because of exported grains and oil seeds. Canada and Mexico are net wood exporters, with Canada at -74 Mt C per year (Environment Canada, 2005) and Mexico at -1 Mt C per year (Masera *et al.*, 1997). The North American export of 61 Mt C per year accounts correctly for the large net transfer of lumber and wood products from Canada to the United States. We know of no analysis of the Canadian or Mexican export-import balance for agricultural products.

3.2.8 River Export
Rivers in the coterminous United States were estimated to export 30-40 Mt C per year to the oceans in the form of dissolved and particulate organic carbon and inorganic carbon derived from the atmosphere (Pacala *et al.*, 2001). An additional 12-20 Mt C per year of inorganic carbon is also exported by rivers but is derived from carbonate minerals. We know of no corresponding estimates for Alaska, Canada, or Mexico.

3.2.9 Coastal Waters
Chapter 15 summarizes the complexity and large uncertainty of the sea-air flux of CO_2 in North American coastal waters. It is important to understand that the source in Mexican coastal waters is not caused by humans and would have been present in pre-industrial times. It is simply the result of the purely physical upwelling of carbon-rich deep waters and is a natural part of the oceanic carbon cycle. It is not yet known how much of the absorption of carbon by United States' and Canadian coastal waters is natural and how much is caused by nutrient additions to the coastal zone by humans. Accordingly, it is essentially impossible to currently assess the potential or costs of carbon management in coastal waters of North America.

3.3 SUMMARY

Fossil-fuel emissions currently dominate the net carbon balance in the United States, Canada, and Mexico (Figure 3.1, Tables 3.1 and 3.2). In 2003, fossil-fuel consumption in the United States emitted 1582 Mt C per year to the atmosphere (confidence ****, see definition of confidence categories in Table 3.1 footnote). This source was partially balanced by a flow of 489 Mt C per year from the atmosphere to land caused by net ecosystem sinks in the United States (***). Canadian fossil-fuel consumption transfered 164 Mt C per year to the atmosphere in 2003 (****), but net ecological sinks capture 64 Mt C per year (**). Mexican fossil-fuel emissions of 110 Mt C per year (****) were supplemented by a net ecosystem source of 48 Mt C per year (*) from tropical deforestation. Each of the three countries has always been a net source of CO_2 emissions to the atmosphere for the past three centuries (Houghton *et al.*, 1999, 2000; Houghton and Hackler, 2000; Hurtt *et al.*, 2002).

What Are the Options That Could Significantly Affect the North American Carbon Cycle?

Coordinating Lead Author: Erik Haites, Margaree Consultants, Inc.

Lead Authors: Ken Caldeira, Carnegie Inst.; Patricia Romero Lankao, NCAR and UAM-Xochimilco; Adam Z. Rose, The Pa. State Univ. and Univ. Southern Calif.; Thomas J. Wilbanks, ORNL

Contributing Authors: Skip Laitner, U.S. EPA; Richard Ready, The Pa. State Univ.; Roger Sedjo, Resources for the Future

For further affiliation details see the Author Team list on pages III and IV

KEY FINDINGS

- Options to reduce energy-related carbon dioxide emissions include improved efficiency, fuel switching (among fossil fuels and non-carbon fuels), and carbon dioxide capture and storage.

- Most energy use, and hence energy-related carbon dioxide emissions, involves equipment or facilities with a relatively long life—5 to 50 years. Many options for reducing these carbon dioxide emissions are most cost-effective, and sometimes only feasible, in new equipment or facilities. This means that cost-effective reduction of energy-related carbon dioxide emissions may best be achieved as existing equipment and facilities are replaced[1]. If emission reductions are implemented over a long time, technological change will have a significant impact on the cost.

- Options to increase carbon sinks include forest growth and agricultural soil sequestration. The amount of carbon that can be captured by these options is significant, but additions to current stocks would be small to moderate relative to carbon emissions. These options can be implemented in the short term, but the amount of carbon sequestered typically is low initially, then rises for a number of years before tapering off again as the total potential is achieved. There is also a significant risk that the carbon sequestered may be released again by natural phenomena or human activities.

- Both policy-induced and voluntary actions can help reduce carbon emissions and increase carbon sinks, but significant changes in the carbon budget are likely to require policy interventions. The effectiveness of a policy depends on the technical feasibility and cost-effectiveness of the portfolio of actions it seeks to promote, on its suitability given the institutional context, and on its interaction with policies implemented to achieve other objectives.

- Policies to reduce atmospheric carbon dioxide concentrations cost effectively in the short- and long-term could include: (1) encouraging adoption of cost-effective emission reduction and sink enhancement actions through such mechanisms as an emissions trading program or an emissions tax; (2) stimulating development of technologies that lower the cost of emissions reduction, carbon capture and sequestration, and sink enhancement; (3) adopting appropriate regulations for sources or actions subject to market imperfections, such as energy efficiency measures and cogeneration; (4) revising existing policies with other objectives that lead to higher carbon dioxide or methane emissions so that the objectives, if still relevant, are achieved with lower emissions; and (5) encouraging voluntary actions.

- Implementation of such policies at a national level, and cooperation at an international level, would reduce the overall cost of achieving a carbon reduction target by providing access to more low-cost mitigation/sequestration options.

---

[1] An emission reduction action is cost-effective if the cost per ton of carbon dioxide reduced is lower than the least-cost alternative.

4.1 INTRODUCTION

This chapter provides an overview of options that can reduce carbon dioxide (CO_2) and methane (CH_4) emissions and those that can enhance carbon sinks, and it attempts to compare them. Finally, it discusses policies to encourage implementation of source reduction and sink enhancement options. No emission reduction or sink enhancement target is proposed, and no policy or option is recommended.

4.2 SOURCE REDUCTION OPTIONS

4.2.1 Energy-Related Carbon Dioxide Emissions

Combustion of fossil fuels is the main source of CO_2 emissions (Chapters 1-3 this report), although some CO_2 is also released in non-combustion and natural processes. Most energy use, and hence energy-related CO_2 emissions, involves equipment or facilities with a relatively long life—5 to 50 years. Many options for reducing these CO_2 emissions are most cost-effective, and sometimes only feasible, in new equipment or facilities (Chapters 6 through 9 this report).

> Canada and the United States use much more energy *per capita* than other high income countries, suggesting considerable potential to reduce energy use and associated CO_2 emissions with little impact on the standard of living.

To stabilize the atmospheric concentration of CO_2 "would require global anthropogenic CO_2 emissions to drop below 1990 levels . . . and to steadily decrease thereafter" (IPCC, 2001)[2]. That entails a transition to a very different energy system, for example, where the major energy carriers are electricity and hydrogen produced by non-fossil sources or from fossil fuels with capture and geological storage of the CO_2 generated. A transition to such an energy system, while also meeting growing energy needs, could take at least several decades. Thus, shorter term (2015–2025) and longer term (post-2050) options are differentiated.

Options to reduce energy-related CO_2 emissions can be grouped into a few categories:
- efficiency improvement,
- fuel switching to fossil fuels with lower carbon content per unit of energy produced or to non-fossil fuels, and
- switching to electricity and hydrogen produced from fossil fuels in processes with CO_2 capture and geological storage.

4.2.1.1 Efficiency Improvement

Energy is used to provide services such as heat, light, and motive power. Any measure that delivers the desired service with less energy is an efficiency improvement[3]. Efficiency improvements reduce CO_2 emissions whenever they reduce the use of fossil fuels at any point between production of the fuel and delivery of the desired service[4]. Energy use can be reduced by improving the efficiency of individual devices (such as refrigerators, industrial boilers, and motors), by improving the efficiency of systems (using the correct motor size for the task), and by using energy that is not currently utilized, such as waste heat[5]. Opportunities for efficiency improvements are available in all sectors.

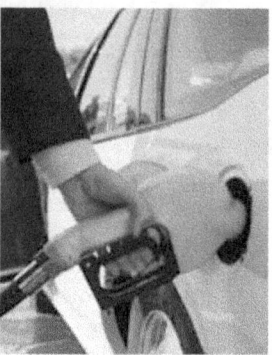

It is useful to distinguish two levels of energy efficiency improvement: (1) the amount consistent with efficient utilization of resources (the economic definition) and (2) the maximum attainable (the engineering definition). Energy efficiency improvement thus covers a broad range, from measures that provide a cost saving to measures that are technically feasible but too expensive under current market conditions to warrant implementation. Market imperfections inhibit adoption of some cost-effective efficiency improvements (NCEP, 2005)[6].

Energy efficiency improvements tend to occur gradually, but steadily, across the economy in response to technological developments, replacement of equipment and buildings, changes in energy prices, and other factors[7]. In the short term, the potential improvement depends largely on greater deployment and use of available efficient equipment and technology. In the long term, it depends largely on tech-

---

[3] In the transportation sector, for example, energy efficiency can be increased by improving the fuel performance of vehicles, shifting to less emissions-intensive modes of transport, and adopting options that reduce transportation demand, such as telecommuting and designing communities so that people live closer to shopping and places of work.
[4] Increasing the fuel economy of vehicles or the efficiency of coal-fired generating units reduces fossil-fuel use directly. Increasing the efficiency of refrigerators or electricity transmission reduces electricity use and hence the fossil fuel used to generate electricity.
[5] For example, 40 to 70% of the energy in the fuel used to generate electricity is wasted. Cogeneration or combined heat and power systems generate electricity and produce steam or hot water. Cogeneration requires a nearby customer for the steam or heat.
[6] Examples of market imperfections include limited foresight, externalities, capital market barriers, and principal/agent split incentive problems. As an example of the principal/agent imperfection, a landlord has little incentive to improve the energy efficiency of the housing unit and its appliances if the tenant pays the energy bills.
[7] The rate of efficiency improvement varies widely across different types of equipment such as lighting, refrigerators, electric motors, and motor vehicles.

---

[2] The later the date at which global anthropogenic CO_2 emissions drop below 1990 levels, the higher the level at which the CO_2 concentration is stabilized.

nological developments. Canada and the United States use much more energy *per capita* than other high-income countries, suggesting considerable potential to reduce energy use and associated CO_2 emissions with little impact on the standard of living[8].

4.2.1.2 Fuel Switching

Energy-related CO_2 emissions are primarily due to combustion of fossil fuels. Thus CO_2 emissions can be reduced by switching to a less carbon-intensive fossil fuel or to a non-carbon fuel.

The CO_2 emissions per unit of energy (carbon intensity) for fossil fuels differ significantly, with coal being the highest, oil and related petroleum products about 25% lower, and natural gas over 40% lower than coal. Oil and/or natural gas can be substituted for coal in all energy uses, mainly electricity generation. However, natural gas is not available everywhere in North America and is much less abundant than coal, limiting the large-scale, long-term replacement of coal with natural gas. Technically, natural gas can replace oil in all energy uses, but to substitute for gasoline and diesel fuel, by far the largest uses of oil, would require conversion of millions of vehicles and development of a gas-refueling infrastructure.

Non-fossil fuels include

- biomass and fuels, such as ethanol and biodiesel, produced from biomass; and
- electricity and hydrogen produced from carbon-free sources.

Biomass can be used directly as a fuel in some situations. Pulp and paper plants and sawmills, for example, can use wood waste and sawdust as fuel. Ethanol, currently produced mainly from corn, is blended with gasoline and biodiesel is produced from vegetable oils and animal fats. Wood residuals and cellulose materials, such as switch grass, can be utilized both for energy and the production of syngases, which can be used to produce biopetroleum (AF&PA, 2006). The CO_2 emission reduction achieved depends on whether the biomass used is replaced, on the emissions associated with production and combustion of the biomass fuel, and the carbon content of the fuel displaced[9].

Carbon-free energy sources include hydro, wind, solar, biomass, geothermal, and nuclear fission[10]. Sometimes they are used to provide energy services directly, such as solar water heating and windmills for pumping water. But they are mainly used to generate electricity, about 35% of the electricity in North America. Currently, generating electricity using any of the carbon free energy sources is usually more costly than using fossil fuels.

Most of the fuel switching options are currently available, and so are viable short-term options in many situations.

4.2.1.3 Electricity and Hydrogen From Fossil Fuels with Carbon Dioxide Capture and Storage

About 65% of the electricity in North America is generated from fossil fuels, mainly coal, but with a rising share for natural gas (EIA, 2003a; Chapter 6 this report). The CO_2 emissions from fossil-fired generating units can be captured and injected into a suitable geological formation for long-term storage.

Hydrogen (H_2) is an energy carrier that emits no CO_2 when burned, but may give rise to CO_2 emissions when it is produced (National Academies, 2004). Currently, most hydrogen is produced from fossil fuels in a process that generates CO_2 (National Research Council, 2004). The CO_2 from this process can be captured and stored in geological formations. Alternatively, hydrogen can be produced from water using electricity, in which case the CO_2 emissions depend on how the electricity is generated. Hydrogen could substitute for

> Carbon-free energy sources include hydro, wind, solar, biomass, geothermal, and nuclear fission. Combined these sources generate about 35% of the electricity in North America.

---

[8] The total primary energy supply *per capita* during 2004, in tons of oil equivalent, was 8.42 for Canada, 7.91 for the United States, 4.43 for France, 4.22 for Germany, 4.18 for Japan, 3.91 for the United Kingdom, and 1.59 for Mexico (IEA, 2006a).

[9] The CO_2 reductions achieved depend on many factors including the inputs used to produce the biomass (fertilizer, irrigation water), whether the land is existing cropland or converted from forests or grasslands, and the management practices used (no-till, conventional till).

[10] Reservoirs for hydroelectric generation produce CO_2 and CH_4 emissions, and production of fuel for nuclear reactors generates CO_2 emissions, so such sources are not totally carbon free.

natural gas in most energy uses and could be used by fuel cell vehicles.

Carbon dioxide can be captured from the emissions of large sources, such as power plants, and pumped into geologic formations for long-term storage, thus permitting continued use of fossil fuels while avoiding CO_2 emissions to the atmosphere[11]. Many variations on this basic theme have been proposed; for example, pre-combustion vs. post-combustion capture, production of hydrogen from fossil fuels, and the use of different chemical approaches and potential storage reservoirs (IPCC, 2005). While most of the basic technology exists, legal, environmental, and safety issues need to be addressed before CO_2 capture and storage can be integrated into our energy system, so this is mainly a long-term option (IPCC, 2005). Carbon dioxide capture and storage could contribute about 30% (15-55%) of the total mitigation effort, mainly after 2025 (IPCC, 2005; IEA, 2006b; Stern, 2006).

> CO$_2$ capture and storage could contribute about 30% of the total mitigation effort, mainly after 2025.

4.2.2 Industrial Processes

The processes used to make cement, lime, and ammonia release CO_2. Because the quantity of CO_2 released is determined by chemical reactions, the process emissions are determined by the output. But the CO_2 could be captured and stored in geological formations. Carbon dioxide also is released when iron ore and coke are heated in a blast furnace to produce molten iron, but alternative steel-making technologies with lower CO_2 emissions are commercially available. Consumption of the carbon anodes during aluminum smelting leads to CO_2 emissions, but good management practices can reduce the emissions. Raw natural gas contains CO_2 that is removed at gas processing plants and could be captured and stored in geological formations.

4.2.3 Methane Emissions

Methane is produced as organic matter decomposes in low-oxygen conditions and is emitted by landfills, wastewater treatment plants, and livestock manure. In many cases, the CH_4 can be collected and used as an energy source. Methane emissions also occur during the transport of natural gas. Such emissions usually can be flared or collected for use as an energy source[12]. Ruminant animals produce CH_4 while digesting their food. Emissions by ruminant farm

> Forest growth and soil sequestration currently offset about 30% of the North American fossil-fuel emissions.

animals can be reduced by measures that improve animal productivity. All of these emission reduction options are currently available.

4.3 TERRESTRIAL SEQUESTRATION OPTIONS

Trees and other plants sequester carbon as biological growth captures carbon from the atmosphere and sequesters it in the plant cells (IPCC, 2000). Currently, very large volumes of carbon are sequestered in the plant cells of the Earth's forests. Increasing the stock of forest through afforestation[13], reforestation, or forest management draws carbon from the atmosphere and increases the carbon sequestered in the forest and the soil of the forested area. Sequestered carbon is released by fire, insects, disease, decay, wood harvesting, conversion of land from its natural state, and disturbance of the soil. Substituting long-lived wood products for steel and cement can reduce emissions and increase the amount of carbon sequestered.

Agricultural practices can increase the carbon sequestered by the soil. Some crops build soil organic matter, which is largely carbon, better than others. Some research shows that crop-fallow systems result in lower soil carbon content than continuous cropping systems (Chapter 10 this report). No-till and low-till cultivation builds soil organic matter.

Conversion of agricultural land to forestry can increase carbon sequestration in soil and tree biomass, but the rate of sequestration depends on environmental factors (such as type of trees planted, soil type, climate, and topography) and management practices (such as thinning, fertilization, and pest control). Conversion of agricultural land to other uses can result in positive or negative net carbon emissions depending upon the land use.

Forest growth and soil sequestration currently offset about 30% (15-45%) of the North American fossil fuel emissions (Chapter 3 this report), and this percentage might be increased to some degree. These options can be implemented in the short term, but the amount of carbon sequestered typically is low initially, then rises for a number of years before tapering off again as the total potential is achieved (Chapters 10-13 this report).

4.4 INTEGRATED COMPARISON OF OPTIONS

As is clear from the previous sections, there are many options to reduce emissions of or to sequester CO_2. To help them decide which options to implement, policy makers need to

---

[11] Since combustion of biomass releases carbon previously removed from the atmosphere, capture and storage of these emissions results in negative emissions (a sink).

[12] Flaring or combustion of CH_4 as an energy source produces CO_2 emissions.

[13] See the *Glossary* for a definition of this term and related terms.

BOX 4.1: Emission Reduction Supply Curve

A tool commonly used to compare emission reduction and sequestration options is an emission reduction supply curve, such as that shown in the figure. It compiles the emission reduction and sequestration options available for a given jurisdiction at a given time. If the analysis is for a future date, a detailed scenario of future conditions is needed. The estimated emission reduction potential of each option is based on local circumstances at the specified time, taking into account the interaction among options, such as improved fuel efficiency for vehicles and greater use of less carbon-intensive fuel. The options are combined into a curve starting with the most cost-effective and ending with the least cost-effective. For each option, the curve shows the cost per metric ton of CO_2 reduced on the vertical axis and the potential emission reduction, tons of CO_2 per year, on the horizontal axis. The curve can be used to identify the lowest cost options to meet a given emission reduction target, the associated marginal cost (the cost per metric ton of the last option included), and total cost (the area under the curve).

An emission reduction supply curve is an excellent tool for assessing alternative emission reduction targets. The best options and cost are easy to identify. The effect on the cost of dropping some options is easy to calculate unless they interact with other options. And the cost impact of having to implement additional options due to underperformance by others is simple to estimate. The drawbacks are that constructing the curve is a complex analytical process and that the curve is out of date almost immediately because fuel prices and the cost or performance of some options change.

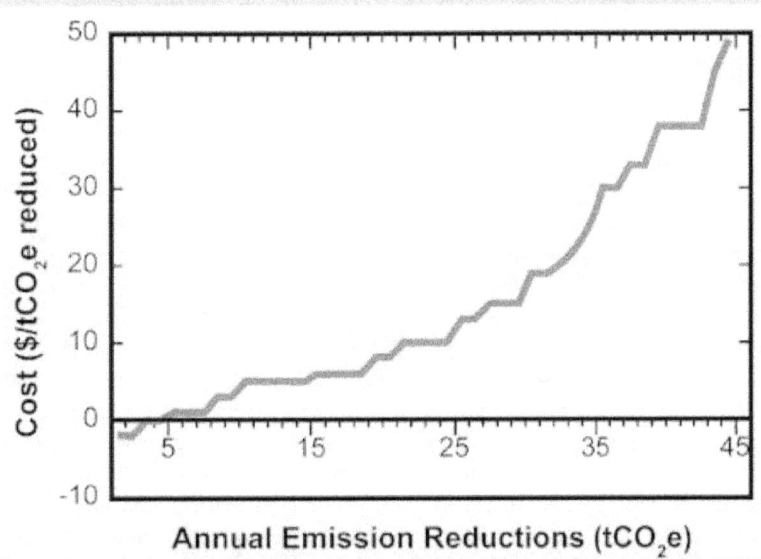

Annual Emission Reductions (tCO₂e)

The curve shows the estimated unit cost ($/t CO_2 equivalent) and annual emission reduction (t CO_2 equivalent) for emission reduction and sequestration options for a given region and date arranged in order of increasing unit cost.

When constructed for a future date, such as 2010 or 2020, the precision suggested by the curve is misleading because the future will differ from the assumed scenario. A useful approach in such cases is to group options into cost ranges, such as less than $5 per metric ton of CO_2, $5 to $15 per metric ton of CO_2, etc., ignoring some interaction effects and the impacts of the policy used to implement the option. This still identifies the most cost-effective options. Comparing the emissions reduction target with the emission reduction potential of the options in each group indicates the most economic strategy.

know the magnitude of the potential emission reduction at various costs for each option so they can select the options that are the most cost-effective—have the lowest cost per metric ton of CO_2 reduced or sequestered.

This involves an integrated comparison of options, which can be surprisingly complex in practice. It is most useful and accurate for short-term options where the cost and performance of each option can be forecast with a high degree of confidence. The performance of many options is interrelated; for example, the emission reductions that can be achieved by blending ethanol in gasoline depend, in addition to the factors relating to ethanol production previously cited, on other options, such as telecommuting to reduce travel demand, the success of modal shift initiatives, and the efficiency of motor vehicles. The prices of fossil fuels affect the cost-effectiveness of many options. Finally, the policy enacted to encourage an option, incentives vs. a regulation for example, can affect its potential.

The emission reduction potential and cost-effectiveness of options also vary by location. Energy sources and sequestration options differ by location; for example, natural gas may not be available, the wind and solar regime vary, hydro potential may be small or large, land suitable for afforestation/reforestation is limited, the agricultural crops may or may not be well suited to low-till cropping. Climate, lifestyles, and consumption patterns also affect the potential of many options; for example, more potential for heating options in a cold climate or air conditioning options in a hot climate. The mix of single-family and multi-residential buildings affects the potential for options focused on those building types, and the scope for public transit options tends to increase with city size. Institutional factors affect the potential of many options as well; for example, the prevalence of rented housing affects the potential to implement residential emission reduction measures, the authority to specify minimum efficiency standards for vehicles, appliances, and equipment may rest with the state/provincial government or the national government, and the ownership and regulatory structure for gas and electric utilities can affect their willingness to offer energy efficiency programs.

The estimated cost and emission reduction potential for the principal short-term CO2 emission reduction and sequestration options are summarized in Table 4.1. All estimates are expressed in 2004 United States dollars per metric ton of carbon . The limitations of emission reduction supply curves noted in the text box apply equally to the cost estimates in Table 4.1.

Most options have a range of costs. The range is due to four factors. First, the cost per unit of emissions reduced varies by location even for a very simple measure. For example, the

emission reduction achieved by installing a more efficient light bulb depends on the hours of use and the generation mix that supplies the electricity. Second, the cost and performance of any option in the future is uncertain. Different assumptions about future costs and performance contribute to the range. Third, most mitigation and sequestration options are subject to diminishing returns, that is, their cost rises at an increasing rate with greater use, as in the power generation, agriculture, and forestry cost estimates[14]. So the estimated scale of adoption contributes to the range. Finally, some categories include multiple options, notably those for the United States economy as a whole, each with its own marginal cost. For example, the "All Industry" category is an aggregation of seven subcategories discussed in Chapter 8 this report. The result again is a range of cost estimates.

The cost estimates in Table 4.1 are the direct costs of the options. A few options, such as the first estimate for power generation in Table 4.1, have a negative annualized cost. This implies that the option is likely to yield cost savings for reasons such as improved combustion efficiency. Some options have ancillary benefits (*e.g.*, reductions in ordinary pollutants, reduced dependence on imported oil, expansion of wildlife habitat associated with afforestation) that reduce their cost from a societal perspective. Indirect (multiplier, general equilibrium, macroeconomic) effects in the economy tend to increase the direct costs (as when the increased cost of energy use raises the price of products that use energy or energy-intensive inputs). Examples of these complicating effects are presented in Chapters 6 through 11 this report, along with some estimates of their impacts on costs.

None of the options listed in Table 4.1 offers the prospect of carbon budget stabilization alone (see below), which indicates a need to consider combinations of options. In any such consideration, costs are the primary driving force (*e.g.*, Table 4.1). Other considerations affecting the choice of options include the magnitudes of their potential contributions, their feasibility, and the time scale of their contribution. Table 4.2 summarizes these characteristics for the main families of emission reduction and sink enhancement options (see also Kauppi *et al.*, 2001).

As indicated in several segments of Table 4.1, costs are sensitive to the policy instruments used to encourage the option. In general, the less restrictive the policy, the lower the cost. That is why the cost estimates for the Feebate[15] are lower than the cost estimate for the Corporate Average Fuel Economy (CAFE) standard. In a similar vein, costs are low-

---

[14] For example, increasing the scale of tree planting to sequester carbon requires more land. Typically, the value of the extra land used rises, so the additional sequestration becomes increasingly costly.
[15] A "Feebate" is a system of progressive vehicle taxes on purchases of less efficient new vehicles and subsidies for more efficient new vehicles.

Table 4.1 Standardized cost estimates for short-term CO$_2$ emission reduction and sequestration options (annualized cost in 2004 constant U.S. dollars per metric ton of carbon [t C]).

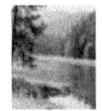

Option/applicable date(s)	Annualized average cost (in $2004 U.S.)	Potential range (Mt C per year) or % reduction	Source
Power generation	-$227 to 1176/tC	N.A.	DOE/EIA (2006)
Transportation/2010 (U.S. permit trading)	$84/t C	N.A.	EIA (2003b)
Transportation/2025 (U.S. permit trading)	$236/t C	22	EIA (2003b)
Transportation/2017 (CAFE standard[a])	$82/t C	39	CBO (2003)
Transportation/2030 (Feebate[b])	$47/t C	67	Greene et al. (2005)
Buildings	N.A.	60% for offices 70% for homes	USGBC (2005) DOE/EERE (2006)
Afforestation/2010-2110	$60 to 120/t C	37 to 224	EPA (2005)
Forest management/2010-2110	$4 to 120/t C	7 to 86	EPA (2005)
Biofuels/2010-2110	$120 to 201/t C	102 to 153	EPA (2005)
Agricultural soil carbon sequestration/2010-2110	$20 to 60/t C	34 to 46	EPA (2005)
All industry			
Reduction of fugitives	$92 to 180/t C	3%	Herzog (1999) Martin et al. (2001) Jaccard et al. (2002, 2003a, 2003b) Worrel et al. (2004) DOE (2006)
Energy efficiency	$0 to 180/t C	8% to 12%	Herzog (1999) Martin et al. (2001) Jaccard et al. (2002, 2003a, 2003b) Worrel et al. (2004) DOE (2006)
Process change	$92 to 180/t C	20%	Herzog (1999) Martin et al. (2001) Jaccard et al. (2002, 2003a, 2003b) Worrel et al. (2004) DOE (2006)
Fuel substitution	$0 to 92/t C	10%	Herzog (1999) Martin et al. (2001) Jaccard et al. (2002, 2003a, 2003b) Worrel et al. (2004) DOE (2006)
CO$_2$ capture and storage	$180 to 367/t C	30%	Herzog (1999) Martin et al. (2001) Jaccard et al. (2002, 2003a, 2003b) Worrel et al. (2004) DOE (2006)
Waste management			
Reduction of fugitives	$0 to 92/t C	90%	Herzog (1999) Jaccard et al. (2002)
CO$_2$ capture and storage	>$367/t C	30%	Herzog (1999) Jaccard et al. (2002)
Entire U.S. economy			
No trading	$102 to 548/t C[c]	Not specified	EMF (2000)
Industrialized country trading	$19 to 299/t C[c]	Not specified	EMF (2000)
Global trading	$7 to 164/t C[c]	Not specified	EMF (2000)

[a] CAFE= Corporate Average Fuel Economy
[b] A "feebate" is a system of progressive vehicle taxes on purchases of less efficient new vehicles and subsidies for more efficient new vehicles.
[c] Annualized marginal cost (cost at upper limit of application, and therefore typically higher than average cost).

ered by expanding the number of participants in an emissions trading arrangement, especially those with a prevalence of low-cost options, such as developing countries. That is why global trading costs are lower than the industrialized country trading case for the United States economy.

The task of choosing the "best" combination of options may seem daunting given the numerous options, their associated cost ranges, and ancillary impacts. This combination will depend on several factors including the emission target, the emitters covered, the compliance period, and the ancillary benefits and costs of the options. The best combination will change over time as locations where cheap options can be implemented are exhausted, and technological change lowers the costs of more expensive options. It is unlikely that decision makers can identify the least-cost combination of options to achieve a given emission target, but they can adopt policies, such as emissions trading or emissions

Table 4.2 Overview of possible contributions of families of options to managing the North American carbon cycle.[a] Note that combining a number of small contributions can add up to a moderate contribution, and combining a number of moderate contributions can sdd up to a large contribution.

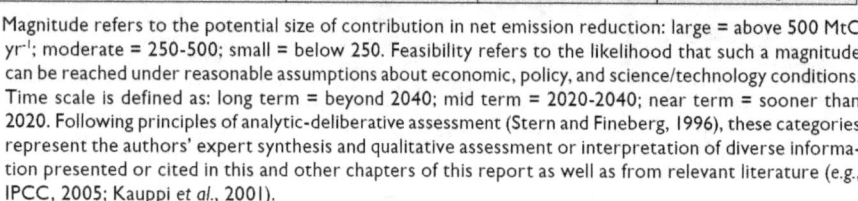

Category of Options	Magnitude of potential contribution	Feasibility of contribution	Time scale of contribution
Emission reduction			
Efficiency improvement	Moderate	High	Near to mid term
Fuel switching:			
- to less carbon-intensive fossil fuels	Small to moderate	High	Near to mid term
- to non-fossil fuels	Moderate to large	Moderate to high	Mid to long term
CO_2 capture and storage	Large[1]	Highly uncertain[2]	Long term[3]
Sink enhancement			
Forests	Small to moderate	Moderate to high	Near to mid term
Soils	Small	Moderate to high	Mid to long term

[a] Magnitude refers to the potential size of contribution in net emission reduction: large = above 500 MtC yr^{-1}; moderate = 250-500; small = below 250. Feasibility refers to the likelihood that such a magnitude can be reached under reasonable assumptions about economic, policy, and science/technology conditions. Time scale is defined as: long term = beyond 2040; mid term = 2020-2040; near term = sooner than 2020. Following principles of analytic-deliberative assessment (Stern and Fineberg, 1996), these categories represent the authors' expert synthesis and qualitative assessment or interpretation of diverse information presented or cited in this and other chapters of this report as well as from relevant literature (e.g., IPCC, 2005; Kauppi et al., 2001).

[1] Depending upon the (uncertain) availability of large geological reservoirs the potential contribution could possibly be very large (much greater than 500 Mt C per year).

[2] Uncertainty in availability of reservoirs, technology, public risk perception and costs among other factors makes the feasibility of large scale applications capable of realizing large potential highly uncertain.

[3] For large-scale or large-magnitude contributions exceeding the small magnitude, near term contributions of pilot-studies or existing oil recovery applications.

energy; expanded use of non-carbon and low-carbon energy technologies; and various changes in forestry, agricultural, and land-use practices. Actions will also be supported by encouraging research and development of technologies that can reduce emissions even further in the long term, such as technologies for removing carbon from fossil fuels and sequestering it in geological formations and possibly other approaches, some of which are currently very controversial, such as certain types of "geoengineering."

Because CO_2 has a long atmospheric residence time[17], immediate action to reduce emissions and increase sequestration allows its atmospheric concentration to be stabilized at a lower level[18]. Policy instruments to promote cost-effective implementation of a portfolio of options covering virtually all emissions sources and sequestration options are available for the short term. Implementation of policy instruments at a national level, and cooperation at an international level, would reduce the overall cost of achieving a carbon reduction target by providing access to more low-cost mitigation/sequestration options.

taxes, that cover a large number of emitters and allow them to use their first-hand knowledge to choose the lowest cost reduction options[16].

4.5 IMPLEMENTATING OPTIONS

4.5.1 Overview

No single technology or approach can achieve a sufficiently large CO_2 emission reduction or sequestration to stabilize the carbon cycle (Hoffert et al., 1998, 2002; Pacala and Socolow, 2004). Decision makers will need to consider a portfolio of options to reduce emissions and increase sequestration in the short term, taking into account constraints on and implications of mitigation strategies and policies. The portfolio of short-term options is likely to include greater efficiency in the production and use of

> No single technology or approach can achieve a sufficiently large CO_2 emission reduction or sequestration to stabilize the carbon cycle.

[16] Swift (2001) finds that emissions trading programs yield greater environmental and economic benefits than regulations. Several other studies of actual policies (Ellerman et al., 2000) and proposed policies (Rose and Oladosu, 2002) have indicated relative cost savings of these incentive-based instruments.

[17] Carbon dioxide has an atmospheric lifetime of 5 to 200 years. A single lifetime can not be defined for CO_2 because of different rates of uptake by different removal processes. (IPCC, 2001, Table 1, p. 38)

[18] IPCC (2001), p. 187.

The effectiveness of such policies is determined by the technical feasibility and cost-effectiveness of the portfolio of options they seek to promote, their interaction with other policies that have unintended impacts on CO_2 emissions, and their suitability given the institutional and socioeconomic context (Raupach *et al.*, 2004). This means that the effectiveness of the portfolio can be limited by factors such as:

- Demographic and social dynamics. Land tenure, population growth, and migration may pose an obstacle to afforestation/reforestation strategies.
- Institutional settings. The acceptability of taxes, subsidies, and regulations to induce the deployment of certain technology may be limited by stakeholder opposition.
- Environmental considerations. The portfolio of options may incur environmental costs such as nuclear waste disposal or biodiversity reduction.
- Institutional and timing aspects of technology transfer. The patent system, for instance, may pose a barrier for some countries and sectors in obtaining the best available technology.

4.5.2 General Considerations

Decisions about the implementation of options for carbon management are made at a variety of geographic scales, by a variety of decision makers, for a variety of reasons. In many cases, they emphasize decentralized voluntary decision-making within market and other institutional conditions that are shaped by governmental policies. Over the past decade in the United States, state and local governments and private firms, motivated by such factors as cost savings, public image, and perceptions of possible future policy directions, have implemented voluntary actions to reduce CO_2 emissions (Kates and Wilbanks, 2003). Although these actions have contributed to a decline in the ratio of CO_2 emissions to GDP (Casler and Rose, 1998), total emissions have continued to increase.

A wide array of policies have been implemented or are under discussion by governments in North America[19]. Policies to encourage reduction and sequestration of CO_2 emissions could include information programs, voluntary programs, conventional regulation, emissions trading, and emissions taxes (Tietenberg, 2000). Working Group III of the Intergovernmental Panel on Climate Change (IPCC) concluded that "[V]oluntary agreements between industry and governments, which vary considerably, are politically attractive, raise awareness among stakeholders, and have played a role

in the evolution of many national policies. . . However, there is little evidence that voluntary agreements have achieved significant emissions reductions beyond business as usual (high agreement/much evidence)." (Gupta *et al.* 2007; see also OECD, 2003b; Harrison, 1999; King and Lenox, 2000; Welch *et al.*, 2000; Darnall and Carmin, 2003; Croci, 2005; Jaccard *et al.*, 2006).

Reducing annual emissions in North America consistently over several decades requires a portfolio of policies across all sectors and gases tailored to fit specific national circumstances. Regulations can require designated sources to keep their emissions below a specified limit, either a quantity per unit of output or an absolute amount per day or year. Regulations can also stipulate minimum or average levels of energy efficiency of appliances, buildings, equipment, and vehicles.

> Although voluntary actions have contributed to a decline in the ratio of CO_2 emissions to GDP, total emissions have continued to increase.

An emissions trading program establishes a cap on the annual emissions of a set of sources. Allowances equal to the cap are issued and can be traded. Each source must monitor its actual emissions and remit allowances equal to its actual emissions to the regulator. An emission trading program creates an incentive for sources with low-cost options to reduce their emissions and sell their surplus allowances. Sources with high-cost options find it less expensive to buy allowances at the market price than to reduce their own emissions enough to achieve compliance.

An emissions tax requires designated sources to pay a specified levy for each unit of its actual emissions. Each emitter will reduce its emissions to the point where the mitigation

---

[19] Policies can be found at: http://www.epa.gov/climatechange/policy/neartermghgreduction.html, http://www.ecoaction.gc.ca/index-eng.cfm, and http://cambio_climatico.ine.gob.mx/ccygob/ccygobingles.html

cost is equal to the tax, but once the mitigation cost exceeds the tax, the emitter will opt to pay the tax.

The framework for evaluating such a policy instrument needs to consider technical, institutional, and socioeconomic constraints that would affect its implementation, such as the ability of sources to monitor their actual emissions, the constitutional authority of national and/or provincial/state governments to impose emissions taxes, regulate emissions and/or regulate efficiency standards. It is also important to consider potential conflicts between carbon reduction policies and policies with other objectives, such as keeping energy costs to consumers as low as possible.

Practically every policy (except cost-saving energy conservation options)[20], no matter what instrument is used to implement it, has a cost in terms of utilization of resources and ensuing price increases that leads to reductions in output, income, employment, or other measures of economic well-being. The total cost is usually higher than the direct cost due to interactions with other segments of the economy and with existing policies ("general equilibrium" effects). Regardless of where the compliance obligation is imposed, the cost ultimately is borne by the general public as consumers, shareholders, employees, taxpayers, and recipients of government services[21]. The cost can have competitiveness impacts if some emitters in other jurisdictions are not subject to similar policies. But societal benefits, such as improved public health and reduced environmental damage, may offset part or all of the cost of implementing the policy.

To achieve a given emission reduction target, regulations that require each affected source to meet a specified emissions limit or implement specified controls are almost always more costly than emissions trading or emissions taxes because they require each affected source to meet the regulation regardless of cost rather than allowing emission reductions to be implemented where the cost is lowest (Bohm and Russell, 1986)[22]. The cost saving available through trading or an emissions tax generally increases with the diversity of sources and share of total emissions covered by the policy (Rose and Oladosu, 2002)[23]. A policy that raises revenue (an emissions tax or auctioned allowances) has a lower cost to the economy than a policy that does not, if the revenue is used to reduce existing distortionary taxes[24] such as sales or income taxes (see, *e.g.*, Parry *et al.*, 1999).

4.5.3 Source Reduction Policies

Historically CO_2 emissions have not been regulated directly. Some energy-related CO_2 emissions have been regulated indirectly through energy policies, such as promotion of renewable energy, and efficiency standards and ratings for equipment, vehicles, and some buildings. Methane emissions from oil and gas production, underground coal mines, and landfills have been regulated, usually for safety reasons.

Policies with other objectives can have a significant impact on CO_2 emissions. Policies to encourage production or use of fossil fuels, such as favorable tax treatment for fossil fuel production, increase CO_2 emissions. Similarly, urban plans and infrastructure that facilitate automobile use rather than public transit increase CO_2 emissions. In contrast, a tax on vehicle fuels reduces CO_2 emissions[25].

Carbon dioxide emissions are suited to emissions trading and emissions taxes. These policies allow considerable flexibility in the location and, to a lesser extent, the timing of the emission reductions[26]. The environmental impacts of

---

[20] These are often called "no regret" options.
[21] The source with the compliance obligation passes on the cost through some combination of higher prices for its products, negotiating lower prices with suppliers, layoffs, and/or lower wages for employees, and lower profits that lead to lower tax payments and lower share prices. Other firms that buy the products or supply the inputs make similar adjustments. Governments raise taxes or reduce services to compensate for the loss of tax revenue. Ultimately, all of the costs are borne by the general public.

[22] As well, regulation is generally inferior to emissions trading or taxes in inducing technological change.
[23] These policies encourage implementation of the lowest cost emission reductions available to the affected sources. They establish a price (the emissions tax or the market price for an allowance) for a unit of emissions and then allow affected sources to respond to the price signal. In principle, these two instruments are equivalent in terms of achievement of the efficient allocation of resources, but they may differ in terms of equity because of how the emission permits are initially distributed and whether a tax or subsidy is used. It is easier to coordinate emissions trading programs than emissions taxes across jurisdictions.
[24] A distortionary tax is one that changes the relative prices of goods or services. For example, income taxes change the relative returns from work, leisure, and savings.
[25] Initially the reduction may be small because demand for gasoline is not very sensitive to price, but over time the tax causes people to adjust their travel patterns and the vehicles they drive, thus yielding larger reductions.
[26] An emissions trading program may allow participants to buy credits issued to entities not covered by the program for emission reductions or increased carbon sequestration. Determination of

CO_2 depend on its atmospheric concentration, which is not sensitive to the location or timing of the emissions. Apart from ground-level safety concerns, the same is true of CH_4 emissions. In addition, the large number and diverse nature of the CO_2 and CH_4 sources means that use of such policies can yield significant cost savings but may also be difficult to implement.

Regulations setting maximum emissions on individual sources or efficiency standards for appliances and equipment might be preferred to emissions trading and taxes. Such regulations may be desirable where monitoring actual emissions is costly or where firms or individuals do not respond well to price signals due to lack of information or market imperfections. Energy efficiency standards for appliances, buildings, equipment, and vehicles tend to fall into this category (OECD, 2003a)[27]. In some cases, such as refrigerators, standards have been used successfully to drive technology development.

4.5.4 Terrestrial Sequestration Policies

To date, policies that explicitly encourage carbon sequestration in terrestrial systems have taken the form of modifying conservation programs aimed at other environmental objectives to include rewards for increasing carbon uptake by forests and agricultural soils. For example, the United States Department of Agriculture modified the enrollment criteria of the Conservation Reserve Program (CRP) and the Environmental Quality Incentives Program to give additional consideration to bids offering to install specific practices and technologies that sequester more carbon. The CRP also was modified to give landowners the right to sell carbon sequestered on lands enrolled in the program in private carbon markets. Policies that affect crop choice (support payments, crop insurance, disaster relief) and farmland preservation (conservation easements, use value taxation, agricultural zoning) may increase or reduce the carbon stock of agricultural soils. And policies that encourage higher agricultural output (support payments) can reduce the carbon stored by agricultural soils if they lead to increased tillage; such policies may increase stored carbon or be neutral with respect to carbon if they do not increase tillage.

A broad suite of policies are potentially available to increase terrestrial carbon stocks:

- Regulations, such as: requirements to limit or offset carbon emissions from land-use practices, requirements to reforest areas that have been logged, good practice standards, and requirements to establish carbon reserves.
- Market-based approaches, including: product labeling,

tradable development rights, markets for terrestrial carbon[28,29], and taxes on carbon emission from terrestrial systems.

- Incentives: tax credits for good management practices, cost-sharing of practice costs, payment of land rents for set-asides, outcome oriented payments based on carbon stored or sequestered (Feng *et al.*, 2003).
- Education and extension: Training, technical assistance, guidance on best management practices, education on impacts of alternative management practices, recommendations, technology pilots, and efforts to address lack of experience, learning costs, and risk aversion (Sedjo, 2001; Sedjo and Swallow, 2002).

Policies to enhance terrestrial carbon sinks have significant potential to store additional carbon more cost effectively than emissions reductions in other sectors, at least for the next few decades (EPA, 2005). The amount of carbon that could be sequestered and the cost-effectiveness of this option would depend on the policies employed and the value placed on terrestrial carbon. (*e.g.*, Marland *et al.*, 2001).

4.5.5 Research and Development Policies

Policies to stimulate research and development of lower emissions technologies can reduce the cost of meeting a long-term reduction target. Policies to reduce CO_2 emissions also influence the rate and direction of technological change (OECD, 2003a; Stern, 2006). By stimulating additional technological change, such policies can reduce the cost of meeting a given reduction target (Goulder, 2004; Grubb *et al.*, 2006; Stern, 2006). Such induced technological change tends to justify earlier and more stringent emission reduction targets (Goulder, 2004; Grubb *et al.*, 2006).

> The environmental impacts of CO_2 depend on its atmospheric concentration, which is not sensitive to the location or timing of the emissions.

Two types of policies are needed to ensure that available technologies can achieve a given cumulative CO_2 reduction or concentration target at least cost. Direct support for research and development produces less emission-intensive technologies and policies to reduce emissions and increase sequestration create a market for those technologies. The combination of "research push" and "market pull" policies is more effective than either strategy on its own (Goulder, 2004; CBO, 2006; Stern, 2006). Policies should encourage research and development for all promising technologies

---

the quantity of credits earned requires resolution of many issues, including the baseline, leakage, and additionality.

[27] The efficiency of standards sometimes can be improved by allowing manufacturers that exceed the standard to earn credits that can be sold to manufacturers that do not meet the standard.

[28] There needs to be a buyer for the credits, such as sources subject to CO_2 emissions trading program or an offset requirement.

[29] Since carbon sequestered in terrestrial plants and soils can be released from these sinks (*e.g.*, through forest fires or a return to tillage), markets for terrestrial carbon may need to address the permanence of the carbon sequestered. A number of options are available to address permanence.

because there is considerable uncertainty about which ones will ultimately prove most useful, socially acceptable, and cost-effective[30].

4.6 CONCLUSIONS

Actions to reduce projected CO_2 and CH_4 concentrations in the atmosphere should recognize the following:

- Emissions are produced by millions of diverse sources, most of which (*e.g.*, power plants, factories, building heating and cooling systems, and large appliances) have lifetimes of 5 to 50 years, and so are likely to adjust only slowly at reasonable cost.
- Potential uptake by agricultural soils and forests is significant but small to moderate relative to emissions (Chapter 11 this report) and can be reversed at any given location by natural phenomena or human activities. Policies to enhance and maintain terrestrial carbon sinks have significant potential to store additional carbon more cost-effectively than emissions reductions in other sectors, at least for the next few decades.
- Technological change will have a significant impact on the cost because emission reductions will be implemented over a long time, and new technologies should lower the cost of future reductions.
- Many policies implemented by national, state/provincial, and municipal jurisdictions and private firms to achieve objectives other than carbon management increase or reduce CO_2/CH_4 emissions.

Under a wide range of assumptions, policies to reduce atmospheric CO_2 and CH_4 concentrations cost-effectively in the short and long term would:

- Encourage adoption of low cost emission reduction and sink enhancement actions. An emission trading program or emissions tax that covers as many sources and sinks as possible, combined with regulations where appropriate, is an example of a way to achieve this. Use of revenues from auctioned allowances and/or emission taxes could reduce the net economic cost of emission reduction policies.
- Stimulate development of technologies that lower the cost of emissions reduction, carbon capture and sequestration, and sink enhancement.
- Adopt appropriate regulations for sources or actions subject to market imperfections, such as energy efficiency measures and cogeneration.
- Revise existing policies at the national, state/provincial, and local level related to objectives other

than carbon management so that the objectives, if still relevant, are achieved with lower CO_2 or CH_4 emissions.

Implementation of such policies at a national level, and cooperation at an international level, would reduce the overall cost of achieving a carbon reduction target by providing access to more low-cost mitigation/sequestration options.

---

[30] In other words, research and development is required for a portfolio of technologies. Because technologies have global markets, international cooperation to stimulate the research and development, as occurs through the International Energy Agency and the Asia-Pacific Partnership on Clean Development and Climate (APP), is appropriate.

How Can We Improve the Usefulness of Carbon Science for Decision Making?

Coordinating Lead Authors: Lisa Dilling, Univ. Colo./NCAR; Ronald Mitchell, Univ. Oreg.

Lead Author: David M. Fairman, Consensus Building Inst., Inc.
Contributing Authors: Myanna Lahsen, IGBP (Brazil) and Univ. Colo.; Susanne Moser, NCAR; Anthony Patt, Boston Univ./IIASA; Chris Potter, NASA; Charles Rice, Kans. State Univ.; Stacy VanDeveer, Univ. N.H.

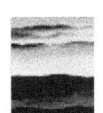

KEY FINDINGS

- Decision makers are seeking more comprehensive information on the carbon cycle and on carbon management options across scales and sectors. Carbon management is a relatively new concept not only for decision makers and members of the public, but also for the science community.
- Improving the usefulness of carbon science in North America will require stronger commitments to generating high quality science that is also decision-relevant.
- Research on the production of policy-relevant scientific information suggests several ways to improve the usefulness of carbon science for decision making, including co-production of knowledge, development of applied modeling tools for decision support, and use of "boundary organizations" that can help carbon scientists and decision makers communicate and collaborate.
- A number of initiatives to improve understanding of decision support needs and options related to the carbon cycle are under way, some as a part of the Climate Change Science Program (CCSP).
- Additional pilot projects should be considered aimed at enhancing interactions between climate change scientists and parties involved in carbon management activities and decisions.

5.1 INTRODUCTION: THE CHALLENGE OF "USABLE" CARBON SCIENCE

This chapter answers two questions:

- How well is the carbon cycle science community doing in "decision support" of carbon cycle management, *i.e.*, in responding to decision makers' demands for carbon cycle management information?
- How can the carbon cycle science community improve such decision support?

Chapters in Parts 2 and 3 of this report identify many research priorities, including assessing the potential for geological storage of carbon dioxide (CO_2), quantifying expansion of the North American carbon sink, and identifying the economic impact of carbon tax systems. This chapter focuses on improving communication and collaboration between scientific researchers and carbon managers, to help researchers be more responsive to decision making, and carbon managers be better informed in making policy, investment, and advocacy decisions.

Humans have been inadvertently altering the Earth's carbon cycle since the dawn of agriculture, and more rapidly since the industrial revolution. These influences have become large enough to cause significant climate change (IPCC, 2007). In response, environmental advocates, business executives, and policy-makers have increasingly recognized the need to manage the carbon cycle deliberately. Effective carbon management requires that the variety of people whose decisions affect carbon emissions and sinks have relevant, appropriate science. Yet, carbon cycle science is rarely organized or conducted to support decision making on managing carbon emissions, uptake and storage (sequestration), and impacts. This reflects that, until recently, scientists have approached carbon cycle science as basic science and only a relatively small, although growing, portion of non-scientist decision makers have demanded carbon cycle information. Consequently, emerging efforts to manage carbon are less informed by carbon cycle science than they could be (Dilling *et al.*, 2003). Applying carbon science to carbon management requires making carbon cycle science more useful to public and private decision-

> Humans have been inadvertently altering the Earth's carbon cycle since the dawn of agriculture, and more rapidly since the industrial revolution. These influences have become large enough to cause significant climate change.

makers at all levels, from national and international policy-makers to the executives and employees of corporations to the millions of individuals whose myriad consumer and household decisions are central to human impacts on the carbon cycle. In particular, scientists and decision makers will need to identify the information most needed in specific sectors for carbon management, adjust research priorities, and develop mechanisms that enhance the credibility of the information generated and the responsiveness of the information-generating process to address stakeholder's views (Lahsen and Nobre, 2007; Mitchell *et al.*, 2006; Cash *et al.*, 2003). Combining some "applied" or "solutions-oriented" research with a portfolio that also includes basic science would make carbon science more directly relevant to decision making.

5.2 TAKING STOCK: WHERE ARE WE NOW IN PROVIDING DECISION SUPPORT TO IMPROVE CAPACITIES FOR CARBON MANAGEMENT?

How effective is the scientific community at providing decision support for carbon management? The Climate Change Science Program (CCSP) Strategic Plan defines decision support as: "the set of analyses and assessments, interdisciplinary research, analytical methods, model and data product development, communication, and operational services that provide timely and useful information to address questions confronting policymakers, resource managers, and other stakeholders" (U.S. Climate Change Science Program, 2003).

Who are the potential stakeholders for information related to the carbon cycle and what are the options and measures for altering human influences on that cycle? Most people constantly, but unconsciously, make decisions that affect the carbon cycle through their use of energy, transportation, living spaces, and natural resources. Increasing attention to climate change has led some policy makers, businesses, advocacy groups, and consumers to begin making choices that consciously limit carbon emissions[1]. Whether carbon emission reductions are driven by political pressures or legal requirements, by economic opportunities, or consumer pressures, or by moral or ethical commitments to averting climate change, people and organizations are seeking information that can help them achieve their specific carbon-related or climate-related goals[2]. Even in countries and economic sectors that lack a consensus on the need to manage carbon, some people and organizations have begun to experiment with carbon-limiting practices and investments in anticipation of a carbon-constrained future.

---

[1] For examples, see Box 5.1

[2] For example, carbon science was presented at recent meetings of the West Coast Governors' Global Warming Initiative and the Climate Action Registry [http://www.climateregistry.org/EVENTS/PastConferences/; http://www.climatechange.ca.gov/events/2005_conference/presentations/]

In designing and producing this report, we engaged individuals from a wide range of sectors and activities, including forestry, agriculture, utilities, fuel companies, carbon brokers, transportation, non-profits, and local and federal governments. Although we did not conduct new research on the informational or decision support needs of stakeholders, a preliminary review suggests that many stakeholders may be interested in carbon-related information (see Box 5.1).

5.3 CURRENT APPROACHES AND TRENDS

Interest in, and attention paid to, carbon information has increased incrementally over the last 20 years. Future levels of interest are likely to depend on perceived risks from carbon emissions as well as on whether and how mandatory and incentive-based policies related to carbon management evolve. As efforts at deliberate carbon management become increasingly common, decision makers from the local to the national level are increasingly open to or actively seeking carbon science information as a direct input to policy and investment decisions (Apps *et al.*, 2003). The government of Canada, having ratified the Kyoto Protocol, has been exploring emission reduction opportunities and offsets and has identified specific needs for applied research (Environment Canada, 2005). For example, Canada's national government recently entered a research partnership with the province of Alberta to assess geological sequestration of CO_2, to develop fuel cell technologies using hydrogen, and to expand the use of vegetative matter (biomass) and biowaste for energy production (Western Economic Diversification Canada, 2006).

Some stakeholders in the United States are actively using carbon science to move forward with voluntary emissions offset programs. For example, the Chicago Climate Exchange brokers agricultural carbon credits in partnership with the Iowa Farm Bureau[3]. Many cities and several states have established commitments to manage carbon emissions, including regional partnerships on the east and west coasts, and non-governmental organizations and utilities have begun to experiment with pilot sequestration projects (Box 5.1). In Europe, for example, mandatory carbon emissions policies have resulted in intense interest in carbon science by those directly affected by such policies (Schröter *et al.*, 2005).

In the United States, federal carbon science has very few mechanisms to assess demand for carbon information across scales and sectors. Thus far, federally-funded carbon science has focused on basic research to clarify fundamental uncertainties in the global carbon cycle and local and regional processes affecting the exchange of carbon (Dilling, in

press). Most federal efforts are organized under the CCSP. The National Aeronautics and Space Administration (NASA) and the National Science Foundation (NSF) manage almost two-thirds of this effort and their missions are limited to basic research, not decision support (CCSP, 2006; Dilling, 2007). Research efforts have also been undertaken at the Department of Energy (DOE), the Department of Agriculture (USDA)[4], and the Department of Interior's Geological Survey (USGS/DOI). Significant technology efforts are underway in the Climate Change Technology Program (CCTP), a sister program to the CCSP focused on technology development. Increasing linkages among these programs may increase the usefulness of CCSP carbon-related research to decision makers. For over a decade, the National Oceanic and Atmospheric Administration (NOAA) Climate Program Office has invested in research and institutions intended to improve the usability of climate science, although that investment is small relative to the investment in climate science itself and has focused on the usability of climate, rather than carbon cycle, science.

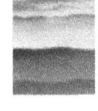

> As efforts at deliberate carbon management become increasingly common, decision makers from the local to the national level are increasingly open to or actively seeking carbon science information as a direct input to policy and investment decisions.

Until recently, the concept of "carbon management" has not been widely recognized—even now, most members of the public do not understand the term "carbon sequestration" or its potential implications (Shackley *et al.*, 2005; Curry *et al.*, 2004). However, the carbon cycle science community is beginning to recognize that it may have information relevant to policy and decision making. Thus prominent carbon scientists have called for "coordinated rigorous, interdisciplinary research that is strategically prioritized to address societal needs" (Sarmiento and Wofsy, 1999) and the North American Carbon Program's (NACP) "Implementation Plan" lists decision support as one of four organizing questions (Denning *et al.*, 2005).

That same plan, however, states that the scientific community knows relatively little about the likely users of information that the NACP will produce. Indeed, the National Academy of Sciences' review of the CCSP stated that "as the decision support elements of the program are implemented, the CCSP will need to do a better job of identifying stakeholders and the types of decisions they need to make" (NRC, 2004). Moreover, they state that "managing risks

---

[3] http://www.iowafarmbureau.com/special/carbon/default.aspx

[4] For example, the Consortium for Agricultural Soil Mitigation of Greenhouse Gases (CASMGS) was recently funded by the USDA to provide information and technology necessary to develop, analyze, and implement carbon sequestration strategies.

This list of sectors is neither exhaustive nor is it based on a systematically rigorous assessment, but is meant to demonstrate the wide variety of stakeholders with a potential interest in carbon-related information.

Agriculture: Tillage and other farming practices significantly influence carbon storage in agricultural soils. Managing these practices presents opportunities both to slow carbon loss and to restore carbon in soils. Farmers have been quite interested in carbon management as a means to stimulate rural economic activity. Since much of the agricultural land in the United States is privately owned, both economic forces and governmental policies will be critical factors in the participation of this sector in carbon management. (Chapter 10 this report).

Forestry: Forests accumulate carbon in above-ground biomass as well as soils. The carbon impact of planting, conserving, and managing forests has been an area of intense interest in international negotiations on climate change (IPCC, 2000). Whether seeking to take advantage of international carbon credits, to offset other emissions, or to simply identify environmental co-benefits of forest actions taken for other reasons, governments, corporations, landowners, and conservation groups may need more information on and insight into the carbon implications of forestry decisions ranging from species selection to silviculture, harvesting methods, and the uses of harvested wood. (Chapter 11 this report).

Utilities and Industries: In the United States, over 85% of energy produced comes from fossil fuels with relatively high carbon intensity. The capital investment and fuel source decisions of utilities and energy-intensive industries thus have major carbon impacts. A small but growing number of companies have made public commitments to reducing carbon emissions, developed business models that demonstrate sensitivity to climate change, and begun exploring carbon capture and storage opportunities. For example, Cinergy, a large Midwestern utility, has experimented with carbon-offset programs in partnership with The Nature Conservancy. (Chapter 6 and 8 this report).

Transportation: Transportation accounts for approximately 37% of carbon emissions in the United States, and about 22% worldwide. Governmental infrastructure investments, automobile manufacturers' decisions about materials, technologies and fuels, and individual choices regarding auto purchases, travel modes, and distances all have significant impacts on carbon emissions. (Chapter 7 this report).

Government: In the United States, national policies currently rely primarily on voluntary measures and incentive structures (U.S. Department of State, 2004; Richards, 2004). Canada, having ratified the Kyoto Protocol, has direct and relatively immediate needs for information that can help it meet its binding targets as cost-effectively as possible (Environment Canada, 2005). The Mexican government appears to be particularly interested in locally relevant research on natural and human influences on the carbon cycle, likely impacts across various regions, and the costs, benefits, and viability of various management options (Martinez and Fernandez-Bremauntz, 2004). Below the national level, more and more states and local governments are taking steps, including setting mandatory policies, to reduce carbon emissions, and may need new carbon cycle science scaled to the state and local level to manage effectively. For example, nine New England and mid-Atlantic states have formed a regional partnership, also observed by Eastern Canadian provinces, to reduce carbon emissions through a cap and trade program combined with a market-based emissions trading system (Regional Greenhouse Gas Initiative—RGGI—www.rggi.org). (Chapters 4 and 14 this report).

Non-Profits and Non-Governmental Organizations (NGOs): Many environmental and business-oriented organizations have an interest in carbon management decision making. Such organizations rely on science to support their positions and to undercut the arguments of opposing advocates. There has been substantial criticism of "advocacy science" in the science-for-policy literature, and new strategies will need to be developed to promote constructive use of carbon cycle science by advocates (Ehrmann and Stinson, 1999; Adler *et al.*, 1999).

and opportunities requires stakeholder support on a range of scales and across multiple sectors, which in turn implies an understanding of the decision context for stakeholders" (NRC, 2004). Successful decision support (*i.e.*, science that improves societal outcomes) requires understanding of who the users are and of the kind of information they are likely to deem relevant and bring to bear on their decision making. Without such knowledge, information runs the risk of being "left on the loading-dock" and not used (Cash et al., 2006; Lahsen and Nobre, 2007).

Some programs within CCSP may shed light on how to link carbon science to user needs. NASA has an Applied Sciences program that seeks to find uses for its data and modeling products using "benchmarking systems," and the USDA and DOE have invested significant resources in science that might inform carbon sequestration efforts and carbon accounting in agriculture and forests. However, these programs have not been integrated into a broader framework self-consciously aimed at making carbon cycle science more useful to decision makers.

Funding agencies, scientists, policy makers, and private sector managers can improve the usefulness of carbon science programs in North America by increasing their commitments to generating decision-relevant carbon cycle information and by integrating those programs more fully into forums and institutions involved in carbon cycle management. The participatory methods and boundary span-

ning institutions identified in the next section help both refine research agendas and accelerate the application of research results to carbon management and societal decision making.

5.4 OPTIONS FOR IMPROVING THE APPLICABILITY OF SCIENTIFIC INFORMATION TO CARBON MANAGEMENT AND DECISION MAKING

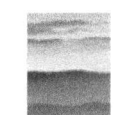

Studies of the creation and use of knowledge for decision making have found that information must be perceived not only as credible, but also as relevant to high priority decisions and as stemming from a process that decision makers view as responsive to their concerns (Mitchell *et al.*, 2006; Cash *et al.*, 2003). Even technically and intellectually rigorous science lacks influence with decision makers if decision makers perceive it as not addressing the decisions they face, as being biased, or as having ignored their views and interests.

Research on the production of policy-relevant scientific information suggests several strategies that can maintain the integrity of the research endeavor while increasing its policy relevance. Although communicating results more effectively is clearly important, generating science that is more applicable to decision making may require deeper changes in the way scientific information is produced. Carbon cycle scientists and carbon decision makers will need to develop methods for interaction that work best in the specific arenas in which they work. At their core, strategies will be effective to the extent that they promote interaction among scientists and stakeholders in the development of research questions, selection of research methods, and review, interpretation, and dissemination of results (Adler *et al.*, 1999; Ehrmann and Stinson, 1999; NRC, 1999; NRC, 2005; Farrell and Jaeger, 2005; Mitchell *et al.*, 2006). Such processes work best when they enhance the usability of the research while preserving the credibility of both scientists and stakeholders. Transparency and expanded participation are important for guarding against politicization and enhancing usability.

Examples of joint scientist-stakeholder development of policy relevant scientific information include:

*   *Co-production of research knowledge (e.g., Regional Integrated Sciences and Assessments)*: In regional partnerships across the United States, university researchers work closely with local operational agencies and others that might incorporate climate information in decision making. New research is developed through ongoing, iterative consultations with all partners (Lemos and Morehouse, 2005). Co-production of research knowledge involves efforts to reach out to, educate, and involve stakeholders in programs that facilitate a dialog

of researchers and stakeholders consulting with and engaging each other in identifying near-term research questions and longer-term research trajectories.

- *Institutional experimentation and adaptive behavior (e.g., adaptive management)*: Adaptive management acknowledges our inherent uncertainty about how natural systems respond to human management, and periodically assesses the outcomes of management decisions and adjusts those decisions accordingly, a form of deliberate "learning by doing" (*cf.*, Holling, 1978). Adaptive management principles have been applied to several resources where multiple stakeholders are involved, including management of river systems and forests (Holling, 1995; Pulwarty and Redmond, 1997; Mitchell *et al.*, 2004; Lemos and Morehouse, 2005).

- *Assessments as policy components (e.g., recovering the stratospheric ozone layer)*: Assessments that were credible, relevant, and responsive played a significant role in the Montreal Protocol's success in phasing out the use of ozone-depleting substances. A highly credible scientific and technical assessment process with diverse academic and industry participation is considered crucial in the Protocol's success (Parson, 2003).

- *Mediated modeling*: Shared tools can facilitate scientist-user interactions, help diverse groups develop common knowledge and understanding of a problem, and clarify common assumptions and differences. In mediated modeling, participants from a wide variety of perspectives jointly construct a computer model to solve complex environmental problems or envision a shared future. The process has been used for watershed management, endangered species management, and other difficult environmental issues (Van den Belt, 2004).

- *Carbon modeling tools as decision support*: Although the United States government has not yet adopted a carbon management policy, some federal agencies have begun to develop online decision support tools, with customizable user interfaces, to estimate carbon sequestration in various ecosystems and under various land-use scenarios (see the NASA Ames Carbon Query and Evaluation Support Tools, http://geo.arc.nasa. gov/sge/casa/cquestwebsite/index.html; the U.S. Forest Service Carbon Online Estimator, http://ncasi.uml. edu/COLE/;and Colorado State's CarbOn Management Evaluation Tool, http://www.cometvr.colostate.edu/).

Over time, well-structured scientist-stakeholder interaction can help both scientists and decision makers (Moser, 2005). Scientists learn to identify research questions that are both scientifically interesting and relevant to decisions, and to present their answers in ways that audiences are more likely to find compelling. Non-scientists learn what questions science can and cannot answer. Such interactions clarify the boundary between empirical questions that scientists can answer (*e.g.*, the sequestration potential of a particular technology) and issues that require political resolution (*e.g.*, the appropriate allocation of carbon reduction targets across firms). Institutional arrangements can convert *ad hoc* successes in scientist-stakeholder interaction into systematic and ongoing networks of scientists, stakeholders, and managers. Such "co-production of knowledge," can enhance both the scientific basis of policy and management and the research agenda for applied science (Lemos and Morehouse, 2005; Gibbons *et al.*, 1994; Patt *et al.*, 2005a).

That said, such interactive approaches have limitations, risks, and costs. Scientists may be reluctant to involve non-scientists who "should" be interested in a given issue, but who can add little scientific value to the research, and whose involvement requires time and effort. Involving private sector firms may require scientists accustomed to working in an open informational environment to navigate in a world of proprietary information. Scientists may also avoid applied, participatory research if they do not see it producing the "cutting edge" (and career enhancing) science most valued by other scientists (Lahsen and Nobre, 2007; Lemos and Morehouse, 2005). Public and private carbon cycle science programs, as well as universities and research institutes, more generally, can help address these obstacles by recognizing that they exist and altering incentive structures to reward innovation in applied research through endowed chairs, fellowships, research grants, and the like.

Some stakeholders may lack the financial resources, expertise, time, or other capacities necessary to meaningful participation. Some will distrust scientists in general, and government-sponsored science in particular, for cultural, institutional, historical, or other reasons. Some may reject the idea of interacting with those with whom they disagree politically or compete economically. Stakeholders may try to manipulate research questions and findings to serve their political or economic interests. In addition, stakeholders often show little interest in diverting their time from other activities to what they perceive as the slow and too-often fruitless pursuit of scientific knowledge (Patt *et al.*, 2005b).

Where direct stakeholder participation proves too difficult, costly, unmanageable, or unproductive, scientists and research managers need other methods to identify the needs of potential users. Science on the one hand, and policy, management, and decision making on the other, often exist as separate social and professional realms, with different traditions, norms, codes of behavior, and reward systems. The boundaries between such realms serve many useful functions but can inhibit the transfer of useful knowledge across those boundaries. A boundary organization is an institution that "straddles the shifting divide" between politics and science (Guston, 2001). Boundary organizations are accountable to both sides of the boundary and involve professionals from each. Boundary spanning individuals and organizations may facilitate the uptake of science by translating scientific findings so that stakeholders find them more useful and by stimulating adjustments in research agendas and approach.

Boundary organizations can exist at a variety of scales and for a variety of purposes. For example, cooperative agricultural extension services and non-governmental organizations (NGOs) successfully convert large-scale scientific understandings of weather, aquifers, or pesticides into locally-tuned guidance to farmers (Cash, 2001). The International Research Institute for Climate Prediction focuses on seasonal-to-interannual scale climate research and modeling to make their research results useful to farmers, anglers, and public health officials (*e.g.*, Agrawala *et al.*, 2001). The Subsidiary Body for Scientific and Technological Advice of the United Nations Framework Convention on Climate Change serves as an international boundary organization that links information and assessments from expert sources (such as the Intergovernmental Panel on Climate Change [IPCC]) to the Conference of the Parties, which focuses on setting policy[5]. The University of California Berkeley Digital Library Project *Calflora* has explicitly designed their database on plants to support environmental planning (Van House *et al.*, 2003).

Though attractive in principle, boundary organizations may not be effective in practice. They may fail to be useful if they are not responsive to both the stakeholders and scientists they seek to engage. They may be captured by one particular stakeholder or science interest. Their usefulness may decline over time if they are unable to keep pace with the salient issues of the principals on either side of the boundary.

Cooperative agricultural extension services and non-governmental organizations (NGOs) successfully convert large-scale scientific understandings of weather, aquifers, or pesticides into locally-tuned guidance to farmers.

Even where boundary organizations do facilitate the translation of scientific expertise for policy, other significant challenges exist in the use of knowledge. People fail to integrate new research and information in their decisions for many reasons. People often are not motivated to use information that supports policies they dislike or that conflicts with pre-existing preferences, interests, or beliefs, or with cognitive, organizational, sociological, or cultural norms (*e.g.*, Douglas and Wildavsky, 1984; Lahsen, 1999; Yaniv, 2004; Lahsen, 2007). These tendencies are important components of a healthy democratic process. Developing processes to make carbon science more useful to decision makers will not guarantee its use, but will make its use more likely.

5.5 RESEARCH NEEDS TO ENHANCE DECISION SUPPORT FOR CARBON MANAGEMENT

The demand for detailed analysis of carbon management issues and options across major economic sectors, nations, and levels of government in North America is likely to grow substantially in the near future. This will be especially true in jurisdictions that place policy constraints on carbon budgets, such as Canada, United States' states comprising the Regional Greenhouse Gas Initiative, or the U.S. State of California. Although new efforts are underway in some federal agencies, carbon cycle science in the United States could be organized and carried out to better and more systematically meet this potential demand. Effective implementation of the goals of the Climate Change Science Program "requires focused research to develop decision support resources and methods" (NRC, 2004). Relevant science could evaluate the impacts, technical feasibility, and economic potential of the wide range of existing and newly-developed options that are likely to be proposed in response to growing regional and national interest in carbon management.

Relevant science could evaluate the impacts, technical feasibility, and economic potential of the wide range of existing and newly-developed options that are likely to be proposed in response to growing interest in carbon management.

[5] http://unfccc.int/2860.php

Creating information for decision support should differ significantly from doing basic science. In such "use-inspired research," societal need is as important as scientific curiosity (Stokes, 1997). Scientists and carbon managers need to improve their joint understanding of the top priority questions facing carbon-related decision making. They need to collaborate more effectively in undertaking research and interpreting results in order to answer those questions.

A first step might involve developing a formal process "for gathering requirements and understanding the problems for which research can inform decision makers outside the scientific community," including forming a decision support working group (Denning *et al.*, 2005). The NRC has recommended that the CCSP's decision support components could be improved by organizing various deliberative activities, including workshops, focus groups, working panels, and citizen advisory groups to: "1) expand the range of decision support options being developed by the program; 2) to match decision support approaches to the decisions, decision makers, and user needs; and 3) to capitalize on the practical knowledge of practitioners, managers, and laypersons" (NRC, 2004).

5.6 SUMMARY AND CONCLUSIONS

The carbon cycle is influenced through both deliberate and inadvertent decisions by diverse and spatially dispersed people and organizations, working in many different sectors and at different scales. To make carbon cycle science more useful to decision makers, we suggest that leaders in the scientific and program level carbon science community initiate the following steps:

- Identify categories of decision makers for whom carbon cycle science is a relevant concern, focusing on policy makers and private sector managers in carbon-intensive sectors (energy, transport, manufacturing, agriculture, and forestry).
- Evaluate existing information about carbon impacts of actions in these arenas, and assess the need and demand for additional information. In some cases, demand may need to be fostered through an interactive process.
- Encourage scientists and research programs to experiment with incremental, as well as major, departures from existing practice with

the goal of making carbon cycle science more credible, relevant, and responsive to carbon managers.

- Involve experts in the social sciences and communication as well as experts in physical, biological, and other natural science disciplines in efforts to produce usable science.
- Consider initiating participatory pilot research projects and identifying existing boundary organizations (or establishing new ones) to bridge carbon management and carbon science.

Energy, Industry, and Waste Management Activities: An Introduction to CO_2 Emissions From Fossil Fuels

Coordinating Lead Author: Gregg Marland, ORNL and Mid Sweden Univ. (Östersund)

Contributing Authors: Robert J. Andres, Univ. N. Dak.; T.J. Blasing, ORNL; Thomas A. Boden, ORNL; Christine T. Broniak, Oreg. State Univ.; Jay S. Gregg, Univ. Md.; London M. Losey, Univ. N. Dak.; Karen Treanton, IEA (Paris)

II.1 THE CONTEXT

Fossil fuels (coal, oil, and natural gas) are used primarily for their concentration of chemical energy, energy that is released as heat when the fuels are burned. Fossil fuels are composed primarily of compounds of hydrogen and carbon, and when the fuels are burned, the hydrogen and carbon oxidize to water and carbon dioxide (CO_2) and heat is released. If the water and CO_2 are released to the atmosphere, the water will soon fall out as rain or snow. The CO_2, however, will increase the concentration of CO_2 in the atmosphere and join the active cycling of carbon that takes place among the atmosphere, biosphere, and hydrosphere. Since humans began taking advantage of fossil-fuel resources for energy, we have been releasing to the atmosphere, over a very short period of time, carbon that was stored deep in the Earth over millions of years. We have been introducing a large perturbation to the active cycling of carbon.

Estimates of fossil-fuel use globally show that there have been significant emissions of CO_2 dating back at least to 1750, and from North America, back at least to 1785. However, this human perturbation of the active carbon cycle is largely a recent process, with the magnitude of the perturbation growing as population grows and demand for energy grows. Over half of the CO_2 released from fossil-fuel burning globally has occurred since 1980 (Figure II.1).

Some CO_2 is also released to the atmosphere during the manufacture of cement. Limestone ($CaCO_3$) is heated to release CO_2 and produce the calcium oxide (CaO) used to manufacture cement. In North America, cement manufacture now releases less than 1% of the mass of CO_2 released by fossil-fuel combustion. However, cement manufacture is the third largest human-caused (anthropogenic) source of CO_2 (after fossil-fuel use and the clearing and oxidation of forests and soils; see Part III this report). The CO_2 emissions from cement manufacture are often included with the accounting of anthropogenic CO_2 emissions from fossil fuels.

> Over half of the CO_2 released from fossil-fuel burning globally has occurred since 1980.

Part II of this report addresses the magnitude and pattern of CO_2 emissions from fossil-fuel consumption and cement manufacture in North America. This introductory section addresses some general issues associated with CO_2 emissions and the annual and cumulative magnitude of total emissions. It looks at the temporal and spatial dis-

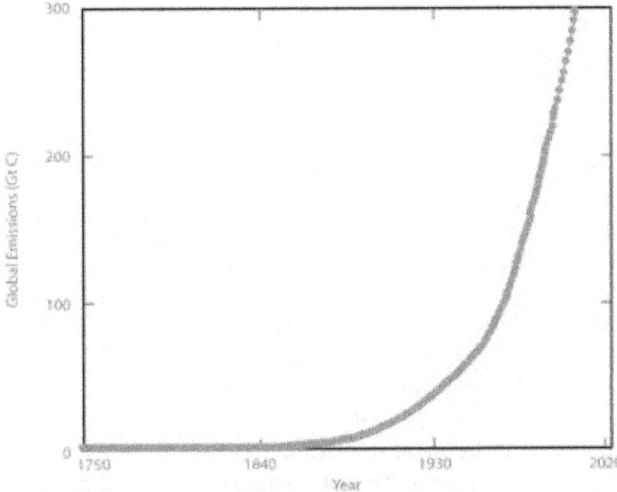

Figure II.1 Cumulative global emmissions of CO_2 from fossil-fuel combustion and cement manufacture from 1751 to 2002. *Source data:* Marland *et al.* (2005).

tribution of emissions and other data likely to be of interest. The following four chapters delve into the sectoral details of emissions so that we can understand the forces that have driven the growth in emissions to date and the possibilities for the magnitude and pattern of emissions in the future. These chapters reveal, for example, that 38% of CO_2 emissions from North America come from enterprises whose primary business is to provide electricity and heat and another 31% come from the transport of passengers and freight. This introduction focuses on the total emissions from the use of fossil fuels and the subsequent chapters provide insight into how these fuels are used and the economic and human factors motivating their use.

II.1.1 Estimating Carbon Dioxide Emissions

It is relatively straightforward to estimate the amount of CO_2 released to the atmosphere when fossil fuels are consumed. Because CO_2 is the equilibrium product of oxidizing the carbon in fossil fuels, we need to know only the amount of fuel used and its carbon content. For greater accuracy, we adjust this estimate to take into consideration the small amount of carbon that is left as ash or soot and is not actually oxidized. We also consider the fraction of fossil fuels that are used for things like asphalt, lubricants, waxes, solvents, and plastics and may not be soon converted to CO_2. Some of these long-lived, carbon-containing products will release their contained carbon to the atmosphere as CO_2 during use or during processing of waste. Other products will hold the carbon in use or in landfills for decades or longer. One of the differences among the various estimates of CO_2 emissions is the way they deal with the carbon in these products.

> It is relatively straightforward to estimate the amount of CO_2 released to the atmosphere when fossil fuels are consumed.

Fossil-fuel consumption is often measured in mass or volume units and, in these terms, the carbon content of fossil fuels is quite variable. However, when we measure the amount of fuel consumed in terms of its energy content, we find that for each of the primary fuel types (coal, oil, and natural gas) there is a strong correlation between the energy content and the carbon content. The rate of CO_2 emitted per unit of useful energy released depends on the ratio of hydrogen to carbon and on the details of the organic compounds in the fuels; but, roughly speaking, the numerical conversion from energy released to carbon released as CO_2 is about 25 kg C per 10^9 joules for coal, 20

Table II.1 A sample of the coefficients used for estimating CO_2 emissions from the amount of fuel burned.

Fuel	Emissions coefficient (kg C/10^9 J net heating value)
Lignite	27.6
Anthracite	26.8
Bituminous coal	25.8
Crude oil	20.0
Residual fuel oil	21.1
Diesel oil	20.2
Jet kerosene	19.5
Gasoline	18.9
Natural gas	15.3

Source: IPCC (1997).

kg C per 10^9 joules for petroleum, and 15 kg C per 10^9 joules for natural gas. Figure PII.2 shows details of the correlation between energy content and carbon content for more than 1000 coal samples. Detailed analysis of the data suggests that hard coal contains 25.16 ± 2.09% kg C per 10^9 joules of coal (measured on a net heating value basis[1]). The value is slightly higher for lignite and brown coal (26.23 kg C ±

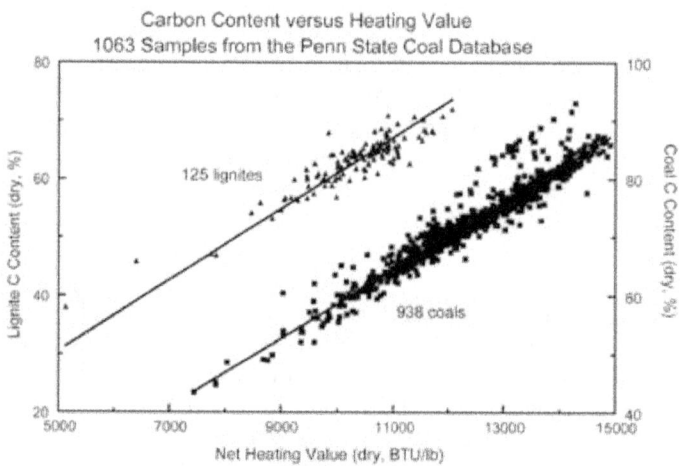

Figure II.2 The carbon content of coal varies with the heat content, shown here as the net heating value. To make them easier to distinguish, data for lignites and brown coals are shown on the left axis, while data for hard coals are offset by 20% and shown on the right axis. Heating value is plotted in the units at which it was originally reported, Btu/lb, where 1 Btu/lb = 2324 J/kg. *Source:* Marland *et al.* (1995).

---

[1] Net heating value (NHV) is the heat release measured when fuel is burned at constant pressure so that the water (H_2O) is released as H_2O vapor. This is distinguished from the gross heating value (GHV), the heat release measured when the fuel is burned at constant volume so that the H_2O is released as liquid H_2O. The difference is essentially the heat of vaporization of the H_2O and is related to the hydrogen content of the fuel.

2.33% per 10^9 joules (also shown in Figure II.2). Similar correlations exist for all fuels and Table PII.1 shows some of the coefficients reported by the Intergovernmental Panel on Climate Change (IPCC) for estimating CO_2 emissions. The differences between the values in Table II.1 and those in Figure II.2 are small, but they begin to explain how different data compilations can end up with different estimates of CO_2 emissions.

Data on fossil-fuel production, trade, consumption, *etc.* are generally collected at the level of some political entity, such as a country, and over some time interval, typically a year. Estimates of national, annual fuel consumption can be based on estimates of fuel production and trade, estimates of actual final consumption, data for fuel sales or some other activity that is clearly related to fuel use, or on estimates and models of the activities that consume fuel (such as vehicle miles driven). In the discussion that follows, some estimates of national, annual CO_2 emissions are based on "apparent consumption" (defined as production + imports – exports +/– changes in stocks), while others are based on more direct estimates of fuel consumption. All of the emissions estimates in this chapter are as the mass of carbon released[2].

The uncertainty in estimates of CO_2 emissions will thus depend on the variability in the chemistry of the fuels, the quality of the data or models of fuel consumption, and on uncertainties in the amount of carbon that is used for non-fuel purposes (such as asphalt and plastics) or is otherwise not burned. For countries like the United States—with good data on fuel production, trade, and consumption—the uncertainty in national emissions of CO_2 is on the order of ±5% or less. In fact, the U.S. Environmental Protection Agency (USEPA, 2005) suggests that their estimates of CO_2 emissions from energy use in the United States are accurate, at the 95% confidence level, within –1 to +6% and Environment Canada (2005) suggests that their estimates for Canada are within –4 to 0%. The Mexican National Report (Mexico, 2001) does not provide estimates of uncertainty, but our analyses with the Mexican data suggest that uncertainty is larger than for the United States and Canada. Emissions estimates for these same three countries, as reported by the Carbon Dioxide Information Analysis Center (CDIAC) and the International Energy Agency (IEA) (see the following section), will have larger uncertainty because these groups are making estimates for all countries. Because they work with data from

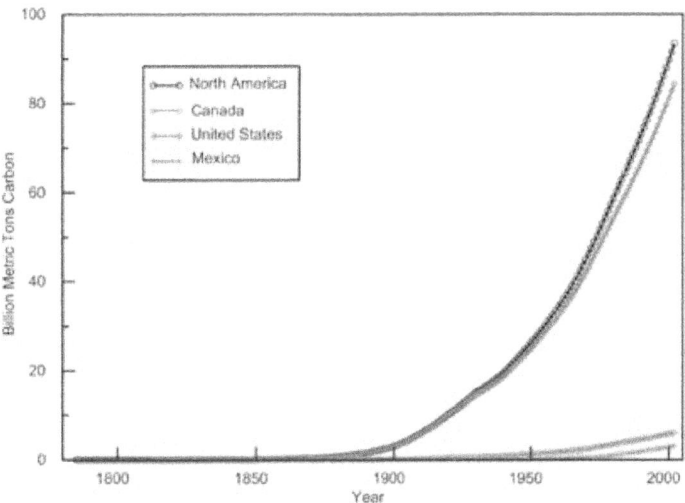

Figure II.3 The cumulative total of CO_2 emissions from fossil-fuel consumption and cement manufacture, as a function of time, for the three countries of North America and for the sum of the three. *Source:* Marland *et al.* (2005).

all countries, they use global average values for things like the emissions coefficients, whereas agencies within the individual countries use values that are more specific to the particular country. When national emissions are calculated by consistent methods it is likely that year-to-year changes can be estimated more accurately than would be suggested by the uncertainties of the individual annual values.

II.1.2 The Magnitude of National and Regional Carbon Dioxide Emissions

Figure II.3 shows that from the beginning of the fossil-fuel era (1751 in these graphs) to the end of 2002, there were 93.5 billion tons of carbon (Gt C) released as CO_2 from fossil-fuel consumption (and cement manufacture) in North America: 84.4 Gt C from the United States, 6.0 from Canada, and 3.1 from Mexico. All three countries of North America are major users of fossil fuels and this 93.5 Gt C was 31.5% of the global total. Among all countries, the United States, Canada, and Mexico ranked as the first, eighth, and eleventh largest emitters of CO_2 from fossil-fuel consumption, respectively (for 2002) (Marland *et al.*, 2005). Figure II.4 shows, for each of these countries and for the sum of the three, the annual total of emissions and the contributions from the different fossil fuels.

The long time series of emissions estimates in Figures II.1, II.3, and II.4 are from the CDIAC (Marland *et al.*, 2005). These estimates are derived from the "apparent consumption" of fuels and are based on data from the United Nations Statistics Office back to 1950 and on data from a mixture of sources for the earlier years (Andres *et al.*, 1999). There are other published estimates (with shorter time series) of national, annual CO_2 emissions. Most notably the IEA (2005) has reported estimates of emissions for many coun-

---

[2] The carbon is actually released to the atmosphere as CO_2 and it is accurate to report (as is often done) either the amount of CO_2 emitted or the amount of C in the CO_2. The numbers can be easily converted back and forth using the ratio of the molecular masses, *i.e.* (mass of C) x (44/12) = (mass of CO_2).

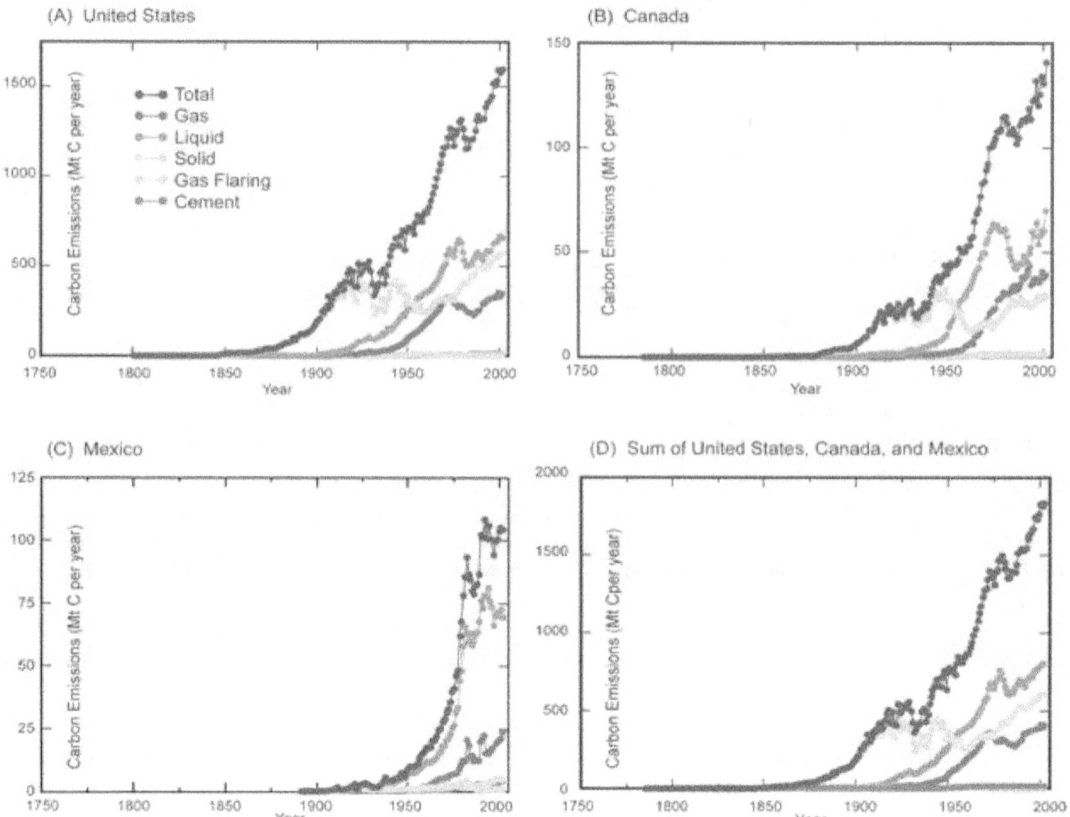

Figure II.4 Annual emissions of CO_2 from fossil-fuel use by fuel type for (A) the United States, (B) Canada, (C) Mexico, and (D) North America, as the sum of the data shown in the other three panels. *Note* that in order to illustrate the contributions of the different fuels, the four plots are not to the same vertical scale. *Source:* Marland *et al.* (2005).

tries for all years back to 1971, and most countries have now provided some estimates of their own emissions as part of their national obligations under the United Nations Framework Convention on Climate Change (UNFCCC, see http://unfccc.int). These latter two sets of estimates are based on data on actual fuel consumption and thus are able to provide details as to the sector of the economy where fuel use is taking place[3].

All three countries of North America are major users of fossil fuels and this 93.5 Gt C was 31.5% of the global total.

Comparing the data from multiple sources can give us some insight into the reliability of the estimates, generally. These different estimates of CO_2 emissions are not, of course, truly independent because they all rely, ultimately, on national data on fuel use; but they do represent

[3] The International Energy Agency provides estimates based on both the reference approach (estimates of apparent consumption) and the sectoral approach (estimates of actual consumption) as described by the IPCC (IPCC, 1997). In the comparison here, we use the numbers that they believe to be the most accurate, those based on the sectoral approach.

different manipulations of this primary data and in many countries there are multiple potential sources of energy data. Many developing countries do not collect or do not report all of the data necessary to precisely estimate CO_2 emissions and in these cases differences can be introduced by how the various agencies derive the basic data on fuel production and use. Because of the way data are collected, there are statistical differences between "consumption" and "apparent consumption" as defined above.

To make comparisons of different estimates of CO_2 emissions we would like to be sure that we are indeed comparing estimates of the same thing. For example, emissions from cement manufacture are not available from all of the sources, so they are not included in the comparisons in Table II.2. All of the estimates in Table II.2, except those from the IEA, include emissions from flaring natural gas at oil production facilities. It is not easy to identify the exact reason the estimates differ, but the differences are generally small. The differences have mostly to do with the statistical difference between consumption and apparent consumption, the way in which correction is made for non-fuel usage of fossil-fuel resources, the conversion from mass or volume to energy

Table II.2 Different estimates (in MtC) of CO2 emissions from fossil-fuel consumption for the United States, Canada, and Mexico.

Country	1990		1998		2002	
United States	CDIAC	1305	CDIAC	1501	CDIAC	1580
	IEA	1320	IEA	1497	IEA	1545
	USEPA	1316	USEPA	1478	USEPA	1534
Canada	CDIAC	112	CDIAC	119	CDIAC	139
	IEA	117	IEA	136	IEA	145
	Canada	117	Canada	133	Canada	144
Mexico	CDIAC	99	CDIAC	96	CDIAC	100
	IEA	80	IEA	96	IEA	100
	Mexico	81	Mexico	96	Mexico	NA

Notes:

Many of these data were published in terms of the mass of CO_2, and these data have been multiplied by 12/44 to get the mass of carbon for the comparison here.

All data except CDIAC include oxidation of non-fuel hydrocarbons.

All data except IEA include flaring of gas at oil and gas processing facilities.

Sources: CDIAC (Marland *et al.*, 2005), IEA (2005), USEPA (2005), Canada (Environment Canada, 2005), and Mexico (2001).

units, and/or the way in which estimates of carbon content are derived. Because the national estimates from CDIAC do not include emissions from the non-fuel uses of petroleum products, we expect them to be slightly smaller than the other estimates shown here, all of which do include these emissions[4]. The comparisons in Table II.2 reveal one number for which there is a notable relative difference among the multiple sources, emissions from Mexico in 1990. Losey (2004) has suggested, based on other criteria, that there is a problem in the United Nations energy data set with the Mexican natural gas data for the three years 1990-1992, and these kinds of analyses result in re-examination of some of the fundamental data.

The IEA (2005, p. 1.4) has systematically compared their estimates with those reported to the UNFCCC by the different countries and they find that the differences for most developed countries are within 5%. The IEA attributes most of the differences to the following: use of the IPCC Tier 1 method that does not take into account different technologies, use of energy data that may have come from different "official" sources within a country, use of average values for net heating value of secondary oil products, use of average emissions values, use of incomplete data on non-fuel uses, different treatment of military emissions, and a different split between what is identified as emissions

from energy and emissions from industrial processes.

With increasing interest in the details of the global carbon cycle there is increasing interest in knowing emissions at spatial and temporal scales finer than countries and years. For the United States, energy data have been collected for many years at the level of states and months and thus estimates of CO_2 emissions can be made by state or by month. Figure II.5 shows the variation in United States' emissions by month and preliminary analyses by Gurney *et al.* (2005) reveal that proper recognition of this variability can be very important in some exercises to model the details of the global carbon cycle.

Because of differences in the way energy data are collected and aggregated, it is not obvious that an estimate of emissions from the United States will be identical to the sum of estimates for the 50 United States' states. Figure II.6 shows that estimates of total annual CO_2 emissions are slightly different if we use data directly from the U.S. Department of Energy (DOE) and sum the estimates for the 50 states or if we sum the estimates for the 12 months of a given year, or if we take United States' energy data as aggregated by the United Nations Statistics Office and calculate the annual total of CO_2 emissions directly. Again,

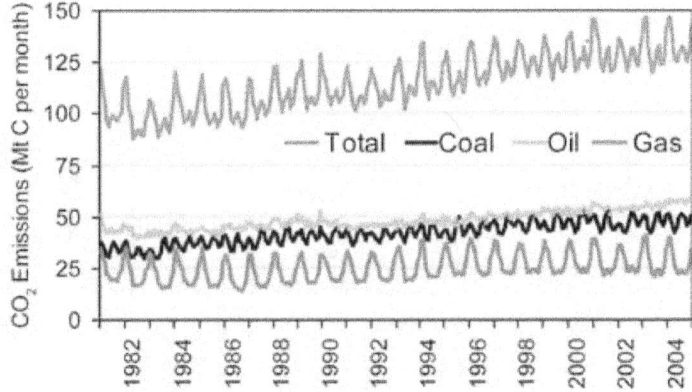

Figure. II.5 Emissions of CO_2 from fossil-fuel consumption in the United States, by month. Emissions from cement manufacturing are not included. *Source:* Blasing *et al.* (2005a).

---

[4] The CDIAC estimate of global total emissions does include estimates of emissions from oxidation from non-fuel use of hydrocarbons.

the state and monthly emissions data are based on estimates of fuel consumption while the national emissions estimates calculated using United Nations' data result from estimates of "apparent consumption." There is a difference between annual values for consumption and annual values of "apparent consumption" (the IEA calls this difference simply "statistical difference") that is related to the way statistics are collected and aggregated. There are also differences in the way values for fuel chemistry and non-fuel usage are averaged at different spatial and temporal scales, but the differences in CO_2 estimates are seen to be within the error bounds generally expected.

Data from DOE permit us to estimate emissions by state or by month (Blasing *et al.*, 2005a and 2005b), but they do not permit us to estimate CO_2 emissions for each state by month directly from the published energy data. Nor do we have sufficiently complete data to estimate emissions from Canada and Mexico by month or province. Andres *et al.* (2005), Gregg (2005), and Losey (2004) have shown that we can disaggregate national total emissions by month or by some national subdivision (such as states or provinces) if we have data on some large fraction of fuel use. Because this approach relies on determining the fractional distribution of an otherwise-determined total, it can be done with incomplete data on fuel use. The estimates will, of course, improve as the fraction of the total fuel use is increased. Figure II.7 is based on sales data for most fossil-fuel commodities and the CDIAC estimates of total national emissions and shows how

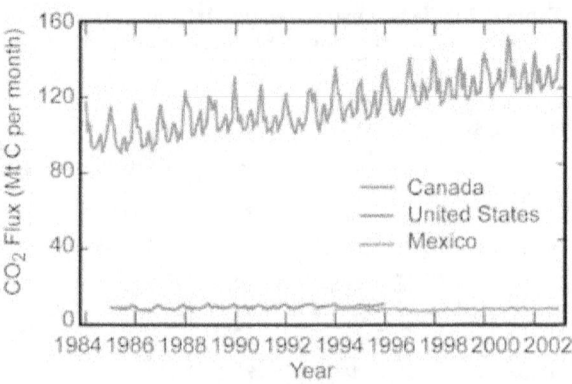

Figure II.7 Carbon dioxide emissions from fossil-fuel consumption in North America, by month. Monthly values are shown where estimates are justified by the availability of monthly data on fuel consumption or sales. *Source:* Andres et al., (2005).

the CO_2 emissions from North America vary at a monthly time scale.

II.1.4 Emissions by Economic Sector

To understand how CO_2 emissions from fossil-fuel use interact in the global and regional cycling of carbon, it is necessary to know the masses of emissions and their spatial and temporal patterns. We have tried to summarize this information here. To understand the trends and the driving forces behind the growth in fossil-fuel emissions, and the opportunities for controlling emissions, it is necessary to look in detail at how the fuels are used. This is the goal of the next four chapters of this report.

Before looking at the details of how energy is used and where CO_2 emissions occur in the economies of North America, however, there are two indices of CO_2 emissions at the national level that provide perspective on the scale and distribution of emissions. These two indices are emis-

> To understand the trends and the driving forces behind the growth in fossil-fuel emissions, and the opportunities for controlling emissions, it is necessary to look in detail at how the fuels are used.

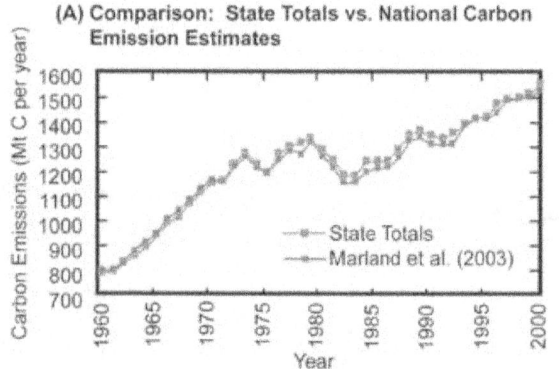

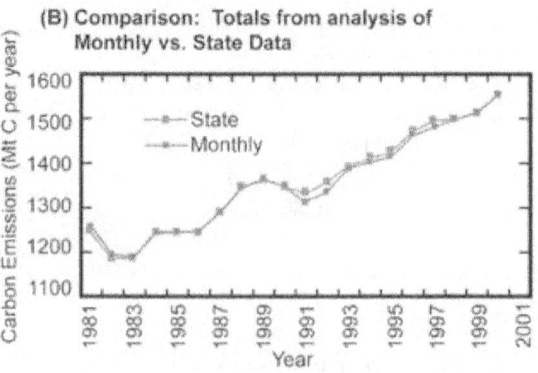

Figure. II.6 A comparison of three different estimates of national annual emissions of CO_2 from fossil-fuel consumption in the United States. (A) Estimates from U.S. Department of Energy data on fuel consumption by state (blue squares) vs. estimates based on UN Statistics Office data on apparent fuel consumption for the full United States (red squares). (B) Estimates based on DOE data on fuel consumption in the 50 U.S. states (blue squares) vs. estimates based on national fuel consumption for each of the 12 months (red squares). The state and monthly data include estimates of oxidation of non-fuel hydrocarbon products; the UN-based estimates do not. *Source:* Blasing et al., (2005b).

Table II.3 Emissions of CO₂ from fossil-fuel consumption (cement manufacture and gas flaring are not included) per unit of GDP for the United States, Canada, Mexico and for the global total.

Country	CO₂ emissions per unit of GDP[a]		
	Year		
	1990	1998	2002
United States	0.19	0.17	0.15
Canada	0.18	0.18	0.16
Mexico	0.13	0.12	0.11
Global Total	0.17	0.15	0.14

[a] Carbon dioxide is measured in kg carbon and GDP is reported in 2000 US$ purchasing power parity.
Source: IEA (2005).

Canada, and Mexico do not cover the same time periods, nor do they present data in the same way. In a discussion of the possibilities for reducing CO₂ emissions in the building sector it is not obvious, for example, whether to include the relevant electricity within the building sector, to leave electric power generation as a separate sector, or to accept some overlap in the discussion. The authors of Chapters 6, 7, 8, and 9 have chosen the system boundaries and data they find most useful for the individual sectors, even though it makes it more difficult to aggregate across sectors.

sions *per capita* and emissions per unit of economic activity, the latter generally represented by CO₂ per unit of gross domestic product (GDP). Figure II.8 shows the 1950–2002 record of CO₂ emissions *per capita* for the three countries of North America and for perspective includes the same data for the Earth as a whole. Similarly, Table II.3 shows CO₂ emissions per unit of GDP for the three countries of North America and for the world total. These are, of course, very complex indices and though they provide some insight they say nothing about the details and the distributions within the means. The data on CO₂ *per capita* for the 50 United States' states (Figure II.9) show that values range over a full order of magnitude, differing in complex ways with the structure of the economies and probably with factors like climate, population density, and access to resources (Blasing *et al.*, 2005b; Neumayer, 2004).

Chapters 6 through 9 of this report discuss the patterns and trends of CO₂ emissions by sector and the driving forces behind the trends that are observed. Estimating emissions by sector brings special challenges in defining sectors and assembling the requisite data. Readers will find that there is consistency and coherence within each of the following chapters but will encounter difficulty in aggregating or summing numbers across chapters. Different experts use different sector boundaries, different data sources, different conversion factors, *etc.* Different analysts and literature sources will find data for different base years and may treat electricity and biomass fuels differently. The national reports of the United States,

Despite these differences in accounting procedures, the four chapters that follow accurately characterize the patterns of emissions and the opportunities for controlling the growth in emissions. They reveal that there are major differences between the countries of North America where, for example, the United States derives 51% of its electricity from coal, Mexico gets 68% from petroleum and natural gas, and Canada gets 58% from hydroelectric stations. Partially as a reflection of this difference, 40% of United States' CO₂ emissions are from enterprises whose primary business is to generate electricity and heat, while this number is only 31% in

> Forty percent of the United States' CO₂ emissions are from enterprises whose primary business is to generate electricity and heat, while this number is only 31% in Mexico and 23% in Canada.

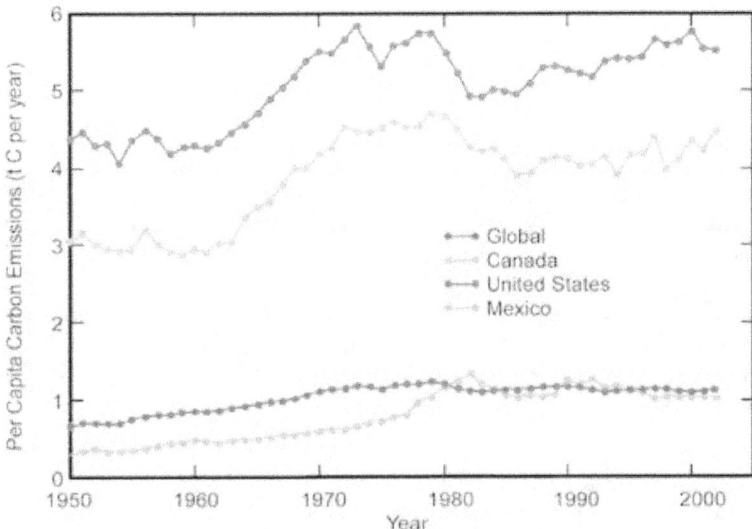

Figure II.8 *Per capita* emissions of CO₂ from fossil-fuel consumption and cement manufacture in the United States, Canada, and Mexico and for the global total of emissions. *Source:* Marland *et al.*, (2005).

Mexico and 23% in Canada (for 2003; from IEA, 2005). Chapter 8 reveals that the sectors are not independent as, for example, a change from fuel burning to electricity in an industrial process will decrease emissions from the industrial sector but increase emissions in the electric power sector. The database of the IEA allows us to summarize CO_2 emissions for the three countries according to sectors that closely correspond to the sectoral division of chapters 6 through 9 (Table II.4).

II.2 CONCLUSION

There are a variety of reasons that we want to know the emissions of CO_2 from fossil fuels, there are a variety of ways of coming up with the desired estimates, and there are a variety of ways of using the estimates. By the nature of the process of fossil-fuel combustion, and because of its economic importance, there are reasonably good data over long time intervals that we can use to make reasonably accurate estimates of CO_2 emissions to the atmosphere. In fact, it is the economic importance of fossil-fuel burning that has assured us of both good data on emissions and great challenges in altering the rate of emissions.

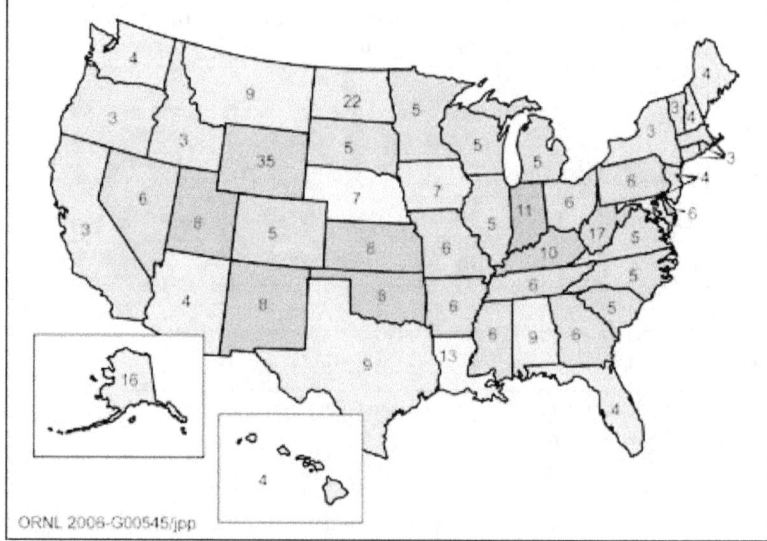

Figure II.9 *Per capita* emissions of CO_2 from fossil-fuel consumption for the 50 United States in 2000. To demonstrate the range, values have been rounded to whole numbers of metric tons carbon *per capita*. A large portion of the range for extreme values is related to the occurrence of coal resources and inter-state transfers of electricity. *Source:* Blasing *et al.* (2005b).

Table II.4 Percentage of CO$_2$ emissions by sector for 2003.

Sector	United States	Canada	Mexico	North America
Energy extraction and conversion[a]	46.2	36.2	47.7	45.4
Transportation[b]	31.3	27.7	30.3	31.0
Industry[c]	11.2	16.8	13.6	11.8
Buildings[d]	11.3	19.3	8.4	11.8

[a] The sum of three IEA categories, "public electricity and heat production," "unallocated autoproducers," and "other energy industries."
[b] IEA category "transport."
[c] IEA category "manufacturing industries and construction."
[d] IEA category "other sectors."
Source: IEA (2005).

Energy Extraction and Conversion

Lead Author: Thomas J. Wilbanks, ORNL

Contributing Authors: Marilyn Brown, Ga. Inst. Tech.; Ken Caldeira, Carnegie Inst.; William Fulkerson, Univ. Tenn.; Erik Haites, Margaree Consultants, Inc; Stephen Pacala, Princeton Univ.; David M. Fairman, Consensus Building Inst., Inc.

KEY FINDINGS

- In recent years, the extraction of primary energy sources and their conversion into energy commodities in North America released on the order of 760 million tons of carbon (2800 million tons of carbon dioxide) per year to the atmosphere, approximately 40% of total North American emissions in 2003 and 10% of total global emissions. Electricity generation is responsible for a very large share of North America's energy extraction and conversion emissions.
- Carbon dioxide emissions from energy supply systems in North America are currently rising.
- Principal drivers behind carbon emissions from energy supply systems are (1) the growing appetite for energy services, closely related to economic and social progress, and (2) the market competitiveness of fossil energy compared with alternatives.
- Emissions from energy supply systems in North America are projected to increase in the future. Projections vary among the countries, but increases approaching 50% or more in coming decades appear likely. Projections for the United States, for example, indicate that carbon dioxide emissions from electricity generation alone will rise to above 900 million tons of carbon (3300 million tons of carbon dioxide) by 2030, an increase of about 45% over emissions in 2004, with three-quarters of the increase associated with greater coal use in electric power plants.
- Prospects for major reductions in carbon dioxide emissions from energy supply systems in North America appear dependent upon (a) the extent, direction, and pace of technological innovation and (b) whether policy conditions favoring carbon emissions reduction that do not now exist will emerge (Figure 6.1). In these regards, the prospects are brighter in the long term (e.g., more than several decades in the future) than in the near term.
- Research and development priorities for managing carbon emissions from energy supply systems include, on the technology side, clarifying and realizing potentials for carbon capture and storage, and on the policy side, understanding the public acceptability of policy incentives for reducing dependence on carbon-intensive energy sources.

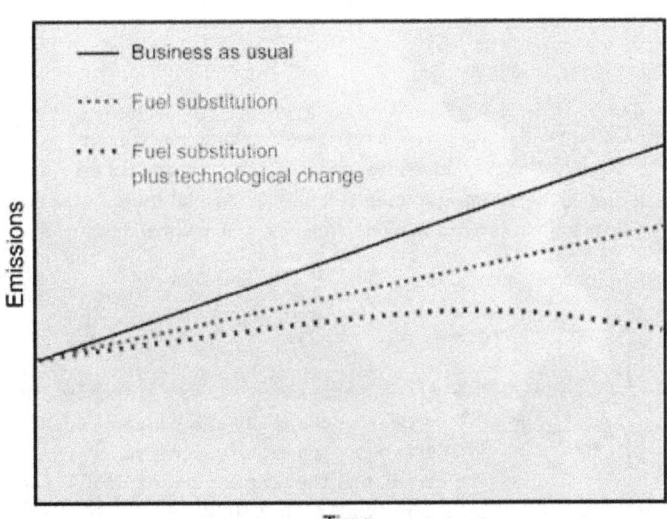

Figure 6.1 Prospects for carbon emissions from energy extraction and conversion in North America, assuming substantial improvement in energy efficiency.

6.1 INTRODUCTION

The energy supply system in North America is a significant part of the North American carbon cycle, because so many of its primary energy resources are fossil fuels associated with extraction and conversion activities that emit greenhouse gases. This chapter summarizes the knowledge bases related to emissions from energy extraction, energy conversion, and other energy supply activities such as energy movement and energy storage, along with options and measures for managing emissions.

Clearly, this topic overlaps the subject matter of other chapters. For instance, the dividing line between energy conversion and other types of industry is sometimes indistinct. One prominent case is emissions associated with electricity and process heat supply for petroleum refining, and other fossil-fuel processing (a large share of their total emissions) included in industrial sector emission totals; another example is industrial co-generation as an energy-efficiency strategy. In addition, biomass energy extraction/conversion is directly related to agriculture and forestry. Moreover, emission-related policy alternatives for energy supply systems are often directed at both supply and demand responses, involving not only emission reductions, but also potential payoffs from efficiency improvements in buildings, industry, and transportation, especially where they reduce the consumption of fossil fuels.

Canada is the world's fifth-largest energy producing country, a significant exporter of both natural gas and electricity to the United States.

6.2 CARBON EMISSIONS INVENTORY

6.2.1 Carbon Emissions From Energy Extraction and Conversion

Carbon emissions from energy resource extraction, conversion into energy commodities, and transmission are one of the "big three" sectors accounting for most of the total emissions from human systems in North America, along with industry and transportation. The largest share of total emissions from energy supply (not including energy end use) is from coal and other fossil-fuel use in producing electricity; fossil-fuel conversion activities such as oil refining and natural gas transmission and distribution also contribute to this total, but in much smaller amounts. Other emission sources are less well defined, but generally small, such as emissions from oil production and methane from reservoirs established partly to support hydropower production (Tremblay *et al.*, 2004), or from materials production (*e.g.*, metals production) associated with other renewable or nuclear energy technologies. Generally, data on emissions have a relatively low level of uncertainty, although the source materials do not include quantitative estimates of uncertainty.

Data on emissions from energy supply systems are unevenly available for the countries of North America, and none are associated with sufficient information to support an assessment of uncertainty. Most emission data sets are organized by fuel consumed rather than by consuming sector, and countries differ in sectors identified and the units of measurement. As a result, inventories are reported in this chapter by country in whatever forms are available rather than constructing a North American inventory that could not be consistent across all three major countries. It is worth noting that Canada and Mexico export energy supplies to the United States, therefore, some emissions from energy supply systems in these countries are associated with energy uses in the United States.

6.2.1.1 CANADA

Canada is the world's fifth-largest energy producing country, a significant exporter of both natural gas and electricity to the United States. In Alberta, which produces nearly two-thirds of Canada's energy, energy accounts for about one-quarter of the province's economic activity; its oil sands are estimated to have more potential energy value than the remaining oil reserves of Saudi Arabia (U.S. Department of Energy, 2004). Although Canada has steadily reduced its energy and carbon intensities since the early 1970s, its overall energy intensity remains high—in part due to its prominence as an energy producer—and total greenhouse gas emissions have grown by 9% since 1990. As of 2003, greenhouse gas emissions were 36.5 million metric tons of carbon (Mt C) equivalents (134 million tons of carbon dioxide [Mt CO_2] equivalents) for electricity and heat generation and 19 Mt C (71 Mt CO_2) for petroleum refining and upgrading and other fossil-fuel production (Environment Canada, 2003). Although the mix of carbon dioxide (CO_2) and methane (CH_4) in these figures is unclear, the carbon emission equivalent is probably within the range of 60-80 Mt C.

6.2.1.2 MEXICO

Mexico is one of the largest sources of energy-related greenhouse gas emissions in Latin America, although its *per capita* emissions are well below the *per capita* average of industrialized countries. The first large oil-producing nation to ratify the Kyoto Protocol, it has promoted shifts to natural gas use to reduce greenhouse gas emissions. The most recent emission figures are from the country's Second National Communication to the United Nations Framework Convention on Climate Change (UNFCCC) in 2001, which included relatively comprehensive data from 1996 and some data from 1998. In 1998, total emissions from "energy industries" were 13 Mt C (47.3 Mt CO_2); from electricity generation they totaled 27.6 Mt C (101.3 Mt CO_2); and "fugitive" emissions from oil and gas production and distribution were between 1.4 and 2.0 Mt C (1.9 and 2.6 Mt of CH_4), depending on the estimated "emission factor" (Government of Mexico, 2001).

6.2.1.3 UNITED STATES

The United States is the largest national emitter of greenhouse gases in the world, and CO_2 emissions associated with electricity generation in 2004 account for 627 Mt C (2299 Mt CO_2), or 39% of a national total of 1600 Mt C (5890 Mt CO_2) (EIA, 2006a). Greenhouse gases are also emitted from oil refining, natural gas transmission, and other fossil energy supply activities, but apart from energy consumption figures included in industry sector calculations, these emissions are relatively small compared with electric power plant emissions. For instance, emissions from petroleum consumed in refining processes in the United States are about 40 Mt C per year (EIA, 2004), while fugitive emissions from gas transmission and distribution pipelines in the United States are about 2.2 Mt C per year[1]**(see Box 6.1 for uncertainty conventions). On the other hand, a study of greenhouse gas emissions from a six-county area in southwestern Kansas found that compressor stations for natural gas pipeline systems are a significant source of emissions at that local scale (AAG, 2003).

6.2.2 Carbon Sinks Associated With Energy Extraction and Conversion

Generally, energy supply in North America is based heavily on mining hydrocarbons from carbon sinks accumulated over millions of years; but current carbon sequestration occurs in plant growth, including the cultivation of feedstocks for bioenergy production. Limited strictly to energy sector applications,

the total contribution of these sinks to the North American carbon cycle is relatively small, while other aspects of bioenergy development are associated with carbon emissions; but the substitution of biomass-derived fuels (approximately emisson-neutral, as stored carbon is released with fuel use) for fossil fuels represents a potentially significant net savings in emissions.

> The substitution of biomass-derived fuels for fossil fuels represents a potentially significant net savings in emissions.

6.3 TRENDS AND DRIVERS

Three principal drivers are behind carbon emissions from energy extraction and conversion:

1. The growing global and national appetite for energy services such as comfort, convenience, mobility, and labor productivity, so closely related to progress with economic and social development and the quality of life (Wilbanks, 1992). Globally, the challenge is to increase total energy services (not necessarily supplies) over the next half-century by a factor of at least three or four—more rapidly than overall economic growth—while reducing environmental impacts from the associated supply systems (NAS, 1999). Mexico shares this need, while increases in Canada and the United States are likely to be more or less proportional to rates of economic growth.

2. The market competitiveness of fossil energy sources compared with supply- and demand-side alternatives. Production costs of electricity from coal, oil, or natural gas at relatively large scales are currently lower than other sources, except large-scale hydropower, and production costs of liquid and gas fuels are currently far lower than other sources, though rising. This is mainly because the energy density and portability of fossil fuels is as yet unmatched by other energy sources, and in some cases policy conditions reinforce fossil-fuel use. These

conditions appear likely to continue for some years. In many cases, the most cost-competitive alternative to fossil-fuel production and use is not alternative supply sources, but efficiency improvement.

3. Enhanced future markets for alternative energy supply sources. In the longer run, however, emissions from energy supply systems may—and in fact, are likely to—begin to decline as alternative technology options are developed and/or improved. Other possible driving forces for attention to alternatives to fossil fuels, at least in the mid to longer term, include the possibility of shrinking oil and/or gas reserves and changes in attitudes toward energy policy interventions.

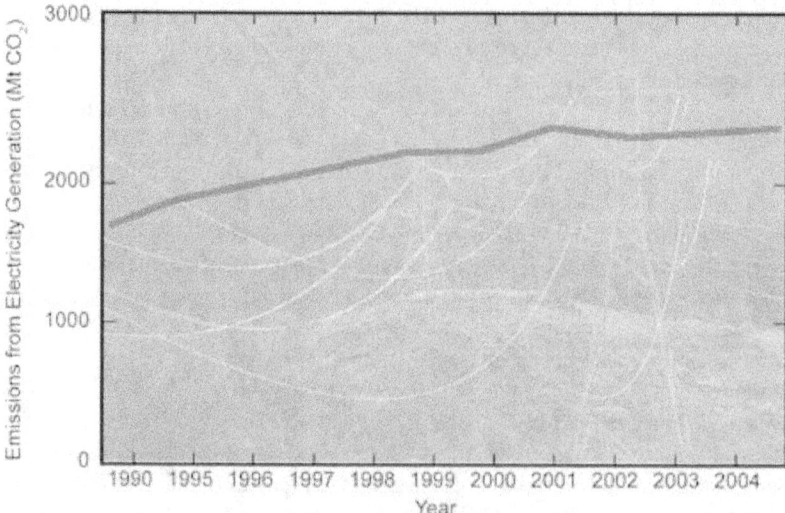

Figure 6.2 U.S. carbon dioxide emissions from electricity generation, 1990-2004. *Source:* EIA, 2004, and the authors' extensions for year 2004.

Total carbon emissions from energy extraction and conversion in North America are currently rising.

Given the power of the first two of these drivers, total carbon emissions from energy extraction and conversion in North America are currently rising (*e.g.*, Figure 6.2). National trends and drivers are as follows. As is always the case, projections of the future involve higher levels of uncertainty than measurements of the present, but source materials do not include quantitative estimates of uncertainties associated with projections of future emissions.

6.3.1 Canada
Canada has ratified the Kyoto Protocol, and it is seeking to meet the Kyoto target of CO_2 emission reduction to 6% below 1990 levels. Of these reductions, 25% are to be through domestic actions and 75% through market mechanisms such as purchases of carbon credits (Government of Canada, 2005). Domestic actions will include a significant reduction in coal consumption. Available projections, however, indicate a total national increase of emissions in CO_2 equivalent of 36.1% by 2020 from 1990 levels (Environment Canada, 2005). Emissions from electricity generation could increase 2000-2020 by as much as two-thirds, while emissions from fossil-fuel production would remain relatively stable (although substantial expansion of oil sands production could be a factor).

It has been estimated that total Mexican CO_2 emissions will grow 69% by 2010, although mitigation measures could reduce this rate of growth by nearly half.

6.3.2 Mexico
It has been estimated that total Mexican CO_2 emissions will grow 69% by 2010, although mitigation measures could reduce this rate of growth by nearly half (Pew Center, 2002). Generally, energy sector emissions in Mexico vary in proportion to economic growth (*e.g.*, declining somewhat with a recession in 2001). However, factors, such as a pressing need for additional electricity supplies (calling for more than doubling production capacity between 1999 and 2008) could increase net emissions, while a national strategy to promote greater use of natural gas (along with other policies related in part to concerns about emissions associated with urban air pollution) could reduce emissions compared with a reference case (EIA, 2005).

6.3.3 United States
The Energy Information Administration (EIA, 2006b) projects that emissions from electricity generation in the United States between 2004 and 2030 will rise from about 627 Mt C (2299 Mt CO_2) to more than 900 Mt C (3300 Mt CO_2) (an increase of about 45%) with three-quarters of the increase associated with greater coal use in electric power plants. EIA projects that technology advances could lower emissions by as much as 9%. Projections of other emissions from energy supply systems appear to be unavailable, but emissions could be expected to rise at a rate just below the rate of change in product consumption in the United States' economy.

6.4 OPTIONS FOR REDUCING EMISSIONS FROM ENERGY EXTRACTION AND CONVERSION

Few aspects of the carbon cycle have received more attention in the past several decades than emissions from fossil energy extraction and conversion. As a result, there is a wide array of technology and policy options, many of which have been examined in considerable detail, although there is not a strong consensus on courses of action.

6.4.1 Technology Options

Technology options for reducing energy-supply-related emissions (other than reduced requirements due to end-use efficiency improvements) consist of:

- reducing emissions from fossil energy extraction, production, and movement (*e.g.*, for electricity generation by improving the efficiency of existing power plants or moving toward the use of lower-emission technologies such as coal gasification-combined cycle generation facilities) and
- shifting from fossil energy sources to other energy sources (*e.g.*, energy from the sun [renewable energy] or from the atom [nuclear energy]).

The most comprehensive description of emission-reducing and fuel switching technologies and their potentials is the U.S. Climate Change Technology Program (CCTP) draft Strategic Plan (U.S. Climate Change Technology Program, 2005), especially Chapters 5 (energy supply) and 6 (capturing and sequestering CO_2)—see also National Laboratory Directors (1997). The CCTP report focuses on five energy supply technology areas: low-emission fossil-based fuels and power, hydrogen as an energy carrier, renewable energy and fuels, nuclear fission, and fusion energy.

There is a widespread consensus that no one of these options, nor one family of options, is a good prospect to stabilize greenhouse gas emissions from energy supply systems, nationally or globally, because each faces daunting constraints (Hoffert *et al.*, 2002). An example is possible physical and/or technological limits to effective global "decarbon-

ization" (*i.e.*, reducing the use of carbon-based energy sources as a proportion of total energy supplies), including renewable or other non-fossil sources of energy use at scales that would dramatically change the global carbon balance between now and 2050. One conclusion is that "the disparity between what is needed and what can be done without great compromise may become more acute."

If many contributions can be combined, the total effect could approach requirements for even relatively ambitious carbon stabilization goals.

Instead, progress with technologies likely to be available in the coming decades may depend on adding together smaller "wedges" of contributions by a variety of resource/technology combinations (Pacala and Socolow, 2004), each of which may be feasible if the demands upon it are moderate. If many such contributions can be combined, the total effect could approach requirements for even relatively ambitious carbon stabilization goals, at least in the first half of the century, although each contribution would need to be economically competitive with current types of fossil energy sources.

A fundamental question is whether prospects for significant decarbonization depend on the emergence of new technologies, in many cases requiring advances in science. For instance, efforts are being made to develop economically affordable and socially acceptable options for large-scale capture of carbon from fossil-fuel streams—with the remaining hydrogen offering a clean energy source—and sequestration of the carbon in the ground or the oceans. This approach is known to be technologically feasible and is being practiced commercially in the North Sea. Recent assessments suggest that it may have considerable promise (*e.g.*, IPCC, 2006). If so, there is at least some chance that fossil energy sources may be used to provide energy services in North America and the world in large quantities in the mid to longer terms without contributing to a carbon cycle imbalance.

What can be expected from technology options over the next quarter to half a century is a matter of debate, partly because the pace of technology development and use depends heavily on policy conditions. Chapter 3 in the CCTP draft Strategic Plan (2005) shows three advanced technology scenarios drawn from work by the Pacific Northwest National Laboratory, varying according to carbon constraints. Potential cumulative contributions to global emission reduction by energy supply technology initiatives between

2000 and 2100 range from about 25 billion tons of carbon (Gt C) equivalent to nearly 350 Gt, which illustrates uncertainties related to both science and policy issues. Carbon capture and storage, along with terrestrial sequestration, could add reductions between about 100 and 325 Gt C. It has been suggested, however, that significantly decarbonizing energy systems by 2050 could require massive efforts on a par with the Manhattan Project or the Apollo Space Program (Hoffert *et al.*, 2002).

Estimated costs of potential technology alternatives for reducing greenhouse gas emissions from energy supply systems are summarized after the following discussion of policy options, because cost estimates are generally based on assumptions about policy interventions.

6.4.2 Policy Options

Policy options for carbon emission reduction from energy supply systems revolve around either incentives or regulatory requirements for such reductions. Generally, interventions may be aimed at (a) shaping technology choice and use or (b) shaping technology development and supply. Many of the policy options are aimed at encouraging end-use efficiency improvement as well as supply-side emission reduction.

Options for intervening to change the relative attractiveness of available energy supply technology alternatives include appealing to voluntary action (*e.g.*, improved consumer information, "green power"), a variety of regulatory actions (*e.g.*, mandated purchase policies such as energy portfolio standards), carbon emission rights trading (where emission reduction would have market value), technology/product standards, production tax credits for non-fossil energy production, tax credits for alternative energy use, and carbon emission taxation or ceilings. Options for changing the relative attractiveness of investing in carbon-emission-reducing technology development and dissemination include tax

credits for certain kinds of energy research and development, public-private sector research and devleopment cost sharing, and electric utility restructuring. For a more comprehensive listing and discussion, see Chapter 6 in IPCC (2001).

In some cases, perceptions that policies and market conditions of the future will be more favorable to emission reduction than at present are motivating private industry to consider investments in technologies whose market competitiveness would grow in such a future. Examples include the CO_2 Capture Project and industry-supported projects at MIT, Princeton, and Stanford (*e.g.*, see http://www.co2captureproject.org/index.htm).

Most estimates of the impacts of energy policy options on greenhouse gas emissions do not differentiate the contributions from energy supply systems from the rest of the energy economy (*e.g.*, IWG, 1997; IWG, 2000; IPCC, 2001; National Commission on Energy Policy, 2004; also see OTA, 1991 and NAS, 1992). For instance the IWG (1997) considered effects of $25 and $50 per ton carbon emission permits on both energy supply and use, while Interlaboratory Working Group (IWG) considered fifty policy/technology options (IWG, 2000; also see IPCC, 2001), most of which would affect both energy supply and energy use decisions.

6.4.3 Estimated Costs of Implementation

Estimating the costs of emission reduction associated with the implementation of various technology and policy options for energy supply and conversion systems is complicated by several realities. First, many estimates are aggregated for the United States or the world as a whole, without separate estimates for the energy extraction and conversion sector. Second, estimates differ in the scenarios considered, the modeling approaches adopted, and the units of measure that are used.

More specifically, estimates of costs of emission reduction vary widely according to assumptions about such issues as how welfare is measured, ancillary benefits, and effects in stimulating technological innovation; and therefore any particular set of cost estimate includes considerable uncertainty. According to IWG (2000), benefits of emission reduction would be comparable to costs, and the National Commission on Energy Policy (2004) estimates that their recommended policy initiatives would be, overall, revenue-neutral with respect to the federal budget. Other participants in energy policymaking, however, are convinced that truly significant carbon emission reductions would have substantial economic impacts (GAO, 2004).

Globally, IPCC (2001) projected that total CO_2 emissions from energy supply and conversion could be reduced in 2020 by 350 to 700 Mt C equivalents per year, based on options that could be adopted using generally accepted policies, at a positive direct cost of less than U.S. $100 per ton of carbon (t C) equivalents. Based on DOE/EIA analyses in 2000, this study includes estimates of the cost of a range of specific emission-reducing technologies for power generation, compared with coal-fired power, although the degree of uncertainty is not clear. Within the United States, the report estimated that the cost of emission reduction per metric ton of carbon emissions reduced would range from -$170 to +$880, depending on the technology used. Marginal abatement costs for the total United States' economy (in 1990 U.S. dollars per metric ton carbon) were estimated by a variety of models compared by the Energy Modeling Forum at $76 to $410 with no emission trading, $14 to $224 with Annex I trading, and $5 to $123 with global trading.

Similarly, the National Commission on Energy Policy (2004) considered costs associated with a tradable emission permit system that would reduce United States' national greenhouse gas emission growth from 44% to 33% from 2002 to 2025, a reduction of 207 Mt C (760 Mt CO_2) in 2025 compared with a reference case. The cost would be a roughly 5% increase in total end-use expenditures compared with the reference case. Electricity prices would rise by 5.4% for residential users, 6.2% for commercial users, and 7.6% for industrial users.

The IWG (2000) estimated that a domestic carbon trading system with a $25/t C permit price would reduce emissions by 13%, or 63 Mt C (230 Mt CO_2), compared with a reference case, while a $50 price would reduce emissions by 17 to 19%, or 83 to 91 Mt C (306 to 332 Mt CO_2). Both cases assume a doubling of United States' government appropriations for cost-shared clean energy research, design, and development.

For carbon capture and sequestration, IPCC (2006) concluded that this option could contribute 15 to 55% to global mitigation between now and 2100 if technologies develop as projected in relatively optimistic scenarios and very large-scale geological carbon sequestration is publicly acceptable. Under these assumptions, the cost is projected to be $110 to $260/t C ($30 to $70/t CO_2). With less optimistic assumptions, the cost could rise above $730/t C ($200/t CO_2).

Net costs to the consumer, however, are balanced in some analyses by benefits from advanced technologies, which are developed and deployed on an accelerated schedule due to policy interventions and changing public preferences. The U.S. Climate Change Technology Program (2005: pp. 3-19) illustrates how costs of achieving different stabilization levels can conceivably be reduced substantially by the use of advanced technologies, and IWG (2000) estimates that net end-user costs of energy can actually be reduced by a domestic carbon trading system if it accelerates the market penetration of more energy-efficient technologies.

> Costs of achieving different stabilization levels can conceivably be reduced substantially by the use of advanced technologies.

In many cases, however, discussions of the promise of technology options are not associated with cost estimates. Economic costs of energy are not one of the drivers of the IPCC Special Report on Emissions Scenarios (SRES) scenarios, and such references as Hoffert *et al.* (2002) and Pacala and Socolow (2004) are concerned with technological potentials and constraints as a limiting condition on market behavior rather than with comparative costs and benefits of particular technology options at the margin.

6.4.4 Summary

In terms of prospects for major emission reductions from energy extraction and conversion in North America, the key issues appear to be the extent, direction, and pace of technological innovation and the likelihood that policy conditions favoring carbon emissions reduction that do not now exist will emerge if concerns about carbon cycle imbalances grow. In these regards, the prospects are brighter in the long term (*e.g.*, more than several decades in the future) than in the near term. History suggests that technology solutions are usually easier to implement than policy solutions, but observed impacts of carbon cycle imbalances might change the political calculus for policy interventions in the future.

6.5 RESEARCH AND DEVELOPMENT NEEDS

If it is possible that truly effective management of carbon emissions from energy supply and conversion systems cannot be realized with the current portfolio of technology alternatives under current policy conditions, then research and development needs and opportunities deserve expanded attention and support (*e.g.*, National Commission on Energy Policy, 2004). If so, the priorities include the following:

Technology. Several objectives seem to be especially relevant to carbon management potentials:
- clarifying and realizing potentials for carbon capture and sequestration;
- clarifying and realizing potentials of affordable renewable energy systems at a relatively large scale;
- addressing social concerns about the nuclear energy fuel cycle, especially in an era of concern about terrorism;
- improving estimates of economic costs and emission reduction benefits of a range of energy technologies

across a range of economic, technological, and policy scenarios; and

- "Blue Sky" research to develop new technology options and families, such as innovative approaches for energy from the sun and from biomass, including possible applications of nanoscience (Caldeira *et al.*, 2005; Lewis, 2005).

Policy. Research and development could also be applied to policy options in order to enlarge their knowledge bases and explore their implications. For instance, research priorities might include learning more about:

- public acceptability of policy incentives for reducing dependence on energy sources associated with carbon emissions;
- possible effects of incentives for the energy industry to increase its support for pathways not limited to fossil fuels;
- approaches toward a more distributed electric power supply enterprise in which certain renewable (and hydrogen) energy options might be more attractive;
- transitions from one energy system/infrastructure to another; and
- interactions and linkage effects among driving forces and responses, along with possible effects of exogenous processes and policy interventions.

In these ways, technology and policy advances might be combined with multiple technologies to transform the capacity to manage carbon emissions from energy supply systems, if that is a high priority for North America.

Transportation

Lead Author: David L. Greene, ORNL

KEY FINDINGS

- The transportation sector of North America released 587 million tons of carbon into the atmosphere in 2003, nearly all in the form of carbon dioxide from combustion of fossil fuels. This comprises 37% of the total carbon dioxide emissions from worldwide transportation activity, which in turn, accounts for about 22% of total global carbon dioxide emissions.

- Transportation energy use in North America and the associated carbon emissions have grown substantially and relatively steadily over the past 40 years. Growth has been most rapid in Mexico, the country most dependent upon road transport.

- Carbon emissions by transport are determined by the levels of passenger and freight activity, the shares of transport modes, the energy intensity of passenger and freight movements, and the carbon intensity of transportation fuels. The growth of passenger and freight activity is driven by population, *per capita* income, and economic output.

- Chiefly as a result of economic growth, energy use by North American transportation is expected to increase by 46% from 2003 to 2025. If the mix of fuels were assumed to remain the same, carbon dioxide emissions would increase from 587 million tons of carbon in 2003 to 859 million tons of carbon in 2025. Canada, the only one of the three countries in North America to have committed to specific greenhouse gas reduction goals, is expected to show the lowest rate of growth in carbon emissions.

- The most widely proposed options for reducing the carbon emissions of the North American transportation sector are increased vehicle fuel economy, increased prices for carbon-based fuels, liquid fuels derived from vegetation (biomass), and in the longer term, hydrogen produced from renewable energy sources (such as hydropower), nuclear energy, or from fossil fuels with carbon capture and storage. Biomass fuels appear to be a promising near- and long-term option, while hydrogen could become an important energy carrier after 2025.

- After the development of advanced energy efficient vehicle technologies and low-carbon fuels, the most pressing research need in the transportation sector is for comprehensive, consistent, and rigorous assessments of carbon emissions mitigation potentials and costs for North America.

7.1 BACKGROUND

Transportation is the largest source of carbon emissions among North American energy end uses (electricity generation is considered energy conversion rather than end use). This fact reflects the vast scale of passenger and freight movements in a region that comprises one-fourth of the global economy, as well as the dominance of relatively energy-intensive road transport and the near total dependence of North American transportation systems on petroleum as a source of energy. If present trends continue, carbon emissions from North American transportation are expected to increase by more than one-half by 2050. Options for mitigating carbon emissions from the transportation sector, like increased vehicle fuel economy and biofuels, could offset the expected growth in transportation activity. However, at present only Canada has committed to achieving a specific reduction in future greenhouse gas (GHG) emissions: 6% below 1990 levels by 2012 (Environment Canada, 2005b).

> Transportation is the largest source of carbon emissions among North American energy end uses (electricity generation is considered energy conversion rather than end use).

7.2 INVENTORY OF CARBON EMISSIONS

Worldwide, transportation produced about 22% (1.5 billion tons of carbon [Gt C]) of total global carbon dioxide (CO_2) emissions from the combustion of fossil fuels (6.6 Gt C) in 2000 (page 3-1 in U.S. EPA, 2005; Marland, Boden, and Andres, 2005). Home to 6.7% of the world's 6.45 billion people and source of 24.8% of the world's $55.5 trillion gross world product (CIA, 2005), North America produces 37% of the total carbon emissions from worldwide transportation activity (Fulton and Eads, 2004).

Transportation activity is driven chiefly by population, economic wealth, and geography. Of the approximately 435 million residents of North America, 68.0% reside in the United States, 24.5% in Mexico, and 7.5% in Canada (CIA, 2005) (these population estimates are judged by the author to have 95% certainty that the actual value is within 10% of the estimate reported, and the gross

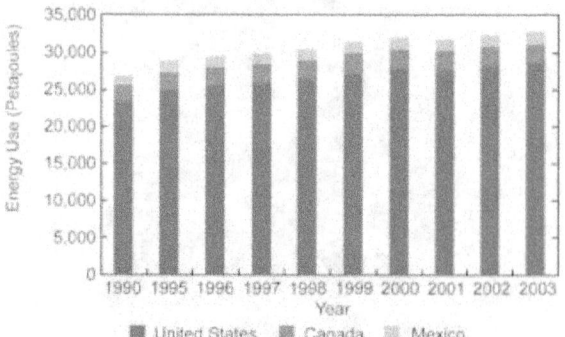

Figure 7.1 Transportation energy use in North America, 1990-2003. *Sources:* NATS (2005), Table 4-1; U.S. DOE/EIA (2005a), Table 2.1e.

domestic product (GDP) estimates are judged to have 95% certainty that the actual value is within 25% of the estimate reported, chiefly because they are not based on triple bottom line accounting). The differences in the sizes of the three countries' economies are far greater. The United States is the world's largest economy, with an estimated GDP of $11.75 trillion in 2004.

Although Mexico has approximately three times the population of Canada, its GDP is roughly the same, $1.006 trillion compared to $1.023 trillion (measured in 2004 purchasing power parity dollars). With the largest population and largest economy, the United States has by far the largest transportation system. The United States accounted for 87% of the energy used for transportation in North America in 2003, Canada for 8%, and Mexico 5% (Figure 7.1) (see Table 4.1 in

Table 7.1 Carbon emissions from transportation in North America in 2003.

North American Carbon Emissions by Country and Mode, 2003/2001 (Mt C)				
	United States 2003	Canada 2003	Mexico 2001	North America 2003/2001
Road	399.4	36.7	26.0	462.0
Domestic Air	46.7	1.9	1.8	50.4
Rail	11.7	1.4	0.4	13.5
Domestic Water	15.7	1.6	0.9	18.1
Pipeline	9.5	2.4		11.9
International Bunker	23.0	3.0	0.5	26.4
Off-Road		4.6		4.6
Total	505.9	51.7	29.4	587.0

Sources: U.S. EPA (2005); Environment Canada (2005a); INE (2003)

Note: Data for Mexico is 2001, United States and Canada are 2003.

Carbon dioxide emissions estimates are considered by the Canadian and Mexican sources to have 95% certainty that the actual value is within 10% of the estimate reported. The United States did not provide quantitative uncertainty estimates for 2003, but these estimates are considered to be equally accurate by the author.

NATS, 2005). These differences in energy use are directly reflected in carbon emissions from the three countries' transportation sectors (Table 7.1)[1].

Transportation is defined as private and public vehicles that move people and commodities (U.S. EPA, 2005, p. 296). This includes automobiles, trucks, buses, motorcycles, railroads and railways (including streetcars and subways), aircraft, ships, barges, and natural gas pipelines. This definition excludes petroleum, coal slurry, and water pipelines, as well as the transmission of electricity, although many countries consider all pipelines part of the transport sector. It also generally excludes mobile sources not engaged in transporting people or goods, such as construction equipment and on-farm agricultural equipment. In addition, carbon emissions from international bunker fuel-use in aviation and waterborne transport, though considered part of transport emissions, are generally accounted for separately from a nation's domestic GHG inventory. In this chapter, however, they are included, as are carbon emissions from military transport operations, because they are real inputs to the carbon cycle. Upstream, or well-to-tank, carbon emissions are not included with transportation end-use, nor are end-of-life emissions produced in the disposal or recycling of materials used in transportation vehicles or infrastructure because these carbon flows are in the domain of other chapters. These two categories of emissions typically comprise 20-30% of total life cycle emissions for transport vehicles (see Table 5.4 in Weiss *et al.*, 2000). In the future, it is likely that upstream carbon emissions will be of greater importance in determining the total emissions due to transportation activities.

In addition to CO_2, the combustion of fossil fuels by transportation produces other GHGs including methane (CH_4), nitrous oxide (N_2O), carbon monoxide (CO), nitrogen oxides (NO_x), and non-CH_4 volatile organic compounds (VOCs). Those containing carbon are generally oxidized in the atmosphere to ultimately produce CO_2. However, the quantities of non-CO_2 gases produced by transportation vehicles are very minor sources of carbon in comparison to the volume of CO_2 emissions. For example, North American emissions of CH_4 by transportation accounted for only 0.03% of total transportation carbon emissions in 2003. This chapter will therefore address primarily the CO_2 emissions from transportation activities (CH_4 emissions are included in the totals presented in Table 7.1, but they are not included in any other estimates presented in this chapter). Estimates of non-CO_2 emissions are also subject to much greater uncertainty. INE (2003) generally put the accuracy of the Mexican 2001 non-CO_2 GHG emissions at 95% certainty that the actual value is within 50% of the estimate reported. However, Environment Canada's 2003 inventory (Environment Canada,

2005a) rates the uncertainty of CH_4 emissions from mobile sources as 95% certain that the actual value is within 10% of the estimate reported.

Four main sources of information on carbon emissions are used in this chapter. The estimates shown in Table 7.1 were obtained from the GHG inventory reports of the three countries, estimated by environmental agencies in accordance with Intergovernmental Panel on Climate Change (IPCC) guidelines. As Annex 1 countries, Canada and the United States are obliged to compile annual inventories under IPCC guidelines. As a non-Annex 1 country, Mexico is not. These inventories are the most authoritative sources for estimates of carbon emissions. The inventory reports, however, do not generally provide estimates of associated energy use and the most recent inventory data available for Mexico are for 2001. Estimates of energy use and carbon emissions produced by the countries' energy agencies are also used in this chapter to illustrate the relationship between energy use and carbon emissions and its historical trends. There are some minor differences between the carbon emissions estimates from the two sources. Finally, future projections of carbon emissions for North America to 2025 were taken from the U.S. Energy Information's Annual Energy Outlook 2005, and projections to 2050 were taken from the World Business Council on Sustainable Development's Sustainable Mobility Project (WBCSD, 2004).

7.2.1 Fuels Used in Transportation

Virtually all of the energy used by the transport sector in North America is derived from petroleum, and most of the remainder comes from natural gas (Table 7.2). In the United States, 96.3% of total transportation energy is obtained by combustion of petroleum fuels (U.S. DOE/EIA, 2005a). Most of the non-petroleum energy is natural gas used to power natural gas pipelines (2.5%, 744 petajoules). During the past two decades, ethanol use (as a blending component for gasoline) has increased from a negligible amount to

Virtually all of the energy used by the transport sector in North America is derived from petroleum, and most of the remainder comes from natural gas.

1.1% of transportation energy use (312 petajoules). Electricity, mostly for passenger rail transport, comprises only 0.1% of United States transport energy use. This pattern of energy use has persisted for more than half a century.

The pattern of energy sources is only a little different in Mexico where 96.2% of transportation energy use is gasoline, diesel, or jet fuel, 3.4% is liquefied petroleum gas, and less than 0.2% is electricity (Rodríguez, 2005). In Canada, natural gas use for natural gas pipelines accounts for 7.5%

---

[1] Uncertainties in these estimates are discussed later in this chapter (see Section 7.5).

Table 7.2 Summary of North American transport energy use and CO₂ emissions in 2003 by energy source or fuel type.

North America energy source	Energy input (Petajoules)	Carbon input (Mt C)
Gasoline	20,923	358.3
Diesel/distillate	7,344	129.5
Jet fuel/kerosene	2,298	68.5
Residual	681	14.5
Other fuels	124	1.3
Natural gas	926	9.7
Electricity	36	0.0
Unallocated/error	466	-
Total	32,798	581.8
United States		
Gasoline	18,520	312.5
Diesel/distillate	6,193	107.1
Jet fuel/kerosene	1,986	62.3
Residual	612	13.1
Other fuels	50	0.2
Natural gas	748	9.7
Electricity	20	0.0
Unallocated/error	466.2	-
Total	28,595.2	504.9

Sources: U.S. EPA (2005), Tables 3-7 and 2-17; Davis and Diegel (2004), Tables 2.6 and 2.7.

Canada		
Gasoline	1,355	26.2
Diesel/distillate	698	13.9
Jet fuel/kerosene	223	4.3
Residual	67	1.3
Other fuels	17	0.2
Natural gas	2	0.0
Electricity	3	0.0
Unallocated/error	0	
Total	2,363	45.9

NRCan (2006), Tables 1 and 8.

Mexico		
Gasoline	1,066	19.5
Diesel/distillate	447	8.5
Jet fuel/kerosene	106	1.9
Residual	4	0.1
Other fuels	57	0.9
Natural gas	1	0.0
Electricity	4	0.0
Unallocated/error		
Total	1,685	31.0

Sources: Transportation energy use by fuel and mode from Rodríguez (2005).

The accuracy of the data in the above table is judged by the author to be 95% certain that the actual value is within 10% of the estimate reported.

Data sources differ somewhat by country with respect to modal, fuel, and greenhouse gas definitions so that the numbers are not precisely comparable. Canadian carbon emissions data include all GHGs produced by transportation in CO₂ equivalents, while the United States' data are CO₂ emissions only. Carbon dioxide emissions for Mexico were estimated by applying U.S. EPA emissions factors to the Mexican energy use data. For Mexico, it is asumed that no transportation carbon emissions result from electricity use.

of transport energy use, 91.8% is petroleum, 0.5% is propane, and only 0.1% is electricity (see Table 1 in NRCan, 2006).

7.2.2 Mode of Transportation

Mode of transportation refers to how people and freight are moved about, whether by road, rail, or air, or in light or heavy vehicles. Carbon dioxide emissions from the North American transportation sector are summarized by mode in Table 7.3, and the distribution of emissions by mode for North America in 2003 is illustrated in Figure 7.2.

7.2.2.1 FREIGHT TRANSPORT

Movement of freight is a major component of the transportation sector in North America. Total freight activity in the United States, measured in metric ton-km, is 20 times that in Mexico and more than 10 times the levels observed in Canada (Figures 7.3A, 7.3B, and 7.3C).

In Mexico, trucking is the mode of choice for freight movements. Four-fifths of Mexican metric ton-km is produced by trucks. Moreover, trucking's modal share has been increasing over time.

In Canada, rail transport accounts for the majority of freight movement (65%). Rail transport is well suited to the approximately linear distribution of Canada's population in close proximity to the United States border, the long-distances from east to west, and the large volumes of raw material flows typical of Canadian freight traffic (see Table 5.2 in NATS, 2005).

Table 7.3 Summary of North American transport energy use and carbon dioxide emissions in 2003 by mode of transportation.

North America transport mode	Energy use (Petajoules)	Carbon emissions (Mt C)
Road	25,830	463.5
Air	2,667	53.0
Rail	751	13.7
Waterborne	1,386	18.4
Pipeline	990	12.3
Internatl./Bunker	0	23.0
Total	31,624	583.9
United States		
Road		
Light vehicles	17,083	303.8
Heavy vehicles	5,505	95.5
Air	2,335	46.7
Rail	655	11.7
Waterborne	1,250	15.7
Pipeline/other	986	9.5
Internatl./Bunker		23.0
Total	27,814	505.8
Source: U.S. EPA (2005), Tables 3-7 and 2-17; Davis and Diegel (2004), Tables 2-6 and 2-7.		
Canada		
Road		
Light vehicles	1,233	23.8
Heavy vehicles	491	12.4
Air	226	4.3
Rail	74	1.6
Waterborne	103	2.1
Pipeline/other		1.8
Total	2,126	46.1
Source: NRCan (2006); Tables 1 and 8.		
Mexico		
Road	1,518	27.9
Light vehicles		
Heavy vehicles		
Air	107	2.0
Rail	22	0.5
Waterborne	33	0.6
Electric	4	-
Total	1,684	32.0
Source: Rodríguez (2005).		

The accuracy of the data in the above table is judged by the author to be 95% certain that the actual value is within 10% of the estimate reported for the larger modes of transportation, and 95% certain that the value is within 25% for the smaller modes.

Data sources differ somewhat by country with respect to modal, fuel, and GHG defintions so that the numbers are not precisely comparable. Canadian carbon emissions data include all GHGs produced by transportation in CO_2 equivalents, while United States data are CO_2 emissions only. Carbon dioxide emissions for Mexico were estimated by applying U.S. EPA emissions factors to the Mexican energy use data. Electricity is assumed to produce no carbon emssions in end use.

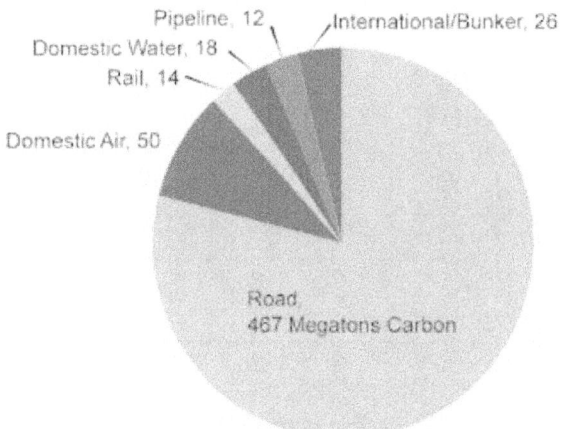

Figure 7.2 North American carbon emissions from transportation by mode; United States and Canada 2003, Mexico 2001. *Sources:* U.S. EPA (2005); Environment Canada (2005a); INE (2003).

In the United States, road freight plays a greater role than in Canada, and rail is less dominant, although rail still carries the largest share of metric ton-km (40%). In none of the countries does air freight account for a significant share of metric ton-km.

7.2.2.2 PASSENGER TRANSPORT

In all three countries, passenger transport is predominantly by road, followed in distant second by air travel. The rate of growth in air travel in North America is more than double that of road transport, so air transport's share of carbon emissions will increase in the future. Nearly complete data are available for passenger-kilometers-traveled (pkt) by mode in the United States and Canada in 2001. Of the more than 8 trillion pkt accounted for by the United States, 86% was by light-duty personal vehicles, most by passenger car but a growing share by light trucks (Figure 7.4A) (motorcycle pkt, about 0.2% of the total, is included with passenger car). Air travel claims 10%; other modes are minor.

> In all three countries, passenger transport is predominantly by road, followed in distant second by air travel.

Canadian passenger travel exhibits a very similar modal structure, but with a smaller role played by light trucks and air and a larger share for buses (Figure 7.4B) (transit numbers for Canada were not available at the time these figures were compiled).

7.3 TRENDS AND DRIVERS

Driven by economic and population growth, transportation energy use has increased substantially in all three countries since 1990. Figures 7.5A and 7.5B illustrate the evolution of transport energy use by mode for Mexico and the United

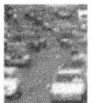

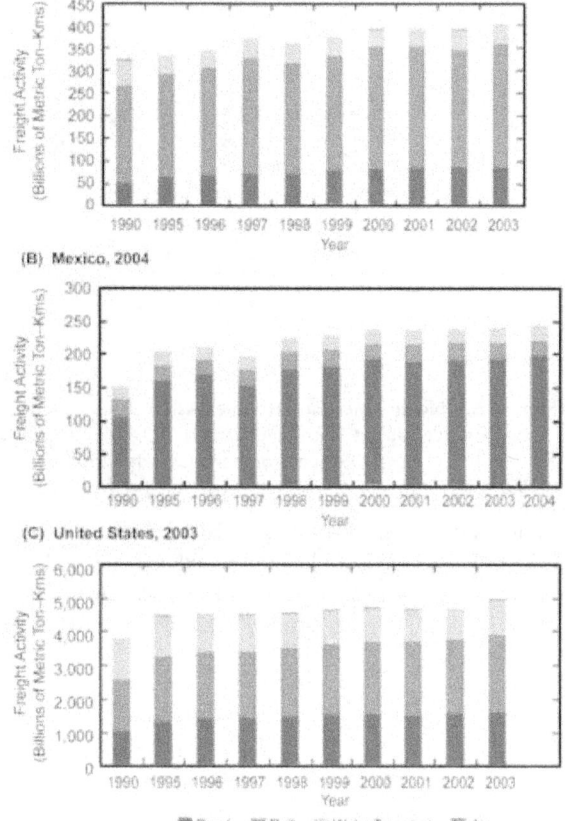

Figure 7.3A Freight activity by mode in Canada.
Figure 7.3B Freight activity by mode in Mexico.
Figure 7.3C Freight activity by mode in the United States.

States. Energy use has grown most rapidly in Mexico, the country most dependent on road transport. In the United States, the steady growth of transportation oil use was interrupted by oil price shocks in 1973-74, 1979-80, and to a much lesser degree in 1991. The impact of the attack on the World Trade Center in 2001 and subsequent changes in air travel procedures had a visible effect on energy use for air travel.

The evolution of transport carbon emissions has closely followed the evolution of energy use. Carbon dioxide emissions by mode are shown for the United States and Canada for the period 1990-2003 in Figures 7.6A and 7.6B. The Canadian data include light-duty commercial vehicles in road freight transport, while all light trucks are included in the light-duty vehicle category in the United States data. These data illustrate the relatively faster growth of freight-transport energy use. Fuel economy standards in both countries restrained the growth of passenger car and light-truck energy use (NAS, 2002). From 1990 to 2003 passenger kilometers traveled by road in Canada increased by 23%, while energy use

increased by only 15%. In 2003, freight activity accounted for more than 40% of Canada's transport energy use. In addition, while passenger transport energy use increased by 15% from 1990 to 2003, freight energy use increased by 40%. The Canadian transport energy statistics do not include natural gas pipelines as a transport mode.

Carbon emissions by transport are determined by the levels of passenger and freight activity, the shares of transport modes, the energy intensity of passenger and freight movements, and the carbon intensity of transportation fuels. In North America, petroleum fuels supply over 95% of transportation's energy requirements and account for 98% of the sector's GHG emissions. Among modes, road vehicles are predominant, producing almost 80% of sectoral GHG emissions. Consequently, the driving forces for transportation GHG emissions have been changes in activity and energy intensity. The principal driving forces of the growth of passenger transportation are population and *per capita* income (WBCSD, 2004). Increased vehicle ownership follows rising *per capita* income, as do vehicle use, fuel consumption, and emissions. In general, energy forecasters expect the greatest growth in vehicle ownership and fossil-fuel use in transpor-

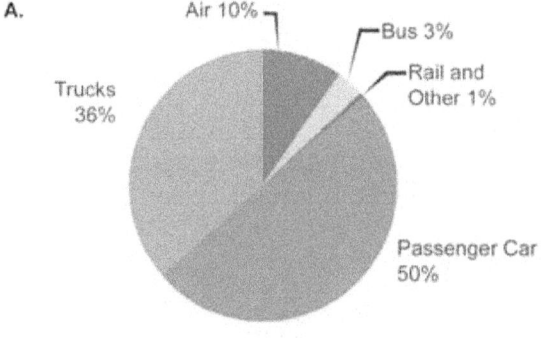

A.

Source: U.S. Bureau of Transportation Statistics, 2006, table 1-37

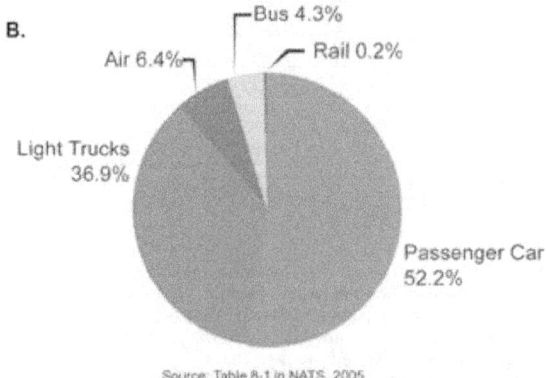

B.

Source: Table 8-1 in NATS, 2005

Figure 7.4A Distribution of passenger travel in the United States by mode.
Figure 7.4B Distribution of passenger travel by mode in Canada.

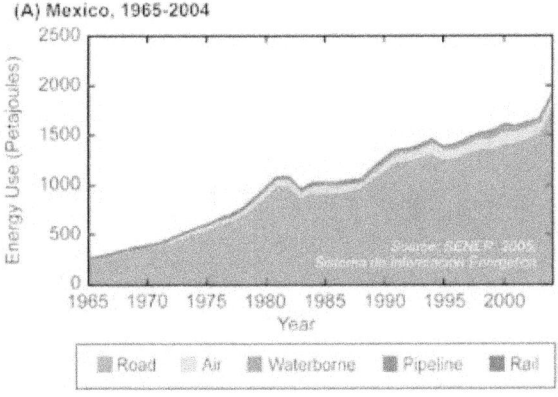

(A) Mexico, 1965-2004

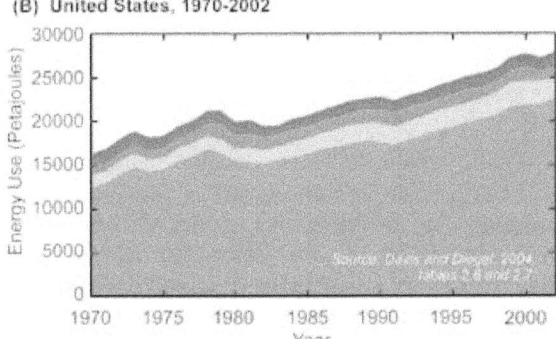

(B) United States, 1970-2002

Figure 7.5A Evolution of transport energy use in Mexico.
Figure 7.5B Evolution of transport energy use in the United States.

tation over the next 25-50 years to occur in the developing economies (U.S. DOE/EIA, 2005b; IEA, 2004; WBCSD, 2004; Nakićenović, Grübler, and McDonald, 1998). The chief driving forces for freight activity are economic growth and the integration of economic activities at both regional and global scales (WBCSD, 2004).

Projections of North American transportation energy use and carbon emissions to 2030 have been published by the U.S. Energy Information Administration (U.S. DOE/EIA, 2005b) and the International Energy Agency (IEA, 2005a). Historical population growth rates are similar in the three countries, 0.92% per year in the United States, 1.17% per year in Mexico, and 0.90% per year in Canada. Recent annual GDP growth rates are 4.4% for the United States, 4.1% for Mexico, and 2.4% for Canada (CIA, 2005). The U.S. Energy Information Administration's Reference Case projection assumes annual GDP growth rates of 3.1% for the United States, 2.4% for Canada, and 3.9% for Mexico (see Table A3 in U.S. DOE/EIA, 2005b). Assumed population growth rates are United States: 0.9%; Canada: 0.6%; Mexico: 1.0% (see Table A14 in U.S. DOE/EIA, 2005b). Chiefly because of economic growth, energy use by North American transportation is expected to increase by 46% from 2003 to 2025 (U.S. DOE/EIA, 2005b). If the mix of fuels is assumed to remain the same, as it nearly does in

the IEO 2005 Reference Case projection, CO_2 emissions would increase from 587 million metric tons of carbon (Mt C) in 2003 to 859 Mt C in 2025 (Figure 7.7). Canada, the only one of the three countries to have committed to specific GHG reduction goals, is expected to show the lowest rate of growth in CO_2 emissions.

The World Business Council for Sustainable Development (WBCSD), in collaboration with the International Energy Agency developed a model for projecting world transport energy use and GHG emissions to 2050 (Table 7.4). The WBCSD's reference case projection foresees the most rapid growth in carbon emissions from transportation occurring in Asia and Latin America (Figure 7.8). Still, in 2050, North America accounts for 26.4% of global CO_2 emissions from transport vehicles (down from a 37.2% share in 2000).

(A) Canada, 1990-2003

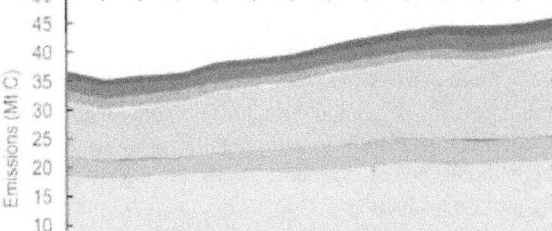

(B) United States, 1990-2003

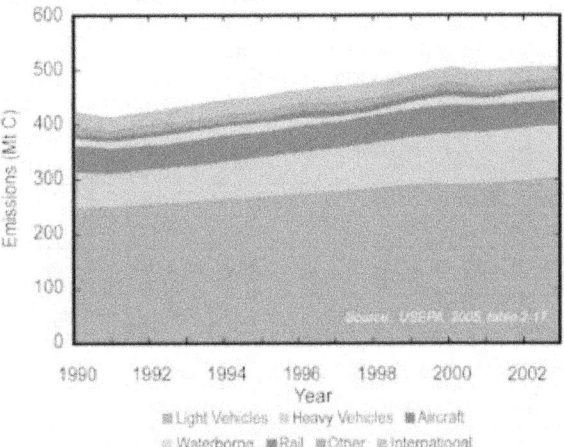

Figure 7.6A Evolution of transport energy use in Mexico.
Source: SENER (2005).
Figure 7.6B Transport CO_2 emissions in the United States.

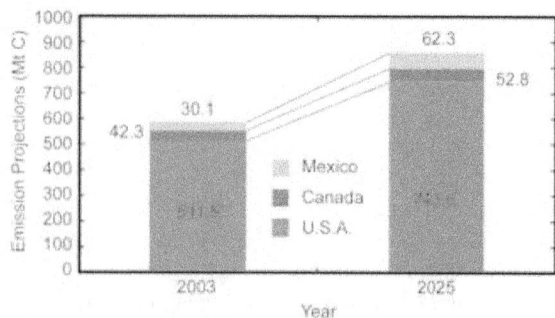

Figure 7.7 Projected CO_2 emissions from the North American transport sector in 2025, based on EIA IEO (2005) reference case. *Source:* NRCan (2006).

7.4 OPTIONS FOR MANAGEMENT

Dozens of policies and measures for reducing petroleum consumption and mitigating carbon emissions from transportation in North America have been identified and assessed (*e.g.*, U.S. DOT, 1998; IEA, 2001; Greene and Schafer, 2003; Greene *et al.*, 2005; CBO, 2003; Harrington and McConnell, 2003; NRTEE, 2005). However, there is no consensus about how much transportation GHG emissions can be reduced and at what cost. In general, top-down models estimating the mitigation impacts of economy-wide carbon taxes or cap-and-trade systems find the cost of mitigation high and the potential modest. On the other hand, bottom-up studies evaluating a wide array of policy options tend to reach the opposite conclusion. Part of the explanation of this paradox may lie in the predominant roles that governments play in constructing, maintaining, and operating the majority of transportation infrastructure and in the strong interrelationship between land-use planning and transportation demand. In addition, top down models typically assume that all markets are efficient, whereas there is evidence of

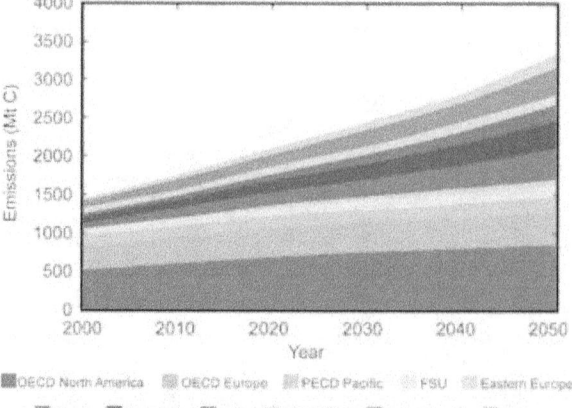

Figure 7.8 World Business Council for Sustainable Development (WBCSD) projections of world transportation vehicle CO_2 emissions to 2050. *Source:* U.S. EPA (2005), Table 2-17.

real-world transportation energy market failures, especially with respect to the determination of light-duty vehicle fuel economy (*e.g.*, Turrentine and Kurani, 2004; Chapter 5 in NAS, 2002). Estimates of the costs and benefits of mitigation policies also vary widely and depend critically on premises concerning (1) the efficiency of transportation energy markets, (2) the values consumers attach to vehicle attributes such as acceleration performance and vehicle weight, and (3) the current and future status of carbon-related technology.

A U.S. Energy Information Administration evaluation of a GHG cap and trade system, expected to result in carbon permit prices of $79/t C in 2010 and $221/t C in 2025, was estimated to reduce 2025 transportation energy use by 4.3 Petajoules (PJ) and to cut transportation's carbon emissions by 10% from 225 Mt C in the reference case to 203 Mt C under this policy (U.S. DOE/EIA, 2003). The average fuel economy of new light-duty vehicles was estimated to increase from 26.4 miles per gallon (mpg, or 8.9 L per 100 km) to 29.0 mpg (8.1 L per 100 km) in the policy case, an improvement of only 10%. A 2002 study by the U.S. National Academy of Sciences (NAS, 2002) estimated that "cost-efficient" fuel economy improvements for United States' light-duty vehicles using proven technologies ranged from 12% for subcompact cars to 27% for large cars, and from 25% for small

Table 7.4 Global carbon emissions from transportation vehicles to 2050 by regions, WBCSD reference case projection (Mt C).

	2000	2010	2020	2030	2040	2050
OECD North America	544	623	708	768	824	882
OECD Europe	313	359	392	412	420	428
OECD Pacific	133	142	153	161	169	179
FSU	48	64	88	109	132	153
Eastern Europe	23	28	36	42	52	66
China	69	108	163	225	308	417
Other Asia	98	131	174	220	283	368
India	38	54	80	108	146	203
Middle East	59	71	88	106	122	138
Latin America	95	127	172	216	275	352
Africa	43	58	80	103	127	158
TOTAL - All Regions	1463	1766	2134	2470	2858	3343

Source: Fulton and Eads (2004).

sport utility vehicles (SUVs) to 42% for large SUVs. The NAS study did not include the potential impacts of diesel or hybrid vehicle technologies and assumed that vehicle size and horsepower would remain constant.

The U.S. Congressional Budget Office (CBO, 2003) estimated that achieving a 10% reduction in United States gasoline use would create total economic costs of approximately $3.6 billion per year if accomplished by means of Corporate Average Fuel Economy (CAFE) standards, $3.0 billion if the same standards allowed trading of fuel economy credits among manufacturers, and $2.9 billion if accomplished via a tax on gasoline. This partial equilibrium analysis assumed that it would take about 14 years for the policies to have their full impact. If one assumes that the United States would consume 22,600 PJ of gasoline in 2017, resulting in 387 Mt of CO_2 emissions, then a 10% reduction amounts to 39 Mt C. At a total cost of $3 billion per year, and attributing the full cost to carbon reduction (vs. other objectives such as reducing petroleum dependence), produces an upper-bound mitigation cost estimate of $77/t C.

The bipartisan National Commission on Energy Policy (NCEP, 2004) surveyed recent assessments of the potential to increase light-duty vehicle fuel economy in the United States. Taking into consideration uncertainties about the costs and technical potential of fuel economy technologies, as well as the future price of fuel, the Commission concluded that future increases in fuel economy of from 40% to 80% could be achieved at a cost that would be fully offset by the value of fuel saved over the life of a vehicle. They estimated that the essentially costless carbon emissions reductions would amount to between 250 and 400 million metric tons per year by 2030.

Systems of progressive vehicle taxes on purchases of less efficient new vehicles and subsidies for more efficient new vehicles ("feebates") are yet another alternative for increasing vehicle fuel economy. A study of the United States market (Greene *et al.*, 2005) examined a variety of feebate structures under two alternative assumptions: (1) consumers consider only the first three years of fuel savings when making new vehicle purchase decisions, and (2) consumers consider the full discounted present value of lifetime fuel savings. The study found that if consumers consider only the first three years of fuel savings, then a feebate of $1000 per 0.01 gal/mile (3.5 L per 100 km), designed to produce no net revenue to the government, would produce net benefits to society in terms of fuel savings and would reduce carbon emissions by 139 Mt C in 2030. If consumers fully valued lifetime fuel savings, the same feebate system would cause a $3 billion loss in consumers' surplus (a technical measure of the change in economic well-being closely approximating

income loss) and reduce carbon emissions by only 67 Mt C, or an implied cost of $44/Mt CO_2.

The most widely proposed options for reducing the carbon content of transportation fuels are liquid fuels derived from biomass and hydrogen produced from renewables, nuclear energy, or from fossil fuels with carbon sequestration. Biomass fuels, such as ethanol from cellulosic feedstocks or liquid hydrocarbon fuels produced via biomass gasification and synthesis, appear to be a promising mid- to long-term option, while hydrogen could become an important energy carrier, but not before 2025 (WBCSD, 2004). The carbon emission reduction potential of biomass fuels for transportation is strongly dependent on the feedstock and conversion processes. Advanced methods of producing ethanol from grain, the predominant feedstock in the United States can reduce carbon emissions by 10% to 30% (Wang, 2005; p. 16 in IEA, 2004). Production of ethanol from sugar cane, as is the current practice in Brazil, or by not-yet-commercialized methods of cellulosic conversion can achieve up to a 90% net reduction over the fuel cycle. Conversion of biomass to liquid hydrocarbon fuels via gasification and synthesis may have a similar potential

> The most widely proposed options for reducing the carbon content of transportation fuels are liquid fuels derived from biomass and hydrogen produced from renewables, nuclear energy, or from fossil fuels with carbon sequestration.

(Williams, 2005). The technical potential for liquid fuels production from biomass is very large and very uncertain; recent estimates of the global potential range from 10 to 400 exajoules per year (see Table 6.8 in IEA, 2004). The U.S. Departments of Energy and Agriculture have estimated that 30% of United States' petroleum use could be replaced by biofuels by 2030 (Perlack *et al.*, 2005). The economic potential will depend on competition for land with other uses, the development of a global market for biofuels, and advances in conversion technologies.

Hydrogen must be considered a long-term option because of the present high cost of fuel cells, technical challenges in hydrogen storage, and the need to construct a new infrastructure for hydrogen production and distribution (NAS, 2004; U.S. DOE, 2005; IEA, 2005b). Hydrogen's potential to mitigate carbon emissions from transport will depend most strongly on how hydrogen is produced. If produced from coal gasification without sequestration of CO_2 emissions in production, it is conceivable that carbon emissions could increase. If produced from fossil fuels with sequestration, or from renewable or nuclear energy, carbon emissions from road and rail vehicles could be virtually eliminated (General Motors *et al.*, 2001).

In a comprehensive assessment of opportunities to reduce GHG emissions from the United States transportation sector, a study published by the Pew Center on Global Climate Change (Greene and Schafer, 2003) estimated that sector-wide reductions in the vicinity of 20% could be achieved by 2015 and 50% by 2030 (Table 7.5). The study's premises assumed no change in the year 2000 distribution of energy use by mode. A wide range of strategies was considered, including research and development, efficiency standards, use of biofuels and hydrogen, pricing policies to encourage efficiency and reduce travel demand, land-use transportation planning options, and public education (Table 7.5). Other key premises of the analysis were that (1) for efficiency improvements the value of fuel saved to the consumer must be greater than or equal to the cost of the improvement, (2) there is no change in vehicle size or performance, (3) pricing

Table 7.5 Potential impacts of transportation GHG reduction policies in the United States by 2015 and 2030[a] based on the 2000 distribution of emissions by mode and fuel (Greene and Shafer, 2003).

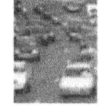

Management option	Carbon emission (Mt C) 2000	Reduction potential per mode/fuel (%)		Transportation sector reduction potential (%)	
		2015	2030	2015	2030
Research, development, and demonstration					
Light-duty vehicles (LDVs)	289	11[b]	38[b]	7[b]	23[b]
Heavy trucks	80	11[b]	24[b]	2[b]	4[b]
Commercial aircraft	53	11[b]	27[b]	1[b]	3[b]
Efficiency standards					
Light-duty vehicles	289	9	31	6	18
Heavy trucks	80	9	20	2	3
Commercial aircraft	53	9	22	1	2
Replacement and alternative fuels					
Low-carbon replacement fuels (~10% of LDV fuel)	27	30	100	2	7
Hydrogen fuel (All LDV fuel)	289	1	6	1	4
Pricing policies					
Low-carbon replacement fuels (~10% of LDV fuel)	27	30	100	2	6
Carbon pricing (All transportation fuel)	489	3	6	3	6
Variabilization (All highway vehicle fuel)	370	8	12	6	9
Behavioral					
Land use and infrastructure (2/3 of highway fuel)	246	5	10	3	5
System efficiency (25% LDV fuel)	72	2	5	0	1
Climate change education (All transportation fuel)	489	1	2	1	2
Fuel economy information (All LDV fuel)	289	1	2	1	1
Total	**489**			**22**	**48**

Notes:

[a] Carbon emissions for the year 2000 are used to weight percent reductions for the respective emissions source and example policy category in calculating total percent reduction potential. The elasticity of vehicle travel with respect to fuel price is −0.15 for all modes. Price elasticity of energy efficiency with respect to fuel price is −0.4.

[b] R&D efficiency improvements have no direct effect on total. Their influence is seen through efficiency standards impacts.

Policies affecting the same target emissions, such as passenger car efficiency, low carbon fuels, and land-use policies are multiplicative, to avoid double counting (e.g. [1−0.1]*[1.0−0.2] = 1−0.28, a 28% rather than a 30% reduction.)

policies shift the incidence but do not increase the overall cost of transportation, and (4) there is a carbon cap and trade system in effect equivalent to a charge of approximately $50/t C. Similar premises underlie the 2030 estimates, except that technological progress is assumed to have expanded the potential for efficiency improvement and lowered the cost of biofuels.

The Pew Center study notes that if transportation demand continues to grow as the IEO 2005 and WBCSD projections anticipate, the potential reductions shown in Table 7.4 would be just large enough to hold United States transportation CO_2 emissions in 2030 to 2000 levels.

A study for the U.S. Department of Energy (ILWG, 2000) produced estimates of carbon mitigation potential for the entire United States economy using a variety of policies generally consistent with carbon taxes of $25-$50/t C. In the study's business as usual case, transportation CO_2 emissions increased from 478 Mt C in 1997 to 700 Mt C in 2020. A combination of technological advances, greater use of biofuel, fuel economy standards, paying for a portion of automobile insurance as a surcharge on gasoline, and others, were estimated to reduce 2020 transportation CO_2 emissions by 155 Mt C to 545 Mt CO_2. The study did not produce cost estimates and did not consider impacts on global energy markets.

A joint study of the U.S. Department of Energy and Natural Resources Canada (Patterson *et al.*, 2003) considered alternative scenarios of highway energy use in the two countries to 2050. The study did not produce estimates of cost-effectiveness for GHG reduction strategies but rather focused on the potential impacts of differing social, economic, and technological trends. Two of the scenarios describe paths that lead to essentially constant GHG emissions from highway vehicles through 2050 through greatly increased efficiency and biofuel and hydrogen use and, in one scenario, reduced demand for vehicle travel.

fuels. The United States and Canada report transport emissions in much greater modal detail, by vehicle type and fuel type within modes. The United States and Mexico report emissions from international bunker fuels in their national inventory reports, while Canada does not. Estimates of international bunker fuel emissions for Canada presented in this chapter were derived by subtracting Air and Waterborne emissions reported by Environment Canada (2005a) which exclude international bunker fuels from total air and waterborne emissions as reported by Natural Resources Canada (2006) which include them. Environment Canada reports off-road emissions from mobile sources separately; in the tables and figures in this chapter, Canadian off-road emissions have been added to road emissions. Both Canada and the United States include emissions from military transport operations in their inventories. It is not clear whether these are included in the estimates for Mexico.

All three countries' GHG inventories discuss uncertainties in estimated emissions. In general, the uncertainties were estimated in accordance with IPCC guidelines. The U.S. EPA provides only an estimate of a 95% confidence

7.5 INCONSISTENCIES AND UNCERTAINTIES

There are some inconsistencies in the way the three North American countries report transportation carbon emissions. The principal source for Mexican emissions data breaks out transportation into four modes (road, air, rail, and waterborne), it does not report emissions for pipelines but does report emissions from use of international bunker

Table 7.6 Uncertainty in estimates of carbon dioxide emissions from energy use in transport: Canada (2003).

Mode	% Below (2.5th Percentile)	% Above (97.5th Percentile)
Total Mobile Sources excluding pipeline	-4	0
Road Transportation	-8	-3
On-Road Gasoline Vehicles	-7	-3
On-Road Diesel Vehicles	-13	-1
Railways	-5	3
Navigation	-3	3
Off-Road Mobile Sources	4	45
Pipeline	-3	3

Source: Environment Canada (2005a), table A7-9.

interval for all CO_2 emissions from the combustion of fossil fuels (-1% to 6%) which can be inferred to apply to transportation. Mexico's INE estimates a total uncertainty for transportation GHG emissions of about ± 10%. For CO_2 emissions from road transport, the uncertainty is put at ± 9% (INE, 2003, Appendix B). The Canadian Greenhouse Gas Inventory provides by far the most extensive and detailed estimates of uncertainty. Given the similarity in methods, the Canadian uncertainty estimates are probably also approximately correct for the United States, and therefore may be considered indicative of the uncertainty of North American carbon emission estimates (Table 7.6). Most significant is the apparent overestimation of carbon emissions from on-road vehicles, offset to a degree by the underestimation of off-road mobile source emissions. Still, total mobile source carbon emissions are estimated to have a 95% confidence interval of (-4% to 0%).

7.6 RESEARCH AND DEVELOPMENT NEEDS

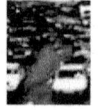

Research needs with respect to the transport sector as a part of the carbon cycle fall into three categories: (1) improved data, (2) comprehensive assessments of mitigation potential, and (3) advances in key mitigation technologies and policies for transportation. The available data are adequate to describe carbon inputs by fuel type and carbon emissions by very broad modal breakdowns by country. Environment Canada (2005a) and the U.S. Environmental Protection Agency (2005) annually publish estimates of transportation's carbon emissions that closely follow IPCC guidelines with respect to methods, data sources, and quantification of uncertainties (GAO, 2003). The Mexican Instituto Nacional de Ecología has published estimates for 2001 that are also based on IPCC methods. However, that report also notes deficiencies in the data available for Mexico's transport sector and recommends establishing an information system for estimating Mexico's transportation GHG emissions on a continuing basis (INE, 2003, p. 21). Knowledge of the magnitude of GHG emissions by type of activity and fuel, and of trends is essential if policies are to be focused on the most important GHG sources.

The most pressing research need is for comprehensive, consistent, and rigorous assessments of the carbon emissions mitigation potential for North American transportation. The lack of such studies for North America parallels a similar dearth of consistent and comprehensive global analyses

The most pressing research need is for comprehensive, consistent, and rigorous assessments of the carbon emissions mitigation potential for North American transportation.

noted by the IPCC (Moomaw and Moreira, 2001). Existing studies focus almost exclusively on a single country, with premises and assumptions varying widely from country to country. Even the best single country studies omit the impacts of carbon reduction policies on global energy markets. Knowledge of how much contribution the transport sector can make to GHG mitigation, at what cost, and what options are capable of achieving those potentials is crucial to the global GHG policy discussion.

Continued research and development of vehicle technologies and fuels that can cost-effectively increase energy efficiency and displace carbon-based fuels is essential to achieving major reductions in transportation carbon emissions. Highly promising technologies for reducing transportation GHG emissions include hybrid vehicles, which are available today, and in the future, plug-in hybrid vehicles capable of accepting electrical energy from the grid, and eventually fuel cell vehicles powered by hydrogen. While hybrids are already in the market and fuel cell vehicles are still years away, all three technologies would benefit from cost reduction. Hydrogen fuel cell vehicles also face significant technological challenges with respect to hydrogen storage and fuel cell durability. Energy-efficient technologies could also greatly reduce GHG emissions from other transport modes. For example, blended wing-body aircraft designs are under development that could reduce fuel burn rates by one-third. Biofuels in the near term and hydrogen in the longer term appear to be the most promising low-carbon fuel options. To achieve the greatest GHG reduction benefits, biofuels must be made from plants' ligno-cellulosic components either by conversion to alcohol or by gasification and synthesis of liquid hydrocarbon fuels. Cost reductions in both feedstock production and fuel conversion are needed.

Industry and Waste Management

Lead Author: John Nyboer, Simon Fraser Univ.

Contributing Authors: Mark Jaccard, Simon Fraser Univ.; Ernst Worrell, LBNL

KEY FINDINGS

- In 2002, North America's industry (not including fossil-fuel mining and processing or electricity generation) contributed 225 million metric tons of carbon (826 million tons of carbon dioxide), 16% of the world's carbon dioxide emissions to the atmosphere from industry. Waste treatment plants and landfill sites in North America accounted for 13.4 million tons of methane (282 million tons of carbon dioxide equivalent; 10 million tons of carbon), roughly 20% of global totals.

- Industrial carbon dioxide emissions from North America decreased nearly 11% between 1990 and 2002, while energy consumption in the United States and Canada increased 8% to 10% during that period. In both countries, a shift in production activity toward less energy-intensive industries and dissemination of more energy efficient equipment kept the rate of energy demand growth lower than industrial gross domestic product growth.

- Changes in industrial carbon dioxide emissions are a consequence of changes in industrial energy demand and changes in the mix of fossil fuels used by industry to supply that demand. Changes in industrial energy demand are themselves a consequence of changes in total industrial output, shifts in the relative shares of industrial sectors, and increases in energy efficiency. Shifts from coal and refined petroleum products to natural gas and electricity contributed to a decline in total industrial carbon dioxide emissions since 1997 in both Canada and the United States.

- An increase in carbon dioxide emissions from North American industry is likely to accompany the forecasted increase in industrial activity (2.3% per year until 2025 for the United States).

- Emissions per unit of industrial activity will likely decline as non-energy intensive industries grow faster than energy intensive industries and with increased penetration of energy efficient equipment. However, continuation of the trend toward less carbon-intensive fuels is uncertain given the rise in natural gas prices relative to coal in recent years.

- Options for reducing carbon dioxide emissions from North American industry can be broadly classified as methods to: (1) reduce process/fugitive emissions or convert currently released emissions; (2) increase energy efficiency, including combined heat and power management; (3) change industrial processes (materials efficiency, recycling, substitution between materials or between materials and energy, and nanotechnology); (4) substitute less carbon intense fuels; and (5) capture and store carbon dioxide.

- Further work on materials substitution holds promise for industrial emissions reduction, such as the replacement of petrochemical feedstocks by feedstocks derived from vegetative matter (biomass), of steel by aluminum in the transport sector, and of concrete by wood in the buildings sector. The prospects for greater usage of energy efficiency technologies are equally substantial.

8.1 INTRODUCTION

This chapter assesses carbon flows through industry (manufacturing and construction including industry process emissions, but excluding fossil-fuel mining and processing)[1] and municipal waste disposal.

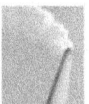

In 2002, industry was responsible for 21% of human-caused (anthropogenic) emissions to the atmosphere.

In 2002, industry was responsible for 1423.8 million metric tons of carbon (Mt C) (5220.6 million tons of carbon dioxide [Mt CO_2]), which is 21% of human-caused (anthropogenic) emissions to the atmosphere (244.8 Mt C [4322.9 Mt CO_2] from fuel combustion and 1179.0 Mt C [897.7 Mt CO_2] from industrial processes). North America's industry contributed 206.9 Mt C (758.7 Mt CO_2) of combustion-sourced emissions and 18.2 Mt C (66.8 Mt CO_2) of process emissions for a total of 225 Mt C (826 Mt CO_2) or 16% of global totals (WRI, 2005; see Figure 8.1A)[2†]. The manufacturing industry contributed 12% of total North American greenhouse gas (GHG) emissions, lower than in many other parts of the world. However, with North America's population at 6.8% of the world's total, industry contributed a proportionally larger share of total industrial emissions *per capita* than the rest of the world[3].

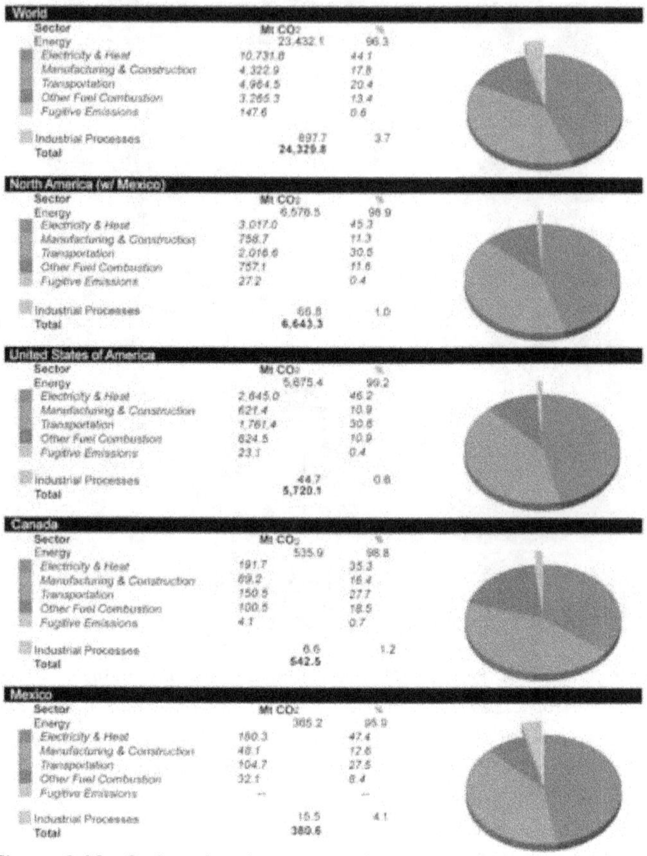

Figure 8.1A Carbon dioxide emissions by sector in 2002. *Source:* WRI (World Resources Institute)(2005). The magnitude and/or range of uncertainty for the given numerical values is not provided in the reference. To convert from Mt CO_2 to MtC, multiply the Mt CO_2 value by 12/44.

World

Sector	Mt CO_2	%
Energy	23,432.1	96.3
Electricity & Heat	10,731.8	44.1
Manufacturing & Construction	4,322.9	17.8
Transportation	4,964.5	20.4
Other Fuel Combustion	3,265.3	13.4
Fugitive Emissions	147.6	0.6
Industrial Processes	897.7	3.7
Total	24,329.8	

North America (w/ Mexico)

Sector	Mt CO_2	%
Energy	6,576.5	98.9
Electricity & Heat	3,017.0	45.3
Manufacturing & Construction	758.7	11.3
Transportation	2,016.6	30.5
Other Fuel Combustion	757.1	11.6
Fugitive Emissions	27.2	0.4
Industrial Processes	66.8	1.0
Total	6,643.3	

United States of America

Sector	Mt CO_2	%
Energy	5,675.4	99.2
Electricity & Heat	2,645.0	46.2
Manufacturing & Construction	621.4	10.9
Transportation	1,761.4	30.8
Other Fuel Combustion	624.5	10.9
Fugitive Emissions	23.1	0.4
Industrial Processes	44.7	0.8
Total	5,720.1	

Canada

Sector	Mt CO_2	%
Energy	535.9	98.8
Electricity & Heat	191.7	35.3
Manufacturing & Construction	89.2	16.4
Transportation	150.5	27.7
Other Fuel Combustion	100.5	18.5
Fugitive Emissions	4.1	0.7
Industrial Processes	6.6	1.2
Total	542.5	

Mexico

Sector	Mt CO_2	%
Energy	365.2	95.9
Electricity & Heat	180.3	47.4
Manufacturing & Construction	48.1	12.6
Transportation	104.7	27.5
Other Fuel Combustion	32.1	8.4
Fugitive Emissions	--	--
Industrial Processes	15.5	4.1
Total	380.6	

Industrial CO_2 emissions decreased nearly 11% between 1990 and 2002 while energy consumption in the United States and Canada increased 8% to 10% (EIA, 2005; CIEED-AC, 2005). In both countries, a shift in production activity toward less energy-intensive industries and dissemination of more energy efficient equipment kept the rate of growth in energy demand lower than industrial gross domestic product (GDP) growth (IEA, 2004)[4]. This slower demand growth, in concert with a shift toward less carbon-intensive fuels, explains the decrease in industrial CO_2 emissions.

The municipal waste stream excludes agricultural and forestry wastes but includes wastewater. Carbon dioxide, generated from aerobic metabolism in waste removal and storage processes, arises from biological material and is considered GHG neutral. Methane (CH_4) released from anaerobic activity at waste treatment plants and landfill sites, forms a substantial portion of carbon emissions to the atmosphere. Given its high global warming potential (GWP) (*i.e.*, the GWP for CH_4 is 21 times that of CO_2), CH_4 plays an important role in the evaluation of possible climate change impacts (WRI, 2005; see Figure 8.1B)[5†]. Globally, CH_4 emissions from waste amount to 66 Mt, or 378 Mt C equivalent (1386 Mt CO_2 equivalent). North American activity accounts for 13.4 Mt of CH_4 (77 Mt C equivalent [282 Mt CO_2 equivalent]) or roughly 20% of global totals.

Substantial sequestration of carbon occurs in landfills[6]. Data on carbon buried there are poor. The Environmental Protection Agency (EPA), using data from Barlaz and Ham (1990) and Barlaz (1994), estimated that 30% of carbon in food waste and up to 80% of carbon in newsprint, leaves, and

---

[1] This includes direct flows only. Indirect carbon flows (*e.g.*, due to electricity generation) are associated with power generation.

[2†] A dagger symbol indicates that the magnitude and/or range of uncertainty for the given numerical value(s) is not provided in the references cited.

[3] North America, including Mexico, was responsible for about 27% of global CO_2 emissions in 2002.

[4] Decomposition analyses can assess changes in energy consumption due to, for example, increases in industry activity, changes in relative productivity to or from more intense industry subsectors, or changes in material or energy efficiency in processes.

[5] While not carbon-based, N_2O from sewage treatment is included in Figure 8.4, below, to show its relative GHG importance.

[6] IPCC guidelines currently do not address landfill sequestration. Such guidelines will be in the 2006 publication.

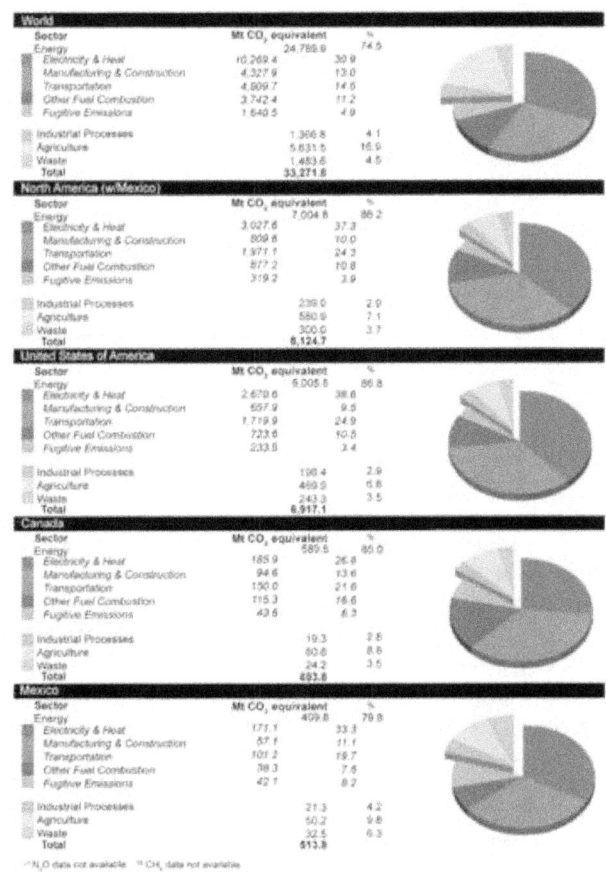

World

Sector	Mt CO₂ equivalent	%
Energy	24,789.9	74.5
Electricity & Heat	10,269.4	30.9
Manufacturing & Construction	4,327.9	13.0
Transportation	4,809.7	14.6
Other Fuel Combustion	3,742.4	11.2
Fugitive Emissions	1,640.5	4.9
Industrial Processes	1,366.8	4.1
Agriculture	5,631.5	16.9
Waste	1,483.6	4.5
Total	33,271.8	

North America (w/Mexico)

Sector	Mt CO₂ equivalent	%
Energy	7,004.8	86.2
Electricity & Heat	3,027.6	37.3
Manufacturing & Construction	809.6	10.0
Transportation	1,971.1	24.3
Other Fuel Combustion	877.2	10.8
Fugitive Emissions	319.2	3.9
Industrial Processes	239.0	2.9
Agriculture	580.9	7.1
Waste	300.0	3.7
Total	8,124.7	

United States of America

Sector	Mt CO₂ equivalent	%
Energy	6,005.6	86.8
Electricity & Heat	2,670.6	38.6
Manufacturing & Construction	657.9	9.5
Transportation	1,719.9	24.9
Other Fuel Combustion	723.6	10.5
Fugitive Emissions	233.6	3.4
Industrial Processes	198.4	2.9
Agriculture	469.9	6.8
Waste	243.3	3.5
Total	6,917.1	

Canada

Sector	Mt CO₂ equivalent	%
Energy	589.5	85.0
Electricity & Heat	185.9	26.8
Manufacturing & Construction	94.6	13.6
Transportation	150.0	21.6
Other Fuel Combustion	115.3	16.6
Fugitive Emissions	43.6	6.3
Industrial Processes	19.3	2.8
Agriculture	60.6	8.8
Waste	24.2	3.5
Total	693.6	

Mexico

Sector	Mt CO₂ equivalent	%
Energy	409.8	79.8
Electricity & Heat	171.1	33.3
Manufacturing & Construction	57.1	11.1
Transportation	101.2	19.7
Other Fuel Combustion	38.3	7.5
Fugitive Emissions	42.1	8.2
Industrial Processes	21.3	4.2
Agriculture	50.2	9.8
Waste	32.5	6.3
Total	513.8	

⁽ᵃ⁾ N₂O data not available. ⁽ᵇ⁾ CH₄ data not available.

Figure 8.1B Greenhouse gas emissions by sector in 2000, CO₂, CH₄, N₂O, PFCs, HFCs, and SF₆. *Source:* WRI (World Resources Institute)(2005). The magnitude and or range of uncertainty for the given numerical values is not provided in the reference. To convert from MtCO₂ equivalent to MtC equivalent, multiply the Mt CO₂ value by 12/44.

related CO_2 emissions occur when carbon-based fuels provide thermal energy to drive industrial processes.

8.2.1 Overview of Carbon Inputs and Outputs

Industry generates about one-third as much emitted carbon as the production of electricity and other fuel supply in North America and only about 55% as much as is generated by the transportation sector.

8.2.1.1 CARBON IN

Carbon-based raw materials typically enter industrial sites as biomass (primarily wood), limestone, soda ash, oil products, coal/coke, natural gas, and natural gas liquids. These inputs are converted to dimension lumber and other wood products, paper and paperboard, cement and lime, glass, and a host of chemical products, plastics, and fertilizers.

While the bulk of the input carbon leaves the industrial site as a product, some leaves as process CO_2 and some is converted to combustible fuel. Waste wood (or hog fuel) and black liquor, generated in the production of chemical pulps, are burned to provide process heat or steam for digesting wood chips or for drying paper or wood products, in some cases providing electricity through cogeneration. Chemical processes utilizing natural gas often generate off-gases that, mixed with conventional fuels, provide process heat. Finally, some of the carbon that enters as a feedstock leaves as solid or liquid waste.

In some industries, carbon is used to remove oxygen from other input materials through "reduction." In most of the literature, such carbon is considered an input to the process and is released as "process" CO_2, even though it acts as a fuel (*i.e.*, it unites with oxygen to form CO_2 and releases heat). For example, in metal smelting and refining processes, a carbon-based reductant separates oxygen from the metal atoms. Coke, from the destructive distillation of coal, enters a blast furnace with iron ore to strip off the oxygen associated with the iron. Carbon anodes in electric arc furnaces in steel mills and specialized electrolytic "Hall-Heroult" cells oxidize to CO_2 as they melt recycled steel or reduce alumina to aluminum.

8.2.1.2 CARBON OUT

Carbon leaves industry as part of the intended commodity or product, as a waste product or as a gas, usually CO_2.

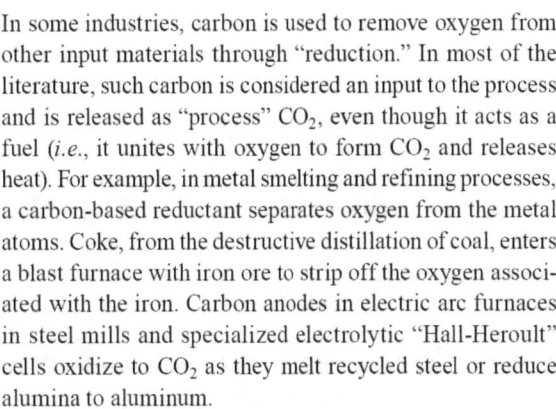

branches remain in the landfill[†]. Plastics show no deterioration. In all, 80% of the carbon entering a landfill site may be sequestered, depending on moisture, aeration, and site conditions. Bogner and Spokas (1993) estimate that "more than 75% of the carbon deposited in landfills remains in sedimentary storage."

8.2 INDUSTRY CARBON CYCLE

Carbon may enter industry as a fuel or as a feedstock where the carbon becomes entrained in the industry's final product. Carbon in the waste stream can be distinguished as atmospheric and non-atmospheric, the former being comprised of process and combustion-related emissions. Process CO_2 emissions, a non-combustive source, are the result of the transformation of the material inputs to the production process. For example, cement production involves the calcination of lime, which chemically alters limestone to form calcium oxide and releases CO_2. Of course, combustion-

Process emissions are CO_2 emissions that occur as a result of the process itself—the calcining of limestone releases about 0.5 tons CO_2 per ton of clinker (unground cement) or about 0.8 tons per ton of lime[7,8]. The oxidation of carbon anodes generates about 1.5 tons CO_2 to produce a ton of aluminum. Stripping hydrogen from CH_4 to make ammonia releases about 1.6 tons CO_2 per ton of ammonia.

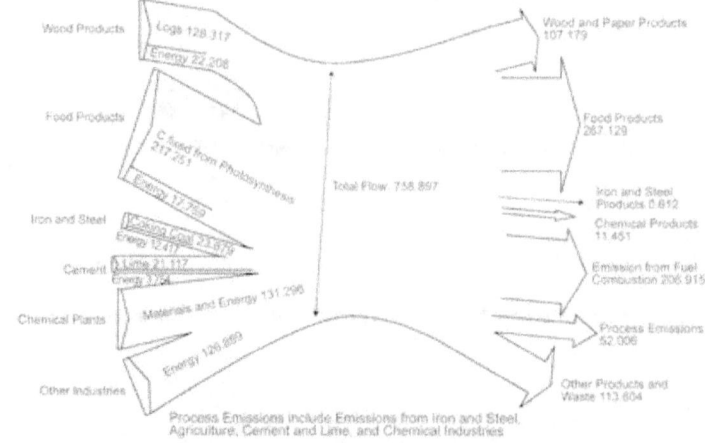

Combustion of carbon-based fuels results in the emission of CO_2. In many cases, the combustion process is not complete and other carbon-based compounds may be released (carbon monoxide, CH_4, volatile organic compounds). These often decompose into CO_2, but their life spans in the atmosphere vary.

Figure 8.2 Carbon flows for Canada, the United States, and Mexico combined. Values in megatons carbon can be converted to megatons CO_2 equivalents by multiplying by 44/12; the ratio of CO_2 mass to carbon mass. Comparable diagrams for the individual countries are in Appendix C. *Source:* Energy data from Statistics Canada Industrial Consumption of Energy survey, Conversion coefficients, IEA Oil Information (2004), IEA Coal Information (2005), IEA Natural Gas Information (2004). Process emissions from Environment Canada, Canada GHG Inventory (2002), EPA, U.S. Emissions Inventory. Production data from Statistics Canada, CANSIM Table 002-0010, Tables 303-0010, -0014 to -0021, -0024, -0060, Pub. Cat. Nos.: 21-020, 26-002, 45-002, Canadian Pulp and Paper Association on forestry products. Production of forestry products: USDA Database; FO-2471000, -2472010, -2482000, -2483040, -6342000, -6342040, U.S. Timber Production, Trade, Consumption, and Price Statistics 1965-2005. Production of organic products (e.g., food): USDA PS&D Official Statistical Results. Steel: International Iron and Steel Institute, World steel in figures (2003). Minerals production: USGS mineral publications.

8.2.1.3 CARBON FLOW

Figure 8.2 illustrates the flows of carbon in and out of industries in North America. Comparable diagrams for individual countries are presented in Appendix C. On the left side of Figure 8.2, all carbon-based material by industry sector is accounted for, whether in fuel or in feedstock. On the right, the exiting arrows portray how much of the carbon leaves as part of the final products from that industry. The carbon in the fossil fuel and feedstock materials leave in the waste stream as emissions from fuel combustion (including biomass), as process emissions, or as other products and waste. Carbon capture and storage potentials are assessed in the industry subsections below.

8.2.2 Sectoral Trends in the Industrial Carbon Cycle

Figure 8.2 shows that energy-intensive industries differ significantly in their carbon cycle dynamics.

8.2.2.1 PULP AND PAPER

While pulp and paper products are quite energy-intensive, much of the energy is obtained from biomass. By using hog fuel and black liquor, some types of pulp mills are energy self-sufficient. Biomass fuels are considered carbon neutral because return of the biomass carbon to the atmosphere completes a cycle that began with carbon uptake from the atmosphere by vegetation[9]. Fuel handling difficulties and air quality concerns can arise from the use of biomass as a fuel.

8.2.2.2 CEMENT, LIME, AND OTHER NONMETALLIC MINERALS

Cement and lime production require the calcination of limestone, which releases CO_2; about 0.78 tons of CO_2 per ton of lime calcined.

$$CaCO_3 \rightarrow CaO + CO_2$$
calcium carbonate   calcium oxide   carbon dioxide

---

[7] In these industries, more CO_2 is generated from processing limestone than from the fossils fuels combusted.
[8] The calcination of limestone also takes place in steel, pulp and paper, glass, and sugar industries.

[9] This is also reflected in the United Nations Framework Convention on Climate Change (UNFCCC) IPCC guidelines to estimate CO_2 emissions.

Outside of the combustion of fossil fuels, lime calcining is the single largest human-caused source of CO_2 emissions. Annual growth in cement production is forecast at 2.4% in the United States for at least the next decade. This industry could potentially utilize sequestration technologies to capture and store CO_2 generated.

The production of soda ash (sodium carbonate) from sodium bicarbonate in the Solvay process releases CO_2, as in glass production, in its utilization. Soda ash is used to produce pulp and paper, detergents, and soft water.

$$2NaHCO_3 \rightarrow Na_2CO_3 + CO_2 + H_2O$$
sodium bicarbonate sodium carbonate carbon dioxide water

8.2.2.3 NONFERROUS METAL SMELTING AND IRON AND STEEL SMELTING

Often metal smelting requires the reduction of metal oxides to obtain pure metal through use of a "reductant", usually coke. Because reduction processes generate relatively pure streams of CO_2, the potential for capture and storage is good.

In electric arc furnaces, carbon anodes decompose to CO_2 as they melt the scrap iron and steel feed in "mini-mills". In Hall-Heroult cells, a carbon anode oxidizes when an electric current forces oxygen from aluminum oxide (alumina) in the production of aluminum[10].

8.2.2.4 METAL AND NONMETAL MINING

Mining involves the extraction of ore and its transformation into a concentrated form. This involves transportation from mine site, milling, and separating mineral-bearing material from the ore. Some transportation depends on truck activity, but the grinding process is driven by electric motors (*i.e.*, indirect release of CO_2). Some processes, like the sintering or agglomeration of iron ore and the liquid extraction of potash, use a considerable amount of fossil fuels directly.

8.2.2.5 CHEMICAL PRODUCTS

This diverse group of industries includes energy-intensive electrolytic processes as well as the consumption of large quantities of natural gas as a feedstock to produce commodities like ammonia, methanol, and hydrogen. Ethylene and propylene monomers from natural gas liquids are used in plastics production. Some chemical processes generate fairly pure streams of CO_2 suitable for capture and storage.

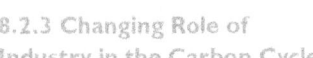

8.2.2.6 FOREST PRODUCTS

This industry uses biomass waste to dry commercial products such as lumber, plywood, and other products. The industry also includes silviculture, the practice of replanting and managing forests.

8.2.2.7 OTHER MANUFACTURING

Most of the remaining industries, while economically important, individually play a relatively minor role in the carbon cycle because they are not energy intensive and use little biomass[11]. In aggregate, however, these various industries contribute significantly to total industrial CO_2 emissions. Industries in this group include the automotive industry, electronic products, leather and allied products, fabricated metals, furniture and related products, and plastics and rubber products.

8.2.3 Changing Role of Industry in the Carbon Cycle

Energy consumption per unit GDP has declined in Canada and the United States by more than 30% since the mid-1970s. In manufacturing, the decline was even greater—more than 50% in the United States since 1974.

The National Energy Modeling System operated by the United States' Energy Information Administration applies growth forecasts from the Global Insight macroeconomic model. While the United States economy is forecast to grow at an average rate of 3.1% per year to 2025, industrial growth is forecast at 2.3% per year—an amalgam of manufactur-

> The shift from coal and refined petroleum products to natural gas and electricity contributed to a decline in total industrial CO_2 emissions since 1997 in both Canada and the United States.

---

[10] Ceramic anodes may soon be available to aluminum producers and significantly reduce process CO_2 emissions.

[11] Except, of course, the food, beverage, and some textile industries.

ing growth of 2.6% per year and non-manufacturing of 1.5% per year. Manufacturing is further disaggregated into energy-intensive industries, growing at 1.5% per year, and non-energy intensive industries at 2.9% per year. The slower growth in

Table 8.1 Energy reductions in recycling.

Recycled material	Energy saved	Recycled material	Energy saved
Aluminum	95%	Glass	31%
Tissue paper	54%	Newsprint	45%
Printing/writing paper	35%	Corrugated cardboard	26%
Plastics	57%–75%	Steel	61%

Source: Hershkowitz (1997)

the energy-intensive industries is reflected in the expected decline in industrial energy intensity of 1.6% per year over the EIA (2005) forecast.

The International Energy Agency reviewed energy consumption and emissions during the last 30 years to identify and project underlying trends in carbon intensity[12]. The review's decomposition analysis (Figure 8.3) attributes changes in industrial energy demand to changes in total industrial output (activity), shifts in the relative shares of industrial sectors (structure), and increases in energy efficiency (intensity).

Changes in carbon emissions result from these three factors, but also from changes in fuel shares—substitution away from or toward more carbon-intensive fuels. The shift from coal and refined petroleum products to natural gas and electricity[13] contributed to a decline in total industrial CO_2 emissions since 1997 in both Canada and the United States. The continuation of this trend is uncertain given the rise in natural gas prices relative to coal in recent years.

8.2.4 Actions and Policies for Carbon Management in Industry

Industry managers can reduce carbon flows through industry by altering the material or energy intensity and character of production (IPCC, 2001). Greater materials efficiency typically reduces energy demands in processing because of reduced materials handling. For example, recycling materials often reduces energy consumption per unit of output by 26 to 95% (Table 8.1). Further work on materials substitution also holds promise for reduced energy consumption and emissions reduction[14].

The prospects for greater energy efficiency are equally substantial. Martin *et al.* (2001) characterized more than 50 key emerging energy efficient technologies, including efficient Hall-Heroult cell retrofits, black liquor gasification in pulp production, and shape casting in steel industries. Worrell *et al.* (2004) covers many of the same technologies and notes that significant potential exists in utilizing efficient motor systems and advanced cogeneration technologies.

At the same time, energy is a valuable production input that, along with capital, can substitute for labor as a means of increasing productivity. Thus overall productivity gains in industry can be both energy-saving and energy-augmenting, and the net impact depends on the nature of technological innovation and the expected long-run cost of energy relative to other inputs. This suggests that, if policies to manage carbon emissions from industry were to be effective, they would need to provide a significant signal to technology innovators and adopters to reflect the negative value that society places on carbon emissions. This in turn suggests the application of regulations or financial instruments, examples being energy efficiency regulations, carbon management regulations, and fees on carbon emissions.

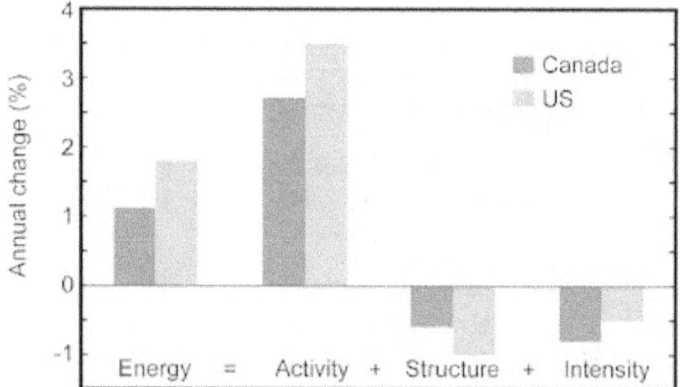

Figure 8.3 Decomposition of energy use, manufacturing section, 1990-1998. *Source:* IEA (2004).

8.3 WASTE MANAGEMENT CARBON CYCLE

The carbon cycle associated with human wastes includes industrial, commercial,

---

[12] Most of the information in this section is obtained from IEA (2004).

[13] As noted earlier, emissions associated with electricity are allocated to the electricity supply sector. Thus a shift to electricity reduces the GHG intensity of the industry using it. If electricity is made in coal-fired plants, however, total CO_2 emissions may actually increase.

[14] For example, substitute petrochemical feedstocks by biomass or concrete by wood in home foundations.

construction, demolition, and residential waste. Municipal solid waste contains significant amounts of carbon. Paper, plastics, yard trimmings, food scraps, wood, rubber, and textiles made up more than 80% of the 236 Mt of municipal solid waste generated in the United States in 2003 (EPA, 2005) and the 25 Mt generated in Canada (Statistics Canada, 2004), as shown in Table 8.2. In Mexico, as much as 20% of wastes are not systematically collected; no disaggregated data are available (EPA, 2005).

Table 8.2 Waste materials flows by region in North America, 2003.

	United States	Canada	Mexico
Total waste (Mt per year)	236.0	24.8	29.2
Recycled	72.0	6.6	–
Carbon-based waste	197.1	19.6	–
Carbon-based waste recycled	47.3[a]	4.3	–
Carbon sequestered (CO_2 equivalents)	10.1	–	–
Methane (kt per year)			
Generated	12,486	1,452	–
Captured, oxidized	6,239	336	–
Emitted	6,247	1,117	–
Emitted (CO_2 equivalents)	131,187	23,453	–

[a] Calculated estimate

Source: EPA (2003b, 2005), Statistics Canada (2004), Mohareb (2004) for Canada methane data, California Evironmental Protection Agency (2003) for Mexico data point.

A portion of municipal solid waste is recycled: 31% in the United States (EPA, 2003b)[†] and 27% in Canada (Statistics Canada, 2004).[†] Up to 14% of the remaining waste is incinerated in the United States, slightly less in Canada. Incineration can reduce the waste stream by up to 80%, but this ensures that more of the carbon reaches the atmosphere as opposed to being sequestered (or subsequently released as CH_4) in a landfill. Incineration, however, can be used to cogenerate electricity and useful heat, which may reduce carbon emissions from stand-alone facilities.

Once in a landfill, carbon in wastes may be acted upon biologically, releasing roughly equal amounts of CO_2 and CH_4 by volume[15] depending on ambient conditions, as well as a trace amount of carbon monoxide and volatile organic compounds. While no direct data on the quantity of CO_2 released from landfills exists, one can estimate the CO_2 released by using this ratio; the estimated amount of CO_2 released from landfills in Canada and the United States (no data from Mexico) would be approximately 38 Mt[16], a relatively small amount compared to the total of other subsectors in this chapter. Also, recall that these emissions are from biomass and, in the context of IPCC assessment guidelines, are considered GHG-neutral.

Depending on the degree to which aerobic or anaerobic metabolism takes place, a considerable amount of carbon remains unaltered and more or less permanently stored in the landfill (75%-80%; see Barlaz and Ham, 1990; Barlaz,

1994; and Bogner and Spokas, 1993). Because data on the proportions of carboniferous material entering landfills can be estimated, approximate carbon contents of these materials can be determined and the degree to which these materials can decompose, it would be possible to estimate the amount of carbon sequestered in a landfill site (see EPIC, 2002; Mohareb *et al.*, 2004; EPA, 2003b; EPA, 2005). While EPA (2005) provides an estimate of carbon sequestered in US landfills (see Table 8.2), no data are available for other regions.

> Municipal solid waste contains significant amounts of carbon.

Anaerobic digestion generates CH_4 gases that can be captured and used in cogenerators. Many of the 1800 municipal solid waste sites in 2003 in the United States captured and combusted landfill-generated CH_4; about half of all the CH_4 produced was combusted or oxidized in some way (EPA, 2005). In Canada, about 23% of the CH_4 emissions were captured and utilized to make energy in 2002 (Mohareb *et al.*, 2004). The resultant CO_2 released from such combustion is considered biological in origin. Thus only CH_4 emissions, at 21 times the CO_2 warming potential, are included as part of GHG inventories. Their combustion greatly alleviates the net contribution to GHG emissions and, if used in cogeneration, may offset the combustion of fossil fuels elsewhere. Figure 8.4 provides an estimate of CH_4 (and nitrous oxide [N_2O] as the other GHG for comparison) released from landfills and waste treatment facilities.

8.4 COSTS RELATED TO CONTROLLING HUMAN-CAUSED IMPACTS ON THE CARBON CYCLE

Defining costs associated with reducing human-caused (anthropogenic) impacts on the carbon cycle is a highly contentious issue. Different approaches to cost assessments (top-down, bottom-up, applicable discount rates, social

---

[15] Based on gas volumes, this means that roughly equivalent amounts of carbon are released as CO_2 as CH_4.
[16] 14 Mt of CH_4 (see Table 8.3) are equivalent, volume wise at standard temperature and pressure, to 38 Mt of CO_2. This derived estimate is highly uncertain and not of the same caliber as other emissions data provided here.

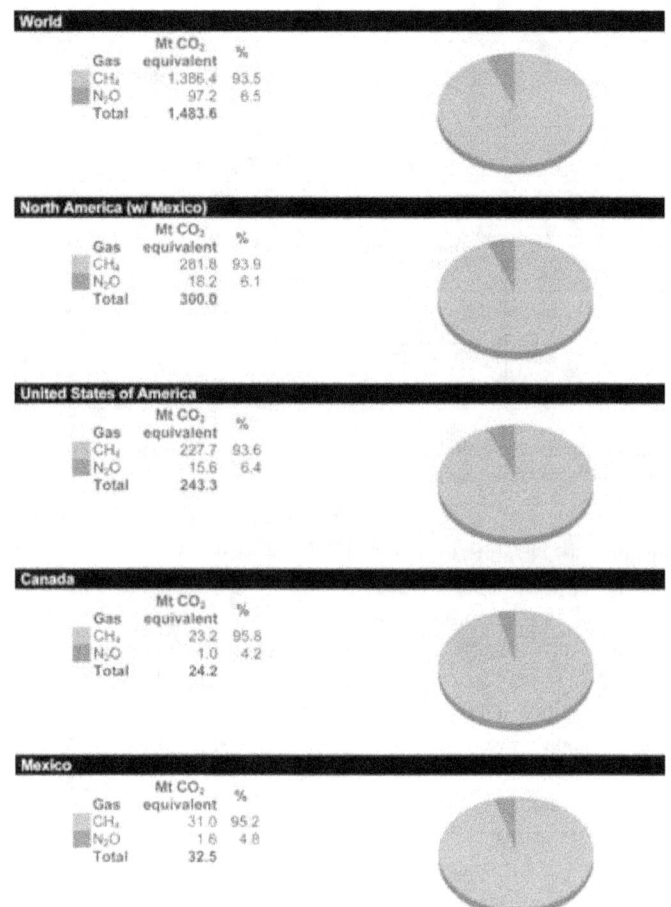

World

Gas	Mt CO$_2$ equivalent	%
CH$_4$	1,386.4	93.5
N$_2$O	97.2	6.5
Total	1,483.6	

North America (w/ Mexico)

Gas	Mt CO$_2$ equivalent	%
CH$_4$	281.8	93.9
N$_2$O	18.2	6.1
Total	300.0	

United States of America

Gas	Mt CO$_2$ equivalent	%
CH$_4$	227.7	93.6
N$_2$O	15.6	6.4
Total	243.3	

Canada

Gas	Mt CO$_2$ equivalent	%
CH$_4$	23.2	95.8
N$_2$O	1.0	4.2
Total	24.2	

Mexico

Gas	Mt CO$_2$ equivalent	%
CH$_4$	31.0	95.2
N$_2$O	1.6	4.8
Total	32.5	

Figure 8.4 Greenhouse gas emissions by gas from waste in 2000. *Source:* WRI (World Resources Institute) (2005). The magnitude and/or range of uncertainty for the given numerical values is not provided in the reference. To convert from Mt CO$_2$ equivalent to Mt C equivalent, multiply the Mt CO$_2$ value by 12/44.

costing, cost effectiveness, no regrets), different understandings of what costs include (risk, welfare, intangibles, capital investment cycles), different values associated with energy demand in different countries (accessibility, availability, infrastructure, resource type and size), actions and technologies included in the analysis, and the perspective on technology development all have an impact on evaluating costs. Should analysts consider only historical responses to energy prices, production and demand elasticities, or income changes? Does one consider only technology options and their strict financial costs or see historic technology investments as sunk costs? Should one include producers' or consumers' welfare? Are there local, national, international issues?

Cost variation within industries is significant. Costs associated with various methods to reduce emissions also vary. Reduction methods can be classified as:

- reducing or altering process/fugitive emissions,
- energy efficiency, including combined heat and power,
- process changes,
- fuel substitution,
- carbon capture and storage.

One can attribute potential reductions over a set time under a range of costs. We suggest the cost-range categories ("A" through "D") shown in Table 8.3. The table contains estimates of the percentage reduction by industry under these cost categories. Costs are not drawn from a single source but are the authors' estimates based on a long history of costs reported in various documents[17]. Some studies focus on technical potential and do not provide the cost of achieving the reductions. As such, achievable reductions are likely overestimated. Others describe optimization models that provide normative costs and likely overestimate potentials and underestimate costs. Still others use top-down approaches where historic data sets are used to determine relationships between emissions and factors of production; costs are often high and emissions reductions underestimated.

When looking at cost numbers like this, one should remember that, for each $10 cost increment per t CO$_2$ (or about $37 per t C), gasoline prices would increase about 2.4¢/L (9¢/U.S. gallon). Diesel fuel cost would be nearly 2.7¢/L (10¢/U.S. gallon). Costs per Gigajoule (GJ)[18] vary by fuel: coal rises about 90¢/GJ, depending on type, heavy fuel oil (HFO) by 73¢, and natural gas by 50¢. At 35% efficiency, coal-fired electricity generation would be about 0.8¢/kWh higher, about 0.65¢/kWh for HFO, and about 0.45¢/kWh for natural gas.

Of course, as the cost of carbon increases, one moves up the carbon supply curve for industrial sectors. However, reductions become marginal or insignificant and so are not included in Table 8.3. If a cell in Table 8.3 shows two cost categories (*e.g.*, A/B) and two reduction levels (%Q$_{red}$ is 15/20), the value associated with the second portrays the additional reduction at that increased expenditure level. Thus spending up to $50/t CO$_2$ to improving efficiency in metal smelting implies a potential reduction of 35% (see Table 8.3). Reductions in each category are not additive for an industry type because categories are not independent.

[17] Studies vary widely in how they define system boundaries, baseline, and time periods, which sectors or subsectors are included, economic assumptions, and many other factors. See `Some Explanatory Notes` in Section 8.4.1 for a list.
[18] A Gigajoule (GJ), or one billion joules, is slightly smaller than 1 MMBtu (1 GJ = 0.948 MMBtu).

Table 8.3 Approximate costs and reductions potential.

Sector	Reduction of fugitives		Energy efficiency		Process change		Fuel substitution		Carbon Capture and Storage	
	Cost category	%Q_{red}	Cost category[a]	%Q_{red}[a]	Cost category	%Q_{red}	Cost category	%Q_{red}	Cost category[a]	%Q_{red}[a]
All industry	B	3	A/B	12/8	B	20	A	10	C	30
P&P	B	5	A/B	10/5	B	40	A	40	D	?
Nonmetal min			A	10	A	40	A	40	C	80
Metal smelt			A/B	15/20	B	10	A	15	C	40
Mining			A	5						
Chemicals	B	10	A/B	10/5	B	25	A	5	C/D	40/20
Forest products	B	5	A	5						
Other man			A	15	A	20	A	5	D	?
Waste	A	90							D	30

[a] If two letters appear, two percent quantities reduced are shown. Each shows the quantity reduced at that cost. That is, if all lesser and higher costs were made, emissions reduction would be the sum of the two values.

Note: The reductions across categories are NOT additive. For example, if "Carbon Capture and Storage" is employed, then fuel switching would have little bearing on the emissions reduction possible. Also, it is difficult to isolate process switching and efficiency improvements.

The "Cost Categories" are as follows:

CO_2-Based: A: \$0–\$25/t CO_2; B: \$25–\$50/t CO_2; C: \$50–\$100/t CO_2; D: >\$100/t CO_2

Carbon-Based: A: \$0–\$92/t C; B: \$92–\$180/t C; C: \$180–\$367/t C; D: >\$367/t C

Because not all reduction methods are applicable to all industries, as one aggregates to an "all industry" level (top line, Table 8.3), the total overall emissions reduction level may be less than any of the individual industries sited.

8.4.1 Some Explanatory Notes

Data come from a variety of sources and do not delineate costs as per the categories described here. Data sources can be notionally categorized into the following groups (with some references listed twice)[19]:

- *General overviews*: Grubb *et al.* (1993), Weyant *et al.* (1999)[20], Grubb *et al.* (2002), Löschel (2002).
- *Top-down analyses*: McKitrick (1996), Herzog (1999), Sands (2002), McFarland *et al.* (2004), Schäfer and Jacoby (2005), Matysek *et al.* (2006).
- *Bottom up analyses*: Martin *et al.* (2001), Humphreys and Mahasenan (2002), Worrell *et al.* (2004), Kim and Worrell (2002), Morris *et al.* (2002), Jaccard *et al.* (2003a), DOE (2006), IEA (2006).
- *Hybrid model analyses*: Böhringer (1998), Jacobsen (1998), Edmonds *et al.* (2000), Koopmans and te Velde (2001), Jaccard (2002), Frei *et al.* (2003), Jaccard *et al.* (2003a), Jaccard *et al.* (2003b), Edenhofer *et al.* (2006).
- *Others*: Newell *et al.* (1999), Sutherland (2000), Jaffe *et al.* (2002).

8.4.1.1 PROCESS AND FUGITIVES

Process and fugitive reductions are only available in certain industries. For example, because wood-products industries burn biomass, fugitives are higher than in other industries and reduction potentials exist.

In the waste sector, the reductions potentials are very large; we have simply estimated possible reductions if we were to trap and burn all landfill CH_4. The costs for this are quite low. EPA (2003a) estimates of between 40% and 60% of CH_4 available for capture may generate net economic benefits.

8.4.1.2 ENERGY EFFICIENCY

The potential for emissions reductions from efficiency improvements is strongly linked with both process change and fuel switching. For example, moving to Cermet-based processes in electric arc furnaces in steel and aluminum smelting industries can significantly improve efficiencies and lower both combustion and process GHG emissions.

A "bottom up" technical analyses tends to show higher potentials and lower costs than when one uses a hybrid or a "top-down" approach to assess reduction potentials due to efficiency improvements; Table 8.3 portrays the outcome of the more conservative hybrid (mix of top-down and bottom-up) approach and provides what some may consider conservative estimates of reduction potential (see particularly Martin *et al.*, 2001; Jaccard *et al.*, 2002; Jaccard *et al.*, 2003a; Jaccard *et al.*, 2003b; and Worrell *et al.*, 2004).

---

[19] Two authors are currently involved with IPCC's upcoming fourth assessment report where estimated costs of reduction are provided. Preliminary reviews of the cost data presented there do not differ substantially from those in table 8.3.

[20] John Weyant of Stanford University is currently editing another analysis similar to this listed publication to be released in the near future.

8.4.1.3 PROCESS CHANGE

Reductions from process change requires not only an understanding of the industry and its potential for change but also an understanding of the market demand for industry products that may change over time. In pulp production, for example, one could move from higher quality kraft pulp to mechanical pulp and increase production ratios (the kraft process only converts one-half the input wood into pulp), but will market acceptability for the end product be unaffected? Numerous substitution possibilities exist in the rather diverse *Other Manufacturing* industries (carpet recycling, alternative uses for plastics, *etc.*).

8.4.1.4 FUEL SUBSTITUTION

It is difficult to isolate fuel substitution and efficiency improvement because fuels display inherent qualities that affect efficiency. Fuel substitution can reduce carbon flow but efficiency may become worse. In wood products industries, shifts to biomass reduces emissions but increases energy use. In terms of higher heating values, shifts from coal or oil to natural gas may worsen efficiencies while reducing emissions[21].

8.4.1.5 CARBON CAPTURE AND STORAGE (CC&S)

In one sense, all industries and landfills could reduce emissions through CC&S but the range of appropriate technologies has not been fully defined and/or the costs are very high. For example, one could combust fuels in a pure oxygen environment such that the exhaust steam is CO_2-rich and suitable for capture and storage. Even so, some industries, like cement production, are reasonable candidates for capture, but cost of transport of the CO_2 to storage may prohibit implementation (see particularly Herzog, 1999; DOE, 2006).

8.5 RESEARCH AND DEVELOPMENT NEEDS

If we assume that carbon management will play a significant role in the future and that fossil fuels are likely to remain an economical energy supply for industries, research and development (R&D) will focus on the control of carbon emissions related to the extraction of this energy. Typical combustion technologies extract and transform fossil fuels' chemical energy relatively efficiently but, outside of further improvements in efficiency, they generally do little to manage the emissions generated. More recently, advanced technologies remove particularly onerous airborne emissions, such as compounds of sulphur and nitrogen, particulates, volatile

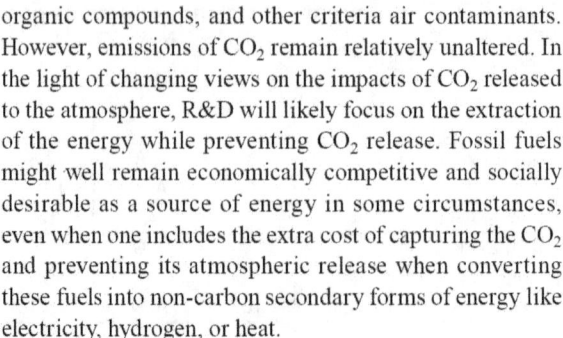

organic compounds, and other criteria air contaminants. However, emissions of CO_2 remain relatively unaltered. In the light of changing views on the impacts of CO_2 released to the atmosphere, R&D will likely focus on the extraction of the energy while preventing CO_2 release. Fossil fuels might well remain economically competitive and socially desirable as a source of energy in some circumstances, even when one includes the extra cost of capturing the CO_2 and preventing its atmospheric release when converting these fuels into non-carbon secondary forms of energy like electricity, hydrogen, or heat.

Some carbon capture and storage processes currently exist; indeed, oil companies have long "sequestered" CO_2 to enhance oil recovery from underground wells simply by injecting it into the oil reservoir. Many newer processes to accomplish CO_2 capture are being investigated, primarily in two categories: pre-combustion and post-combustion processes. Pre-combustion alternatives include gasification processes where, for example, coal's energy is entrapped in hydrogen and the CO_2 stream is subsequently sequestered. Post-combustion alternatives include carbon combustion in pure oxygen atmospheres and then trapping the resultant CO_2 for sequestration, and flue stack devices designed to extract the CO_2 from the flue gases for delivery to sequestration systems. Research has also been conducted on devices that can extract CO_2 directly from the atmosphere (Keith *et al.*, 2003).

---

[21] As the ratio of hydrogen to carbon rises in a fossil fuel, more of the total heat released upon combustion is caught up in the latent heat of vaporization of water and is typically lost to process. This loss is equivalent to the difference between a fuel's higher heating value and its lower heating value.

CHAPTER 9

Buildings

Lead Author: James E. McMahon, LBNL

Contributing Authors: Michael A. McNeil, LBNL, Itha Sánchez Ramos, Instituto de Investigaciones Eléctricas, Mexico

KEY FINDINGS

- The buildings sector of North America was responsible for annual carbon dioxide emissions of 671 million tons of carbon in 2003, which is 37% of total North American carbon dioxide emissions and 10% of global emissions. United States buildings alone are responsible for more carbon dioxide emissions than total carbon dioxide emissions of any other country in the world, except China.

- Carbon dioxide emissions from energy use in buildings in the United States and Canada increased by 30% from 1990 to 2003, an annual growth rate of 2.1% per year.

- Carbon dioxide emissions from buildings have grown with energy consumption, which in turn is increasing with population and income. Rising incomes have led to larger residential buildings and increased household appliance ownership.

- These trends are likely to continue in the future, with increased energy efficiency of building materials and equipment and slowing population growth, especially in Mexico, only partially offsetting the general growth in population and income.

- Options for reducing the carbon dioxide emissions of new and existing buildings include increasing the efficiency of equipment and implementing insulation and passive design measures to provide thermal comfort and lighting with reduced energy. Current best practices can reduce emissions from buildings by at least 60% for offices and 70% for homes. Technology options could be supported by a portfolio of policy options that take advantage of cooperative activities, avoid unduly burdening certain sectors, and are cost effective.

- Because reducing carbon dioxide emissions from buildings is currently secondary to reducing building costs, continued improvement of energy efficiency in buildings and reduced carbon dioxide emissions from the building sector will require a better understanding of the total societal cost of carbon dioxide emissions as an externality of building costs, including the costs of mitigation compared to the costs of continued emissions.

9.1 BACKGROUND

In 2003, buildings were responsible for 615 million metric tons of carbon (Mt C)[1] emitted in the United States (DOE/EIA, 2005), 40 Mt C in Canada (Natural Resources Canada, 2005a), and 17 Mt C in Mexico (SENER México, 2005), for a total of 671 Mt C in North America[2†]. According to the International Energy Agency, total energy-related emissions in North America in this year were 1815 Mt C (IEA, 2005). Therefore, buildings were responsible for 37% of energy-related emissions in North America. North American buildings accounted for 10% of global energy emissions, which totaled 6814 Mt C. United States' buildings alone are responsible for more carbon dioxide (CO_2) emissions than total CO_2 emissions of any other country in the world, except China (Kinzey *et al.*, 2002). Significant carbon emissions are due to energy consumption during the operation of the buildings; other emissions, not well quantified, may occur from water use in and around the buildings and from land-use impacts related to buildings. Buildings are responsible for 72% of United States electricity consumption and 54% of natural gas consumption (DOE/EERE, 2005)[3]. The discussions in this chapter include an accounting of CO_2 emissions from electricity consumed in the buildings sector; however, this represents a potential double counting of the CO_2 emissions from fossil fuels that are used to generate that electricity (Chapter 6, this report). This chapter provides a description of how energy, including electrical energy, is used within the buildings sector. Following the discussion of such end uses of energy, this chapter then describes the opportunities and potential for reducing energy consumption within the sector.

Many options are available for reducing the carbon impacts of new and existing buildings, including increasing equipment efficiency and implementing alternative design, construction, and operational measures to provide thermal comfort and lighting with reduced energy. Current best practices can reduce carbon emissions for buildings by at least 60% for offices[4] and up to 70% for homes[5]. Residential and commercial buildings in the United States and

> North American buildings accounted for 10% of global energy emissions, 2003.

> Current best practices can reduce carbon emissions for buildings by at least 60% for offices and up to 70% for homes.

Canada occupy 27 billion m² (2.7 million hectares)[†] of floor space, providing a large area available for siting non-carbon-emitting on-site energy supplies (*e.g.*, photovoltaic panels on roofs)[6]. With the most cutting-edge technology, at the least, emissions can be dramatically reduced, and at best, buildings can produce electricity without carbon emissions by means of on-site renewable electricity generation.

9.2 CARBON FLUXES

Carbon fluxes from energy emissions in buildings are well understood, since primary energy inputs from the source of production are tracked, their emissions rates are known, and the total end user consumption data are gathered and reported by energy utilities, typically monthly. The quantity of energy consumed by each particular end use is slightly less well known because attribution requires detailed data on use patterns in a wide variety of contexts. The governments of North America have invested in detailed energy consumption surveys, which allow researchers to identify opportunities for reducing energy use.

The largest contribution to carbon emissions from buildings is through the operation of energy-using equipment. The energy consumed in the average home accounts for 2.9 metric tons[7] of carbon per year in the United States, 1.7 metric tons[8] per year in Canada, and 0.6 metric tons[9] in Mexico (DOE/EIA, 2005; Natural Resources Canada, 2005b; SENER México, 2004)[†]. Energy consumption in a 500 m² commercial, government, or public-use building in the United States produces 1.9 metric tons of carbon (DOE/EIA, 2005)[10†]. Energy consumption includes electricity as

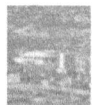

---

[1] Carbon dioxide emissions only.
[2†] A dagger symbol indicates that the magnitude and/or range of uncertainty for the given numerical value(s) is not provided in the references cited.
[3] See Tables 1.1.6 and 1.1.7 in DOE/EERE (2005).
[4] Leadership in Energy and Environment Design (LEED) Gold Certification (USGBC, 2005).
[5] U.S. DOE Building America Program (DOE/EERE, 2006).

[6] A recent study estimates a potential of 711 GW generation capacity from rooftop installation of photovoltaic systems (Chaudhari *et al.*, 2004).
[7] United States' residential sector emissions of 334 Mt C divided by 114 million households in 2004; the numerical value given for "tons of carbon" is for carbon dioxide emissions only.
[8] Canada residential sector emissions of 20.6 Mt C divided by 12.2 million households in 2003.
[9] Mexico residential sector emissions of 13.2 Mt C divided by 23.8 million households in 2004.
[10] United States' commercial sector emissions per m² in 2003 times 500 m².

well as the direct combustion of fossil fuels (natural gas, bottled gas, and petroleum distillates) and the burning of wood. Because most electricity in North America is produced from fossil fuels, each kilowatt-hour consumed in a building contributed about 180 g of carbon to the atmosphere in 2003 (DOE/EIA, 2005)[11]. The equivalent amount of energy from natural gas or other fuels contributed about 52 g of carbon (DOE/EIA, 2005)[12]. Renewable energy accounted for 9% of electricity production in 2003, down from 12% in 1990. Renewable site energy use in buildings also decreased in that time, from 4% to 2%, mostly due to decreasing use of wood as a household fuel (DOE/EERE, 2005)[13].

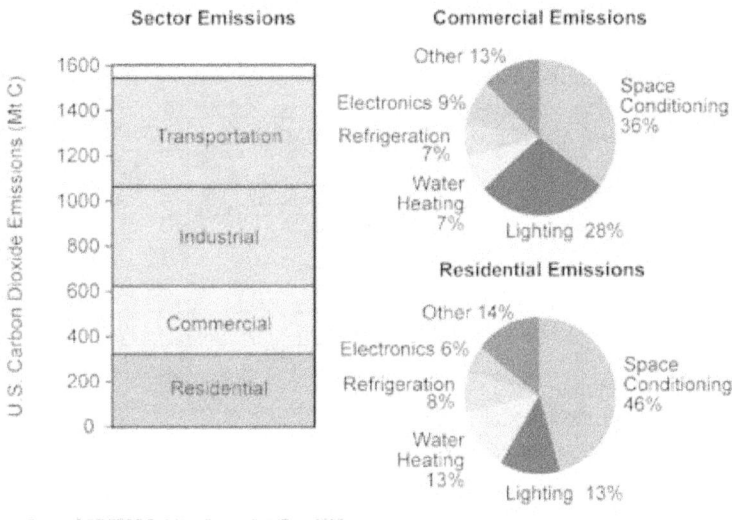

Figure 9.1 United States' carbon emissions by sector and (for commercial and residential buildings) by end use.

Buildings-sector CO_2 emissions and the relative contribution of each end use are shown in Figure 9.1. In the United States, five end uses account for 87% of primary energy consumption in buildings: space conditioning (including space heating, cooling, and ventilation), 40.9%; lighting, 19.8%; water heating, 10.5%; refrigeration, 7.9%; and electronics (including televisions, computers, and office equipment), 7.7% (DOE/EERE, 2005)[14†]. Space heating and cooling are the largest single uses for residences, commercial, and public-sector buildings, accounting for 46% and 35% of primary energy, respectively, in the United States (DOE/EERE, 2005)[15]. Water heating is the second-highest energy consumer in the United States and Canada in terms of site energy, while lighting is the second-highest source of CO_2 emissions, due to the higher emissions per unit of electricity compared to natural gas.

Heating and cooling loads are highly climate dependent; colder regions use heating during much of the year (primarily with natural gas), while warm regions seldom use heating. The majority of United States households own an air conditioner; and although air-conditioner ownership has been historically low in Mexico[16], sales of this equipment are now growing significantly, 14% per year over the past 10 years[17]. Space-conditioning energy end use depends

significantly on building construction (*e.g.*, insulation, air infiltration) and operation (thermostat settings). Water heating is a major consumer of energy in the United States and Canada, where storage-tank systems are common.

Aside from heating and cooling, lighting, and water heating, energy is consumed by a variety of appliances, mostly electrical. Most homes in the United States and Canada own all of the major appliances, including refrigerators, freezers, clothes washers, clothes dryers, dishwashers, and at least one color television. The remainder of household energy consumption comes from small appliances (blenders and microwaves, for example) and increasingly from electronic devices such as entertainment equipment and personal computers. In Mexico, 96.6% of households used electricity in 2005, and recent years have shown a marked growth in appliance ownership: ownership rates in 2000 were 85.9% for televisions, 68.5% for refrigerators, 52% for washing machines, and only 9.3% for computers. By the end of 2005 ownership rates had grown to 91% for televisions, 79% for refrigerators, 62.7% for washing machines, and 19.6% for computers (INEGI, 2005).

Many end uses—such as water heating and space heating, cooling, and ventilation—occur in most commercial sector buildings. Factors such as climate and building construction influence the carbon emissions by these buildings. In addition, commercial buildings contain specialized equipment, such as large-scale refrigeration units in supermarkets, cooking equipment in food preparation businesses, and computers, printers, and copiers in office buildings. Office equipment is the largest component of electricity use

---

[11] United States' emissions from electricity divided by delivered energy.

[12] United States'emissions from natural gas and other fuels divided by delivered energy.

[13] See Table 1.1.2 and Summary Table 2 in DOE/EERE (2005).

[14] Does not include the adjustment EIA uses to relieve differences between data sources.

[15] Table 1.2.3 and Table 1.3.3 in DOE/EERE (2005); available at http://buildingsdatabook.eere.energy.gov (2003 data).

[16] Air conditioners have typically been used only in the northern and coastal areas of Mexico.

[17] Air conditioner sales 1995–2004 from Asociacion Nacional de

Fabricantes de Aparatos Domesticos, A.C. (ANFAD).

aside from cooling and lighting. Due to heat from internal loads, many commercial buildings use air-conditioning year round in most climates in North America.

Residential and commercial buildings in the United States are responsible for 37% of CO_2 emissions from energy nationally and 34% of emissions from energy in North America as a whole. Total emissions from buildings in the United States are ten times as high as in the other two countries combined, due to a large population compared to Canada, and high *per capita* consumption compared to Mexico. On a *per capita* basis, building energy consumption in the United States (65 Gigajoules [GJ] per person per year) is comparable with that of Canada (75 GJ per person per year).[†] This is about seven to eight times higher than in Mexico, where 9 GJ is consumed per person per year[18][†].

In general, contributions from the residential sector are roughly equal to that of the commercial sector, except in Mexico, where the commercial sector contributes less. Electricity contributes more emissions than all other fuels combined in the United States and Mexico (2.6 and 1.8 times as much, respectively). In Canada, natural gas is on par with electricity (0.85 times as many emissions) due to high heating loads resulting from the cold climate. Fuel oil represents most of Canada's "other fuels" for the commercial sector. Firewood (*leña*) remains an important fuel for many Mexican households for heating, water heating, and cooking. Table 9.1 summarizes CO_2 emissions by country, sector, and fuel type.

The energy consumed during building operation is the most important input to the carbon cycle from buildings; but it is not the only one. The construction, renovation, and demolition of buildings also generate a significant flux of wood and other materials. Construction of a typical 204 m^2 (2200 ft^2) house requires about 20 metric tons of wood and creates 2 to 7 metric tons of construction waste (DOE/

> Emissions from energy use in buildings in the United States and Canada increased 30% from 1990 to 2003.

Table 9.1 Carbon dioxide emissions from energy consumed in buildings.

	2003 Carbon Dioxide Emissions (Mt C)			
	Electricity	Natural Gas	Other Fuels	All Fuels
United States	**445.8**	**122.1**	**46.5**	**614.5**
Residential	229.2	75.6	29.3	334.1
Commercial	216.6	46.5	17.2	280.4
Canada	**17.7**	**15.8**	**6.1**	**39.5**
Residential	9.4	8.7	2.5	20.6
Commercial	8.2	7.1	3.5	18.9
Mexico	**10.7**	**0.5**	**5.6**	**16.9**
Residential	7.3	0.4	5.5	13.2
Commercial [a]	3.5	0.1	0.1	3.7

[a] Mexican commercial building emissions include electricity statistics provided by the National Energy Balance (SENER, 2004). Recent investigations suggest that these may be significantly underestimated, since the methodology used categorizes most large commercial and public sector buildings in the category "medium industry" (Odón de Buen Rodríguez, President, Energía Technología y Educación SC, Puente de Xoco, Mexico, personal communication to James McMahon, Lawrence Berkeley National Laboratory, Berkeley, California, November 23, 2006).

EERE, 2005)[19][†]. Building lifetimes are many decades and, especially for commercial buildings, may include several cycles of remodeling and renovation. In the United States as a whole, water supplied to residential and commercial customers accounts for about 6% of total national fresh water consumption. This water consumption also impacts the carbon cycle because water supply, treatment, and waste disposal require energy.

9.3 TRENDS AND DRIVERS

Several factors influence trends in carbon emissions in the buildings sector. Some driver variables tend to increase emissions, while others decrease emissions. Emissions from energy use in buildings in the United States and Canada increased 30% from 1990 to 2003 (DOE/EERE, 2005; Natural Resources Canada, 2005a)[20], corresponding to an annual growth rate of 2.1%.

Carbon emissions from buildings have grown with energy consumption, which in turn is increasing with population and income. Demographic shifts therefore have a direct influence on residential energy consumption. Rising incomes have led to larger residential buildings and the amount of living area *per capita* is increasing in all three countries in North America. On one hand, total population growth is slowing, especially in Mexico, as families are having fewer children than in the past. Annual population growth during the 1990s was 1.1% in the United States, 1.0% in Canada,

---

[18] Total building energy in 1999 (Source: IEA) divided by population (Source: UN Department of Economic and Social Affairs) United States, 18296 million GJ divided 282 million; Canada 2280 million GJ divided by 30.5 million; Mexico 855 million GJ divided by 97.4 million.

[19] Construction data from Table 2.1.7 in DOE/EERE (2005); wood content estimated from lumber content. Construction waste from Table 3.4.1 in DOE/EERE (2005).
[20] Data from Table 3.1.1 in DOE/EERE (2005).

Table 9.2 Principal drivers of buildings emissions trends.

Driver	United States		Canada		Mexico	
	Total 2000	Growth Rate 1990-2000	Total 2000	Growth Rate 1990-2000	Total 2000	Growth Rate 1990-2000
Population (millions)	288	1.1%	31.0	1.0%	100	1.7%
Household Size (persons per household)	2.5	-0.6%	2.6	-0.9%	5.3	-0.1%
Per capita GDP (thousand $US 1995)	31.7	2.0%	23.0	1.8%	3.8	1.8%
Residential Floor space (billion m²)	15.7	2.4%	1.5	2.4%	0.85	N/A
Commercial Floor space (million m²)	6.4	0.6%	0.5	1.6%	N/A	N/A
Building Energy Emissions per GDP (g C/$US)	70	-0.5%	59	-0.9%	N/A	N/A

Source: Population - United Nations Department of Economic and Social Affairs (UNDESA); Household Size - United Nations Development Programme (UNDP); gross domestic product (GDP) - World Bank

Source: Floor space - EIA-EERE (2005), U.S. residential floor space estimated from 2001 Residential Energy Consumption Survey (DOE-EIA), Natural Resources Canada (2005a). Mexican residential floor space estimated from Table 1.8 in CONAFOVI (2001)

Source: Emissions - EIA-EERE (2005), Natural Resources Canada (2005b)

and 1.7% in Mexico. In the period from 1970 to 1990, it was 1.0%, 1.2%, and 2.5%, respectively[21] [†]. By 2005, annual population growth in Mexico declined to 1% (INEGI, 2005). On the other hand, a shift from large, extended-family households to nuclear-family and single-occupant households means an increase in the number of households per unit population[22], each with its own heating and cooling systems and appliances.

The consumption of energy on a *per capita* basis or per unit economic activity (gross domestic product [GDP]) is also not constant but depends on several underlying factors. Economic development is a primary driver of overall *per capita* energy consumption and influences the mix of fuels used[23]. *Per capita* energy consumption generally grows with economic development, since wealthier people live in larger dwellings and use more energy[24]. Recently, computers, printers, and other office equipment have become com-

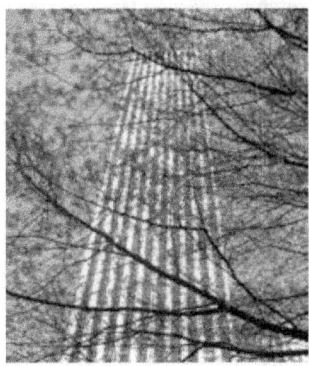

monplace in nearly all businesses and in most homes. These end uses now constitute 7% of primary household energy consumption. Because of these growing electricity uses, the ratio of electricity to total household primary energy has increased. This is significant to emissions because of the large

emissions associated with the combustion of fossil fuels in power plants. Electricity can be generated from renewable sources such as solar or wind, but their full potential has yet to be realized.

In the United States, the major drivers of energy consumption growth are growth in commercial floor space and an increase in the size of the average home. The size of an average United States single-family home has grown from 160 m² (1720 ft²) for a house built in 1980 to 216 m² (2330 ft²) in 2003[†]. In the same time, commercial floor space *per capita* has increased from 20 to 22.6 m² (215 to 240 ft²) (DOE/EERE, 2005)[25] [†]. Certain end uses once considered luxuries have now become commonplace. Only 56% of United States' homes in 1978 used mechanical space-cooling equipment (DOE/EIA, 2005). By 2001, ownership grew to 83% driven by near total saturation in warmer climates and a demographic shift in new construction to these regions. Table 9.2 shows emissions trends as well as the underlying drivers.

> In the United States, the major drivers of energy consumption growth are growth in commercial floor space and an increase in the size of the average home.

Although the general trend has been toward growth in *per capita* emissions, emissions per unit of GDP have decreased in past decades due to improvements in efficiency. Efficiency performance of most types of equipment has generally increased, as has the thermal insulation of buildings, due to influences such as technology improvements and voluntary and mandatory efficiency standards and building codes. The energy crisis of the 1970s was followed by

---

[21] *Source:* U.N. Department of Economic and Social Affairs.
[22] See household size statistics in Table 9.2.
[23] For example, whether biomass, natural gas, or electricity is used for space heating and cooking.
[24] See Table 4.2.6 in DOE/EERE (2005).

[25] See Tables 2.1.6 and 2.2.1 in DOE/EERE (2005). Residential data are from 1981.

99

BOX 9.1: Electricity Consumption in the United States and in California

Since the mid-1970s, the state of California has pursued an aggressive set of efficiency regulations and utility programs. As a result, *per capita* electricity consumption has stabilized in that state, while it continues to grow in the United States as a whole.

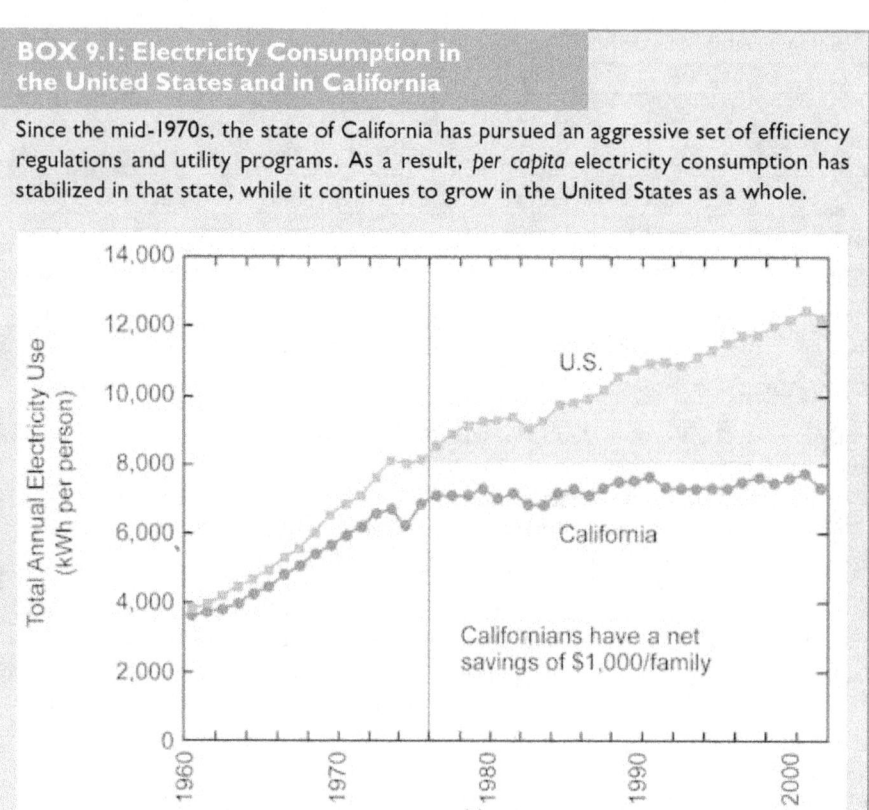

Source: California Energy Commission— Available at http://www.energy.ca.gov/2005publications/CEC-999-2005-007/CEC-999-2005-007.PDF, Slide 5

metering, incentives and financing, establishment of voluntary guidelines, procurement programs, energy audits and retrofits, and mandatory regulation. The most effective approaches will likely include more than one of these options in a policy portfolio that takes advantage of synergies, avoids unduly burdening certain sectors, and is cost effective. Major participants include not only federal agencies, but also state and local governments, energy and water utilities, private research and development firms, equipment manufacturers and importers, energy services companies (ESCOs), nonprofit organizations, and building owners and occupants. An ESCO is a company that offers to reduce a client's utility costs, often with the cost savings being split with the client through an energy performance contract or a shared savings agreement.

a sharp decline in economic energy intensity. Increases in efficiency were driven both by market-related technology improvements and incentives and by the establishment of federal and state/provincial government policies designed to encourage or require energy efficiency.

9.4 OPTIONS FOR MANAGEMENT

A variety of alternatives exists for reducing emissions from the buildings sector. Technology- and market-driven improvements in efficiency are expected to continue for most equipment, but this will probably not be sufficient to curtail emissions growth adequately without government intervention. The government has many different ways in which it can manage emissions that have been proven effective in influencing the flow of products from manufacturers to users (Interlaboratory Working Group, 2000). That flow may involve six steps: advancing technologies; product development and manufacturing; supply, distribution, and wholesale purchasing; retail purchasing; system design and installation; and operation and maintenance (Wiel and Mc-Mahon, 2005). Options for specific products or packages include government investment in research and development, information and education programs, energy pricing and

- **Technology adoption supported by research and development:** Government has the opportunity to encourage development and adoption of energy-efficient technologies through investment in research and development, which can advance technologies and bring down prices, therefore enabling a larger market. Successful programs have contributed to the development of high-efficiency lighting, heating, cooling, and refrigeration. Research and development has also had an impact on the improvement of insulation, ducting, and windows. Finally, government support of research and development has been critical in the reduction of costs associated with development of renewable energy.

- **Voluntary Programs:** By now, there are a wide range of efficiency technologies and best practices available and if the most cost-effective among them were widely utilized, carbon emissions would be reduced. Voluntary measures can be effective in overcoming some market barriers. Government has been active with programs to educate consumers with endorsement labels or ratings (such as the U.S. Environmental Protection Agency's [EPA's] and U.S. Department of Energy's [DOE's] Energy Star Appliances and Homes) and public-private partnerships (such as DOE's "Building America Program"). Government is not the only player, however. Energy utilities can offer rebates for efficient appliances and ESCOs can facilitate best practices at the firm level. Finally, nongovernmental organizations and professional societies (such as the U.S. Green Building Council and the American Institute of Architects) can play a role in establishing benchmarks and ratings.

- **Regulations:** Governments can dramatically impact energy consumption through well-considered regulations that address market failures with cost-effective measures. Regulations facilitate best practices in two ways: they eliminate the lowest-performing equipment from the market, and they boost the market share of high-efficiency technologies. Widely used examples are mandatory energy efficiency standards for appliances, equipment, and lighting, mandatory labeling programs, and building codes. Most equipment standards are instituted at a national level, whereas most states have their own set of prescriptive building codes (and sometimes energy performance standards for equipment) to guarantee a minimum standard for energy-saving design in homes and businesses.

Although large strides in efficiency improvement have been made over the past three decades, significant improvements are still possible. They will involve continued improvement in equipment technology and will increasingly take a whole-building approach that integrates the design of the building and the energy consumption of the equipment inside it. The improvements may also involve alternative ways to provide energy services, such as cogeneration of heat and electricity and thermal energy storage units (Public Technology Inc. and U.S. Green Building Council, 1996).

Whole-building certification standards evaluate a package of efficiency and design options. An example is the Leadership in Energy and Environmental Design (LEED) certification system developed by the U.S. Green Building Council, a non-profit organization. In existence for five years, the LEED program has certified 36 million m² (390 million ft²) of com-

mercial and public-sector buildings and has recently implemented a certification system for homes. The LEED program includes a graduated rating system (Certified, Silver, Gold, or Platinum) for environmentally friendly design, of which energy efficiency is a key component (USGBC, 2005).

On the government side, the EPA's Energy Star Homes program awards certification to new homes that are independently verified to be at least 30% more energy-efficient than homes built to the 1993 national Model Energy Code, or 15% more efficient than state energy code, whichever is more rigorous. Likewise, the DOE's Building America program partners with homebuilders, providing research and development toward goals to decrease primary energy consumption by 30% for participating projects by 2007, and by 50% by 2015.

BOX 9.2: Impact of Efficiency Improvements

Between 1974 and 2001, the energy consumption of the average refrigerator sold in the United States dropped by 74%, a change driven by market forces and regulations. From 1987 to 2005, the U.S. Congress and DOE promulgated labels or minimum efficiency standards for over 40 residential and commercial product types. Canada and Mexico also have many product labels and efficiency standards, and a program is under way to harmonize standards throughout North America in connection with the North American Free Trade Agreement (NAFTA).

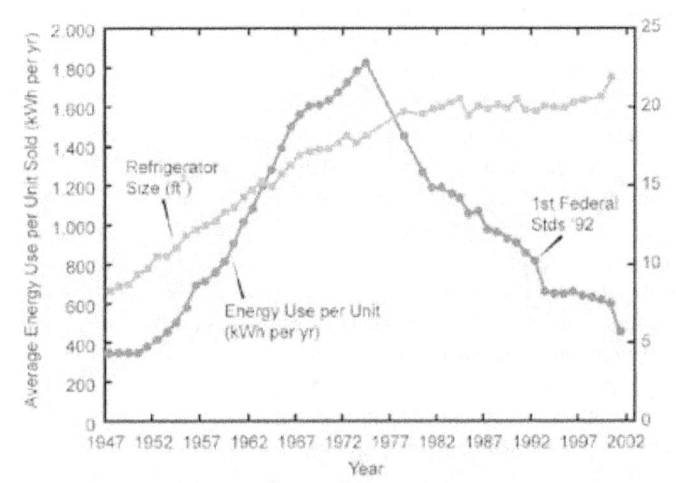

Source: California Energy Commission—Available at http://www.energy.ca.gov/2005publications/CEC-999-2005-007/CEC-999-2005-007.PDF, Slide 7

9.5 RESEARCH AND DEVELOPMENT NEEDS

Research, development, demonstration, and deployment of technologies and programs to improve energy efficiency in buildings and to produce energy with fewer carbon emissions have involved significant effort over the last 30 years. These efforts have contributed options toward carbon management. Technologies and markets continue to evolve, representing new crops of "low-hanging fruit" available for harvesting. However, in most buildings-related decisions in North America, reducing carbon emissions remains a secondary objective to other goals, such as reducing first costs (DeCanio, 1993 and 1994). The questions for which answers could significantly change the discussion about options for carbon management include the following:

- What is the total societal cost of environmental externalities[26], including carbon emissions? Energy resources in North America have been abundant and affordable, but external costs have not been completely accounted for. Most economic decisions are weighted toward the short term and do not consider the complete costs. Total societal costs of carbon emissions are unknown and because it is a global issue, difficult to allocate. Practical difficulties notwithstanding, this is a key issue, answers to which could influence priorities for research and development as well as policies such as energy pricing, carbon taxes, or credits.
- What cost-effective reduced-carbon-emitting equipment and building systems—including energy demand (efficient equipment) and supply (renewable energy)— are available in the short, medium, and long term? Policymakers must have sufficient information to be confident that particular new technology types or programs will be effective and affordable. For consumers to consider a set of options seriously, the technologies must be manifested as products that are widely available and competitive in the marketplace. Therefore, economic and market analyses are necessary before attractive options for managing carbon can be proposed.
- How do the costs of mitigation compare to the costs of continued emissions? The answers to the previous two questions can be compared in order to develop a supply curve of conserved carbon comprising a series of least-cost options, whether changes to energy demand or to supply, for managing carbon emissions. The supply curve of conserved carbon will need to be updated at regular intervals to account for changes in technologies, production practices, and market acceptance of competing solutions.

[26] External costs are the costs borne by society beyond those included in the market prices of goods. For example, carbon emissions may cause environmental damage not reflected in the market transactions associated with the buying and selling of energy (Rabl and Spadaro, 2007).

PART III OVERVIEW

The Carbon Cycle in Land and Water Systems

Lead Author: R.A. Houghton, Woods Hole Research Center

The six chapters (Chapters 10-15) in Part III consider the current and future carbon balance of terrestrial and aquatic ecosystems in North America. Although the amount of carbon exchanged between these ecosystems and the atmosphere each year through photosynthesis and plant and microbial respiration is large, the net balance for all of the ecosystems combined is currently a net sink of 370-505 million tons of carbon (Mt C) per year[1]. This net sink offsets only about 20-30% of current fossil-fuel emissions from the region (1856 Mt C per year in 2003) (Chapter 3 this report). The cause of this terrestrial carbon sink is uncertain. Although management has the potential for removing carbon from the atmosphere and storing it in vegetation and soil, most of the current sink is not the result of current management practices. Instead, most of it may be attributed to a combination of past management and the response of terrestrial ecosystems to environmental changes.

The large sink in the forests of Canada and the United States, for example, is, in some measure, the consequence of continued forest growth following agricultural abandonment that occurred in the past. This is partly the result of past and current management practices (*e.g.,* fire suppression), and partly the result of forest responses to a changing environment

(climatic change, carbon dioxide [CO_2] fertilization, and the increased mobilization of nutrients). The relative importance of these broad factors in accounting for the current sink is unknown. Estimates vary from attributing nearly 100% of the sink in United States forests to regrowth (Caspersen *et al.*,

2000; Hurtt *et al.*, 2002) to attributing nearly all of it to CO_2 fertilization (Schimel *et al.*, 2000). The attribution question is critical because the current sink may be expected to increase in the future if the important mechanism is CO_2 fertilization, for example, but may be expected to decline if the important mechanism is forest regrowth (forests accumulate carbon more slowly as they age). Understanding the history of land use, management, and disturbance is critical because disturbance and recovery are major determinants of the net terrestrial carbon flux.

Land-use change and management have been, and will be, important in the carbon balance of other ecosystems besides forests. The expansion of cultivated lands in

> Understanding the history of land use, management, and disturbance is critical because disturbance and recovery are major determinants of the net terrestrial carbon flux.

Canada and the United States in the 1800s released large amounts of carbon to the atmosphere (Houghton *et al.*, 1999), leaving those lands with the potential for recovery (*i.e.*, a future carbon sink), if managed properly. For example, recent changes in farming practice may have begun to recover the carbon that was lost decades ago. Recovery of carbon in soil, however, generally takes longer than its loss through cultivation. Grazing lands, although not directly affected by cultivation, have, nevertheless, been managed in the United States through fire suppression. The combined effects of grazing and fire suppression are believed to have promoted the invasion of woody vegetation, possibly a carbon sink at present. Wetlands are also a net carbon sink, but the magnitude of the sink was larger in the past than it is today, again, as a result of land-use change (draining of wetlands for agriculture and forestry). The only lands that seem to have escaped management are those lands overlying permafrost (perennially frozen ground), and they are clearly subject to change in the future as a result of global warming. Settled lands, by definition, are managed, and are dominated by fossil-fuel emissions. Nevertheless, the accumulation of carbon in urban and

---

[1] The lower estimate is from this overview, the larger estimate from Chapter 3, with most of the difference attributable to uncertainty in the sink from woody encroachment. See Table III.1, footnote h, for discussion of this range.

suburban trees suggests a net sequestration of carbon in the biotic component of long-standing settled lands. Residential lands recently cleared from forests, on the other hand, are sources of carbon (Wienert and Hamburg, 2006).

From the perspective of carbon and climate, ecosystems are important if (1) they are currently large sources or sinks of carbon or (2) they have the potential to become large sources or sinks of carbon in the future through either management or environmental change, where "large" sources or sinks, in this context, are determined by the product of area (hectares) times flux per unit area (or flux density) (megagrams of carbon [Mg C] per hectare per year).

The largest carbon sink in North America (270 Mt C per year) is associated with forests (Chapter 11 this report) (Table III-1). The sink includes the carbon accumulating in wood products (*e.g.*, in increasing numbers of houses and landfills) as well as in the forests themselves. A sink is believed to exist in wetlands (Chapter 13 this report), including the wetlands overlying permafrost (Chapter 12 this report), although the magnitude of this sink is uncertain. More certain is the fact that the current sink is considerably smaller than it was before wetlands were drained for agriculture and forestry. The other important aspect of wetlands is that they hold more than half of the carbon in North America. Thus, despite the current net sink in these systems, their potential for future emissions is large.

Although management has the potential to increase the carbon sequestered in agricultural (cultivated) lands, these lands today are nearly in balance with respect to carbon (Chapter 10 this report). The carbon lost to the atmosphere from cultivation of organic soils (soils dominated by organic matter) is approximately balanced by the carbon accumulated in mineral soils (soils consisting of more inorganic material, such as sand or clay). In the past, before cultivation, these soils held considerably more carbon than they do today, but 25-30% of that carbon was lost soon after the lands were initially cultivated. In large areas of grazing lands, there is the possibility that the invasion and spread of woody vegetation (woody encroachment) is responsible for a significant net carbon sink at present (Chapter 10 this report). The magnitude (and even sign) of this flux is uncertain, however, in part because some ecosystems lose carbon below-ground (soils) as they accumulate it aboveground (woody vegetation), and in part because the invasion and spread of exotic grasses into semi-arid lands of the western United States are increasing the frequency of fires, reversing woody encroachment, and releasing carbon (Bradley *et al.*, 2006).

Wetlands hold more than half of the carbon in North America. Thus, despite the current net sink in these systems, their potential for future emissions is large.

The emissions of carbon from settled lands are largely considered in the chapters in Part II and in Chapter 14 of this report. Non-fossil carbon seems to be accumulating in trees in these lands, but the net changes in soil carbon are uncertain.

The only ecosystems that appear to release carbon to the atmosphere at present are the coastal waters. The estimated flux of carbon is close to zero (and difficult to determine) because the gross fluxes (from river transport, photosynthesis, and respiration) are large and variable in both space and time.

The average net fluxes of carbon expressed as Mg C per hectare per year in Table III-1 are for comparative purposes. They show the relative flux density for different types of ecosystems. These annual fluxes of carbon are rarely determined with direct measurements of flux, however, because of the extreme variability of fluxes in time and space, even within a single ecosystem type. Extrapolating from a few isolated measurements to an estimate for the whole region's flux is difficult. Rather, the net changes are more often based on differences in measured stocks over intervals of 10 years, or longer (Chapter 3 this report), or are based on the large and rapid changes per hectare that are reasonably well documented for certain forms of management, such as the changes in carbon stocks that result from the conversion of forest to cultivated land. Thus, most of the flux estimates in Table III-1 are long-term and large-area estimates.

Nevertheless, average flux density is one factor important in determining an ecosystem's role as a net source or sink for carbon. The other important factor is area. Permafrost wetlands, for example, are currently a small net sink for carbon. They cover a large area, however, hold large stocks of carbon, and, thus, have the potential to become a significant net source of carbon if the permafrost thaws with global warming (Smith *et al.*, 2001; Smith *et al.*, 2005a; Osterkamp and Romanovsky, 1999; Osterkamp *et al.*, 2000). Forests clearly dominate the net uptake and storage of carbon in North America, although wetlands and settled lands have mean flux densities that are above average.

The two factors (flux density and area) demonstrate the level of management required to remove a significant amount of carbon from the atmosphere and keep it on land. Under current conditions, sequestration of 100 Mt C per year, for example (about 7% of fossil-fuel emissions from North America), requires nearly half the forest area (Table III-1). As discussed above, the cause of this sequestration is uncertain, but enhancing it through management over a few hundred million hectares would require considerable effort. Nevertheless, the cost (in $/metric ton CO_2) may be low relative to other options for managing carbon. For example, forestry activities are estimated to have the potential to sequester 100-200 Mt C per year in the United States at prices ranging from less than $10/ton of CO_2 for improved forest management, to $15/ton for afforestation, to $30-50/ton for production of biofuels (Chapter 11 this report). Somewhat smaller sinks of 10-70 Mt C per year might be stored in agricultural soils at low to moderate costs ($3-30/ton CO_2) (Chapter 10 this report). The maximum amounts of carbon that might be accumulated in forests and agricultural soils are not known, thus, the number of years these rates of sequestration might be expected to continue is also unknown. It seems unlikely that the amount of carbon currently held in forests and agricultural lands could double. Changes in climate will also affect carbon storage, but the net effect of management and climate is uncertain.

Table III.1 Ecosystems in North America: their areas, net annual fluxes of carbon (negative values are sinks), and carbon stocks (including both vegetation and soils).

Type of ecosystem	Area (10⁶ ha)	Current mean flux density (Mg C per ha per year)	Current flux (Mt C per year)	Carbon stocks (Mt C)	Mean carbon stocks (Mt C per ha)
Agriculture	231	0.0	0±15[a]	18,500	80
Grass, shrub and arid	558	−0.01	−6[b]	59,950	107
Forests	771	−0.35	−269[c]	171,500	222
Permafrost lands					
Peatlands	51	−0.13	−6.7	57,700	1130
Mineral soils[d]	517	−0.03	−14	98,780	191
Non-permafrost wetlands					
Peatlands	86	−0.12	−10	126,400	1470
Mineral soils	105	−0.21	−22.3	38,100	363
Estuarine	4.5	−2.3	−10.2	900	200
Settled lands[e]	104	−0.31	−32	~1,000	10
Coastal waters	384	0.05	19		
Sum	2427[f]	−0.15[g]	−370[h]	572,830[f]	
Total	2126[i]			480,000[j]	225[g]

a. Fossil-fuel inputs to crop management are not included. Some of the carbon sequestration is occurring on grasslands as well as croplands, but the inventories do not separate these fluxes. The near-zero flux is for Canada and the United States only. Including Mexican croplands would likely change the flux to a net source because croplands are expanding in Mexico, and the carbon in biomass and soil is released to the atmosphere as native ecosystems are cultivated.

b. Fossil-fuels are not included. The small net sink results from the Conservation Reserve Program in the United States. Including Mexico is likely to change the net sink to a source because forests are being converted to grazing lands. Neither woody encroachment nor woody elimination is included in this estimate of flux because the uncertainties are so large.

c. Includes an annual sink of 68 Mt C per year in wood products as well as a sink of 201 Mt C per year in forested ecosystems.

d. Includes zones with continuous, discontinuous, sporadic, and isolated permafrost; that is, not all of the lands are strictly over permafrost.

e. Urban trees only (does not include soil carbon). Note that this sink is accounted for as part of the forest sink in Chapter 3 (Table 3.1).

f. Sum does not include coastal waters. The summed area is larger than the total area (note i) because of double counting. For example, an estimated 75 × 10⁶ hectares (ha) of permafrost lands in Canada are forested (and may be included in forest area as well as permafrost area), 26 x 10⁶ ha of wetlands in the United States are forested, and 54 x 10⁶ ha of wetlands are shrublands. In addition, an estimated 75 x 10⁶ ha of other wooded lands are included as both forests and rangelands, and ~70 x 10⁶ ha of grasslands and shrublands are counted also as non-permafrost lands within areas defined as sporadic or isolated permafrost (see note d).

g. Weighted average; does not include coastal waters.

h. Does not include coastal waters. The total annual sink of 370 Mt C is lower than the estimate of 505 Mt C presented in Chapter 3 (Table 3.1). The largest difference results from the flux of carbon attributed to woody encroachment. Chapter 3 includes a sink of 120 Mt C per year; Table III-1, above, presents a net flux of zero (see note b). Other differences between the two estimates include: (1) an additional sink in Table III-1 of 14 Mt C per year in permafrost mineral soils and (2) a sink of 25 Mt C per year in rivers and reservoirs that is included in Table 3.1 but not in Table III-1. In addition, there are small differences in the estimates for agricultural lands and grasslands.

i. Areas (10⁶ ha) (*The Times Atlas of the World*, 1990)

Globe	North America	Canada	United States	Mexico
14,900	2,126	992	936	197

j. Total carbon stocks are reduced by the areas double counted (see note f).

Despite the limited nature of carbon uptake and storage in offsetting the global emissions of carbon from fossil fuels, local and regional activities may, nevertheless, offset local and regional emissions of fossil carbon. This offset, as well as other co-benefits, may be particularly successful in urban and suburban systems (Chapter 14 this report).

The effects and cost of managing aquatic systems are less clear. Increasing the area of wetlands, for example, would presumably increase the sequestration of carbon; but it would also increase emissions of methane (CH_4), countering the effect of carbon storage. Fertilization of coastal waters with iron has been proposed as a method for increasing oceanic uptake of CO_2, but neither the amount of carbon that might be sequestered nor the side effects are known (Chapter 15 this report).

A few studies have estimated the potential magnitudes of future carbon sinks as a result of management (Chapters 10, 11 this report). However, the contribution of management, as opposed to the environment, in today's sink is unclear (Chapter 3 this report), and for the future, the relative roles of management and environmental change are even less clear. The two drivers might work together to enhance terrestrial carbon sinks, as seems to have been the case during recent decades (Prentice *et al.*, 2001) (Chapter 2 this report). On the other hand, they might work in opposing directions. A worst-case scenario, quite possible, is one in which management will become ineffective in the face of large natural sources of carbon not previously experienced in the modern world. In other words, while management is likely to be essential for sequestering carbon, it may not be sufficient to preserve the current terrestrial carbon sink over North America, let alone to offset fossil-fuel emissions.

At least one other observation about storing carbon in terrestrial and aquatic ecosystems should be mentioned. In contrast to the hundreds of millions of hectares that must be managed to sequester 100 Mt C annually, a few million hectares of forest fires can release an equivalent amount of carbon in a single year. This disparity in flux densities underscores the fact that a few million hectares are disturbed each year, while hundreds of millions of hectares are recovering from past disturbances. The natural fluxes of carbon are large in comparison to net fluxes. The observation is relevant for carbon management, because the cumulative effects of managing small net sinks to mitigate fossil-fuel emissions will have to be understood, analyzed, monitored, and evaluated in the context of larger, highly variable, and uncertain sources and sinks in the natural cycle.

The major challenge for future research is quantification of the mechanisms responsible for current (and future) fluxes of carbon. In particular, what are the relative effects of man-agement (including land-use change), environmental change, and natural disturbance in determining sources and sinks of carbon for today and tomorrow? Will the current natural sinks continue, grow in magnitude, or reverse to become net sources? What is the role of soils in the current (and future) carbon balance (Davidson and Janssens, 2006)? What are the most cost-effective means of managing carbon?

Answering these questions will require two scales of measurement: (1) an expanded network of intensive research sites dedicated to understanding basic processes (*e.g.*, the effects of management and environmental effects on carbon stocks), and (2) extensive national-level networks of monitoring sites, through which uncertainties in carbon stocks (inventories) would be reduced and changes, directly measured. Elements of these measurements are underway, but the effort has not yet been adequate for resolving these questions.

KEY UNCERTAINTIES AND GAPS IN UNDERSTANDING THE CARBON CYCLE OF NORTH AMERICA

- As mentioned above, the net flux of carbon resulting from woody encroachment and its inverse, woody elimination, is highly uncertain. Even the sign of the flux is in question.
- Rivers, lakes, dams, and other inland waters are mentioned in Chapter 15 as being a source of carbon, but they are claimed elsewhere to be a sink (Chapter 3 this report). The sign of the net carbon flux attributable to erosion, transport, deposition, accumulation, and decomposition is uncertain (*e.g.*, Stallard, 1998; Lal, 2001; Smith *et al.*, 2005b).
- Several chapters cite studies that have attempted to quantify the potential for management to increase carbon sinks in the future, but no studies have yet attempted to estimate the potential future sources of carbon for North America as they have for the globe (*e.g.*, Friedlingstein *et al.*, 2006; Jones *et al.*, 2005). Global models that include the feedbacks between climatic change and the carbon cycle have all shown decreased carbon sinks over the next century. In North America, warming of wetlands and thawing of permafrost, in particular, are likely to increase emissions of carbon to the atmosphere, CH_4 as well as CO_2; and periods of unusually low rainfall, combined with warming trends, are likely to release carbon from the ecosystems of the Mountain West and the southwestern United States through increasing their vulnerability to wildfires and insect outbreaks (Potter *et al.*, 2003 and 2005).

CHAPTER 10

Agricultural and Grazing Lands

Lead Authors: Richard T. Conant, Colo. State Univ.; Keith Paustian, Colo. State Univ.

Contributing Authors: Felipe García-Oliva, UNAM; H. Henry Janzen, Agriculture and Agri-Food Canada; Victor J. Jaramillo, UNAM; Donald E. Johnson, Colo. State Univ. (deceased); Suren N. Kulshreshtha, Univ. Saskatchewan

KEY FINDINGS

- Agricultural and grazing lands (cropland, pasture, rangeland, shrublands, and arid lands) occupy 789 million hectares (1.95 billion acres), which is 47% of the land area of North America, and contain 78.5 ± 19.5[1] billion tons of organic carbon (17% of North American terrestrial carbon) in the soil alone.

- The emissions and uptake and storage of carbon on agricultural lands are mainly determined by two conditions: management and changes in the environment. The effects of converting forest and grassland to agricultural lands and of agricultural management (e.g., cultivation, conservation tillage) are reasonably well known and have been responsible for historic losses of carbon in Canada and the United States (and for current losses in Mexico); the effects of climate change or of elevated concentrations of atmospheric carbon dioxide are uncertain.

- Conservation-oriented management of agricultural lands (e.g., use of conservation tillage, improved cropping and grazing systems, reduced bare fallow, set-asides of fragile lands, and restoration of degraded soils) can significantly increase soil carbon stocks.

- Agricultural and grazing lands in the United States and Canada are currently near neutral with respect to their soil carbon balance, but agricultural and grazing lands in Mexico are likely losing carbon due to land-use change. Although agricultural soils are estimated to currently uptake about 19-20 million tons of carbon per year, the cultivation of organic soils releases approximately 6-12 million tons of carbon per year. On-farm fossil-fuel use (around 31 million tons of carbon per year), agricultural liming (1.2 million tons of carbon per year), and manufacture of agricultural inputs including fertilizer (approximately 6 million tons of carbon per year) yields a net source from the agricultural sector of about 25-30 million tons of carbon per year.

- As much as 120 million tons of carbon per year may be accumulating through woody encroachment of arid and semi-arid lands of North America; this value is highly uncertain. Woody encroachment is generally accompanied by decreased forage production, and ongoing efforts to reestablish forage species are likely to reverse carbon accumulation by vegetation.

- Projections of future trends in agricultural land area and soil carbon stocks are unavailable or highly uncertain because of uncertainty in future land-use change and agricultural management practice.

- Annualized prices of $15/metric ton carbon dioxide, could yield mitigation amounts of 46 million tons of carbon per year captured in agricultural soils and 14.5 million tons of carbon per year from reductions in fossil-fuel use. At lower prices of $5/metric ton carbon dioxide, the corresponding values would be 34 million tons of carbon per year and 9 million tons of carbon per year, respectively.

- Policies designed to suppress emissions of one greenhouse gas need to consider complex interactions to ensure that *net* emissions of total greenhouse gases are reduced. For example, increased use of fertilizer or irrigation may increase crop residues and carbon uptake and storage, but may stimulate emissions of methane or nitrous oxide.

---

[1] The uncertainty in this value is given as one standard error of the mean.

- Many of the practices that lead to carbon capture and storage or to reduced carbon dioxide and methane emissions from agricultural lands not only increase production efficiencies, but lead to environmental co-benefits, for example, improved soil fertility, reduced erosion, and pesticide immobilization.
- An expanded network of intensive research sites would allow us to better understand the effects of management on carbon cycling and storage in agricultural systems. An extensive national-level network of soil monitoring sites in which changes in carbon stocks are directly measured would allow us to reduce the uncertainty in the inventory of agricultural and grazing land carbon. Better information about the spatial extent of woody encroachment, the amount and growth of woody vegetation, and variation in impacts on soil carbon stocks would help reduce the large uncertainty of the carbon impacts of woody encroachment.

10.1 INVENTORY

10.1.1 Background

Agricultural and grazing lands (cropland, pasture, rangeland, shrublands, and arid lands)[2] occupy 47% of the land area in North America (59% in the United States, 70% in Mexico, and 11% in Canada), and contain 17% of the terrestrial carbon. Most of the carbon in these ecosystems is held in soils. Live vegetation in cropland generally contains less than 5% of total carbon, whereas vegetation in grazing lands contains a greater proportion (5–30%), but still less than that in forested systems (30–65%). Agricultural and grazing lands in North America contain 78.5 ± 19.5 (± 1 standard error) billion tons of organic carbon (Gt C) in the soil (Table 10.1). Significant increases in vegetation carbon stocks in some grazing lands have been observed and, together with soil carbon stocks from croplands and grazing lands, likely contribute significantly to the large North American terrestrial carbon sink (Houghton *et al.*, 1999; Pacala *et al.*, 2001; Eve *et al.*, 2002; Ogle *et al.*, 2003). These lands also emit greenhouse gases: fossil-fuel use for on-farm machinery and buildings, for manufacture of agricultural inputs, and for transportation account for 3–5% of total carbon dioxide (CO_2) emissions in developed countries (Enquete Commission, 1995); activities on agricultural and grazing

> Agricultural and grazing lands are actively managed and have the capacity to take up and store carbon. Thus improving management could lead to substantial reductions in CO_2 and CH_4 emissions.

BOX 10.1: Nitrous Oxide Emissions From Agricultural and Grazing Lands

Nitrous oxide (N_2O) is the most potent greenhouse gas in terms of global warming potential, with a radiative forcing 296 times that of CO_2 (IPCC, 2001). Agricultural activities that add mineral or organic nitrogen (fertilization, plant N_2 fixation, manure additions, *etc.*) augment naturally occurring N_2O emissions from nitrification and denitrification by 0.0125 kg N_2O per kg nitrogen applied (Mosier *et al.*, 1998a). Agriculture contributes significantly to total global N_2O fluxes through soil emissions (35% of total global emissions), animal waste handling (12%), nitrate leaching (7%), synthetic fertilizer application (5%), grazing animals (4%), and crop residue management (2%). Agriculture is the largest source of N_2O in the United States (78% of total N_2O emissions), Canada (59%), and Mexico (76%).

lands, like livestock production, animal waste management, biomass burning, and rice cultivation emit 35% of global anthropogenic methane (CH_4) (27% of United States', 31% of Mexican, and 27% of Canadian CH_4 emissions) (Mosier *et al.*, 1998b; CISCC, 2001; Ministry of the Environment, 2006; EPA, 2006); and agricultural and grazing lands are the largest anthropogenic source of nitrous oxide (N_2O) emissions (CAST, 2004; see Box 10.1). However, agricultural and grazing lands are actively managed and have the capacity to take up and store carbon. Thus improving management could lead to substantial reductions in CO_2 and CH_4 emissions and could sequester carbon to offset emissions from other lands or sectors.

10.1.2 Carbon Dioxide Fluxes From Agricultural and Grazing Land

The main processes governing the carbon balance of agricultural and grazing lands are the same as for other ecosystems: the photosynthetic uptake and assimilation of CO_2 into organic compounds, the release of gaseous carbon through respiration (primarily CO_2 but also CH_4), and fire. Like other terrestrial ecosystems in general, for which CO_2 emissions are approximately two orders of magnitude greater than CH_4 emissions, carbon cycling in most agricultural and grazing lands is dominated by fluxes of CO_2 rather than CH_4. In agricultural lands, carbon assimilation is directed towards production of food, fiber, and forage by manipulating species composition and growing conditions (soil fertility, irrigation, *etc.*). Biomass, being predominantly herbaceous (*i.e.*, non-woody), is a small, transient carbon pool (compared to forests) and hence soils constitute the dominant carbon stock. Cropland systems can be among the most productive ecosystems, but in some cases restricted growing season length, fallow periods, and grazing-induced shifts in species

Table 10.1 Soil organic carbon pools in agricultural and grazing lands in Canada, Mexico, and the United States. The data values are given in Gt C. The area (in millions of hectares) for each climatic zone is in parentheses. Current soil carbon stocks are secondary quantities derived from an initial starting point of undisturbed native ecosystems carbon stocks, which were quantified using the intersection of (Moderate Resolution Imaging Spectroradiometer-International Geosphere-Biosphere Programme) MODIS-IGBP[a] land cover types (Friedl et *al.*, 2002) and mean soil carbon contents to 1-m depth from Sombroek et *al.* (1993), spatially arrayed using Food and Agriculture Organization soil classes (ISRIC, 2002), and summed by climate zone. These undisturbed native ecosystem carbon stock values were then multiplied by soil carbon loss factors for tillage- and overgrazing-induced losses (Nabuurs et *al.*, 2004; Ogle et *al.*, 2004) to estimate current soil carbon stocks (see Figure 10.2). Uncertainties (± one standard error) were derived from uncertainty associated with soil carbon stocks and soil carbon loss factors.

Practice	Temperate dry[b,c]	Temperate wet	Tropical dry	Tropical wet	Total
Agricultural lands					
Canada	1.79±0.35 (17.3)	1.77±0.36 (22.1)	–	–	3.60±0.77 (39.4)
Mexico	–	–	0.24±0.06 (3.9)	0.53±0.14 (10.2)	0.81±0.22 (14.1)
United States	3.31±0.74 (34.8)	8.66±2.18 (108.4)	0.35±0.08 (5.6)	1.53±0.33 (28.4)	14.05±3.20 (177.1)
Total	**5.16±1.07 (52.1)**	**10.57±2.42 (130.5)**	**0.61±0.14 (9.5)**	**2.18±0.54 (38.6)**	**18.5±4.16 (230.6)**
Grazing lands					
Canada	2.17±0.55 (18.4)	9.49±1.27 (40.8)	–	–	11.66±4.88 (59.2)
Mexico	–	–	7.20±1.62 (99.1)	2.19±0.58 (20.3)	9.99±2.60 (119.4)
United States	16.89±3.62 (209.9)	5.67±1.39 (55.0)	4.26±0.98 (68.1)	4.30±0.89 (46.7)	32.88±7.18 (379.7)
Total	**19.34±4.27 (228.3)**	**21.07±5.80 (95.8)**	**12.59±2.73 (167.1)**	**6.94±1.86 (67.0)**	**59.95±14.65 (558.2)**

[a] Cropland area was derived from the IGBP cropland land cover class plus the area in the cropland/natural vegetation IGBP class in Mexico and one-half of the area in the cropland/natural vegetation IGBP class in Canada and the United States. Grazing land area includes IGBP woody savannas, savannas, and grasslands in all three countries, plus open shrubland in Mexico and open shrublands (not in Alaska) in the United States.

[b] Temperate zones are those located above 30° latitude. Tropical zones (below 30° latitude) include subtropical regions.

[c] Dry climates were defined as those where the ratio of mean annual precipitation (MAP) to potential evapotranspiration (PET) is less than one; in wet areas, MAP/PET is greater than one.

Inorganic carbon in the soil is comprised of primary carbonate minerals, such as calcite ($CaCO_3$) or dolomite ($CaMg[CO_3]_2$), or secondary minerals formed when carbonate (CO_3^{2-}), derived from soil CO_2, combines with base cations (e.g., Ca^{2+}, Mg^{2+}) and precipitates within the soil profile in arid and semi-arid ecosystems. Weathering of primary carbonate minerals in humid regions can be a source of CO_2, whereas formation of secondary carbonates in drier areas is a sink for CO_2; however, the magnitude of either flux is highly uncertain. Agricultural liming involves addition of primary carbonate minerals to the acid soils to increase the pH. In Canada and the United States, about 0.1 and 1.1 Mt C per year is emitted from liming (Sobool and Kulshreshtha, 2005; EPA, 2006). Inorganic carbon stocks in North America have been estimated at 66.8 Gt C (Sombroek et *al.*, 1993).

composition or production can reduce carbon uptake relative to that in other ecosystems. These factors, along with tillage-induced soil disturbances and removal of plant carbon through harvest, have depleted soil carbon stocks by 20-40% (or more) from pre-cultivated conditions (Davidson and Ackerman, 1993; Houghton and Goodale, 2004). Soil organic carbon stocks in grazing lands (see Box 10.2 for information on inorganic soil carbon stocks) have been depleted to a lesser degree than for cropland (Ogle *et al.*, 2004), and in some regions biomass has increased due to suppression of disturbance and subsequent woody encroachment (see Box 10.3). Woody encroachment is potentially a significant sink for atmospheric CO_2, but the magnitude of the sink is poorly constrained (Houghton *et al.*, 1999; Pacala *et al.*, 2001). Since woody encroachment leads to decreased forage production, management practices are aimed at reversing it, with consequent reductions in biomass carbon. Dis-

> Much of the carbon lost from agricultural soil and biomass pools can be recovered with changes in management practices.

BOX 10.3: Impacts of Woody Encroachment Into Grasslands on Ecosystem Carbon Stocks

Encroachment of woody species into grasslands—caused by overgrazing-induced reduction in grass biomass and subsequent reduction or elimination of grassland fires—is widespread in the United States and Mexico, decreases forage production, and is unlikely to be reversed without costly mechanical intervention (Van Auken, 2000). Encroachment of woody species into grassland tends to increase biomass carbon stocks by one million grams of carbon (1 Mg C) per hectare per year (Pacala et al., 2001), with estimated net sequestration of 120–130 Mt C per year in encroaching woody biomass (Houghton et al., 1999; Pacala et al., 2001). In response to woody encroachment, soil organic carbon stocks can significantly increase or decrease, thus predicting impacts on soil carbon or ecosystem carbon stocks is very difficult (Jackson et al., 2002). Invasion of grass species into native shrublands tends to lead to the release of soil organic carbon (Bradley et al., 2006).

turbance-induced increases in decomposition rates of above-ground litter and harvest removal of some (30–50% of forage in grazing systems, 40–50% in grain crops) or all (*e.g.*, corn for silage) of the above-ground biomass, have drastically altered carbon cycling within agricultural lands and thus the sources and sinks of CO_2 to the atmosphere.

Much of the carbon lost from agricultural soil and biomass pools can be recovered with changes in management practices that increase carbon inputs, stabilize carbon within the system, or reduce carbon losses, while still maintaining outputs of food, fiber, and forage. Increased production, increased residue carbon inputs to the soil, and increased organic matter additions have reversed historic soil carbon losses in long-term experimental plots (*e.g.*, Buyanovsky and Wagner, 1998). However, the management practices that promote soil carbon sequestration would need to be maintained over time to avoid subsequent losses of sequestered carbon. Across Canada and the United States, mineral soils have been sequestering 2.5[†] and 17.0 ± 0.45 million metric tons of carbon (Mt C) per year[3] (Ministry of the Environment, 2006; Ogle *et al.*, 2003; EPA, 2006), respectively, largely through increased production and improved management practices on annual cropland (Figure 10.1, Table 10.2). Conversion of agricultural land to grassland, like under the Conservation Reserve Program in the United States (7.6–11.5 Mt C per year on 31.5 million acres [12.5 million hectares] of land), and afforestation have also sequestered carbon in agricul-

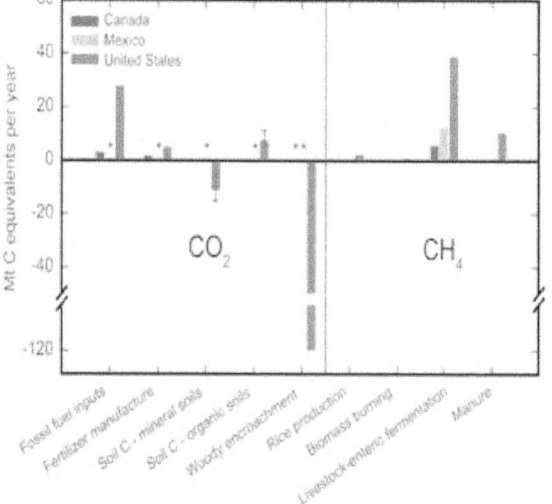

Figure 10.1 North American agricultural and grazing land CO_2 (left side) and CH_4 (right side), adjusted for global warming potential. All units are in Mt C-equivalent per year for years around 2000. Negative values indicate net flux from the atmosphere to soil and biomass carbon pools (*i.e.*, sequestration). All data are from Canadian (Matin *et al.*, 2004) and U.S. (EPA, 2006) National Inventories and from the second Mexican National Communication (CISCC, 2001), except for Canadian (from Kulshreshtha *et al.*, 2000) and U.S. fossil-fuel inputs (from Lal *et al.*, 1998) and woody encroachment (from Houghton *et al.*, 1999). Values are for 2003 for Canada, 1998 for Mexico, and 2004 for the United States. A global warming potential of 23 for methane was used to convert emissions of CH_4 to CO_2 equivalents (IPCC, 2001) and a factor of 12/44 to convert from CO_2 to carbon. Asterisks indicate unavailable data. Data ranges are indicated by error bars where available.

---

[3] † A dagger symbol indicates that the magnitude and/or range of uncertainty for the given numerical value(s) is not provided in the references cited.

Table 10.2 North American agricultural and grazing land carbon fluxes for the years around 2000. All units are in Mt C per year. Negative numbers (in parentheses) indicate net flux from the atmosphere to soil and biomass carbon pools. Unless otherwise noted, data are from Canadian (Matin et al., 2004) and United States' National Inventories (EPA, 2006), and from the Second Mexican National Communication (CISCC, 2001). Values are for 2003 for the United States and Canada, and 1998 for Mexico. A factor of 12/44 was used to convert from CO_2 to carbon and a factor of 12/16 to convert CH_4 to carbon

	Canada	Mexico	United States	Total
CO_2				
On-farm fossil-fuel use	2.9[a]	ND	28[b]	30.9
Fertilizer manufacture	1.7	ND	4.7	6.4
Mineral soil carbon sequestration	(2.5)	ND	(17±0.45)	(19.1) – (20.0)
Organic soil cultivation	0.1	ND	8.3±3.2	5.6 – 11.9
Agricultural liming	0.1	ND	1.1	1.2
Woody encroachment	ND	ND	(120)[c]	(120)
Total	2.3	ND	(114.7) – (120.1)	(117) – (122.4)
CH_4				
Rice production	0	0.011	0.25±0.28	0.26
Biomass burning	<0.01	<0.01	0.03±0.02	0.05
Livestock	0.62	1.48	3.67±0.53	5.93
Manure	0.18	0.05	1.28±0.24	1.60
Total	0.80	1.54	5.23	7.84

ND = no data reported.
[a] From Kulshreshtha et al. (2000).
[b] From Lal et al. (1998).
[c] From Houghton et al. (1999).

largest CO_2 emitters within the agricultural sector (Enquete Commission, 1995).

Much of the ammonia production and urea application (United States: 4.3 Mt C per year; Mexico: 0.4 Mt C per year; Canada: 1.7 Mt C per year) and phosphoric acid manufacture (United States: 0.4 Mt C per year; Mexico: 0.2 Mt C per year; Canada: not reported) are devoted to agricultural uses.

10.1.3 Methane Fluxes From Agricultural and Grazing Lands

Cropland and grazing land soils act as both sources and sinks for atmospheric CH_4. Methane formation is an anaerobic process and is most significant in waterlogged soils, like those under paddy rice cultivation (United States: 0.25 ± 0.28 Mt CH_4-C per year; Mexico: 0.01 Mt CH_4-C

tural and grazing lands (Follett et al., 2001a). In contrast, cultivation of organic soils (e.g., peat-derived soils) is releasing an estimated 0.1 and 8.3 ± 3.2 Mt C per year[†] from soils in Canada and the United States (Matin et al., 2004; Ministry of the Environment, 2006; Ogle et al., 2003; EPA, 2006). Compared with other systems, the high productivity and management-induced disturbances of agricultural systems promote movement and redistribution (through erosion, runoff, and leaching) of organic and inorganic carbon, sequestering potentially large amounts of carbon in sediments and water (Raymond and Cole, 2003; Smith et al., 2005; Yoo et al., 2005). However, the net impact of soil erosion on carbon emissions to the atmosphere remains highly uncertain.

Production, delivery, and use of field equipment, fertilizer, seed, pesticides, irrigation water, and maintenance of animal production facilities contribute 3–5% of total fossil-fuel CO_2 emissions in developed countries (Enquete Commission, 1995). On-farm fossil-fuel emissions together with manufacture of fertilizers and pesticides contribute emissions of 32.7 Mt C per year[†] within the United States (Lal et al., 1998) and 4.6 Mt C per year in Canada (Kulshreshtha et al., 2000) (Table 10.2). Energy consumption for heating and cooling high intensity animal production facilities is among the

per year[†]; Canada: negligible, not reported; Table 10.2). Methane is also formed by incomplete biomass combustion of crop residues (United States: 0.03 ± 0.02 Mt CH_4-C per year; Mexico: <0.01 Mt CH_4-C per year; Canada: negligible, not reported; Table 10.2). Methane oxidation in soils is a global sink for about 5% of CH_4 produced annually and is mainly limited by CH_4 diffusion into the soil. However, intensive cropland management tends to reduce soil CH_4 consumption relative to forests and extensively managed grazing lands (CAST, 2004). Management-induced changes in CH_4-C fluxes have a smaller impact on terrestrial carbon cycling than changes in CO_2-C fluxes (Table 10.2), but relatively greater radiative forcing for CH_4 amplifies the impact of increasing atmospheric CH_4 concentrations on net radiative forcing (Figure 10.1). Recent research has shown that live plant biomass and litter produce substantial amounts of CH_4, potentially making plants as large a source of CH_4 as livestock (Keppler et al., 2006). If this is the case, activities that increase plant biomass (and sequester CO_2) may lead to increased CH_4 production (Keppler et al., 2006).

10.1.4 Methane Fluxes From Livestock

Enteric fermentation (the process of organic matter breakdown by gut flora within the gastrointestinal tract of animals, particularly ruminants) allows for the digestion of fibrous

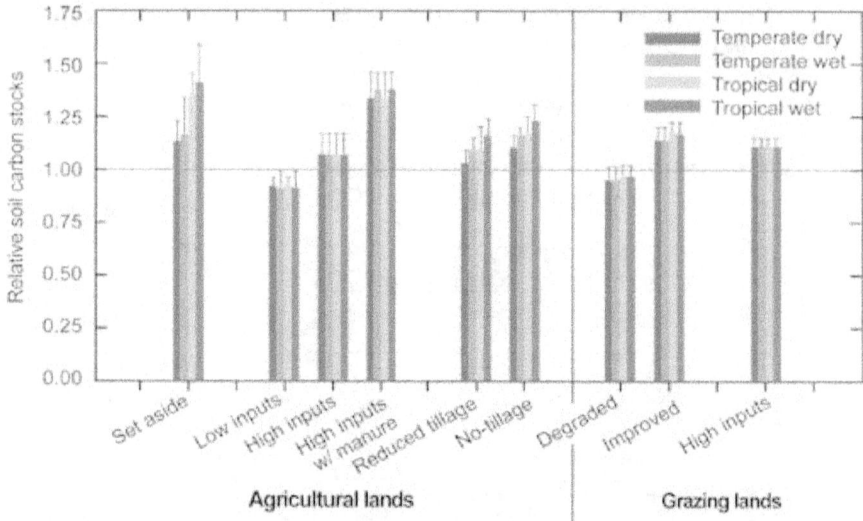

Figure 10.2 Relative soil carbon following implementation of new agricultural or grassland management practices. Conventionally tilled, medium-input cultivated land and moderately grazed grasslands with moderate inputs are defaults for agricultural and grazing lands, respectively. Default soil carbon stocks (like those in Table 10.1) can be multiplied by one or more stock change factors to estimate carbon sequestration rates (over a 20-year time period). The dashed horizontal line indicates default soil carbon stocks (*i.e.*, those under conventional-tillage cropland or undegraded grazingland, with medium inputs). Temperature/precipitation divisions are the same as those described in Table 10.1. Data are from Nabuurs *et al.* (2004) and Ogle *et al.* (2004).

storage temperature, and duration of storage. Unlike enteric CH_4, the major sources of manure CH_4 emissions in the United States are from swine (44%) and dairy cattle (39%). Manure CH_4 production is greater for production systems with anoxic lagoons, largely anoxic pits, or manure handled or stored as slurry. Between 1990 and 2002, CH_4 emissions from manure management increased 25% in the United States and 21% in Canada (EPA, 2000; Matin *et al.*, 2004).

10.2 DRIVERS AND TRENDS

The extent to which agriculture will contribute to greenhouse gas mitigation will largely depend on government policy decisions, but mitigation opportunities will also be constrained by technological advances and changing environmental conditions (see discussion below). Estimates from national inventories suggest that United States' and Canadian agricultural soils are currently near neutral or small net sinks for CO_2, which has occurred as a consequence of changing management (*e.g.*, reduced tillage intensity) and government programs designed for purposes other than greenhouse gas mitigation (*e.g.*, soil conservation, commodity regulation). However, to realize the much larger potential for soil carbon sequestration (see section below) and for significant reductions in CH_4 (and N_2O) emissions, specific policies targeted at greenhouse gas reductions are required. It is generally recognized that farmers (and other economic actors) are, as a group, "profit-maximizers," which implies that to change from current practices to ones that reduce net emissions, farmers will incur additional costs (termed "opportunity costs"). Hence, where the incentives (*e.g.*, carbon offset market payments, government subsidies) to adopt new practices exceed the opportunity costs, farmers will adopt new practices. Crop productivity, production input expenses, marketing costs, *etc.* (which determine profitability) vary widely within (and between) countries. Thus, the payment needed to achieve a unit of emission reduction will vary, among and within regions. In general, each successive increment of carbon sequestration or emission reduction comes at a progressively higher cost

materials by livestock, but the extensive fermentation of the ruminant diet requires 5–7% of the dietary gross energy to be belched out as CH_4 to sustain the anaerobic processes (Johnson and Johnson, 1995). Methane emissions from livestock contribute significantly to total CH_4 emissions in the United States (3.7 ± 0.53 Mt CH_4-C per year, 20% of total United States' CH_4 emissions), Canada (0.78 ± 0.14 Mt CH_4-C per year, 22% of total) (Ministry of the Environment, 2006; Sobool and Kulshreshtha, 2005), and Mexico (1.5 Mt CH_4-C per year, 27% of total)[†] with the vast majority of enteric CH_4 emissions from beef (72%) and dairy cattle (23%) (Table 10.2). Emissions from ruminants are tightly coupled to feed consumption, since CH_4 emission per unit of feed energy is relatively constant, except for feedlot cattle with diets high in cereal grain contents, for which the fractional loss falls to one-third to one-half of normal rates (Johnson and Johnson, 1995). Between 1990 and 2002, CH_4 emissions from enteric fermentation fell 2% in the United States but increased by 20% in Canada (EPA, 2000; Matin *et al.*, 2004).

Where the incentives (*e.g.*, carbon offset market payments, government subsidies) to adopt new practices exceed the opportunity costs, farmers will adopt new practices.

Methane emissions during manure storage (United States: 1.3 ± 0.24 Mt CH_4 per year; Mexico: 0.06 Mt CH_4 per year [†]; Canada: 0.3 ± 0.05 Mt CH_4 per year) are governed by the amount of degradable organic matter, degree of anoxia, (this relationship is often shown in the form of an upward bending marginal cost curve).

The interaction of changes in technological and environmental conditions, including crop growth improvements, impacts of CO_2 increase, nitrogen deposition, and climate change, will shape future trends in greenhouse gas emissions and mitigation from agricultural and grazing lands. A continuation of the yield increases seen in the past several decades for agricultural crops (Reilly and Fuglie, 1998) would tend to enhance the potential for soil carbon sequestration (CAST, 2004). Similarly, increased plant growth due to higher concentrations of CO_2 (and nitrogen deposition) has been projected to boost carbon uptake on agricultural (and other) lands, offsetting some or all of the climate-change induced reductions in productivity projected in some regions of North America (NAS, 2001). However, recent syntheses from field-scale FACE (Free-Air Carbon dioxide Enrichment) studies of croplands (Long et al., 2006) and grasslands (Nowak et al., 2004) suggest that the growth enhancement from CO_2 fertilization may be much less than previously thought. Feedbacks between temperature and soil carbon stocks could counteract efforts to reduce greenhouse gases via carbon sequestration within agricultural ecosystems. Increased temperatures tend to increase the rate of biological processes—including plant respiration and organic matter decay, and CO_2 release by soil organisms—particularly in temperate climates that prevail across most of North America. Because soil carbon stocks, including those in agricultural lands, contain such large amounts of carbon, small percentage increases in the rate of soil organic matter decomposition could lead to substantially increased emissions (Jenkinson et al., 1991; Cox et al., 2000). There is currently a scientific debate about the relative temperature sensitivity of the different constituents making up soil organic matter (e.g., Kätterer et al., 1998; Giardina and Ryan, 2000; Ågren and Bosatta, 2002; Knorr et al., 2005), reflecting uncertainty in the possible degree and magnitude of climate change feedbacks. Despite this uncertainty, the potential for climate and other environmental feedbacks to influence the carbon balance of agricultural systems by perturbing productivity (and carbon input rates) and organic

matter turnover, and potentially soil N_2O and CH_4 fluxes, cannot be overlooked.

10.3 OPTIONS FOR MANAGEMENT

10.3.1 Carbon Sequestration

Agricultural and grazing land management practices capable of increasing carbon inputs or decreasing carbon outputs, while still maintaining yields, can be divided into two classes: those that impact carbon inputs, and those that affect carbon release through decomposition and disturbance. Reversion to native vegetation or setting agricultural land aside as grassland, such as in the Canadian Prairie Cover Program and the U.S. Conservation Reserve Program, can increase the proportion of photosynthesized carbon retained in the system and sequester carbon in the soil[4] (Conant et al., 2001; Post and Kwon, 2000; Follett et al., 2001b) (Figure 10.2). In annual cropland, improved crop rotations, yield enhancement measures, organic amendments, cover crops, improved fertilization and irrigation practices, and reduced bare fallow tend to increase productivity and carbon inputs, and thus soil carbon stocks (Lal et al., 1998; Paustian et al., 1998; Vanden-Bygaart et al., 2003) (Figure 10.2). Tillage, traditionally used for soil preparation and weed control, disturbs the soil and stimulates decomposition and loss of soil carbon. Practices that substantially reduce (reduced-till) or eliminate (no-till) tillage-induced disturbances are being increasingly adopted and generally increase soil carbon stocks while maintaining or enhancing productivity levels (Paustian et al., 1997; Ogle et al., 2003) (Figure 10.2). Estimates of the technical potential for annual cropland soil carbon sequestration are on the order of 50–100 Mt C per year in the United States (Lal et al., 2003; Sperow et al., 2003) and 3.3–6.4 Mt C per year in Canada (Boehm et al., 2004).

Within grazing lands, historical overgrazing has substantially reduced productive capacity in many areas, leading to loss of soil carbon stocks (Conant and Paustian, 2002) (Figure 10.2). Conversely, improved grazing management and production inputs (like fertilizer, adding (nitrogen-fixing) legumes, organic amendments, and irrigation) can increase productivity, carbon inputs, and soil carbon stocks (Conant et al., 2001), potentially storing 0.44 Mt C per year[†] in Canada (Lynch et al., 2005) and as much as 16–54 (mean = 33.2) Mt C per year in the United States (Follett et al., 2001a). Such improvements will carry a carbon cost, par-

---

[4] The bulk of carbon sequestration potential in agricultural and grazing lands is restricted to soil carbon pools, though carbon can be sequestered in woody biomass in agroforestry systems (Sheinbaum and Masera, 2000). Woody encroachment on grasslands can also store substantial amounts of carbon in biomass, but the phenomenon is neither well-controlled nor desirable from the standpoint of livestock production, since it results in decreased forage productivity, and the impacts on soil carbon pools are highly variable and poorly understood.

ticularly fertilization and irrigation, since their production and implementation require the use of fossil fuels.

10.3.2 Fossil-Fuel Derived Emission Reductions

> Converting from conventional plowing to no-tillage can reduce on-farm fossil-fuel emissions by 25–80% and total fossil-fuel emissions by 14–25%.

The efficiency with which on-farm (from tractors and machinery) and off-farm (from production of agricultural input) energy inputs are converted to agricultural products varies several-fold (Lal, 2004). Where more energy-efficient practices can be substituted for less efficient ones, fossil-fuel CO_2 emissions can be reduced (Lal, 2004). For example, converting from conventional plowing to no-tillage can reduce on-farm fossil-fuel emissions by 25–80% (Frye, 1984; Robertson et al., 2000) and total fossil-fuel emissions by 14–25% (West and Marland, 2003). Substitution of legumes for mineral nitrogen can reduce energy input by 15% in cropping systems incorporating legumes (Pimentel et al., 2005). More efficient heating and cooling (e.g., better building insulation) could reduce CO_2 emissions associated with housed animal facilities (e.g., dairy). Substitution of crop-derived fuels for fossil fuels could decrease net emissions.

Energy intensity (energy per unit product) for the United States' agricultural sector has declined since the 1970s (Paustian et al., 1998). Between 1990 and 2000, fossil-fuel emissions on Canadian farms increased by 35%[†] (Sobool and Kulshreshtha, 2005).

10.3.3 Methane Emission Reduction

Reducing flood duration and decreasing organic matter additions to paddy rice fields can reduce CH_4 emissions. Soil amendments such as ammonium sulfate and calcium carbide inhibit CH_4 formation. Coupled with adoption of new rice cultivars that favor lower CH_4 emissions, these management practices could reduce CH_4 emission from paddy rice systems by 16–70% (mean = 40%) of current emissions (Mosier et al., 1998b).

Biomass burning is uncommon in most Canadian and United States' crop production systems; less than 3% of crop residues are burned annually in the United States (EPA, 2006). Biomass burning in conjunction with land clearing and with subsistence agriculture still occurs in Mexico, but these practices are declining. The primary path for emission reduction is reducing residue burning (CAST, 2004).

> Practices that sequester carbon in agricultural and grazing land soils improve soil fertility, buffering capacity, and pesticide immobilization.

Refinement of feed quality, feed rationing, additives, and livestock production efficiency chains can all reduce CH_4 emissions from ruminant livestock with minimal impacts on productivity or profits (CAST, 2004). Boadi et al. (2004) review several examples of increases in energy intensity. Wider adoption of more efficient practices could reduce CH_4 production from 5–8% to 2–3% of gross feed energy (Agriculture and Agri-Food Canada, 1999), reducing CH_4 emissions by 20–30% (Mosier et al., 1998b).

Methane emissions from manure storage are proportional to duration of storage under anoxic conditions. Handling solid rather than liquid manure, storing manure for shorter periods of time, and keeping storage tanks cool can reduce emissions from stored manure (CAST, 2004). More important, capture of CH_4 produced during anaerobic decomposition of manure (in covered lagoons or small- or large-scale digesters) can reduce emissions by 70–80% (Mosier et al., 1998b). Use of digester systems is spreading in the United States, with 50 digesters currently in operation and 60 systems in construction or planned (NRCS, 2005). Energy production using CH_4 captured during manure storage will reduce energy demands and associated CO_2 emissions.

10.3.4 Environmental Co-benefits From Carbon Sequestration and Emission Reduction Activities

Many of the practices that lead to carbon sequestration and reduced CO_2 and CH_4 emissions not only increase production efficiencies but also lead to environmental co-benefits. Practices that sequester carbon in agricultural and grazing land soils improve soil fertility, buffering capacity, and pesticide immobilization (Lal, 2002; CAST, 2004). Increasing soil carbon content makes the soil more easily workable and reduces energy requirements for field operations (CAST, 2004). Decreasing soil disturbance and retaining more surface crop residues enhance water infiltration and prevent wind and water erosion, improving air quality. Increased water retention plus improved fertilizer management reduces nitrogen losses and subsequent nitrate (NO_3^-) leaching and downstream eutrophication.

10.3.5 Economics and Policy Assessment

Policies for agricultural mitigation activities can range from transfer payments (such as subsidies, tax credits, etc.) to encourage greenhouse gas mitigating practices or taxes or penalties to discourage practices with high emissions, to emission offset trading in a free market-based system with governmental sanction. Currently the policy context of the three North American countries differs greatly. Canada and the United States are both Annex 1 (developed countries) within the United Nations Framework Convention on Climate Change (UNFCCC), but Canada is obligated to mandatory emission reductions as a party to the Kyoto Protocol, while the United States currently maintains a national, voluntary

emission reduction policy outside of Kyoto. Mexico is a non-Annex 1 (developing) country and thus is not currently subject to mandatory emission reductions under Kyoto.

At present, there is relatively little practical experience upon which to judge the costs and effectiveness of agricultural mitigation activities. Governments are still in the process of developing policies and, moreover, the economics of various mitigation activities will only be known when there is a significant economic incentive for emission reductions, *e.g.*, through regulatory emission caps or government-sponsored bids and contracts. However, several economic analyses have been performed in the United States, using a variety of models (*e.g.*, McCarl and Schneider, 2001; Antle *et al.*, 2003; Lewandrowski *et al.*, 2004). Most studies have focused on carbon sequestration, and less work has been done on the economics of reducing CH_4 and N_2O emissions. While results differ between models and for different parts of the country, some preliminary conclusions have been drawn (see Boehm *et al.*, 2004; CAST, 2004).

- Additional carbon (10–70 Mt C per year), above current rates, could be sequestered in soils at low to moderate costs ($10–100 per metric ton of carbon).
- Mitigation practices that maintain the primary income source (*i.e.*, crop/livestock production), such as conservation tillage and pasture improvement, have a lower cost per ton sequestered carbon compared with practices where mitigation would be a primary income source (*i.e.*, foregoing income from crop and/or livestock production), such as land set-asides, even if the latter have a higher biological sequestration potential.
- With higher energy prices, major shifts in land use in favor of energy crops and afforestation may occur at the expense of annual cropland and pasture.
- Policies based on per-ton payments (for carbon actually sequestered) are more economically efficient than per-hectare payments (for adopting specific practices, see Antle *et al.*, 2003), although the former have a higher verification cost (*i.e.*, measuring actual carbon sequestered versus measuring adoption of specific farming practices on a given area of land).

A recent study commissioned by the U.S. Environmental Protection Agency (EPA, 2005), evaluated some agricultural mitigation options for different policy scenarios, including constant CO_2 price scenarios for 2010–2110, where the price represents the incentive required for the mitigation activity. Annualized prices of $15/ton of CO_2 would yield mitigation amounts of 46 Mt C per year through agricultural soil carbon sequestration and 14.5 Mt C per year from fossil-fuel use reduction (compared with the estimated United States' national ecosystem carbon sink of 480 Mt C per year). At lower prices of $5/ton CO_2, the corresponding values would be 34 Mt C per year (for soil sequestration) and 9 Mt C per year (for fossil-fuel reduction), respectively, reflecting the effect of price on the supply of mitigation activities[5].

10.3.6 Other Policy Considerations

Agricultural mitigation of CO_2 through carbon sequestration and emission reductions for CH_4 (and N_2O), differ in ways that impact policy design and implementation. Direct emission reductions of CH_4 and CO_2 from fossil-fuel use are considered "permanent" reductions, while carbon sequestration is a "non-permanent" reduction, in that carbon stored through conservation practices could potentially be re-emitted if management practices revert back to the previous state or otherwise change so that the stored carbon is lost. This *permanence* issue applies to all forms of carbon sinks. In addition, soil carbon storage, with a given change in management (*e.g.*, tillage reduction, pasture improvement, afforestation), will tend to level off at a new steady state level after 15–30 years, after which there is no further accumulation of carbon (West *et al.*, 2004). Enhanced management practices must be sustained to maintain these higher carbon stocks. Key implications for policy are that the value of sequestered carbon could be discounted compared to direct emission reductions to compensate for the possibility of future emissions. Alternatively, long-term contracts will be needed to build and maintain carbon stocks, which will tend to increase the price per unit of sequestered carbon. However, even temporary storage of carbon has economic value (CAST, 2004), and various proposed concepts of leasing carbon storage or applying discount rates could accommodate carbon sequestration as part of a carbon offset trading system (CAST, 2004). In addition, switching to practices that increase soil carbon (and hence, improve soil fertility) could be more profitable to farmers in the long-run, so that additional incentives to maintain the practices once they become well established may not be necessary (Paustian *et al.*, 2006).

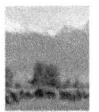

Another policy issue relating to carbon sequestration is *leakage* (also termed "slippage" in economics), whereby mitigation actions in one area (*e.g.*, geographic region, pro-

---

[5] These estimates were produced using a national-scale economic sector model which estimates the linkage between CO_2 prices and the supply of mitigation activities, for specified price scenarios. Hence, the model can produce a range of CO_2 mitigation amounts as a function of price, but the model was not used to estimate the uncertainty of mitigation amounts at a given price level.

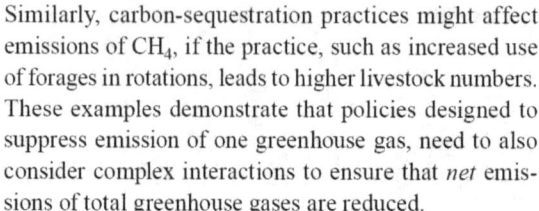

BOX 10.4: Agricultural and Grazing Land N₂O Emission Reductions

When mineral soil nitrogen content is increased by nitrogen additions (*i.e.*, fertilizer), a portion of that nitrogen can be transformed to N_2O as a byproduct of two microbiological processes (nitrification and denitrification) and lost to the atmosphere. Coincidental introduction of large amounts of easily decomposable organic matter and NO_3^- from either a plow down of cover crop or manure addition greatly stimulates denitrification under wet conditions (Peoples *et al.*, 2004). Some practices intended to sequester atmospheric carbon in soil could prompt increases in N_2O fluxes. For example, reducing tillage intensity tends to increase soil moisture, leading to increased N_2O fluxes, particularly in wetter environments (Six *et al.*, 2004). Synchronizing organic amendment applications with plant nitrogen uptake and minimizing manure storage under anoxic conditions can reduce N_2O emissions by 10–25% and will increase nitrogen use efficiency which can decrease indirect emissions (in waterways) by 5–20% (CAST, 2004).

Similarly, carbon-sequestration practices might affect emissions of CH_4, if the practice, such as increased use of forages in rotations, leads to higher livestock numbers. These examples demonstrate that policies designed to suppress emission of one greenhouse gas, need to also consider complex interactions to ensure that *net* emissions of total greenhouse gases are reduced.

A variety of other factors will affect the willingness of farmers to adopt greenhouse gas reducing practices and the efficacy of agricultural policies, including perceptions of risk, information and extension efforts, technological developments, and social and ethical values (Paustian *et al.*, 2006). Many of these factors are difficult to incorporate into traditional economic analyses. Pilot mitigation projects, along with additional research using integrated ecosystem and economic assessment approaches (*e.g.*, Antle *et al.*, 2001), will allow us to get a clearer picture of the actual potential of agriculture to contribute to greenhouse gas mitigation efforts.

10.4 RESEARCH AND DEVELOPMENT NEEDS

Expanding the network of intensive research sites dedicated to understanding basic processes, coupled with national-level networks of soil monitoring/validation sites, could reduce inventory uncertainty and contribute to attributing changes in ecosystem carbon stocks to changes in land management (see Bellamy *et al.*, 2005). Expansion of both networks should be informed about how different geographic areas and ecosystems contribute to uncertainty and the likelihood that reducing uncertainty could inform policy decisions. For example, changes in ecosystem carbon stocks due to woody encroachment on grasslands constitute one of the largest, but least certain, aspects of terrestrial carbon cycling in North America (Houghton *et al.*, 1999; Pacala *et al.*, 2001). Better information about the spatial extent of woody encroachment, the amount and growth of woody biomass, and variation in the impacts on soil carbon stocks would help reduce that uncertainty. Identifying location, cause, and size of this sink could help identify practices that may promote continued sequestration of carbon and would constrain estimates of carbon storage in other lands, possibly helping to identify other policy options. Uncertainty in land use, land-use change, soil carbon responses to management (*e.g.*, tillage) on particular soils, and impacts of cultivation on soil carbon stocks (*e.g.*, impacts of erosion) are the largest contributors to uncertainty in the Canadian and United States' national agricultural greenhouse gas inventories (Ogle *et al.*, 2003; VandenBygaart *et al.*, 2003). Finally, if the goal of a policy instrument is to reduce greenhouse gas emissions, net impacts on CO_2, CH_4, and N_2O emissions, which are not as well understood, should be considered.

duction system) stimulate additional emissions elsewhere. For forest carbon sequestration, leakage is a major concern. For example, reducing harvest rates in one area (thereby maintaining higher biomass carbon stocks) can stimulate increased cutting and reduction in stored carbon in other areas, as was seen with the reduction in harvesting in the Pacific Northwest during the 1990s (Murray *et al.*, 2004). Preliminary studies suggest that leakage is of minor concern for agricultural carbon sequestration, since most practices would have little or no effect on the supply and demand of agricultural commodities. However, there are uncertain and conflicting views on whether land-set asides in which land is taken out of agricultural production, such as the Conservation Reserve Program in the United States, might be subject to significant leakage.

A further question, relevant to policies for carbon sequestration, is how practices for conserving carbon affect emissions of other greenhouse gases. Of particular importance is the interaction of carbon sequestration with N_2O emissions, because N_2O is such a potent greenhouse gas (Robertson and Grace, 2004; Six *et al.*, 2004; Gregorich *et al.*, 2005). (See Box 10.4). In some environs, carbon-sequestration practices, such as reduced tillage, can stimulate N_2O emissions, thereby offsetting part of the benefit; elsewhere, carbon-conserving practices may suppress N_2O emissions, amplifying the net benefit (Smith *et al.*, 2001; Smith and Conen, 2004; Conant *et al.*, 2005; Helgason *et al.*, 2005).

CHAPTER 11

North American Forests

Lead Authors: Richard A. Birdsey, USDA Forest Service; Jennifer C. Jenkins, Univ. Vt.; Mark Johnston, Saskatchewan Research Council; Elisabeth Huber-Sannwald, Instituto Potosino de Investigación Científica y Tecnológica

Contributing Authors: Brian Amiro, Univ. Manitoba; Ben de Jong, ECOSUR; Jorge D. Etchevers Barra, Colegio de Postgraduado; Nancy French, Altarum Inst.; Felipe García-Oliva, UNAM; Mark Harmon, Oreg. State Univ.; Linda S. Heath, USDA Forest Service; Victor J. Jaramillo, UNAM; Kurt Johnsen, USDA Forest Service; Beverly E. Law, Oreg. State Univ.; Erika Marín-Spiotta, Univ. Calif. Berkeley; Omar Masera, UNAM; Ronald Neilson, USDA Forest Service; Yude Pan, USDA Forest Service; Kurt S. Pregitzer, Mich. Tech. Univ.

KEY FINDINGS

- North American forests contain roughly 170 ± 40 billion tons of carbon, of which approximately 28% is in live vegetation and 72% is in dead organic matter.
- North American forests were a net carbon sink of -270 ± 130 million tons of carbon per year over the last 10 to 15 years.
- Deforestation continues in Mexico where forests are a source of carbon dioxide to the atmosphere. Forests of the United States and parts of Canada have become a carbon sink as a consequence of the recovery of forests following the abandonment of agricultural land.
- Carbon dioxide emissions from Canada's forests are highly variable because of interannual changes in area burned by wildfire.
- The size of the carbon sink in United States' forests appears to be declining based on inventory data from 1952 to the present.
- Many factors that cause changes in carbon stocks of forests have been identified, including land-use change, timber harvesting, natural disturbance, increasing atmospheric carbon dioxide, climate change, nitrogen deposition, and ozone in the lower atmosphere. There is a lack of consensus about how these different natural and human-caused factors contribute to the current sink, and the relative importance of factors varies geographically.
- There have been several continental- to sub continental-scale assessments of future changes in carbon and vegetation distribution in North America, but the resulting projections of future trends for North American forests are highly uncertain. Some of this is due to uncertainty in future climate, but there is also considerable uncertainty in forest response to climate change and in the interaction of climate with other natural and human-caused factors.
- Forest management strategies can be adapted to manipulate the carbon sink strength of forest systems. The net effect of these management strategies will depend on the area of forests under management, management objectives for resources other than carbon, and the type of disturbance regime being considered.
- Decisions concerning carbon storage in North American forests and their management as carbon sources and sinks will be significantly improved by (1) filling gaps in inventories of carbon pools and fluxes, (2) a better understanding of how management practices affect carbon in forests, (3) a better estimate of potential changes in forest carbon under climate change and other factors, and (4) the increased availability of decision support tools for carbon management in forests.

11.1 INTRODUCTION

The forest area of North America totals 771 million hectares (ha), 36% of the land area of North America and about 20% of the world's forest area (Food and Agriculture Organization, 2001)† (see Table 11.1 and Box 11.1 for estimates and uncertainty conventions, respectively). About 45% of this forest area is classified as boreal, mostly in Canada and some in Alaska. Temperate and tropical forests constitute the remainder of the forest area.

North American forests are critical components of the global carbon cycle, exchanging large amounts of carbon dioxide (CO_2) and other gases with the atmosphere and oceans. In this chapter, we present the most recent estimates of the role of forests in the North American carbon balance, describe the main factors that affect forest carbon stocks and fluxes, describe how forests affect the carbon cycle through CO_2 sequestration and emissions, and discuss management options and research needs.

11.2 CARBON STOCKS AND FLUXES

11.2.1 Ecosystem Carbon Stocks and Pools

North American forests contain more than 170 billion tons of carbon (Gt C), of which 28% is in live biomass and 72% is in dead organic matter (Table 11.2). Among the three countries, Canada's forests contain the most carbon and Mexico's forests the least.

Carbon density (the amount of carbon stored per unit of land area) is highly variable. In Canada, the majority of carbon storage occurs in boreal and cordilleran forests (Kurz and

BOX 11.1: CCSP SAP 2.2 Uncertainty Conventions

***** = 95% certain that the actual value is within 10% of the estimate reported,

**** = 95% certain that the estimate is within 25%,

*** = 95% certain that the estimate is within 50%,

** = 95% certain that the estimate is within 100%, and

* = uncertainty greater than 100%.

† = The magnitude and/or range of uncertainty for the given numerical value(s) is not provided in the references cited.

Apps, 1999). In the United States, forests of the Northeast, Upper Midwest, Pacific Coast, and Alaska (with 14 Gt C) store the most carbon. In Mexico, temperate forests contain 4.5 Gt C, tropical forests contain 4.1 Gt C, and semiarid forests contain 5.0 Gt C.

11.2.2 Net North American Forest Carbon Fluxes

According to nearly all published studies, North American lands are a net carbon sink (Pacala *et al.*, 2001). A summary of currently available data from greenhouse gas inventories and other sources suggests that the magnitude of the North American forest carbon sink was approximately -269 million metric tons of carbon (Mt C) per year over the last decade or so, with United States' forests accounting for most of the sink (Table 11.3). This estimate is likely to be within 50% of the true value.

Canadian forests were estimated to be a net sink of -17 Mt C per year from 1990-2004 (Environment Canada, 2006) (Table 11.3). These estimates pertain to the area of forest considered to be "managed" under international reporting guidelines, which is 83% of the total area of Canada's forests. The estimates also include the carbon changes that result from land-use change. Changes in forest soil carbon are included; urban forests are excluded (Chapter 14 this report). High interannual variability is averaged into this estimate—the annual change varied from approximately -50 to +40 between 1990 and 2004. Years with net emissions were generally years with high forest fire activity (Environment Canada, 2005) (Figure 11.1).

Most of the net sink in United States' forests is in aboveground carbon pools, which account for -146 Mt C per year (Smith and Heath, 2005). The net sink for the below-ground carbon pool is estimated at -90 Mt C (Pacala *et al.*, 2001) (Table 11.3). The size of the carbon sink in United States' forest ecosystems appears to have declined slightly over the last decade (Smith and Heath, 2005). In

Table 11.1 Area of forest land by biome and country, 2000 (1000 ha)ᵃ. See Box 11.1 for uncertainty conventions.

Ecological zone:	Canadaᵇ	U.S.ᶜ	Mexicoᵈ	Total
Tropical/subtropical	0*****	115,200*****	30,700*****	145,900*****
Temperate	101,100*****	142,400*****	32,900*****	276,400*****
Boreal	303,000*****	45,500*****	0*****	348,500*****
Total	404,100*****	303,100*****	63,600*****	770,800*****

ᵃ The certainty for estimates in this table are listed in Box 11.1. See sources for estimates (e.g., see Bechtold and Patterson, 2005 for the United States).
ᵇ Canadian Forest Service (2005)
ᶜ Smith *et al.* (2004)
ᵈ Palacio-Prieto *et al.* (2000)

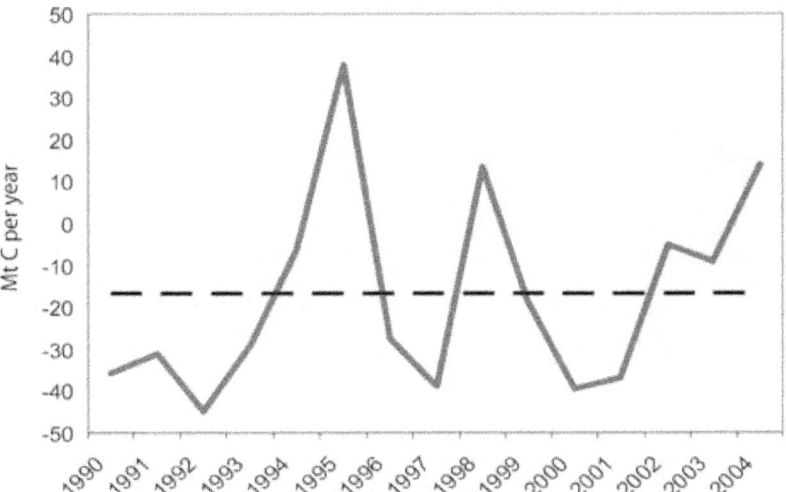

Figure 11.1 Average and annual estimates of change in carbon stocks for forest ecosystems of Canada, 1990-2004. Interannual variability is high because of changes in rates and impacts of disturbances such as fire and insects (from Environment Canada, 2006).

Table 11.2 Carbon stocks in forests by ecosystem carbon pool and country (Mt C)[a]. See Box 11.1 for uncertainty conventions.

Ecosystem carbon pool:	Canada[b]	U.S.[c]	Mexico[d]	Total
Biomass	14,500****	24,900****	7,700****	47,100****
Dead organic matter[e]	71,300****	41,700****	11,400****	124,400****
Total	85,800****	66,600****	19,100****	171,500****

[a] The certainty for estimates in this table are listed in Box 11.1. See sources for estimates (Heath and Smith, 2000; Smith and Heath, 2000). The estimated carbon stock in North American forests is thus 171,500 ± 43,000 Mt C.
[b] Kurz and Apps (1999)
[c] Heath and Smith (2004), Birdsey and Heath (1995)
[d] Masera et al. (2001)
[e] Includes litter, coarse woody debris, and soil carbon.

for forest stands at various stages of recovery after disturbance reveal patterns and causes of sink or source strength, which is highly dependent on time since disturbance. Representative estimates for North America are summarized in Appendix D. As forests are planted or regrow on abandoned farmland, or as they recover from fire, harvest, or other disturbance, there is an initial period of slow (or negative) carbon sequestration followed by a period of rapid carbon sequestration. Many forests continue sequestering significant amounts of carbon for 125 years or more after establishment (Smith et al., 2006). Eventually, the rate of sequestration slows as forests reach a new balance of carbon uptake and release, and in old growth forests processes of carbon uptake are very nearly balanced by processes of release (Chapter 3, this report).

11.3 TRENDS AND DRIVERS

11.3.1 Overview of Trends and Drivers of Change in Carbon Stocks

Many factors that cause changes in carbon stocks of forests and wood products have been identified, but the relative importance of each remains difficult to quantify (Barford et al., 2001; Caspersen et al., 2000; Goodale et al., 2002; Körner, 2000; Schimel et al., 2000). Land-use

contrast, a steady or increasing supply of timber products now and in the foreseeable future (Haynes, 2003) means that the rate of increase in the wood products carbon pool is likely to remain steady.

For Mexico, the most comprehensive available estimate for the forest sector suggests a source of +52 Mt C per year in the 1990s (Masera et al., 1997) (Table 11.3). This estimate does not include changes in the wood products carbon pool. The main cause of the estimated source is deforestation, which is offset to a much lesser degree by restoration and recovery of degraded forestland.

Landscape-scale estimates of ecosystem carbon fluxes reflect the dynamics of individual forest stands that respond to unique combinations of disturbance history, management intensity, vegetation, and site characteristics. Extensive land-based measurements of forest/atmosphere carbon exchange

Table 11.3 Change in carbon stocks for forests and wood products by country (Mt C per year). See Box 11.1 for uncertainty conventions.

Carbon pool:	Canada[a]	U.S.[b]	Mexico[c]	Total
Forest ecosystem	−17**	−236***	+52**	−201
Wood products	−11**	−57***	ND[d]	−68
Total	−28**	−293***	+52**	−269

[a] Data for 1990-2004, taken from Environment Canada (2006), Goodale et al. (2002).
[b] From Smith and Heath (2005) (excluding soils), and Pacala et al. (2001) (soils). Estimates do not include urban forests.
[c] From Masera (1997)
[d] Estimates are not available.

and timber harvesting seem to be dominant factors according to repeated forest inventories from 1952 to 1997 that show forest carbon stocks (excluding soils) increasing by about 175 Mt C per year. The most recent inventories show a decline in the rate of carbon uptake by forests, which appears to be mainly the result of changing growth and harvest rates following a long history of land-use change and management (Birdsey *et al.*, 2006; Smith and Heath, 2005). The factors behind net emissions from Mexico's forests are deforestation, forest degradation, and forest fires that are not fully offset by forest regeneration (Masera *et al.*, 1997; De Jong *et al.*, 2000).

change, timber harvesting, natural disturbance, increasing atmospheric CO_2, climate change, nitrogen deposition, and tropospheric ozone all have effects on carbon stocks in forests, with their relative influence depending on geographic location, the type of forest, and specific site factors. It is important for policy implementation and management of forest carbon to separate the effects of direct human actions from natural factors.

The natural and human-caused (anthropogenic) factors that significantly influence forest carbon stocks are different for each country, and still debated in the scientific literature. Natural disturbances are significant in Canada, but estimates of the relative effects of different kinds of disturbance are uncertain. One study estimated that impacts of wildfire and insects caused emissions of about +40 Mt C per year[†] of carbon to the atmosphere over the two decades (Kurz and Apps, 1999). Another study concluded that the positive effects of climate, CO_2, and nitrogen deposition outweighed the effects of wildfire and insects, making Canada's forests a net carbon sink in the same period (Chen *et al.*, 2003). In the United States, land-use change

The most recent inventories for the U.S. show a decline in the rate of carbon uptake by forests.

11.3.2 Effects of Land-use Change

Since 1990, approximately 549,000 ha of former cropland or grassland in Canada have been abandoned and are reverting to forest, while 71,000 ha of forest have been converted to cropland, grassland, or settlements, for a net increase in forest area of 478,000 ha (Environment Canada, 2005)[†]. In 2004, approximately 25,000 ha were converted from forest to cropland, 19,000 ha from forest to settlements, and approximately 3,000 ha converted to wetlands. These land-use changes resulted in emissions of about 4 Mt C (Environment Canada, 2005)[†].

In the last century more than 130 million ha of land in the conterminous United States were either afforested (62 million ha)[†] or deforested (70 million ha)[†] (Birdsey and Lewis, 2003). Houghton *et al.* (1999) estimated that cumulative changes in forest carbon stocks for the period from 1700 to 1990 in the United States were about +25 Gt C,[†] primarily from conversion of forestland to agricultural use and reduction of carbon stocks for wood products.

Emissions from Mexican forests to the atmosphere are primarily due to the impacts of deforestation to pasture and degradation of 720,000 to 880,000 ha per year[†] (Masera *et al.*, 1997; Palacio-Prieto *et al.*, 2000). The highest deforestation rates occur in the tropical deciduous forests (304,000 ha in 1990)[†] and the lowest in temperate broadleaf forests (59,000 ha in 1990)[†].

11.3.3 Effects of Forest Management

The direct human impact on North American forests ranges from very minimal for protected areas to very intense for plantations (Table 11.4). Between these extremes is the vast majority of forestland, which is impacted by a wide range of human activities and government policies that influence harvesting, wood products, and regeneration.

Table 11.4 Area of forestland by management class and country, 2000 (1000 ha)[a]. See Box 11.1 for uncertainty conventions.

Management class:	Canada	U.S.	Mexico	Total
Protected	19,300*****	66,700*****	6,000*****	92,000*****
Plantation	4,500*****	16,200*****	200*****	20,900*****
Other	380,300*****	220,200*****	57,400*****	657,900*****
Total	404,100*****	303,100*****	63,600*****	770,800*****

[a]From Food and Agriculture Organization (2001), Natural Resources Canada (2005). The certainty for estimates in this table are listed in Box 11.1. See sources for estimates (*e.g.*, for the United States, see Bechtold and Patterson, 2005).

Forests and other wooded land in Canada occupy about 402 million ha. Approximately 310 million ha is considered forest of which 255 million ha (83%) are under active forest management (Environment Canada, 2005)[†]. Managed forests are considered to be under the direct influence of human activity and not reserved. Less than 1% of the area under active management is harvested annually. Apps et al. (1999) used a carbon budget model to simulate carbon in harvested wood products (HWP) for Canada. Approximately 800 Mt C were stored in the Canadian HWP sector in 1989, of which 50 Mt C were in imported wood products, 550 Mt C in exported products, and 200 Mt C in wood products produced and consumed domestically[†].

Between 1990 and 2000, about 4 million ha per year were harvested in the United States, two-thirds by partial-cut harvest and one-third by clear-cut (Birdsey and Lewis, 2003). Between 1987 and 1997, about 1 million ha per year were planted with trees, and about 800,000 ha were treated to improve the quality and/or quantity of timber produced (Birdsey and Lewis, 2003). Harvesting in United States' forests accounts for substantially more tree mortality than natural causes such as wildfire and insect outbreaks (Smith et al., 2004). The harvested wood resulted in -57 Mt C added to landfills and products in use, and an additional 88 Mt C were emitted from harvested wood burned for energy (Skog and Nicholson, 1998)[†].

About 80% of the forested area in Mexico is socially owned by communal land grants (ejidos) and rural communities. About 95% of timber harvesting occurs in native temperate forests (SEMARNAP, 1996). Illegal harvesting involves 13.3 million cubic meters of wood every year (Torres, 2004). The rural population is the controlling factor for changes in carbon stocks from wildfire, wood extraction, shifting agriculture practices, and conversion of land to crop and pasture use.

11.3.4 Effects of Climate and Atmospheric Chemistry

Environmental factors, including climate variability, nitrogen deposition, tropospheric ozone, and elevated CO_2, have been recognized as significant factors affecting the carbon cycle of forests (Aber et al., 2001; Ollinger et al., 2002). Some studies indicate that these effects are significantly smaller than the effects of land management and land-use change (Caspersen et al., 2000; Schimel et al., 2000). Recent reviews of ecosystem-scale studies known as Free Air CO_2 Exchange (FACE) experiments suggest that rising CO_2 increases net primary productivity by 12-23% over all species studied (Norby et al., 2005; Nowak et al., 2004). However, it is uncertain whether this effect results in a lasting increase in sequestered carbon or causes a more rapid cycling of carbon between the ecosystem and the atmosphere

(Körner et al., 2005; Lichter et al., 2005). Experiments have also shown that the effects of rising CO_2 are significantly moderated by increasing tropospheric ozone (Karnosky et al., 2003; Loya et al., 2003). When nitrogen availability is also considered, reduced soil fertility limits the response to rising CO_2, but nitrogen deposition can increase soil fertility to counteract that effect (Finzi et al., 2006; Johnson et al., 1998; Oren et al., 2001). Observations of photosynthetic activity from satellites suggest that productivity changes due to lengthening of the growing season depend on whether areas were disturbed by fire (Goetz et al., 2005). Based on these conflicting and complicated results from different studies and approaches, a definitive assessment of the relative importance, and interactions, of natural and anthropogenic factors is a high priority for research (U.S. Climate Change Science Program, 2003).

11.3.5 Effects of Natural Disturbances

Wildfire, insects, diseases, and weather events are common natural disturbances in North America. These factors impact all forests but differ in magnitude by geographic region. Wildfires were the largest disturbance in the twentieth century in Canada (Weber and Flannigan, 1997). In the 1980s and 1990s, the average total burned area was 2.6 million ha per year in Canada's forests, with a maximum 7.6 million ha per year in 1989[†]. Carbon emissions from forest fires range from less than +1 Mt C per year in the interior of British Columbia to more than +10 Mt C per year in the western boreal forest. Total emissions from forest fires in Canada averaged approximately +27 Mt C per year between 1959 and 1999 (Amiro *et al.*, 2001)[†]. Estimated carbon emissions from four major insect pests in Canadian forests (spruce budworm, jack pine budworm, hemlock looper, and mountain pine beetle) varied from +5 to 10 Mt C per year in the 1970s to less than +2 Mt C per year in the mid-1990s[1]. Much of the Canadian forest is expected to experience increases in fire severity (Parisien *et al.*, 2005) and burn areas (Flannigan *et al.*, 2005), and continued outbreaks of forest pests are also likely (Volney and Hirsch, 2005).

In United States' forests, insects, diseases, and wildfire combined, affect more than 30 million ha per decade (Birdsey and Lewis, 2003). Damage from weather events (hurricanes, tornadoes, and ice storms) may exceed 20 million ha per decade (Dale *et al.*, 2001). Although forest inventory data reveal the extent of tree mortality attributed to all causes combined, estimates of the impacts of individual categories of natural disturbance on carbon pools of temperate forests are scarce. The impacts of fire are clearly significant. According to one estimate, the average annual carbon emissions from biomass burning in the contemporary United States ranges from 9 to 59 Mt C (Leenhouts, 1998). McNulty (2002) estimated that large hurricanes in the United States could convert 20 Mt C of live biomass into detrital carbon pools.

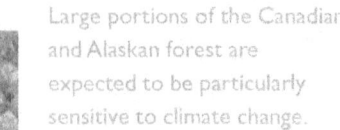

Large portions of the Canadian and Alaskan forest are expected to be particularly sensitive to climate change.

The number and area of sites affected by forest fires in Mexico have fluctuated considerably between 1970 and 2002, with a clear tendency of an increasing number of fire events (4,000-7,000 in the 1970s and 1,800-15,000 in the 1990s), and overall, larger areas are being affected (0.08-0.25 million ha in the 1970s and 0.05-0.85 million ha in the 1990s). During El Nino years, increasing drought increases fire frequencies (Torres, 2004). Between 1995 and 2000, an average of 8,900 fire events occurred per year and affected about 327,000 ha of the forested area. Currently, no estimates are available on the contribution of these fires to CO_2 emissions. Pests and diseases are important natural disturbance agents in temperate forests of Mexico; however, no statistics exist on the extent of the affected land area.

11.3.6 Projections of Future Trends

11.3.6.1 CANADA

Large portions of the Canadian and Alaskan forest are expected to be particularly sensitive to climate change (Hogg and Bernier, 2005). Climate change effects on forest growth could be positive (*e.g.*, increased rates of photosynthesis and increased water use efficiency) or negative (decreased water availability, higher rates of respiration) (Baldocchi and Amthor, 2001). It is difficult to predict the direction of these changes and they will likely vary by species and local conditions of soils and topography (Johnston and Williamson, 2005). Because of the large area of boreal forests and expected high degree of warming in northern latitudes, Canada and Alaska require close monitoring over the next few decades as these areas will likely be critical to determining the carbon balance of North America.

11.3.6.2 UNITED STATES

Assessments of future changes in carbon and vegetation distribution in the United States suggest that under most future climate conditions, net primary production (NPP) would respond positively to changing climate but total carbon storage would remain relatively constant (VEMAP Members, 1995; Pan *et al.*, 1998; Neilson *et al.*, 1998; Joyce *et al.*, 2001). Some climate scenarios indicate that much of the Northwest U.S. will receive more annual precipitation. When coupled with higher CO_2 and longer growing seasons, simulations show woody expansion and increased sequestration of carbon as well as increases in fire (Bachelet *et al.*, 2001). However, recent scenarios from the Hadley climate model show drying in the Northwest, which produces some forest decline (Price *et al.*, 2004). Many simulations show continued growth in eastern forests through the end of the twenty-first century, but some show the opposite, especially in the Southeast. Eastern forests could experience a period of enhanced growth in the early stages of warming, due to elevated CO_2, increased precipitation, and a longer growing

---

[1] These estimates are the product of regional carbon density values, the proportion of mortality in defoliated stands given in Kurz and Apps (1999), data on area affected taken from NFDP (2005), and the proportion of carbon in insect-killed stands that is emitted directly to the atmosphere (0.1) from the disturbance matrix for insects used in the CBM-CFS (Kurz et al., 1992).

season. However, further warming could bring on increasing drought stress, reducing the carrying capacity of the ecosystem and causing carbon losses through drought-induced dieback and increased fire and insect disturbances. North American boreal forests are of particular concern due to substantial increases in fire activity projected under most future climate scenarios (Flannigan *et al.*, 2005).

11.3.6.3 MEXICO

For Mexican forests, deforestation will continue to cause large carbon emissions in the years to come. However, government programs (since 2001) are trying to reduce deforestation rates and forest degradation, implement sustainable forestry in native forests, promote commercial plantations and diverse agroforestry systems, and promote afforestation and protection of natural areas (Masera *et al.*, 1997).

11.4 OPTIONS FOR MANAGEMENT

Forest management strategies can be adapted to increase the amount of carbon uptake by forest systems. Alternative strategies for wood products are also important in several ways: how long carbon is retained in use, how much wood is used for biofuel, and substitution of wood for other materials that use more energy to produce. The net effect of these management and production strategies on carbon stocks and emissions will depend on emerging government policies for greenhouse gas management, the area of forests under management, management objectives for resources other than carbon, and the type of management and production regime being considered.

The forest sector includes a variety of activities that can contribute to increasing carbon sequestration, including: afforestation, mine land reclamation, forest restoration, agroforestry, forest management, biomass energy, forest preservation, wood products management, and urban forestry (Birdsey *et al.*, 2000). Although the science of managing forests specifically for carbon sequestration is not well developed, some ecological principles are emerging to guide management decisions (Appendix E). The prospective role of forestry in helping to stabilize atmospheric CO_2 depends on government policy, harvesting and disturbance rates, expectations of future forest productivity, the fate and longevity of forest products, and the ability to deploy technology and forest practices to increase the retention of sequestered CO_2. Market factors are also important in guiding the behavior of the private sector.

For Canada, Price *et al.* (1997) examined the effects of reducing natural disturbance, manipulating stand density, and changing rotation lengths for a forested landscape in northwest Alberta. By replacing natural disturbance (fire) with a simulated harvesting regime, they found that long-term equilibrium carbon storage increased from 105 to 130 Mt C. Controlling stand density following harvest had minimal impacts in the short term but increased landscape-level carbon storage by 13% after 150 years. Kurz *et al.* (1998) investigated the impacts on landscape-level carbon storage of the transition from natural to managed disturbance regimes. For a boreal landscape in northern Quebec, a simulated fire disturbance interval of 120 yr was replaced by a harvest cycle of 120 yr. The net impact was that the average age of forests in the landscape declined from 110 yr to 70 yr, and total carbon storage in forests declined from 16.3 to 14.8 Mt C (including both ecosystem and forest products pools).

Market approaches and incentive programs to manage greenhouse gases, particularly CO_2, are under development in the United States, the European Union, and

> Substantial increases in fire activity for North American boreal forests are projected under most future climate scenarios.

elsewhere (Totten, 1999). Since forestry activities have highly variable costs because of site productivity and operational variability, most recent studies of forestry potential develop "cost curves", *i.e.*, estimates of how much carbon will be sequestered by a given activity for various carbon prices (value in a market system) or payments (in an incentive system). There is also a temporal dimension to the analyses because the rate of change in forest carbon stocks is variable over time, with forestry activities tending to have a high initial rate of net carbon sequestration followed by a lower or even a negative rate as forests reach advanced age.

In the United States, a bundle of forestry activities could potentially increase carbon sequestration from -100 to -200 Mt C per year according to several studies (Birdsey *et al.*, 2000; Lewandrowski *et al.*, 2004; Environmental Protection Agency, 2005; Stavins and Richards, 2005). The rate of annual mitigation would likely decline over time as low-cost forestry opportunities become scarcer, forestry sinks become saturated, and timber harvesting takes place.

Table 11.5 Illustrative emissions reduction potential of various forestry activities in the United States under a range of prices and sequestration rates[a].

Forestry activity	Carbon sequestration rate (t CO_2 per ha per year)	Price range ($/t CO_2)	Emissions reduction potential (Mt CO_2 per year)
Afforestation	5.4–23.5	15–30	137–823
Forest management	5.2–7.7	1–30	25–314
Biofuels	11.8–13.6	30–50	375–561

[a] Adapted from Environmental Protection Agency (2005). Maximum price analyzed was $50/t CO_2.

Economic analyses of the U.S. forestry potential have focused on three broad categories of activities: afforestation (conversion of agricultural land to forest), improved management of existing forests, and use of woody biomass for fuel. Improved management of existing forest lands may be attractive to landowners at carbon prices below $10 per ton of CO_2; afforestation requires a moderate price of $15 per ton of CO_2 or more to induce landowners to participate; and biofuels become dominant at prices of $30-50 per ton of CO_2 (Lewandrowski *et al.*, 2004; Stavins and Richards, 2005; Environmental Protection Agency, 2005). Table 11.5 shows a simple scenario of emissions reduction below baseline, annualized over the time period 2010-2110, for forestry activities as part of a bundle of reduction options for the land base.

Production of renewable materials that have lower life-cycle emissions of greenhouse gases than non-renewable alternatives is a promising strategy for reducing emissions. Lippke *et al.* (2004) found that wood components used in residential construction had lower emissions of CO_2 from energy inputs than either concrete or steel.

Co-benefits are vitally important for inducing good forest carbon management. For example, conversion of agricultural land to forest will generally have positive effects on water, air, and soil quality and on biodiversity. In practice, some forest carbon sequestration projects have already been initiated even though sequestered carbon has little current value (Winrock International, 2005). In many of the current projects, carbon is a secondary objective that supports other landowner interests, such as restoration of degraded habitat. But co-effects may not all be beneficial. Water quantity may decline because of increased transpiration by trees relative to other vegetation. And taking land out of crop production may affect food prices—at higher carbon prices, nearly 40 million ha may be converted from cropland to forest (Environmental Protection Agency, 2005). Implementation of a forest carbon management policy will need to carefully consider co-effects, both positive and negative.

11.5 DATA GAPS AND INFORMATION NEEDS FOR DECISION SUPPORT

Decisions concerning carbon storage in North American forests and their management as carbon sources and sinks will be significantly improved by (1) filling gaps in inventories of carbon pools and fluxes, (2) a better understanding of how management practices affect carbon in forests, and (3) the increased availability of decision support tools for carbon management in forests.

11.5.1 Major Data Gaps in Estimates of Carbon Pools and Fluxes

Effective carbon policy and management to increase carbon sequestration and/or reduce emissions requires thorough understanding of current carbon stock sizes and flux rates, and responses to disturbance. Data gaps complicate analyses of the potential for policies to influence natural, social, and economic drivers that can change carbon stocks and fluxes. Forests in an area as large as North America are quite diverse, and comprehensive data sets that can be used to analyze forestry opportunities, such as spatially explicit historical

management and disturbance rates and effects on the carbon cycle, would enable managers to change forest carbon stocks and fluxes. Although this report provides aggregate statistics on forest carbon by biome and country, users could benefit from spatially explicit estimates of forest carbon. Such an analysis might involve matching estimates based on forest inventories as presented by political unit and general forest type (Birdsey and Lewis, 2003) with data developed using remote sensing techniques (Running *et al.*, 2004). Research at the level of individual sites has proven the feasibility of this combination (*e.g.*, Van Tuyl *et al.*, 2005; Turner *et al.*, 2006). This kind of analysis could facilitate development of a forest carbon map for North America.

In the United States, the range of estimates of the size of the land carbon sink is between -0.30 and -0.58 Mt C per year (Pacala *et al.*, 2001). Significant data gaps among carbon pools include carbon in wood products, soils, woody debris, and water transport (Birdsey, 2004; Pacala *et al.*, 2001). Geographic areas that are poorly represented in the available data sets include much of the Intermountain Western United States and Alaska, where forests of low productivity have not been inventoried as intensively as more productive timberlands (Birdsey, 2004). Accurate quantification of the relative magnitude of various causal mechanisms at large spatial scales is not yet possible, although research is ongoing to combine various approaches and data sets: large-scale observations, process-based modeling, ecosystem experiments, and laboratory investigations (Foley and Ramankutty, 2004).

Data gaps exist for Canada, particularly regarding changes in forest soil carbon and forest lands that are considered "unmanaged" (17% of forest lands). Aboveground biomass is better represented in forest inventories; however, the information needs to be updated and made more consistent among provinces. The new Canadian National Forest Inventory, currently under way, will provide a uniform coverage at a 20 × 20 km grid that will be the basis for future forest carbon inventories. Data are also lacking on carbon fluxes, particularly those due to insect outbreaks and forest stand senescence. The ability to model forest carbon stock changes has considerably improved with the release of the Carbon Budget Model of the Canadian Forest Sector (CBM-CFS3)(Kurz *et al.*, 2002); however, the CBM-CFS3 was not designed to incorporate climate change impacts (Price *et al.*, 1999; Hogg and Bernier, 2005).

For Mexico, there is very little data about measured carbon stocks for all forest types. Information on forest ecosystem carbon fluxes is primarily based on deforestation rates, while

fundamental knowledge of carbon exchange processes in almost all forest ecosystems is missing. That information is essential for understanding the effects of both natural and human-induced drivers (hurricanes, fires, insect outbreaks, climate change, migration, and forest management strategies), which all strongly impact the forest carbon cycle. Current carbon estimates are derived from studies in preferred sites in natural reserves with species-rich tropical forests. Therefore, inferences made from the studies on regional and national carbon stocks and fluxes probably give biased estimates on the carbon cycle.

11.5.2 Major Data Gaps in Knowledge of Forest Management Effects

There is insufficient information available to guide land managers in specific situations to change forest management practices to increase carbon sequestration, and there is some uncertainty about the longevity of effects (Caldeira *et al.*, 2004). This reflects a gap in the availability of inexpensive techniques for measuring, monitoring, and predicting changes in ecosystem carbon pools at the smaller scales appropriate for managers. There is more information available about management effects on live biomass and woody debris, and less about effects on soils and wood products. This imbalance in data has the potential to produce unintended consequences if predicted results are based on incomplete carbon accounting.

In the tropics, agroforestry systems offer a promising economic alternative to slash-and-burn agriculture, including highly effective soil conservation practices and mid-term and long-term carbon mitigation options (Soto-Pinto *et al.*, 2001; Nelson and de Jong, 2003; Albrecht and Kandji, 2003). However, a detailed assessment of current implementations of agroforestry systems in different regions of Mexico is missing. Agroforestry also has potential in temperate agricultural landscapes, but as with forest management, there

is a lack of data about how specific systems affect carbon storage (Nair and Nair, 2003).

Refining management of forests to realize significant carbon sequestration, while at the same time continuing to satisfy the needs of forests and the services they provide (*e.g.*, timber harvest, recreational value, watershed management) will require a multi-criteria decision support framework for a holistic and adaptive management program of the carbon cycle in North American forests. For example, methods should be developed for enhancing the efficiency of forest management, increasing the carbon storage per acre from existing forests, or even increasing the acreage devoted to forest systems that provide carbon sequestration. Currently there is little information about how appropriate incentives might be applied to accomplish these goals effectively, but given the importance of forests in the global carbon cycle, success in this endeavor could have important long-term and large-scale effects on global atmospheric carbon stocks.

> Given the importance of forests in the global carbon cycle, success in enhancing the efficiency of forests as a renewable energy source could have important long-term and large-scale effects on global atmospheric carbon stocks.

11.5.3 Availability of Decision Support Tools

Few decision support tools for land managers that include complete carbon accounting are available; one example is the CBM-CFS3 carbon accounting model (Kurz *et al.*, 2002). Some are in development or have been used primarily in research studies (Proctor *et al.*, 2005; Potter *et al.*, 2003). As markets emerge for trading carbon credits, and if credits for forest management activities have value in those markets, then the demand for decision support tools will encourage their development.

12

PTER

Carbon Cycles in the Permafrost Region of North America

Lead Author: Charles Tarnocai, Agriculture and Agri-Food Canada

Contributing Authors: Chien-Lu Ping, Univ. Alaska; John Kimble, USDA NRCS (retired)

KEY FINDINGS

- Much of northern North America (more than 6 million square kilometers) is characterized by the presence of permafrost (soils or rocks that remain frozen for at least two consecutive years). This permafrost region contains approximately 25% of the world's total soil organic carbon, a massive pool of carbon that is vulnerable to release to the atmosphere as carbon dioxide in response to an already detectable polar warming.

- The soils of the permafrost region of North America contain 213 billion tons of organic carbon, approximately 61% of the carbon in all soils of North America.

- The soils of the permafrost region of North America are currently a net sink of approximately 11 million tons of carbon per year.

- The soils of the permafrost region of North America have been slowly accumulating carbon for the last 5000–8000 years. More recently, increased human activity in the region has resulted in permafrost degradation and at least localized loss of soil carbon.

- Patterns of climate, especially the region's cool and cold temperatures and their interaction with soil hydrology to produce wet and frozen soils, are primarily responsible for the historical accumulation of carbon in the region. Non-climatic drivers of carbon change include human activities, including flooding associated with hydroelectric development, that degrade permafrost and lead to carbon loss. Fires, increasingly common in the region, also lead to carbon loss.

- Projections of future warming of the polar regions of North America lead to projections of carbon loss from the soils of the permafrost region, with upwards of 78% (34 billion tons) and 41% (40 billion tons) of carbon stored in soils of the Subarctic and northern-most coniferous (Boreal) regions, respectively, being severely or extremely severely affected by future climate change.

- Options for management of carbon in the permafrost region of North America, including construction methods that cause as little disturbance of the permafrost and surface as possible, are primarily those which avoid permafrost degradation and subsequent carbon losses.

- Most research needs for the permafrost region are focused on reducing uncertainties in knowing how much carbon is vulnerable to a warming climate and how sensitive that carbon loss is to climate change. Development and adoption of measures that reduce or avoid the negative impact of human activities on permafrost are also needed.

12.1 INTRODUCTION

It is especially important to understand the carbon cycle in the permafrost region of North America because the soils in this area contain large amounts of organic carbon that is vulnerable to release to the atmosphere as carbon dioxide (CO_2) and methane (CH_4) in response to climate warming. It is predicted that the average annual air temperature in the permafrost region will increase 3–4°C by 2020 and 5–10°C by 2050 (Hengeveld, 2000, see Box 12.1)[†]. The soils in this region contain approximately 61%[***] of the organic carbon occurring in all soils in North America (Lacelle *et al.*, 2000) even though the permafrost area covers only about 21%[***] of the soil area of the continent. Release of even a fraction of this carbon in greenhouse gases could have global consequences.

Permafrost is defined, on the basis of temperature, as soils or rocks that remain below 0°C for at least two consecutive years (van Everdingen, 1998 revised May 2005). Permafrost terrain often contains large quantities of ground ice in the upper section of the permafrost. If this terrain is well protected by forests or peat, this ground ice is generally in equilibrium with the current climate. If this insulating layer is not sufficient, however, even small temperature changes, especially in the southern part of the permafrost region, could cause degradation and result in severe thermal erosion (thawing). For example, some of the permafrost that formed in central Alaska during the Little Ice Age is now degrading in response to warming during the last 150 years (Jorgenson *et al.*, 2001).

> Some of the permafrost that formed in central Alaska during the Little Ice Age is now degrading in response to warming during the last 150 years.

The permafrost region in North America is divided into four zones on the basis of the percentage of the land area underlain by permafrost (Figure 12.1). These zones are the Continuous Permafrost Zone (≥90 to 100%), the Discontinuous

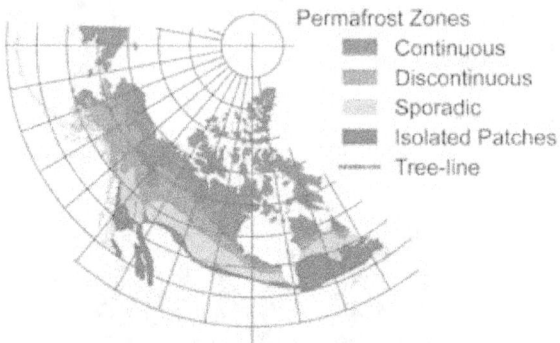

Figure 12.1 Permafrost zones in North America (Brown *et al.*, 1997).

Permafrost Zone (≥50 to <90%), the Sporadic Permafrost Zone (≥10 to <50%), and the Isolated Patches Permafrost Zone (0 to <10%) (Brown *et al.*, 1997).

These permafrost zones encompass three major ecoclimatic provinces (ecological regions) (Figure 12.2): the Arctic (north of the arctic tree line), the Subarctic (open canopy coniferous forest), and the Boreal (closed canopy forest, either

Figure 12.2 Arctic, Subarctic, and Boreal ecoclimatic provinces (ecological regions) in North America (Ecoregions Working Group, 1989; Baily and Cushwa, 1981).

BOX 12.1: CCSP SAP 2.2 Uncertainty Conventions

*****	= 95% certain that the actual value is within 10% of the estimate reported,
****	= 95% certain that the estimate is within 25%,
***	= 95% certain that the estimate is within 50%,
**	= 95% certain that the estimate is within 100%, and
*	= uncertainty greater than 100%.
†	= The magnitude and/or range of uncertainty for the given numerical value(s) is not provided in the references cited.

coniferous or mixed coniferous and deciduous). Peatlands (organic wetlands characterized by more than 40 cm of peat accumulation) cover large areas in the Boreal, Subarctic, and southern part of the Arctic ecoclimatic provinces.

Although northern ecosystems (Arctic, Subarctic, and Boreal) in North America cover approximately 14% of the global land area, they contain approximately 25% of the world's total soil organic carbon (Oechel and Vourlitis, 1994) [†]. In addition, Oechel and Vourlitis (1994) indicate that the tundra (Arctic) ecosystems alone contain approximately 12% of the global soil carbon pool, even though they account for only 6% of the total global land area[†]. Based on direct measure of the carbon density to one meter depth, the soil carbon pool should be doubled (Michaelson *et al.*, 1996). The soils of the permafrost region of North America are currently a carbon sink and are unique because they are able to actively sequester carbon and store it for thousands of years.

The objectives of this chapter are to give the below-ground carbon stocks and to explain the mechanisms associated with the carbon cycle (sources and sinks) in the soils of the permafrost region of North America.

12.2 PROCESSES AFFECTING THE CARBON CYCLE IN A PERMAFROST ENVIRONMENT

12.2.1 Soils of the Permafrost Region

Soils cover approximately 6,211,340 square kilometers (km^2) [***] of the area of the North American permafrost region (Tables 12.1 and 12.2), with approximately 58%[***] of the land area being occupied by permafrost-affected (perennially frozen) soils (Cryosols/Gelisols) and the remainder by non-permafrost soils (Soil Carbon Database Working Group, 1993). Approximately 17%[***] of this area is associated with organic soils (peatlands), the remainder with mineral soils (Soil Carbon Database Working Group, 1993). It is important to distinguish between mineral soils and organic soils in the region because different processes are responsible for the carbon cycle in these two types of soils.

12.2.2 Mineral Soils

Table 12.1 Areas of mineral soils in the various permafrost zones.

Permafrost zones	Area (10^3 km^2)		
	Canada[a]	Alaska[b]	Total
Continuous	2001.80	353.46	2355.26
Discontinuous	636.63	479.15	1115.78
Sporadic	717.63	110.98	828.61
Isolated Patches	868.08	0.73	868.81
Total	4224.14	944.32	5168.46

[a] Calculated using the Soil Carbon of Canada Database (Soil Carbon Database Working Group, 1993).
[b] Calculated using the Northern and Mid Latitudes Soil Database (Cryosol Working Group, 2001).

Table 12.2 Areas of peatlands (organic soils) in the various permafrost zones.

Permafrost zones	Area (10^3 km^2)		
	Canada[a]	Alaska[b]	Total
Continuous	176.70	51.31	228.01
Discontinuous	243.51	28.74	272.25
Sporadic	307.72	0.62	308.34
Isolated Patches	221.23	13.05	234.28
Total	949.16	93.72	1042.88

[a] Calculated using the Peatlands of Canada Database (Tarnocai *et al.*, 2005).
[b] Calculated using the Northern and Mid Latitudes Soil Database (Cryosol Working Group, 2001).

The schematic diagram in Figure 12.3 provides general information about the carbon sinks and sources in mineral soils. Most of the permafrost-affected mineral soils are carbon sinks because of the slow decomposition rate due to cold and wet conditions and the process of cryoturbation, which moves surface organic matter into the deeper soil layers. Other processes, such as decomposition, wildfires, and thermal degradation, release carbon into the atmosphere and, thus, act as carbon sources.

For unfrozen soils and noncryoturbated frozen soils in the permafrost region, the carbon cycle is similar to that in soils occurring in temperate regions. In these soils, organic matter is deposited on the soil surface. Some soluble organic matter may move downward, but because these soils are not affected by cryoturbation, they have no mechanism for moving organic matter from the surface into the deeper soil layers and preserving it from decomposition and wildfires. Most of their below-ground carbon originates from roots and its residence time is relatively short.

The role of cryoturbation: Although permafrost-affected ecosystems pro-

Carbon Sinks

Permafrost-affected soil with a thick surface organic layer, dark-colored organic intrusions in the brown soil layer, and an underlying frozen, high-ice-content layer. The organic intrusions were translocated from the surface by cryoturbation. (Mackenzie Valley, Canada)

Carbon sources

Eroding high-ice-content permafrost soil composed of a dark frozen soil layer with an almost pure ice layer below. The thawing process generated a flow slide in which high-organic-content soil materials slumped into the water-saturated environment. (Mackenzie Delta area, Canada)

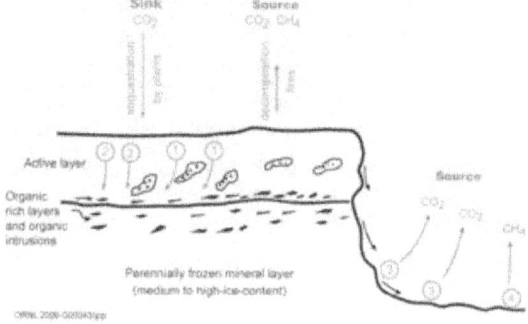

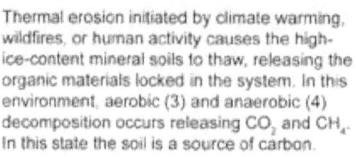

Perennially frozen deposit composed of an active layer that freezes and thaws annually and an underlying perennially frozen layer that has a high ice content.

Organic material deposited annually on the soil surface builds up as an organic soil layer. Some of this surface organic material is translocated into the deeper soil layers by cryoturbation (1). In addition, soluble organic matter is translocated into the deeper soil layers by movement of water to the freezing front and by gravity (2). Because these deeper soil layers have low temperatures (0 to −15°C), the organic material decomposes very slowly. Thus, more organic material accumulates as long as the soil is frozen. In this state, the permafrost soil acts as a carbon sink.

Thermal erosion initiated by climate warming, wildfires, or human activity causes the high-ice-content mineral soils to thaw, releasing the organic materials locked in the system. In this environment, aerobic (3) and anaerobic (4) decomposition occurs releasing CO_2 and CH_4. In this state the soil is a source of carbon.

Figure 12.3 Carbon cycle in permafrost-affected upland (mineral) soils, showing below-ground organic carbon sinks and sources.

decomposing as a result of biological activity. A large portion of this litter, however, builds up on the soil surface, forming an organic soil horizon. Cryoturbation causes some of this organic material to move down into the deeper soil layers (Bockheim and Tarnocai, 1998). Soluble organic materials move downward because of the effect of gravity and the movement of water along the thermal gradient toward the freezing front (Kokelj and Burn, 2005). Once the organic material has moved down to the cold, deeper soil layers where very little or no biological decomposition takes place, it may be preserved for many thousands of years. Radiocarbon dates from cryoturbated soil materials ranged between 490 and 11,200 yr BP (Zoltai *et al.*, 1978). These dates were randomly distributed within the soil and did not appear in chronological sequence by depth (the deepest material was not necessarily the oldest), indicating that cryoturbation is an ongoing process.

The permafrost table (top of the permafrost) is very dynamic and is subject to deepening due to factors such as removal of vegetation and/or the insulating surface organic layer, wildfires, global climate change, and other natural or human activities. When this occurs, the seasonally thawed layer (active layer) becomes deeper and the organic material is able to move even deeper into the soil (translocation). However, if such factors cause thawing of the soil and melting of the ground ice, some or all of the organic materials locked in the system could be exposed to the atmosphere. This change in soil environment gives rise to both aerobic and anaerobic decomposition, releasing carbon into the atmosphere as CO_2 and CH_4, respectively (Figure 12.3). At this stage, the soil can become a major carbon source.

If, however, the permafrost table rises (and the active layer becomes shallower) because of reestablishment of the vegetation or buildup of the surface organic layer, this deep organic material becomes part of the permafrost and is,

duce much less biomass than do temperate ecosystems, permafrost-affected soils that are subject to cryoturbation (frost-churning), a cryogenic process, have a unique ability to sequester a portion of this organic matter and store it for thousands of years. A number of models have been developed to explain the mechanisms involved in cryoturbation (Mackay, 1980; Van Vliet-Lanoë, 1991; Vandenberghe, 1992). The most recent model involves the process of differential frost heave (heave-subsidence), which produces downward and lateral movement of materials (Walker *et al.*, 2002; Peterson and Krantz, 2003).

Part of the organic matter produced annually by the vegetation is deposited as litter on the soil surface, with some thus, more securely preserved. This is the main reason that permafrost-affected soils contain high amounts of organic carbon not only in the upper (0–100 cm) layer, but also in the deeper layers. These cryoturbated, permafrost-affected soils are effective carbon sinks.

12.2.3 Peatlands (Organic Soils)

The schematic diagram in Figure 12.4 provides general information about the processes driving the carbon sinks and sources in peatland soils. The water-saturated conditions, low soil temperatures, and acidic conditions of northern peatlands provide an environment in which very little decomposition occurs; hence, the litter is converted to peat and preserved. This gradual buildup process has been ongoing in peatlands during the last 5000–8000 years, resulting in peat deposits that are an average of 2–3 meters (m) thick

and, in some cases, up to 10 m thick. At this stage, peatlands can act as very effective carbon sinks for many thousands of years (Figure 12.4).

Carbon dynamics: Data for carbon accumulation in various peatland types in the permafrost regions are given in Table 12.3. Although some values for the rate of peat accumulation are higher (associated with unfrozen peatlands), the value for frozen peatlands, which are more widespread, is 13.31 ± 2.20 grams of carbon per square meter (g C per m²) per year (Robinson and Moore, 1999). Peat accumulations in the various ecological regions were calculated on the basis of the thickness of the deposit and the date of the basal peat. The rate of peat accumulation is generally highest in the Boreal region and decreases northward (Table 12.3). Note, however, that if the surface of the peat deposit has eroded, the calculated rate of accumulation (based on the age of the basal peat and a decreased deposit thickness) will appear to be higher than it should be. This is probably the reason for some of the high rates of peat accumulation found for the Arctic region, which likely experienced a rapid rate of accumulation during the Hypsithermal Maximum with subsequent erosion of the surface of some of the deposits reducing their thicknesses. Wildfires, decomposition, and leaching of soluble organic compounds release approximately one-third of the carbon input, causing most of the carbon loss in these peatlands.

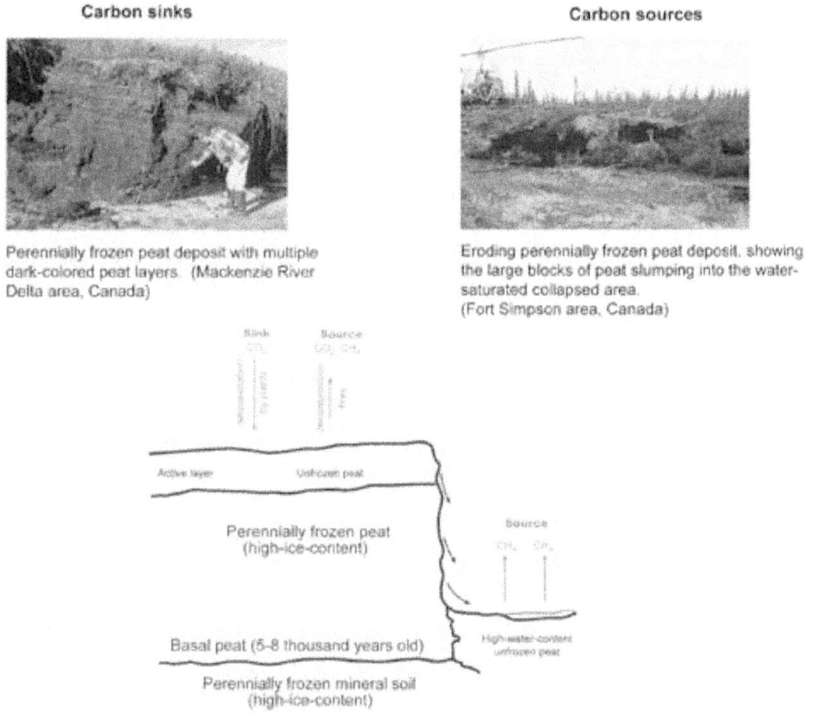

12.3 BELOW-GROUND CARBON STOCKS

The carbon content of mineral soils to a 1-m depth is 49–61 kilograms (kg) per m² for permafrost-affected soils and 12–17 kg per m² for unfrozen soils (Tables 12.4 and 12.5). The carbon content of organic soils (peatlands) for the total depth of the deposit is 81–129 kg per m² for permafrost-affected soils and 43–144 kg per

Carbon sinks

Perennially frozen peat deposit with multiple dark-colored peat layers. (Mackenzie River Delta area, Canada)

Carbon sources

Eroding perennially frozen peat deposit, showing the large blocks of peat slumping into the water-saturated collapsed area. (Fort Simpson area, Canada)

Perennially frozen peat deposits consist of an active layer that freezes and thaws annually and an underlying perennially frozen layer composed of ice-rich frozen peat and mineral materials.

Organic material is deposited annually on the peatland surface. Although a large portion (≥ 90%) of this organic material decomposes, the remainder is added to the peat deposit, producing an annual peat accumulation. The low soil temperatures (0 to -15°C) and the water-saturated and acid conditions cause this added organic carbon to be preserved and stored. This has been occurring for the last 5–8 thousand years. In this state, the peatland is a carbon sink.

Thermal erosion (thawing) of frozen peat deposits occurs as a result of climate change, wildfires, or human disturbances, releasing large amounts of water from the melting ice. This is mixed with the slumped peat material, initiating anaerobic decomposition in the much warmer environment. Anaerobic decomposition produces CH₄, which is expelled into the atmosphere. In this state, the peatland is a source of carbon.

Figure 12.4 Carbon cycle in permafrost peatlands, showing below-ground organic carbon sinks and sources.

Table 12.3 Organic carbon accumulation and loss in various Canadian peatlands. Positive values indicate net flux into the atmosphere (source); negative values indicate carbon sequestration (land sinks).

Peatlands	Amount of carbon
Boreal peatlands	–9.8 Mt per year[a]
All Canadian peatlands	–30 Mt per year[b]
All mineral and organic soils	–18 mg per m² per year[c]
Rich fens	–13.58 ± 1.07 g per m² per year[d]
Poor fens (unfrozen, Discontinuous Permafrost Zone)	–20.34 ± 2.86 g per m² per year[d]
Peat plateaus (frozen, Discontinuous Permafrost Zone)	–13.31 ± 2.20 g per m² per year[d]
Collapse fens	–13.54 ± 1.50 g per m² per year[d]
Bogs (unfrozen, Discontinuous Permafrost Zone)	–21.81 ± 3.25 g per m² per year[d]
Dissolved organic carbon (DOC)	+2 g per m² per year[e]
Arctic peatlands	–0 to –16 cm/100 yr[f]
Subarctic peatlands	–2 to –5 cm/100 yr[f]
Boreal peatlands	–2 to –11 cm/100 yr[f]
Carbon release by each fire in northern boreal peatlands	+1.46 kg C per m²[g]
Carbon release by fires in all terrain	+27 Mt per year[h]
Carbon release by fires in Western Canadian peatlands	+5.9 Mt per year[h]

[a] Zoltai et al. (1988).

[b] Gorham (1988).

[c] Liblik et al. (1997).

[d] Robinson and Moore (1999).

[e] Moore (1997).

[f] Calculated based on the thickness of the deposit and the date of the basal peat (National Wetlands Working Group, 1988).

[g] Robinson and Moore (2000).

[h] Turetsky et al. (2004).

Note: Except as explicitly indicated otherwise, no estimates of the confidence, certainty, or uncertainty of the numerical values in this table are available.

m² for unfrozen soils (Tables 12.4 and 12.5) (Tarnocai, 1998 and 2000).

Soils in the permafrost region of North America contain 213 billion tons (Gt) of organic carbon (Tables 12.6 and 12.7), which is approximately 61% of the organic carbon in all soils on this continent (Lacelle et al., 2000). Mineral soils contain approximately 99 Gt of organic carbon in the 0- to 100-cm depth (Table 12.6). Although peatlands (organic soils) cover a smaller area than mineral soils (17% vs. 83%), they contain approximately 114 Gt of organic carbon in the total depth of the deposit, or more than half (54%) of the soil organic carbon of the region (Table 12.7).

12.4 CARBON FLUXES

12.4.1 Mineral Soils

Very little information is available about carbon fluxes in both unfrozen and perennially frozen mineral soils in the permafrost regions. For unfrozen upland mineral soils, Trumbore and Harden (1997) report a carbon accumulation of 60-100 g C per m² per year (Table 12.4). They further indicate that the slow decomposition results in rapid organic matter accumulation, but the turnover time due to wildfires (every 500–1000 years) eliminates the accumulated carbon except for the deep carbon derived from roots in the subsoil. The turnover time for this deep carbon is 100-1600 years. Therefore, the carbon stocks in these unfrozen soils are low, and the turnover time of this carbon is 100 to 1600 years.

As with unfrozen mineral soils, very little information has been published on the carbon cycle in perennially frozen mineral soils. The carbon cycle in these soils differs from that in unfrozen soils in that, because of cryogenic activities, these soils are able to move the organic matter deposited on the soil surface into the deeper soil layers. Assuming that cryoturbation was active in these soils during the last six thousand years (Zoltai et al., 1978), an average of 9 million tons of carbon (Mt C)** have been added annually to these soils. Most of this carbon has been cryoturbated into the deeper soil layers, but some of the carbon in the surface organic layer is released by decomposition and, periodically, by wildfires. The schematic diagram in Figure 12.5 shows the carbon cycle in these soils.

12.4.2 Peatlands (Organic Soils)

Peatland vegetation deposits various amounts of organic material (litter) annually on the peatland surface. Reader and Stewart (1972) found that the amount of litter (dry biomass) deposited annually on the bog surface in boreal peatlands in Manitoba, Canada was 489-1750 g per m². Approximately 25% of the original litter fall was found to have decomposed during the following year. In the course of the study, they found that the average annual accumulation rate was 10% of the annual net primary production. Robinson and Moore (1999) and Robinson et al. (2003) found that, in the Sporadic Permafrost Zone, mean carbon accumulation rates over the past 100 years for unfrozen bogs and frost mounds were 88.6 ± 4.4 and 78.5 ± 8.8 g per m² per year, respectively. They also found that, in the Discontinuous Permafrost Zone, the mean carbon accumulation rate during the past 1200 years in frozen peat plateaus was 13.31 ± 2.20 g per m² per year, while in unfrozen fens

Table 12.4 Soil carbon pools and fluxes for the permafrost areas of Canada. Positive flux numbers indicate net flux into the atmosphere (source); negative values indicate carbon sequestration (land sinks).

Type	Peatlands		Mineral soils		Total
	Perennially frozen	Unfrozen	Perennially frozen	Unfrozen	
Current area (× 10³ km²)	422[a]	527[a]	2088[b]	2136[b]	5173
Current pool (Gt)	47[c]	65[a]	56[c]	28[b]	196
Current atm. flux (g per m² per year)	−5.7[d]	−15.2[e]			
Carbon accumulation (g per m² per year)	−13.3 ± 2.20[f]	−20.34 ± 2.86 to −21.81 ± 3.25[f]		−60 to −100[g]	
Carbon release by fires (g per m² per year)[h]	+7.57[i]				
Methane flux (g per m² per year)		+2.0[j]			

[a] Calculated using the Peatlands of Canada Database (Tarnocai et al., 2005).

[b] Calculated using the Soil Carbon of Canada Database (Soil Carbon Database Working Group, 1993).

[c] Tarnocai (1998).

[d] Using C accumulation rate of 0.13 mg per ha per year (this report).

[e] Using C accumulation rate of 0.194 mg per ha per year (Vitt et al., 2000).

[f] Robinson and Moore (1999).

[g] Trumbore and Harden (1997).

[h] Fires recur every 150–190 years (Kuhry, 1994; Robinson and Moore, 2000).

[i] Robinson and Moore (2000).

[j] Moore and Roulet (1995).

Note: Except as explicitly indicated otherwise, no estimates of the confidence, certainty, or uncertainty of the numerical values in this table are available.

The schematic diagram presented in Figure 12.6 summarizes the carbon cycle in peatlands in the permafrost region. Based on average values for the rate of peat accumulation, approximately 17 g C per m² per year, or 18 Mt C, is added annually to peatlands in this region of North America. Approximately 1.46 kg C per m² is released to the atmosphere every 600 years by wildfires in the northern boreal peatlands. In addition, decomposition of unfrozen peatlands releases approximately 2.0 g C per m² per year, and a further 2.0 g C per m² per year is released by leaching of dissolved organic carbon (DOC), leading to a carbon decrease of approximately 4 Mt** annually, not including that released by wildfires (Figure 12.6). Note that these values are based on current measurements. However, rates of peat accumulation have varied during the past 5000–8000 years, with periods during which the rate of peat accumulation was much higher than at present.

and bogs the comparable rates were 20.34 ± 2.86 and 21.81 ± 3.25 g per m² per year, respectively.

Because peatlands cover large areas in the permafrost region of North America, their contribution to the carbon stocks is significant (Table 12.5). Zoltai et al. (1988) estimated that the annual carbon accumulation capacity of boreal peatlands is approximately 9.8 Mt[†]. Gorham (1988), in contrast, estimated that Canadian peatlands accumulate approximately 30 Mt C[†] annually.

Currently, wildfires are probably the greatest natural force in converting peatlands to a carbon source. Ritchie (1987) found that the western Canadian boreal forests have a fire return interval of 50-100 years, while Kuhry (1994) indicated that, for wetter Sphagnum bogs, the interval is 400-1700 years. For peat plateau bogs, each fire resulted in an average decrease in carbon mass of 1.46 kg per m² and an average decrease in height of 2.74 cm, which represents about 150 years of peat accumulation (Robinson and Moore, 2000). In recent years, the number of these wildfires has increased, as has the area burned, releasing increasing amounts of carbon into the atmosphere.

The soils in the permafrost region of North America currently act as a net carbon sink.

Table 12.5 Average organic carbon content for soils in the various ecological regions (Tarnocai, 1998 and 2000).

Ecological regions	Average carbon content (kg per m²)			
	Mineral soils[a]		Organic soils (peatlands)[b]	
	Frozen	Unfrozen	Frozen	Unfrozen
Arctic	49	12	86	43
Subarctic	61	17	129	144
Boreal	50	16	81	134

[a] For the 1-m depth.

[b] For the total depth of the peat deposit.

Table 12.6 Organic carbon mass in mineral soils in the various permafrost zones.

Permafrost zones	Carbon mass^a (Gt)		
	Canada^b	Alaska^c	Total
Continuous	51.10	9.04	60.14
Discontinuous	10.33	4.82	15.15
Sporadic	9.15	0.75	9.90
Isolated Patches	13.59	0	13.59
Total	84.17	14.61	98.78

^{a} Calculated for the 0–100 cm depth.
^{b} Calculated using the Soil Carbon of Canada Database (Soil Carbon Database Working Group, 1993).
^{c} Calculated using the Northern and Mid Latitudes Soil Database (Cryosol Working Group, 2001).

Table 12.7 Organic carbon mass in peatlands (organic soils) in the various permafrost zones.

Permafrost zones	Carbon mass^a (Gt)		
	Canada^b	Alaska^c	Total
Continuous	21.82	1.46	23.28
Discontinuous	26.54	0.84	27.38
Sporadic	30.66	0.27	30.93
Isolated Patches	32.95	0	32.95
Total	111.97	2.57	114.54

^{a} Calculated for the total depth of the peat deposit.
^{b} Calculated using the Peatlands of Canada Database (Tarnocai et al., 2005).
^{c} Calculated using the Northern and Mid Latitudes Soil Database (Cryosol Working Group, 2001).

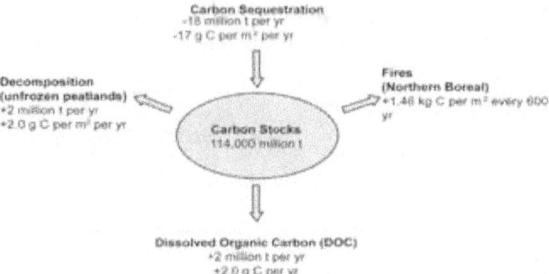

Figure 12.6 Carbon cycle in peatlands in the permafrost region.

12.4.3 Total Flux

Based on the limited data available for this vast, and largely inaccessible, area of the continent, approximately 27 Mt C per year** is deposited on the surface of mineral soils and peatlands (organic soils) in the permafrost region of North America. Approximately 8 Mt per year** of surface carbon (excluding vegetation) is released by decomposition and wildfires, and by leaching into the water systems. Thus, the soils in the permafrost region of North America currently act as a sink for approximately 19 Mt C per year** and as a source for approximately 8 Mt C per year** and are, therefore, a net carbon sink (Figures 12.5 and 12.6).

12.5 POSSIBLE EFFECTS OF GLOBAL CLIMATE CHANGE

The permafrost region is unique because the soils in this vast area contain large amounts of organic materials and much of the carbon has been actively sequestered by peat accumulation (organic soils) and cryoturbation (mineral soils) and stored in the permafrost for many thousands of years. Historical patterns of climate are responsible for the large amount of carbon found in the soils of the region today, but cryoturbation is a consequence of the region's current cool to cold climate and the effects of that climate on soil hydrology. As a result, patterns of climate and climate change are dominant drivers of carbon cycling in the region. Future climate change will determine the fate of that carbon and whether the region will remain a slow but significant carbon sink, or whether it will reverse and become a source, rapidly releasing large amounts of CO_2 and CH_4 to the atmosphere.

12.5.1 Peatlands

A model for estimating the sensitivity of peatlands to global climate change was developed using current climate (1x CO_2), vegetation, and permafrost data together with the changes in these variables expected in a 2x CO_2 environment (Kettles and Tarnocai, 1999). The data generated by this model were used to produce a peatland sensitivity map. Using geographic information system (GIS) techniques, this map was overlaid on the peatland map of Canada to

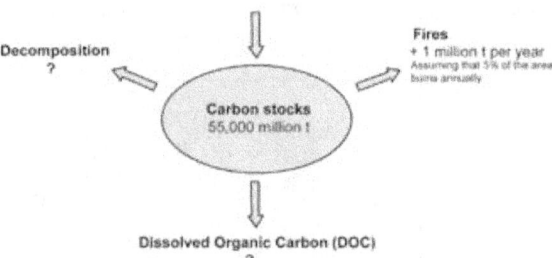

Figure 12.5 Carbon cycle in perennially frozen mineral soils in the permafrost region. Question marks represent data values that cannot be quantified.

determine both the sensitivity ratings of the various peatland areas and the associated organic carbon masses. The sensitivity ratings, or classes, used are no change, very slight, slight, moderate, severe, and extremely severe. Because global climate change is expected to have the greatest impact on the ecological processes and permafrost distribution in peatlands in the severe and extremely severe categories (Kettles and Tarnocai, 1999), the areas and carbon masses of peatlands in these two sensitivity classes are considered to be most vulnerable to climate change. The sensitivity ratings are determined by the degree of change in the ecological zonation combined with the degree of change in the permafrost zonation, with the greater the change, the more severe the sensitivity rating. For example, if a portion of the Subarctic becomes Boreal in ecology and the associated sporadic permafrost disappears (no permafrost remains in the region), the sensitivity of this region is rated as extremely severe. If however, a portion of the Boreal remains Boreal in ecology, but the discontinuous permafrost disappears (no permafrost remains in the region), the sensitivity of this region is rated as severe.

The peatland sensitivity model (Tarnocai, 2006) indicates that the greatest effect of global climate change will occur in the Subarctic region, where about 85% (314,270 km²) of the peatland area and 78% (33.96 Gt) of the organic carbon mass will be severely or extremely severely affected by climate change, with 66% of the area and 57% of the organic carbon mass being extremely severely affected (Figure 12.7) [†]. The second largest effect will occur in the Boreal region, where about 49% (353,100 km²) of the peatland area and 41% (40.20 Gt) of the organic carbon mass will be severely or extremely severely affected, with 10% of both the area and organic carbon mass being extremely severely affected.

These two regions contain almost all (99%) of the Canadian peatland area and organic carbon mass that is predicted to be severely or extremely severely affected (Figure 12.7) (Tarnocai, 2006).

In the Subarctic region and the northern part of the Boreal region, where most of the perennially frozen peatlands occur, the increased temperatures are expected to cause increased thawing of the perennially frozen peat. Thawing of the ice-rich peat and the underlying mineral soil will initially result in water-saturated conditions. These water-saturated conditions, together with the higher temperatures, result in anaerobic decomposition, leading to the production of CH_4.

In the southern part of the Boreal region, where the peatlands are generally unfrozen, the main impact is expected to be drought conditions resulting from higher summer temperatures and higher evapotranspiration. Under such conditions, peatlands become a net source of CO_2 because the oxygenated conditions lead to aerobic decomposition (Melillo *et al.*, 1990; Christensen, 1991). These dry conditions will likely also increase wildfires and, eventually, burning of peat, leading to the release of CO_2 to the atmosphere.

> The greatest effect of global climate change will occur in the Subarctic region, where about 85% of the peatland area and 78% of the organic carbon mass will be severely or extremely severely affected by climate change.

12.5.2 Permafrost-affected Mineral Soils

The same model described above was used to determine the effect of climate change on mineral permafrost-affected soils. The model suggests that approximately 21% (11.9 Gt)[†] of the total organic carbon in these soils could be severely or extremely severely affected by climate warming (Tarnocai, 1999). The model also suggests that the permafrost will probably disappear from the soils (the soils will become unfrozen) in the Sporadic and Isolated Patches permafrost zones. The main reason for the high sensitivity of mineral soils in these zones is that soil temperatures at both the 100- and 150-cm depths are only slightly below freezing (-0.3°C). The slightest disturbance or climate warming could initiate rapid thawing in these soils, with resultant loss of carbon (Tarnocai, 1999).

Figure 12.7 The organic carbon mass in the various sensitivity classes for the Subarctic and Boreal ecoclimatic provinces (ecological regions) (Tarnocai, 2006).

12.6 NON-CLIMATIC DRIVERS

Wildfires are an important part of the ecology of Boreal and Subarctic forests and are probably the major non-climatic drivers of carbon change in the permafrost region. There has been a rapid increase in both the frequency of fires and the area burned as a result of warmer and drier summers and increased human activity in the region. According to observations of natives, not only has the frequency of lightning strikes increased in the more southerly areas, but they have now appeared in more northerly areas where they were previously unknown. Because lightning is the major cause of wildfires in areas of little habitation, it is likely largely responsible for the increase in wildfires now being observed.

There has been a rapid increase in both the frequency of fires and the area burned as a result of warmer and drier summers and increased human activity in the region.

Increased human activity as a result of the construction of pipelines, roads, airstrips, and mines, expansion of agriculture, and development and expansion of town sites has disturbed the natural soil cover and exposed the organic-rich soil layers, leading to increased soil temperatures and, hence, decomposition of the exposed organic materials. Burgess and Tarnocai (1997), studying the Norman Wells Pipeline, provide some examples of the effect of pipeline construction on frozen peatlands and permafrost in Canada.

Shoreline erosion along rivers, lakes, and oceans and thermal erosion (thermokarst) are also common processes in the permafrost region, exposing the carbon-rich frozen soil layers to the atmosphere and making the organic materials available for decomposition. Along the 1957 km of the Beaufort Sea coast of Arctic Alaska, an estimated 1.8×10^5 Mg C per year erodes into the Arctic Ocean due to thawing of permafrost. As a result, CO_2 and CH_4 are released directly into the atmosphere, but most of this carbon goes into the ocean as particulate organic matter, and a small fraction as dissolved organic carbon (Jorgenson and Brown, 2005; Ping *et al.*, 2006).

Large hydroelectric projects in northern areas, such as Southern Indian Lake in Manitoba and the James Bay region of Quebec, have flooded vast areas of peatlands and initiated permafrost degradation and decomposition of organic carbon, some of which is released into the atmosphere as CH_4. Of greater immediate concern, however, is the carbon that has entered the water system as dissolved organic carbon. These compounds include contaminants such as persistent organic pollutants (*e.g.*, Polychlorinated biphenyls [PCBs], Dichloro-Diphenyl-Trichloroethane [DDT], Hexachlorocyclohexanes [HCH], and chlorobenzene [AMAP, 2004]) that have been widely distributed in northern ecosystems over many years, much of it deposited by snowfalls, concentrated by cryoturbation, and stored in the organic soils. Of particular concern is the release of methylmercury because peatlands are net producers of this compound (Driscoll *et al.*, 1998; Suchanek *et al.*, 2000), which is a much greater health hazard than inorganic or elemental mercury. Natives in the regions where these hydroelectric developments have taken place have developed mercury poisoning after ingesting fish contaminated by this mercury, leading to serious health problems for many of the people. This is an example of what can happen when permafrost degrades as a result of human activities. When climate warming occurs, the widespread degradation of permafrost, with the resulting release of such dangerous pollutants into the water systems, could cause serious health problems for fish, animals, and humans that rely on such waters.

12.7 OPTIONS FOR MANAGEMENT OF CARBON IN THE PERMAFROST REGION

Although wildfires are the most effective mechanism for releasing carbon into the atmosphere, they are also an important factor in maintaining the integrity of northern ecosystems. Therefore, such fires are allowed to burn naturally and are controlled only if they are close to settlements or other man-made structures.

The construction methods currently used in permafrost terrain are designed to cause as little surface disturbance as possible and to preserve the permafrost. Thus, the construction of pipelines, airstrips, and highways is commonly carried out in the winter so that the heavy equipment used will cause minimal surface disturbance.

The greatest threat to the region is a warmer (and possibly drier) climate, which would drastically affect not only the carbon cycle, but also the biological systems, including human life. Unfortunately, we know very little about how to manage the natural systems in this new environment.

12.8 DATA GAPS AND UNCERTAINTIES

The permafrost environment is a very complex system, and the data available for it are very limited with numerous gaps and uncertainties. Information on the distribution of soils in the permafrost region is based on small-scale maps, and the carbon stocks calculated for these soils are derived from a relatively small number of datasets. Although there is some understanding of the carbon sinks and sources in these soils, the limited amount of data available make it very difficult, or impossible, to assign reliable values. Only limited amounts of flux data have been collected for the permafrost-affected soils and, in some cases, it has been collected on sites that are not representative of the overall landscape. This makes it very difficult to scale this information up for a larger area. As Davidson and Janssens (2006) state:

> "...the unresolved question regarding peatlands and permafrost is not the degree to which the currently constrained decomposition rates are temperature sensitive, but rather how much permafrost is likely to melt and how much of the peatland area is likely

to dry significantly. Such regional
changes in temperature, precipitation,
and drainage are still difficult to predict
in global circulation models. Hence,
the climate change predictions, as
much as our understanding of carbon
dynamics, limit our ability to predict
the magnitude of likely vulnerability of
peat and permafrost carbon to climate
change."

To obtain more reliable estimates of the carbon sinks and
sources in permafrost-affected soils, we need much more
detailed data on the distribution and characteristics of these
soils. Carbon stock estimates currently exist only for the upper 1 m of the soil. Limited data from the Mackenzie River
Valley in Canada, Arctic coast of Alaska and the Kolyma
Lowland of NE Russia indicate that a considerable amount
of soil organic carbon occurs below the 1-m depth, even at
the 3-m depth. Future estimates of carbon stocks should be
extended to cover a depth of 0-2 m or, in some cases, even
greater depths. More measurements of carbon fluxes and
inputs are also needed if we are to understand the carbon
sequestration process in these soils in the various permafrost
zones. Our understanding of the effect that rapid climate
warming will have on the carbon sinks and sources in these
soils is also very limited. Future research should focus in
greater detail on how the interactions of climate with the
biological and physical environments will affect the carbon
balance in permafrost-affected soils.

The changes that are occurring, and will occur, in the permafrost region are almost totally driven by natural forces
and so are almost impossible for humans to manage on a
large scale. Human activities, such as they are, are aimed at
protecting the permafrost and, thus, preserving the carbon.
Perhaps we humans should realize that there are systems
(*e.g.*, glaciers, ocean currents, droughts, and rainfall) that
will be impossible for us to manage. We simply must learn
to accept them, and if possible, adapt.

13

CHAPTER

Wetlands

Lead Author: Scott D. Bridgham, Univ. Oreg.

Contributing Authors: J. Patrick Megonigal, Smithsonian Environmental Research Center; Jason K. Keller, Smithsonian Environmental Research Center; Norman B. Bliss, SAIC, USGS Center for Earth Resources Observation and Science; Carl Trettin, USDA Forest Service

KEY FINDINGS

- North America is home to approximately 40% of the global wetland area, encompassing about 2.5 million square kilometers (965,000 square miles) with a carbon pool of approximately 223 billion tons, mostly in peatland soils.

- North American wetlands currently are a carbon dioxide sink of approximately 49 million tons of carbon per year, but that estimate has an uncertainty of greater than 100%. North American wetlands are also a source of approximately 9 million tons of methane, a more potent atmospheric heat-trapping gas. The uncertainty in that flux is also greater than 100%.

- Historically, the destruction of North American wetlands through land-use change has reduced carbon storage in wetlands by 15 million tons of carbon per year, primarily through the oxidation of carbon in peatland soils as they are drained and a more general reduction in carbon uptake and storage capacity of wetlands converted to other land uses. Methane emissions have also declined with the loss of wetland area.

- Projections of future carbon storage and methane emissions of North American wetlands are highly uncertain and complex, but the large carbon pools in peatlands may be at risk for oxidation and release to the atmosphere as carbon dioxide if they become substantially warmer and drier. Methane emissions may increase with warming, but the response will likely vary with wetland type and with changes in precipitation.

- Because of the potentially significant role of North American wetlands in methane production, the activities associated with the restoration, creation, and protection of wetlands are likely to focus on the ecosystem services that wetlands provide, such as filtering of toxics, coastal erosion protection, wildlife habitat, and havens of biological diversity, rather than on carbon sequestration, *per se*.

- Research needs to reduce the uncertainties in carbon storage and fluxes in wetlands to provide information about management options in terms of carbon uptake and storage and trace gas fluxes.

13.1 INTRODUCTION

While there are a variety of legal and scientific definitions of a wetland (National Research Council, 1995; National Wetlands Working Group, 1997), most emphasize the presence of waterlogged conditions in the upper soil profile during at least part of the growing season, and plant species and soil conditions that reflect these hydrologic conditions. Waterlogging tends to suppress microbial decomposition more than plant productivity, so wetlands are known for their ability to accumulate large amounts of soil carbon, most spectacularly seen in large peat deposits that are often many meters deep. Thus, when examining carbon dynamics, it is important to distinguish between freshwater wetlands with surface soil organic matter deposits greater than 40 cm thick (*i.e.*, peatlands) and those with lesser amounts of soil organic matter (*i.e.*, freshwater mineral-soil wetlands [FWMS]). Some wetlands have permafrost (fluxes and pools in wetlands with and without permafrost are discussed separately in Appendix F). We also differentiate between freshwater wetlands and estuarine wetlands (salt marshes, mangroves, and mud flats) with marine-derived salinity.

Peatlands occupy about 3% of the terrestrial global surface, yet they contain 16–33% of the total soil carbon pool (Gorham, 1991; Maltby and Immirzi, 1993)[1]. Most peatlands occur between 50 and 70° N, although significant areas occur at lower latitudes (Matthews and Fung, 1987; Aselmann and Crutzen, 1989; Maltby and Immirzi, 1993). Large areas of peatlands exist in Alaska, Canada, and in the northern midwestern, northeastern, and southeastern United States (Bridgham *et al.*, 2000). Because this peat formed over thousands of years, these areas represent a large carbon pool, but with relatively slow rates of accumulation. By comparison, estuarine wetlands and some freshwater mineral-soil wetlands rapidly sequester carbon as soil organic matter due to rapid burial

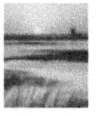

5.5% of the land area of the contiguous United States is wetlands. This represents just 48% of the original wetland area in the conterminous United States.

in sediments. Large areas of wetlands have been converted to other land uses, globally and in North America (Dugan, 1993; OECD, 1996), which may have resulted in a net flux of carbon to the atmosphere (Armentano and Menges, 1986; Maltby and Immirzi, 1993). Additionally, wetlands emit 92–237 million tons of methane (Mt CH_4) per year[1], which is a large fraction of the total annual global flux of about 600 Mt CH_4 per year (Ehhalt *et al.*, 2001). This is important because CH_4 is a potent greenhouse gas (GHG), second in importance only to carbon dioxide (CO_2) (Ehhalt *et al.*, 2001).

---

[1] The uncertainties for the numerical values cited in this chapter are presented and explained in Table 13.1 and Appendix F.

A number of previous studies have examined the role of peatlands in the global carbon balance (reviewed in Mitra *et al.*, 2005), and Roulet (2000) focused on the role of Canadian peatlands in the Kyoto process. Here we augment these previous studies by considering all types of wetlands (not just peatlands) and integrate new data to examine the carbon balance in the wetlands of Canada, the United States, and Mexico. We also briefly compare these values to those from global wetlands. We limit this review to those components of the carbon budget that result in a net gaseous exchange with the atmosphere on an interannual basis and do not consider other internal carbon fluxes. We do not consider dissolved organic carbon (DOC) fluxes from wetlands, although they may be substantial (Moore, 1997), because the oxidation of the DOC would be counted as atmospheric carbon emissions in the receiving ecosystems downstream and we do not want to double-count fluxes.

Given that many undisturbed wetlands are a natural sink for CO_2 and a source of CH_4, a note of caution in interpretation of our data is important. Using the Intergovernmental Panel on Climate Change (IPCC) terminology, a radiative forcing denotes "an externally imposed perturbation in the radiative energy budget of the Earth's climate system" (Ramaswamy *et al.*, 2001). Thus, it is the change from a baseline condition in GHG fluxes in wetlands that constitute a radiative forcing that will impact climate change, and carbon fluxes in unperturbed wetlands are important only in establishing a baseline condition. For example, historical steady state rates of CH_4 emissions from wetlands have zero net radiative forcing, but an increase in CH_4 emissions due to climatic warming would constitute a positive radiative forcing. Similarly, steady state rates of soil carbon sequestration in wetlands have zero net radiative forcing, but the lost sequestration capacity and the oxidation of the soil carbon pool in drained wetlands are both positive radiative forcings.

13.2 INVENTORIES

13.2.1 Current Wetland Area and Rates of Loss

The current and original wetland area and rates of loss are the basis for all further estimates of pools and fluxes in this chapter. The loss of wetlands has caused the oxidation of their soil carbon, particularly in peatlands, reduced their ability to sequester carbon, and reduced their emissions of CH_4. The strengths and weakness of the wetland inventories of Canada, the United States, and Mexico are discussed in Appendix F.

The conterminous United States has 312,000 km² of FWMS wetlands, 93,000 km² of peatlands, and 25,000 km² of estuarine wetlands, which encompass 5.5% of the land area (Table 13.1). This represents just 48% of the original wetland area in the conterminous United States (Table F.1 in Appendix F).

Table 13.1 The area, carbon pool, net carbon balance, and methane flux from wetlands in North America and the world. Positive fluxes indicate net fluxes to the atmosphere, whereas negative fluxes indicate net fluxes into an ecosystem. Citations and assumptions in calculations are in the text and in Appendix F (see Box 13.1 for uncertainty conventions).

	Area[a] (km²)	Carbon Pool[b] (Gt C)	Net Carbon Balance[c] (Mt C per year)	Historical Loss in Sequestration Capacity (Mt C per year)	Methane Flux (Mt CH₄ per year)
Canada					
Peatland	1,135,608****	152****	-19***	0.3*	3.2**
Freshwater Mineral	158,720**	4.9**	-2.7*	3.4*	1.2*
Estuarine	6,400***	0.1***	-1.3**	0.5*	0.0***
Total	1,300,728****	157****	-23**	4.2*	4.4*
Alaska					
Peatland	132,196****	15.9**	-2.0**	0.0****	0.3*
Freshwater Mineral	555,629****	27.1**	-9.4*	0.0****	1.4*
Estuarine	8,400****	0.1***	-1.9**	0.0****	0.0***
Total	696,224*****	43.2**	-13*	0.0****	1.7*
Conterminous United States					
Peatland	93,477*****	14.4***	5.7*	1.2*	0.7**
Freshwater Mineral	312,193*****	6.2***	-9.8*	7.6*	2.4**
Estuarine	25,000*****	0.6*****	-5.4**	0.5*	0.0***
Total	430,670*****	21.2***	-9.5*	9.4*	3.1**
U.S. Total	1,126,895*****	64.3**	-23*	9.4*	4.8**
Mexico					
Peatland	10,000*	1.5*	-1.6*	ND*	0.1*
Freshwater Mineral	20,685*	0.4*	-0.4*	ND*	0.2*
Estuarine	5,000*	0.2*	-1.6*	1.0*	0.0*
Total	35,685*	2.0*	-3.6*	ND*	0.2*
North America					
Peatland	1,371,281****	184****	-17*	1.5*	4.3**
Freshwater Mineral	1,047,227****	39***	-22*	11*	5.1*
Estuarine	44,800***	0.9***	-10**	2.0*	0.1**
Total	2,463,308****	223****	-49*	15*	9.4*
Global					
Peatland	3,443,000***	462***	150**	16*	37**
Freshwater Mineral	2,315,000***	46***	-39*	45*	68**
Estuarine	203,000*	5.4*	-43*	21*	0.2**
Total	5,961,000***	513***	68*	82*	105**

[a] Estuarine includes salt marsh, mangrove, and mudflat, except for Mexico and global for which no mudflat estimates were available.

[b] Includes soil carbon and plant carbon, but overall soil carbon is 98% of the total pool.

[c] Includes soil carbon sequestration, plant carbon sequestration, and loss of carbon due to drainage of wetlands. Plant carbon sequestration and soil oxidative flux due to drainage are either unknown or negligible for North American wetlands except for the conterminous United States (see Appendix F).

ND indicates that no data are available.

BOX 13.1: CCSP SAP 2.2 Uncertainty Conventions

*****	=	95% certain that the actual value is within 10% of the estimate reported,
****	=	95% certain that the estimate is within 25%,
***	=	95% certain that the estimate is within 50%,
**	=	95% certain that the estimate is within 100%, and
*	=	uncertainty greater than 100%.

However, wetland losses in the United States have declined from 1855 km² per year in the 1950s–1970s to 237 km² per year in the 1980s–1990s (Dahl, 2000). Such data mask large differences in loss rates among wetland classes and conversion of wetlands to other classes (Dahl, 2000), with potentially large effects on carbon stocks and fluxes. For example, the majority of wetland losses in the United States have occurred in FWMS wetlands. As of the early 1980s, 84% of United States' peatlands were unaltered (Armentano and Menges, 1986; Maltby and Immirzi, 1993; Rubec, 1996), and, given the current regulatory environment in the United States, recent rates of loss are likely small.

Canada has 1,301,000 km² of wetlands, covering 14% of its land area, of which 87% are peatlands (Table 13.1). Canada has lost about 14% of its wetlands, mainly due to agricultural development of FWMS wetlands (Rubec, 1996), although the ability to estimate wetland losses in Canada is limited by the lack of a regular wetland inventory.

The wetland area in Mexico is estimated at 36,000 km² (Table 13.1), with an estimated historical loss of 16,000 km² (Table F.1 in Appendix F). However, given the lack of a nationwide wetland inventory and a general paucity of data, this number is highly uncertain.

> North America currently has about 43% of the global wetland area.

Problems with inadequate wetland inventories are even more prevalent in lesser developed countries (Finlayson *et al.*, 1999). We estimate a global wetland area of 6.0 × 10⁶ km² (Table 13.1); thus, North America currently has about 43% of the global wetland area. It has been estimated that about 50% of the world's original wetlands have been converted to other uses (Moser *et al.*, 1996).

13.2.2 Carbon Pools

We estimate that North American wetlands have a current soil and plant carbon pool of 223 billion tons (Gt), of which approximately 98% is in the soil (Table 13.1). The majority of this carbon is in peatlands, with FWMS wetlands contributing about 18% of the carbon pool. The large amount of soil carbon (27 Gt) in Alaskan FWMS wetlands had not been identified in previous studies (see Appendix F).

13.2.3 Soil Carbon Fluxes

North American peatlands currently have a net carbon balance of about -17 million metric tons of carbon (Mt C)

per year (Table 13.1), but several large fluxes are incorporated into this estimate. (Negative numbers indicate net fluxes into the ecosystem, whereas positive numbers indicate net fluxes into the atmosphere). Peatlands sequester -29 Mt C per year (Table F.2 in Appendix F). However, this carbon sink is partially offset by a net oxidative flux of 18 Mt C per year as of the early 1980s in peatlands in the conterminous United States that have been drained for agriculture and forestry (Armentano and Menges, 1986). Despite a substantial reduction in the rate of wetland loss since the 1980s (Dahl, 2000), drained organic soils continue to lose carbon over many decades, so the actual flux to the atmosphere is probably close to the 1980s estimate. There has also been a loss in sequestration capacity in drained peatlands of 1.5 Mt C per year (Table 13.1), so the overall soil carbon sink of North American peatlands is about 20 Mt C per year smaller than it would have been in the absence of disturbance.

Very little attention has been given to the role of FWMS wetlands in North American or global carbon balance estimates, with the exception of CH₄ emissions. Carbon sequestration associated with sediment deposition is a potentially large, but poorly quantified, flux in wetlands (Stallard, 1998; Smith *et al.*, 2001). We estimate that North American FWMS wetlands sequester -18 Mt C per year in sedimentation (Table F.2 in Appendix F). However, as discussed in Appendix F, wetland sedimentation rates are extremely variable. Moreover, almost no studies have placed sediment carbon sequestration in FWMS wetlands in a landscape context, considering allochthonous-derived (from on-site plant production) versus autochthonous-derived (imported from outside the wetland) carbon, replacement of carbon in terrestrial source areas, and differences in decomposition rates between sink and source areas (Stallard, 1998; Harden *et al.*, 1999; Smith *et al.*, 2001). However, it is clear that sedimentation in FWMS wetlands

is a potentially substantial carbon sink and an important unknown in carbon budgets. For example, agriculture typically increases sedimentation rates by 10- to 100-fold and 90% of sediments are stored within the watershed, amounting to about -40 Mt C per year in the conterminous United States (Stallard, 1998; Smith *et al.*, 2001). Our estimate of sediment carbon sequestration in FWMS wetlands seems quite reasonable in comparison to within-watershed sediment storage in North America. Moreover, Stallard (1998) and Smith *et al.* (2001) estimated a global sediment sink on the order of -1 Gt C per year.

Decomposition of soil carbon in FWMS wetlands that have been converted to other land uses appears to be responsible for only a negligible loss of soil carbon, currently (Table F.2 in Appendix F). However, due to the historical loss of FWMS wetland area, we estimate that they currently sequester 11 Mt C per year less than they did prior to disturbance (Table 13.1). This estimate has the same unknowns described in the previous paragraph on current sediment carbon sequestration in extant FWMS wetlands.

We estimate that estuarine wetlands currently sequester -10.2 Mt C per year (Table F.2 in Appendix F), with a historical reduction in sequestration capacity of 2.0 Mt C per year due to loss of area (Table 13.1). However, the reduction is almost certainly greater because our "original" area is only from the 1950s. Despite the relatively small area of estuarine wetlands, they currently contribute about 31% of total wetland carbon sequestration in the conterminous United States and about 18% of the North American total. Estuarine wetlands sequester carbon at a rate about 10 times higher on an area basis than other wetland ecosystems due to high sedimentation rates, high soil carbon content, and constant burial due to sea level rise. Estimates of sediment deposition rates in estuarine wetlands are reasonably robust, but the same 'landscape' issues of allochthonous versus autochthonous inputs of carbon, replenishment of carbon in source area soils, and differences in decomposition rates between sink and source areas exist as for FWMS wetlands. Another large uncertainty in the estuarine carbon budget is the area and carbon content of mud flats, particularly in Canada and Mexico.

Overall, North American wetland soils appear to be a substantial carbon sink with a net flux of -49 Mt C per year (with very large error bounds because of FWMS wetlands) (Table 13.1). The large-scale conversion of wetlands to upland uses has led to a reduction in the wetland soil carbon sequestration capacity of 15 Mt C per year from the likely historical rate (Table 13.1), but this estimate is driven by large losses of FWMS wetlands with their highly uncertain sedimentation carbon sink. Adding in the current net oxidative flux of 18 Mt C per year from conterminous United States' peatlands,

we estimate that North American wetlands currently sequester 33 Mt C per year less than they did historically (Table F.2 in Appendix F). Furthermore, North American peatlands and FWMS wetlands have lost 2.6 Gt and 0.8 Gt of soil carbon, respectively, and collectively they have lost 2.4 Gt of plant carbon since approximately 1800. Very little data exist to estimate carbon fluxes for freshwater Mexican wetlands, but because of their small area, they will not likely have a large impact on the overall North American estimates.

The global wetland soil carbon balance has only been examined in peatlands, which currently are a moderate source of atmospheric carbon of about 150 Mt C per year (Table 13.1), largely due to the oxidation of peat drained for agriculture and forestry and secondarily due to peat combustion for fuel (Armentano and Menges, 1986; Maltby and Immirzi, 1993). The cumulative historical shift in soil carbon stocks has been estimated to be 5.5 to 7.1 Gt C (Maltby and Immirzi, 1993). Although we are aware of no previous evaluation of the carbon balance of global FWMS and estuarine wetlands, using the assumption noted above, we estimate that they are a sink of approximately -39 and -43 Mt per year, respectively.

13.2.4 Methane and Nitrous Oxide Emissions

We estimate that North American wetlands emit 9.4 Mt CH_4 per year (Table 13.1). For comparison, a mechanistic CH_4 model yielded emissions of 3.8 and 7.1 Mt CH_4 per year for Alaska and Canada, respectively (Zhuang *et al.*, 2004). A regional inverse atmospheric modeling approach estimated total CH_4 emissions (from all sources) of 16 and 54 Mt CH_4 per year for boreal and temperate North America, respectively (Fletcher *et al.*, 2004b).

> Despite the relatively small area of estuarine wetlands, they currently contribute about 31% of total wetland carbon sequestration in the conterminous United States and about 18% of the North American total.

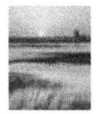

Methane emissions are currently about 5 Mt CH_4 per year less than they were historically in North American wetlands (see Table F.4 in Appendix F) because of the loss of wetland area. We do not consider the effects of conversion of wetlands from one type to another (Dahl, 2000), which may have a significant impact on CH_4 emissions. Similarly, we estimate that global CH_4 emissions from natural wetlands are only about half of what they were historically due to loss of area (Table F.4 in Appendix F). However, this may be an overestimate because wetland losses have been higher in more developed countries than less developed countries (Moser *et al.*, 1996), and wetlands at lower latitudes have higher emissions on average (Bartlett and Harriss, 1993).

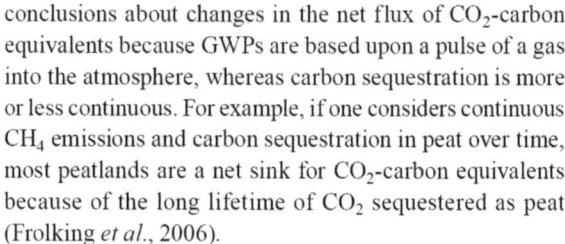

When we multiplied the very low published estimates of nitrous oxide emissions from natural and disturbed wetlands (Joosten and Clarke, 2002) by North American wetland area, the flux was insignificant (data not shown). However, nitrous oxide emissions have been measured in few wetlands, particularly in FWMS wetlands and wetlands with high nitrogen inputs (*e.g.*, from agricultural run-off) where emissions might be expected to be higher.

We use global warming potentials (GWPs) as a convenient way to compare the relative contributions of CO_2 and CH_4 fluxes in North American wetlands to the Earth's radiative balance. The GWP is the radiative effect of a pulse of a substance into the atmosphere relative to CO_2 over a particular time horizon (Ramaswamy *et al.*, 2001). However, it is important to distinguish between *radiative balance*, which refers to the static radiative effect of a substance, and *radiative forcing*, which refers to an externally imposed perturbation on the Earth's radiative energy budget (Ramaswamy *et al.*, 2001). Thus, changes in radiative balance lead to a radiative forcing, which subsequently leads to a change in the Earth's surface temperature. For example, wetlands have a large effect on the Earth's radiative balance through high CH_4 emissions, but it is only to the extent that emissions change through time that they represent a positive or negative radiative forcing and impact climate change.

> Historically, the destruction of wetlands through land-use changes has had the largest effect on the carbon fluxes.

Methane has GWPs of 1.9, 6.3, and 16.9 CO_2-carbon equivalents on a mass basis across 500-year, 100-year, and 20-year time frames, respectively (Ramaswamy *et al.*, 2001)[2]. Depending upon the time frame and within the large confidence limits of many of our estimates in Table 13.1, the *net radiative balance* of North American wetlands as a whole currently are approximately neutral in terms of net CO_2-carbon equivalents to the atmosphere (note that we discuss *net radiative forcing* in *Trends and Drivers of Wetland Carbon Fluxes,* Section 13.3). The exception is estuarine wetlands, which are a net sink for CO_2-carbon equivalents because they support both rapid rates of carbon sequestration and low CH_4 emissions. However, caution should be exercised in using GWPs to draw

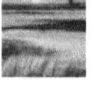

> Wetland ecosystems provide havens for biodiversity, recharge of groundwater, and reduction in flooding and fish nurseries.

conclusions about changes in the net flux of CO_2-carbon equivalents because GWPs are based upon a pulse of a gas into the atmosphere, whereas carbon sequestration is more or less continuous. For example, if one considers continuous CH_4 emissions and carbon sequestration in peat over time, most peatlands are a net sink for CO_2-carbon equivalents because of the long lifetime of CO_2 sequestered as peat (Frolking *et al.*, 2006).

13.2.5 Plant Carbon Fluxes

We estimate that wetland forests in the conterminous United States currently sequester -10.3 Mt C per year as increased plant biomass (see Table F.3 in Appendix F). Sequestration in plants in undisturbed wetland forests in Alaska, many peatlands, and estuarine wetlands is probably minimal, although there may be substantial logging of Canadian forested peatlands that we do not have the data to account for.

13.3 TRENDS AND DRIVERS OF WETLAND CARBON FLUXES

While extensive research has been done on carbon cycling and pools in North American wetlands, to our knowledge, this is the first attempt at an overall carbon budget for all of the wetlands of North America, although others have examined the carbon budget for North American peatlands as part of global assessments (Armentano and Menges, 1986; Maltby and Immirzi, 1993; Joosten and Clarke, 2002). Historically, the destruction of wetlands through land-use changes has had the largest effect on the carbon fluxes and, consequently, the radiative forcing of North American wetlands. The primary effects have been a reduction in their ability to sequester carbon (a small to moderate increase in radiative forcing depending on carbon sequestration by sedimentation in FWMS and estuarine wetlands), oxidation of their soil carbon reserves upon drainage (a small increase in radiative forcing), and a reduction in CH_4 emissions (a small to large decrease in radiative forcing depending on actual emissions) (Table F.1 and Appendix F). Globally, the disturbance of peatlands appears to have shifted them into a net source of carbon to the atmosphere. Any positive effect of wetland loss due to a reduction in their CH_4 emissions, and hence radiative forcing, will be more than negated by the loss of the many ecosystem services they provide, such as havens for biodiversity, recharge of groundwater, reduction in flooding, fish nurseries, *etc.* (Zedler and Kercher, 2005).

A majority of the effort in examining future global change impacts on wetlands has focused on northern peatlands because of their large soil carbon reserves, although under current climate conditions they have modest CH_4 emissions (Moore and Roulet, 1995; Roulet, 2000; Joosten and Clarke, 2002, and references therein). The effects of global change

---

[2] GWPs in Ramaswamy *et al.* (2001) were originally reported in CO_2-mass equivalents. We have converted them into CO_2-carbon equivalents so that the net carbon balance and CH_4 flux columns in Table 13.1 can be directly compared by multiplying CH_4 fluxes by the GWPs given here.

on carbon sequestration in peatlands are probably of minor importance as a global flux because of the relatively low rate of peat accumulation. However, losses of soil carbon stocks in peatlands drained for agriculture and forestry (Table F.2 in Appendix F) attest to the possibility of large losses from the massive soil carbon deposits in northern peatlands if they become substantially drier in a future climate. Furthermore, Turetsky *et al.* (2004) estimated that up to 5.9 Mt C per year are released from western Canadian peatlands by fire and predicted that increases in fire frequency may cause these systems to become net atmospheric carbon sources. We did not add this flux to our estimate of the net carbon balance of North American wetlands because historical oxidation of peat by fire should be integrated in the peat sequestration estimates and recent changes due to anthropogenic effects are highly uncertain.

Our compilation shows that attention needs to be directed toward understanding climate change impacts to FWMS wetlands, which collectively emit similar amounts of CH_4 and potentially sequester an equivalent amount of carbon than peatlands. The effects of changing water table depths are somewhat more tractable in FWMS wetlands than peatlands because FWMS wetlands have less potential for oxidation of soil organic matter. In forested FWMS wetlands, increased precipitation and runoff may increase radiative forcing by simultaneously decreasing wood production and increasing methanogenesis (Megonigal *et al.*, 2005). The influence of changes in hydrology on CH_4 emissions, plant productivity, soil carbon preservation, and sedimentation will need to be addressed in order to fully anticipate climate change impacts on radiative forcing in these systems.

The effects of global change on estuarine wetlands is of concern because sequestration rates are rapid, and they can be expected to increase in proportion to the rate of sea level rise provided estuarine wetland area does not decline. Because CH_4 emissions from estuarine wetlands are low, this increase in sequestration capacity could represent a net decrease in radiative forcing, depending on how much of the sequestered carbon is autochthonous. Changes in tidal wetland area with sea-level rise will depend on rates of inland migration, erosion at the wetland-estuary boundary, and wetland elevation change. The rate of loss of tidal wetland area has declined in past decades due to regulations on draining and filling activities (Dahl, 2000). However, rapid conversion to open water is occurring in coastal Louisiana (Bourne, 2000) and Maryland (Kearney and Stevenson, 1991), suggesting that marsh area will decline with increased rates of sea level rise (Kearney *et al.*, 2002). A multitude of human and climate factors are contributing to the current losses (Turner, 1997; Day Jr. *et al.*, 2000; Day Jr. *et al.*, 2001). Although it is un-

certain how global changes in climate, eutrophication, and other factors will interact with sea level rise (Najjar *et al.*, 2000), it is likely that increased rates of sea level rise will cause an overall decline in estuarine marsh area and soil carbon sequestration.

One of the greatest concerns is how climate change will affect future CH_4 emissions from wetlands because of their large GWP. Wetlands emit about 105 Mt CH_4 per year (Table 13.1), or 20% of the global total. Increases in atmospheric CH_4 concentrations over the past century have had the second largest radiative forcing (after CO_2) in human-induced climate change (Ehhalt *et al.*, 2001). Moreover, CH_4 fluxes from wetlands have provided an important radiative feedback on climate over the geologic past (Chappellaz *et al.*, 1993; Blunier *et al.*, 1995; Petit *et al.*, 1999). The large global warming observed since the 1990s may have resulted in increased CH_4 emissions from wetlands (Fletcher *et al.*, 2004a; Wang *et al.*, 2004; Zhuang *et al.*, 2004).

> It is likely that increased rates of sea level rise will cause an overall decline in estuarine marsh area and soil carbon sequestration.

Data (Bartlett and Harriss, 1993; Moore *et al.*, 1998; Updegraff *et al.*, 2001) and modeling (Gedney *et al.*, 2004; Zhuang *et al.*, 2004) strongly support the contention that water table position and temperature are the primary environmental controls over CH_4 emissions. How this generalization plays out with future climate change is, however, more complex. For example, most climate models predict much of Canada will be warmer and drier in the future. Based upon this prediction, Moore *et al.* (1998) proposed a variety of responses to climate change in the carbon fluxes from different types of Canadian peatlands. Methane emissions may increase in collapsed former-permafrost bogs (which will be warmer and wetter) but decrease in fens and other types of bogs (warmer and drier). A CH_4-process model predicted that modest warming will increase global wetland emissions,

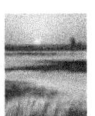

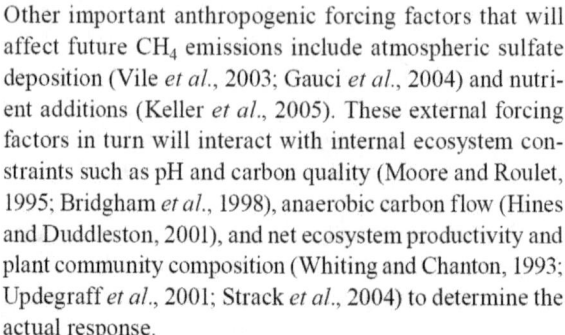

but larger increases in temperature will decrease emissions because of drier conditions (Cao *et al.*, 1998).

The direct, non-climatic effects of increasing atmospheric CO_2 on carbon cycling in wetland ecosystems has received far less attention than upland systems. Field studies have been done in tussock tundra (Tissue and Oechel, 1987; Oechel *et al.*, 1994), bog-type peatlands (Hoosbeek *et al.*, 2001), rice paddies (Kim *et al.*, 2001), and a salt marsh (Rasse *et al.*, 2005); and a somewhat wider variety of wetlands have been studied in small scale glasshouse systems. Temperate and tropical wetland ecosystems consistently respond to elevated CO_2 with an increase in photosynthesis and/or biomass (Vann and Megonigal, 2003). By comparison, the response of northern peatland plant communities has been inconsistent. A hypothesis that remains untested is that the elevated CO_2 response of northern peatlands will be limited by nitrogen availability. In an *in situ* study of tussock tundra, complete photosynthetic acclimation occurred when CO_2 was elevated, but acclimation was far less severe with both elevated CO_2 and a 4°C increase in air temperature (Oechel *et al.*, 1994). It was hypothesized that soil warming relieved a severe nutrient limitation on photosynthesis by increasing nitrogen mineralization.

A consistent response to elevated CO_2-enhanced photosynthesis in wetlands is an increase in CH_4 emissions ranging from 50 to 350% (Megonigal and Schlesinger, 1997; Vann and Megonigal, 2003). It is generally assumed that the increased supply of plant photosynthate stimulates anaerobic microbial carbon metabolism, of which CH_4 is a primary end product. An increase in CH_4 emissions from wetlands due to elevated CO_2 constitutes a positive feedback on radiative forcing because CO_2 is rapidly converted to a more effective GHG (CH_4).

An elevated CO_2-induced increase in CH_4 emissions may be offset by an increase in carbon sequestration in soil organic matter or wood. Although there are very little data to evaluate this hypothesis, a study on seedlings of a wetland-adapted tree species reported that elevated CO_2 stimulated photosynthesis and CH_4 emissions, but not growth, under flooded conditions (Megonigal *et al.*, 2005). It is possible that elevated CO_2 will stimulate soil carbon sequestration, particularly in tidal wetlands experiencing sea level rise, but a net loss of soil carbon is also possible due to priming effects (*i.e.*, increased labile carbon inputs from elevated CO_2 enhance decomposition of the overall soil carbon pool) (Hoosbeek *et al.*, 2004; Lichter *et al.*, 2005). Elevated CO_2 has the potential to influence the carbon budgets of adjacent aquatic ecosystems by increasing export of dissolved organic carbon (Freeman *et al.*, 2004) and dissolved inorganic carbon (Marsh *et al.*, 2005).

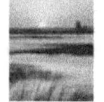

Other important anthropogenic forcing factors that will affect future CH_4 emissions include atmospheric sulfate deposition (Vile *et al.*, 2003; Gauci *et al.*, 2004) and nutrient additions (Keller *et al.*, 2005). These external forcing factors in turn will interact with internal ecosystem constraints such as pH and carbon quality (Moore and Roulet, 1995; Bridgham *et al.*, 1998), anaerobic carbon flow (Hines and Duddleston, 2001), and net ecosystem productivity and plant community composition (Whiting and Chanton, 1993; Updegraff *et al.*, 2001; Strack *et al.*, 2004) to determine the actual response.

13.4 OPTIONS FOR MANAGEMENT

Wetland policies in the United States and Canada are driven by a variety of federal, state or provincial, and local laws and regulations in recognition of the many wetland ecosystem services and large historical loss rates (Lynch-Stewart *et al.*, 1999; National Research Council, 2001; Zedler and Kercher, 2005). Thus, any actions to enhance the ability of wetlands to sequester carbon, or reduce their CH_4 emissions, must be implemented within the context of the existing regulatory framework. The most important option in the United States has already been largely achieved, and that is to reduce the historical rate of peatland losses with their accompanying large oxidative losses of the stored soil carbon. Decreases in the rates of loss of all wetlands have helped to maintain their soil carbon sequestration potential.

There has been strong interest expressed in using carbon sequestration as a rationale for wetland restoration and creation in the United States, Canada, and elsewhere (Wylynko, 1999; Watson *et al.*, 2000). However, high CH_4 emissions from conterminous United States' wetlands suggest that creating and restoring wetlands may increase net radiative forcing, although adequate data do not exist to fully evaluate this possibility. Roulet (2000) came to a similar conclusion concerning the restoration of Canadian wetlands. Net radiative forcing from restoration will likely vary among different kinds of wetlands and the specifics of their carbon budgets. The possibility of increasing radiative forcing by creating or restoring wetlands does not apply to estuarine wetlands, which emit relatively little CH_4 compared to the carbon they sequester. Restoration of drained peatlands may stop the rapid loss of their soil carbon, which may compensate for increased CH_4 emissions. However, Canadian peatlands restored from peat extraction operations increased their net emissions of carbon because of straw addition during the restoration process, although it was assumed that they would eventually become a net sink (Cleary *et al.*, 2005).

Regardless of their internal carbon balance, the area of restored wetlands is currently too small to form a significant carbon sink at the continental scale. Between 1986 and 1997,

only 4157 km² of uplands were converted into wetlands in the conterminous United States (Dahl, 2000). Using the soil carbon sequestration rate of 3.05 Mg C per hectare per year found by Euliss *et al.* (2006) for restored prairie pothole wetlands[3], we estimate that wetland restoration in the United States would have sequestered 1.3 Mt C over this 11-year period. However, larger areas of wetland restoration may have a significant impact on carbon sequestration. A simulation model of planting 20,000 km² into bottomland hardwood trees as part of the Wetland Reserve Program in the United States showed a sequestration of 4 Mt C per year through 2045 (Barker *et al.*, 1996). Euliss *et al.* (2006) estimated that if all cropland on former prairie pothole wetlands in the United States and Canada (162,244 km²) were restored that 378 Mt C would be sequestered over 10 years in soils and plants. However, neither study accounted for the GWP of increased CH_4 emissions.

Potentially more significant is the conversion of wetlands from one type to another; for example, 8.7% (37,200 km²) of the wetlands in the conterminous United States in 1997

were in a previous wetland category in 1986 (Dahl, 2000). The net effect of these conversions on wetland carbon fluxes is unknown. Similarly, Roulet (2000) argued that too many uncertainties exist to include Canadian wetlands in the Kyoto Protocol.

In summary, North American wetlands form a very large carbon pool, primarily because of storage as peat, and are a small-to-moderate carbon sink (excluding CH_4 effects). The largest unknown in the wetland carbon budget is the amount and significance of sedimentation in FWMS and estuarine wetlands, and CH_4 emissions in freshwater wetlands. With the exception of estuarine wetlands, CH_4 emissions from wetlands may largely offset any positive benefits of carbon sequestration in soils and plants. Given these conclusions, it is probably unwarranted to use carbon sequestration as a rationale for the protection and restoration of FWMS wetlands, although the many other ecosystem services that they provide justify these actions. However, protecting and restoring peatlands will stop the loss of their soil carbon (at least over the long term) and estuarine wetlands are an important carbon sink given their limited areal extent and low CH_4 emissions.

The most important areas for further scientific research in terms of current carbon fluxes in the United States are to establish an unbiased, landscape-level sampling scheme to determine sediment carbon sequestration in FWMS and estuarine wetlands and additional measurements of annual CH_4 emissions to better constrain these important fluxes. It would also be beneficial if the approximately decadal National Wetland Inventory (NWI) status and trends data were collected in sufficient detail with respect to the Cowardin *et al.* (1979) classification scheme to determine changes among mineral-soil wetlands and peatlands.

> Larger areas of wetland restoration may have a significant impact on carbon sequestration, but may also increase methane emissions offsetting any positive greenhouse gas effects.

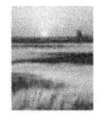

Canada lacks any regular inventory of its wetlands, and thus, it is difficult to quantify land-use impacts upon their carbon fluxes and pools. While excellent scientific data exists on most aspects of carbon cycling in Canadian peatlands, Canadian FWMS and estuarine wetlands have been relatively poorly studied, despite having suffered large proportional losses to land-use change. Wetland data for Mexico is almost entirely lacking. Thus, anything that can be done to improve upon this would be helpful. All wetland inventories should consider the area of estuarine mud flats, which have the potential to sequester considerable carbon and are poorly understood with respect to carbon sequestration.

---

[3] Euliss *et al.* (2006) regressed surface soil carbon stores in 27 restored semi-permanent prairie pothole wetlands against years since restoration to derive this estimate ($r^2 = 0.31$, $P = 0.002$). However, there was no significant relationship in seasonal prairie pothole wetlands ($r^2 = 0.04$, $P = 0.241$).

The greatest unknown is how global change will affect the carbon pools and fluxes of North American wetlands. We will not be able to accurately predict the role of North American wetlands as potential positive or negative feedbacks to anthropogenic climate change without knowing the integrative effects of changes in temperature, precipitation, atmospheric CO_2 concentrations, and atmospheric deposition of nitrogen and sulfur within the context of internal ecosystem drivers of wetlands. To our knowledge, no manipulative experiment has simultaneously measured more than two of these perturbations in any North American wetland, and few have been done at any site. Modeling expertise of the carbon dynamics of wetlands has rapidly improved in the last few years (Frolking *et al.*, 2002; Zhuang *et al.*, 2004, and references therein), but this needs even further development in the future, including for FWMS and estuarine wetlands.

ACKNOWLEDGMENTS

Steve Campbell (U.S. Department of Agriculture [USDA] Natural Resource Conservation Service [NRCS], OR) synthesized the National Soil Information database so that it was useful to us. Information on wetland soils within specific states was provided by Joseph Moore (USDA NRCS, AK), Robert Weihrouch (USDA NRCS, WI), and Susan Platz (USDA NRCS, MN). Charles Tarnocai provided invaluable data on Canadian peatlands. Thomas Dahl (U.S. Fish and Wildlife Service) explored the possibility of combining NWI data with United States' soils maps. Nigel Roulet (McGill University) gave valuable advice on recent references. R. Kelman Wieder provided useful initial information on peatlands in Canada. Comments of two anonymous reviewers and Shuguang Liu (USGS Center for Earth Resources Observation and Science) greatly improved this manuscript.

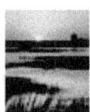

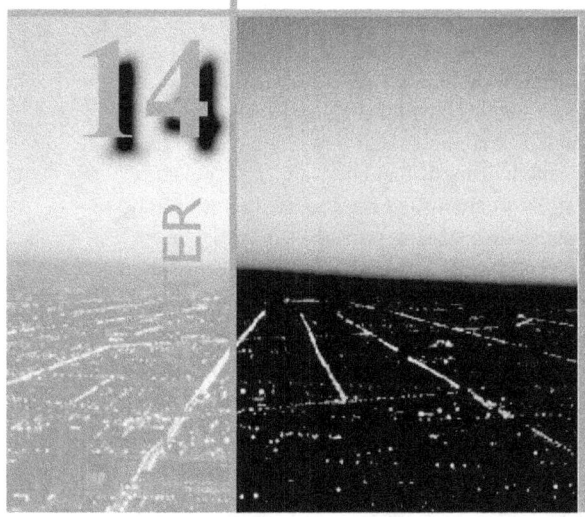

14

Human Settlements and the North American Carbon Cycle

Lead Authors: Diane E. Pataki, Univ. Calif., Irvine

Contributing Authors: Alan S. Fung, Dalhousie Univ.; David J. Nowak, USDA Forest Service; E. Gregory McPherson, USDA Forest Service; Richard V. Pouyat, USDA Forest Service; Nancy Golubiewski, Landcare Research; Christopher Kennedy, Univ. Toronto; Patricia Romero Lankao, NCAR and UAM-Xochimilco; Ralph Alig, USDA Forest Service

KEY FINDINGS

- Human settlements occupy almost 5% of the North American land area.
- There is currently insufficient information to determine the complete carbon balance of human settlements in North America. Fossil-fuel emissions, however, very likely dominate carbon fluxes from settlements.
- An estimated 410 to 1679 million tons of carbon are currently stored in the urban tree component of North American settlements. The growth of urban trees in North America produces a sink of approximately 16 to 49 million tons of carbon per year, which is 1 to 3% of the fossil-fuel emissions from North America in 2003.
- Estimates of historical trends of the net carbon balance of North American settlements are not available. Fossil-fuel emissions have likely gone up with the growth of urban lands, but the net balance of carbon loss during conversion of natural to urban or suburban land cover and subsequent uptake by lawns and urban trees is highly uncertain.
- The density and development patterns of human settlements are drivers of fossil-fuel emissions, especially in the residential and transportation sectors. Biological carbon gains and losses are influenced by type of predevelopment land cover, post-development urban design and landscaping choices, soil and landscape management practices, and the time since land conversion.
- Projections of future trends in the net carbon balance of North American settlements are not available. However, the projected expansion of urban areas in North America will strongly impact the future North American carbon cycle as human settlements affect (1) the direct emission of carbon dioxide from fossil-fuel combustion, (2) plant and soil carbon cycling in converting wildlands to residential and urban land cover.
- A number of municipalities in Canada, Mexico, and the United States have made commitments to voluntary greenhouse gas emission reductions under the Cities for Climate Protection program of International Governments for Local Sustainability (formerly the International Council for Local Environmental Initiatives [ICLEI]). Reductions have in some cases been associated with improvements in air quality.

- Research is needed to improve comprehensive carbon inventories for settled areas, to improve understanding of how development processes relate to driving forces for the carbon cycle, and to improve linkages between understanding of human and environmental systems in settled areas.

14.1 BACKGROUND

Activities in human settlements form the basis for much of North America's contribution to global carbon dioxide (CO_2) emissions. Settlements such as cities, towns, and suburbs vary widely in density, form, and distribution. Urban settlements, as they have been defined by the census bureaus of the United States, Canada, and Mexico, make up approximately 75 to 80% of the population of the continent, and this proportion is projected to continue to increase (United Nations, 2004). The density and forms of new development will strongly impact the future trajectory of the North American carbon cycle as human settlements affect the carbon cycle by (1) direct emission of CO_2 from fossil-fuel combustion, (2) alterations to plant and soil carbon cycles in conversion of wildlands to residential and urban land cover, and (3) indirect effects of residential and urban land cover on energy use and ecosystem carbon cycling.

14.2 CARBON INVENTORIES OF HUMAN SETTLEMENTS

Conversion of agricultural and wildlands to settlements of varying densities is occurring at a rapid rate in North America, faster, in fact, than the rate of population growth. For example, according to U.S. Census Bureau estimates, urban land in the coterminous United States increased by 23% in the 1990s (Nowak *et al.*, 2005) while the population increased by 13%.[1†] Given these trends, it is important to determine the carbon balance of different types of settlements and how future urban policy and planning may impact the magnitude of CO_2 sources and sinks at regional, continental, and global scales. However, unlike many other types of common land cover, complete carbon inventories including fossil-fuel emissions and biological sources and sinks of carbon have been conducted only rarely for settlements as a whole. Assessing the carbon balance of settlements is challenging, as they are characterized by large CO_2 emissions from fuel combustion and decomposition of organic waste as well as transformations to vegetation and soil that affect carbon sources and sinks.

> Conversion of agricultural and wildlands to settlements of varying densities is occurring at a rapid rate in North America, faster, in fact, than the rate of population growth.

Determining the extent of human settlements across North America also presents a challenge, as definitions of "developed," "built-up," and "urban" land vary greatly, particularly among nations. The U.S., Canadian, and Mexican census definitions are not consistent; in addition, several other classification schemes for defining and mapping settlements have been developed, such as the U.S. Department of Agriculture's National Resource Inventory categorization of developed land, which uses a variety of methods based on satellite imagery and ground-based information. One method of classifying settled land cover that has been consistently applied at a continental scale is the Global Rural-Urban Mapping Project conducted by a consortium of institutions, including Columbia University and the World Bank (CIESIN *et al.*, 2004). This estimate, which is based on nighttime lights satellite imagery, is 1,039,450 km², almost 5% of the total continental land area (Figure 14.1).[†]

Currently, there is insufficient information to determine the complete current or historical carbon balance of total continental land area. Fossil-fuel emissions very likely dominate carbon fluxes from settlements, just as settlement-related emissions likely dominate total fossil-fuel consumption in North America. However, specific estimates of the proportion of total fossil-fuel emissions directly attributable to settlements are difficult to make given current inventory methods, which are often conducted on a state or province-wide basis. In addition, the biological component of the carbon balance of settlements is highly uncertain, particularly with regard to the influence of urbanization on soil carbon pools and biogenic greenhouse gas emissions.

For the urban tree component of the settlement carbon balance, carbon stocks and sequestration have been estimated for urban land cover (as defined by the U.S. Census Bureau) in the coterminous United States to be on the order of 700 million tons (Mt) (335-980 million metric tons of carbon [Mt C]) with sequestration rates of 22.8 Mt C per year (13.7-25.9 Mt C per year) (Nowak and Crane, 2002). These estimates

[1†] A dagger symbol indicates that the magnitude and/or range of uncertainty for the given numerical value(s) is not provided in the references cited.

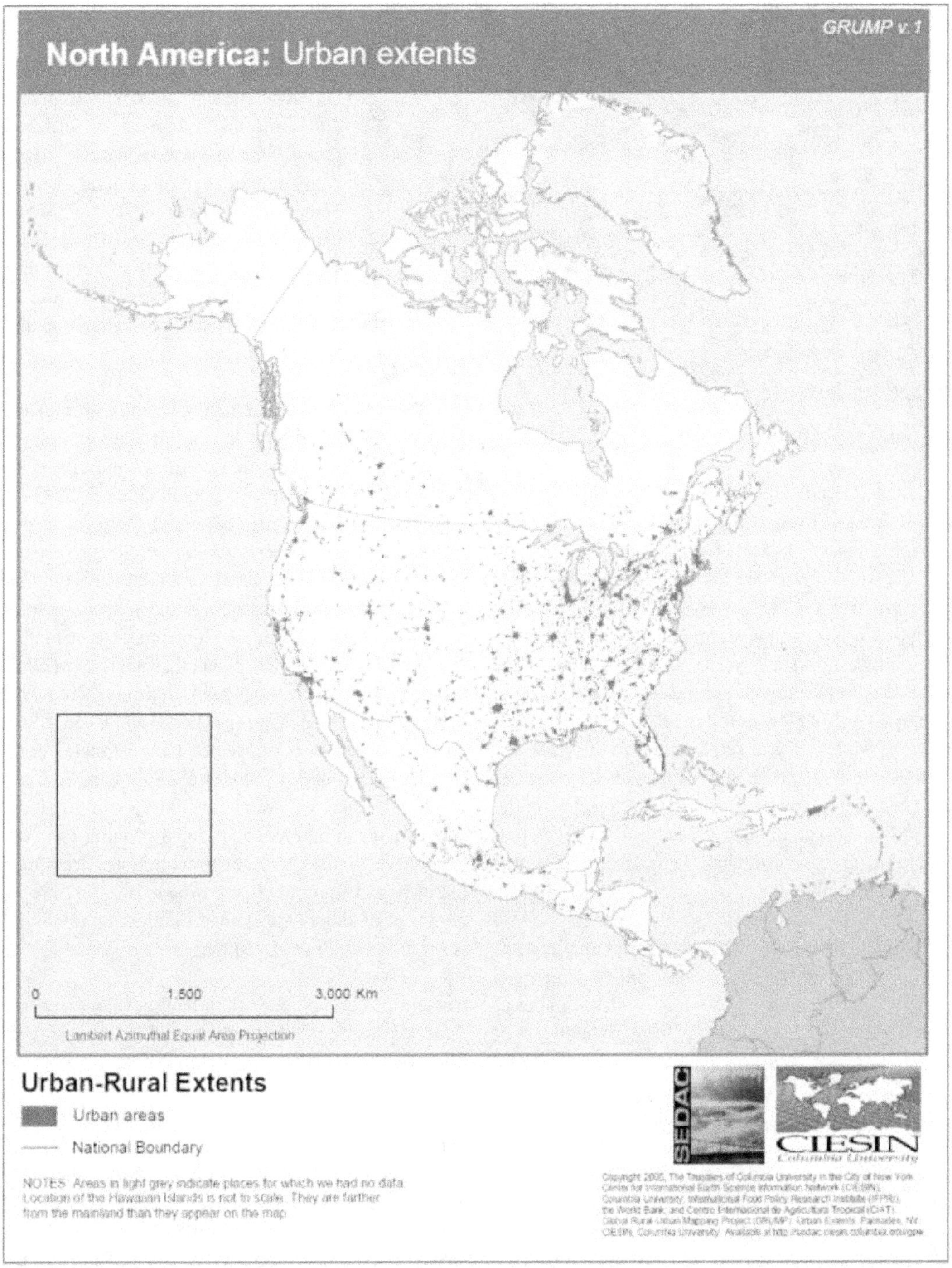

Figure 14.1 North America urban extents.

encompass a great deal of regional variability and contain some uncertainty about differences in carbon allocation between urban and natural trees, as urban trees have been less studied. However, to a first approximation, these estimates can be used to infer a probable range of urban tree carbon stocks and gross sequestration on a continental basis.

Nowak and Crane (2002) estimated that urban tree carbon storage in the Canadian border states (excluding semi-arid Montana, Idaho, and North Dakota) ranged from 24 to 45 tons of carbon per hectare (t C per ha), and carbon sequestration ranged from 0.8 to 1.5 t C per ha per year. Applying these values to a range of estimates of the extent of urban

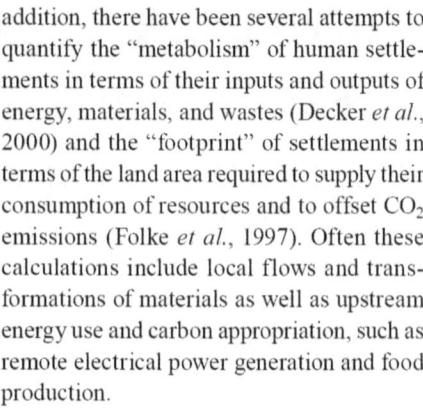

addition, there have been several attempts to quantify the "metabolism" of human settlements in terms of their inputs and outputs of energy, materials, and wastes (Decker *et al.*, 2000) and the "footprint" of settlements in terms of the land area required to supply their consumption of resources and to offset CO_2 emissions (Folke *et al.*, 1997). Often these calculations include local flows and transformations of materials as well as upstream energy use and carbon appropriation, such as remote electrical power generation and food production.

To conduct metabolic and footprint analyses of specific settlements, energy and fuel use statistics are needed for individual munici-

land in Canada (28,045 km² from the 1996 Canadian Census and 131,560 km² from CIESIN *et al.*, 2004), Canadian urban forest carbon stocks are between 67 and 592 Mt while carbon sequestration rates are between 2.2 and 19.7 Mt C per year. Similarly, for Mexico, Nowak and Crane (2002) estimated that urban carbon storage and sequestration in the U.S. southwestern states varied from 4.4 to 10.5 t per ha and 0.1 to 0.3 t per ha per year, respectively, leading to estimates of 10 to 107 Mt C stored in urban trees in Mexico and 0.2 to 3.1 Mt C per year sequestered. In this analysis, urban "trees" were defined as vegetation with woody stems greater than 1 inch diameter as measured 4.5 feet from the ground; carbon storage of other types of urban vegetation is not included in these estimates. Estimates of historical trends are not available.

palities, and these data are seldom made available at that scale. Consequently, metabolic and footprint analyses of carbon flows and conversions associated with metropolitan regions have been conducted for a relatively small number of cities. A metabolic analysis of the Toronto metropolitan region showed per capita net CO_2 emissions of 14 t CO_2 per year[†] (Sahely *et al.*, 2003), higher than analyses of other large metropolitan areas in developed countries (Newman, 1999; Pataki *et al.*, 2006a; Warren-Rhodes and Koenig, 2001). In contrast, an analysis of Mexico City estimated per capita CO_2 emissions of 3.4 t CO_2 per year[†] (Romero Lankao *et al.*, 2004). Local emissions inventories can provide useful supplements to national and global inventories in order to ensure that emissions reductions policies are applied effectively and equitably (Easterling *et al.*, 2003). A detailed review of methodological uncertainties and research needs is given in Pataki *et al.* (2006b).

Projections for increases in the extent of developed, nonfederal land cover in the United States in the next 25 years would increase the proportion of developed land from 5.2% to 9.2% of total land cover.

While complete national or continental-scale estimates of the carbon budget of settlements including fossil fuels, vegetation, and soils are not available, several methods are available to assess the full carbon balance of individual settlements and can be applied in the next several years toward constructing larger-scale inventories. Atmospheric measurements can be used to determine the net losses of carbon from settlements and urbanizing regions (Grimmond *et al.*, 2002; Grimmond *et al.*, 2004; Nemitz *et al.*, 2002; Soegaard and Moller-Jensen, 2003). Specific sources of CO_2 can be determined from unique isotopic signatures (Pataki *et al.*, 2003; Pataki *et al.*, 2006b) and from the relationship between CO_2 and carbon monoxide (Lin *et al.*, 2004). Many of these techniques have been commonly applied to natural ecosystems and may be easily adapted for settled regions. In

Current projections for urban land development in North America highlight the importance of improving carbon inventories of settlements and assessing patterns and impacts of future urban and rural development. Projections for increases in the extent of developed, nonfederal land cover in the United States in the next 25 years are as high as 79%, which would increase the proportion of developed land from 5.2% to 9.2% of total land cover (Alig *et al.*, 2004). The potential consequences of this increase for the carbon cycle are significant in terms of CO_2 emissions from an expanded housing stock and transportation network as well as from conversion of agricultural land, forest, rangeland, and other ecosystems to urban land cover. Because the dynamics of carbon cycling in settled areas encompass a range of physical, biological, social, and economic processes, studies of the potential impacts of future development on the carbon cycle must be interdisciplinary. Large-scale research on what has been called the study "of cities as ecosystems" (Pickett *et al.*, 2001) has begun only relatively recently, pioneered

Table 14.1 Increases in number of households and the total population of the United States, Canada, and Mexico between 1985 and and 2000. (United Nations, 2002; United Nations Habitat, 2003).

	Total Population (%)	Households (%)
Canada	19	39
Mexico	33	60
United States	15	25

by interdisciplinary studies such as the National Science Foundation's Long-Term Ecological Research sites in the central Arizona-Phoenix area and in Baltimore (Grimm *et al.*, 2000). Although there is not yet sufficient data to construct a complete carbon inventory of settlements across North America, it is a feasible research goal to do so in the next several years if additional studies in individual municipalities are conducted in a variety of urbanizing regions.

14.3 TRENDS AND DRIVERS

Drivers of change in the carbon cycle associated with human settlements include (1) factors that influence the rate of land conversion and urbanization, such as population growth and density, household size, economic growth, and transportation infrastructure; (2) additional factors that influence fossil-fuel emissions, such as climate, residence and building characteristics, transit choices, and affluence; and (3) factors that influence biological carbon gains and losses, including the type of predevelopment land cover, post-development urban design and landscaping choices, soil and landscape management practices, and the time since land conversion.

14.3.1 Fossil-fuel Emissions
The density and patterns of development of human settlements (*i.e.*, their "form") are drivers of the magnitude of the fossil-fuel emissions component of the carbon cycle. The size and number of residences and households influence CO_2 emissions from the residential sector, and the spatial distribution of residences, commercial districts, and transportation networks is a key influence in the vehicular and transportation sectors. Many of the attributes of urban form that influence the magnitude of fossil-fuel emissions are linked to historical patterns of economic development, which have differed in Canada, the United States, and Mexico. The future trajectory of development and associated levels of affluence and technological and social change will strongly influence key aspects of urban form such as residence size, vehicle miles traveled, and investment in urban infrastructure, along with associated fossil-fuel emissions. Whereas emissions from the transportation and residential sectors are discussed in detail in Chapters 7 and 9, respectively, this chapter discusses specific aspects of the form of human settlements that affect the current continental carbon balance and its possible future trajectories.

> Although there is not yet sufficient data to construct a complete carbon inventory of settlements across North America, it is a feasible research goal to do so in the next several years.

Household size in terms of the number of occupants per household has been declining in North America (Table 14.1) while the average size of new residences has been increasing. For example, the average size of new, single family homes in the United States increased from 139 m² (1500 ft²) to more than 214 m² (2300 ft²) between 1970 and 2004 (NAHB, 2005). These trends have contributed to increases in per capita CO_2 emissions from the residential sector as well as increases in the consumption of land for residential and urban development (Alig *et.al.*, 2003; Ironmonger *et al.*, 1995; Liu *et al.*, 2003; MacKellar *et al.*, 1995). In addition, when considering total emissions from settlements, the trajectory of the transportation and residential sectors may be linked. There have been a number of qualitative discussions of the role of "urban sprawl" in influencing fossil-fuel and pollutant emissions from cities (CEC, 2001; Gonzalez, 2005), although definitions of urban sprawl vary (Ewing *et al.*, 2003). Quantitative linkages between urban form and energy use have been attempted by comparing datasets for a variety of cities, but the results have been difficult to interpret due to the large number of factors that may affect transportation patterns and energy consumption (Anderson *et al.*, 1996). For example, in a seminal analysis of data from a variety of cities, Kenworthy and Newman (1990) found a negative correlation between population density and per capita energy use in the transportation sector. However, their data have been reanalyzed and reinterpreted in a number of subsequent studies that

have highlighted other important driving variables, such as income levels, employment density, and transit choice (Gomez-Ibanez, 1991; Gordon and Richardson, 1989; Mindali *et al.*, 2004).

Quantifying the nature and extent of the linkage between development patterns of human settlements and greenhouse gas emissions is critical from the perspective of evaluating the potential impacts of land-use policy.

Quantifying the nature and extent of the linkage between development patterns of human settlements and greenhouse gas emissions is critical from the perspective of evaluating the potential impacts of land-use policy. One way forward is to further the application of integrated land-use and transportation models that have been developed to analyze future patterns of urban development in a variety of cities (Agarwal *et al.*, 2000; EPA, 2000; Hunt *et al.*, 2005). Only a handful have been applied to date for generating fossil-fuel emissions scenarios from individual metropolitan areas (Jaccard *et al.*, 1997; Pataki *et al.*, 2006a), such that larger-scale national or continental projections for human settlements are not currently available. However, there is potential to add a carbon cycle component to these models that would assess the linkages between land-use and land-cover change, residential and commercial energy use and emissions, emissions from the transportation sector, and net carbon gains and losses in biological sinks following land conversion. A critical feature of these models is that they may be used to evaluate future scenarios and the potential impacts of policies to influence land-use patterns and transportation networks in individual settlements and developing regions.

14.3.2 Vegetation and Soils in Human Settlements

Human settlements contain vegetation and soils that are often overlooked in national inventories, as they fall outside common classification schemes. Nevertheless, patterns of development affect the carbon balance of biological systems, both in the replacement of natural ecosystems with rural, residential, or urban land cover and in processes within settlements that affect constructed and managed land cover. In the United States, satellite data and ecosystem modeling for the mid-1990s suggested that urbanization occurred largely on productive agricultural land and therefore caused a net loss of carbon fixed by photosynthesis of 40 Mt C per year [†] (Imhoff *et al.*, 2004).

Urban trees generally result in net reductions in energy use.

Urban forests and vegetation sequester carbon directly as described under carbon inventories. In addition, urban trees influence the carbon balance of municipalities indirectly through their effects on energy use. Depending on their placement relative to buildings, trees may cause shading and windbreak effects, as well as evaporative cooling due to transpiration (Akbari, 2002; Oke, 1989; Taha, 1997). These effects have been estimated in a variety of studies, mostly involving model calculations that suggest that urban trees generally result in net reductions in energy use (Akbari, 2002; Akbari and Konopacki, 2005; Akbari *et al.*, 1997; Akbari and Taha, 1992; Huang *et al.*, 1987). Taking into account CO_2 emissions resulting from tree maintenance and decomposition of removed trees, "avoided" emissions from energy savings were responsible for approximately half of the total net reduction in CO_2 emissions from seven municipal urban forests, with the remainder attributable to direct sequestration of CO_2 (McPherson *et al.*, 2005). Direct measurements of meteorological fluxes that quantify the contribution of vegetation are needed to validate these estimates.

Like natural ecosystems, soils in human settlements contain carbon, although rates of sequestration are much more uncertain in urban soils than in natural soils. In general, soil carbon is lost following disturbances associated with conversion from natural to urban or suburban land cover (Pouyat *et al.*, 2002). Soil carbon pools may subsequently increase at varying rates, depending on the soil and land cover type, local climate, and management intensity (Golubiewski, 2006; Pouyat *et al.*, 2002; Qian and Follet, 2002). In ecosystems with low rates of carbon sequestration in native soil such as arid and semi-arid ecosystems, conversion to highly managed, settled land cover can result in higher rates of carbon sequestration and storage than pre-settlement due to large inputs of water, fertilizer, and organic matter (Golubiewski, 2006). Pouyat *et al.* (2006) used urban soil organic carbon measurements to estimate the total above- and below-ground carbon storage, including soil carbon, in U.S. urban land cover to be 2,640 Mt (1,890 to 3,300 Mt). This range does not include the uncertainty in classifying urban land cover,

but applies the range of uncertainty in above-ground urban carbon stocks reported in Nowak and Crane (2002) and the standard deviation of urban soil carbon densities reported in Pouyat *et al.* (2006). In addition, irrigated and fertilized urban soils have been associated with higher emissions of CO_2 and the potent greenhouse gas nitrous oxide (N_2O) relative to natural soils, offsetting some potential gains of sequestering carbon in urban soils (Kaye *et al.*, 2004; Kaye *et al.*, 2005; Koerner and Klopatek, 2002). Finally, full carbon accounting that incorporates fossil-fuel emissions associated with soil management (*e.g.*, irrigation and fertilizer production and transport) has not yet been conducted. In general, additional data on soil carbon balance in human settlements are required to assess the potential for managing urban and residential soils for carbon sequestration.

14.4 OPTIONS FOR MANAGEMENT

A number of municipalities in Canada, the United States, and Mexico have committed to voluntary programs of greenhouse gas emissions reductions. Under the Cities for Climate Protection program (CCP) of International Governments for Local Sustainability (ICLEI, formerly the International Council of Local Environmental Initiatives) 269 towns, cities, and counties in North America have committed to conducting emissions inventories, establishing a target for reductions, and monitoring the results of reductions initiatives (the current count of the number of municipalities participating in voluntary greenhouse gas reduction programs may be found on-line at http://www.iclei.org). Emissions reductions targets vary by municipality, as do the scope of reductions, which may apply to the municipality as a whole or only to government operations (*i.e.*, emissions related to operation of government-owned buildings, facilities, and vehicle fleets).

Kousky and Schneider (2003) interviewed representatives from 23 participating CCP municipalities in the United States who indicated that cost savings and other co-benefits of greenhouse gas reductions in cities and towns were the most commonly cited reasons for participating in voluntary greenhouse gas reductions programs. Potential cost savings include reductions in energy and fuel costs from energy efficiency programs in buildings, street lights, and traffic lights; energy cogeneration in landfills and sewage treatment plants; mass transit programs; and replacement of municipal vehicles and buses with alternative fuel or hybrid vehicles (ICLEI, 1993; 2000). Other perceived co-benefits include reductions in emissions of particulate and oxidant pollutants, alleviation of traffic congestion, and availability of lower-income housing in efforts to curb urban sprawl. These co-benefits are often "perceived" because many municipalities have not attempted to quantify them as part of their emissions reductions programs (Kousky and Sch-

neider, 2003); however, it has been suggested that they play a key role in efforts to promote reductions of municipal-scale greenhouse gas emissions because local constituents regard them as an issue of interest (Betsill, 2001).

Of the co-benefits of municipal programs to reduce CO_2 emissions, improvements in air quality are perhaps the most well studied. Cifuentes (2001) analyzed the benefits of reductions in atmospheric particulate matter measuring less than 10 micrometers (μm) in diameter (PM10) and ozone concentrations in four cities in North and South America. Using a greenhouse gas reduction of 13% of 2000 levels by 2020 from energy efficiency and fuel substitution programs, Cifuentes (2001) estimated that PM10 and ozone concentrations would decline by 10% of 2000 levels. Estimated health benefits from such a reduction included avoidance of 64,000 (18,000-116,000) premature deaths associated with air quality-related heath problems as well as avoidance of 91,000 (28,000-153,000) hospital admissions and 787,000 (136,000-1,430,000) emergency room visits. However, using calculations for co-control of CO_2 and air pollutants in Mexico City, West *et al.* (2004) found that in practice, if electrical energy is primarily generated in remote locations relative to the urban area, cost-effective energy efficiency programs may have a relatively small effect on air quality. In that case, options for reducing greenhouse gas emissions would have to be implemented primarily in the transportation sector to appreciably affect air quality.

> Two hundred and sixty nine towns, cities, and counties in North America have committed to conducting emissions inventories, establishing a target for reductions, and monitoring the results of reductions initiatives.

14.5 RESEARCH NEEDS

Additional studies of the carbon balance of settlements of varying densities, geographical location, and patterns of development are needed to quantify the potential impacts of various policy and planning alternatives on net greenhouse gas emissions. While it may seem intuitive that policies to curb urban sprawl or enhance tree planting programs will result in emissions reductions, different aspects of urban form (*e.g.*, housing density, availability of public transportation, type and location of forest cover) may have different net effects on carbon sources and sinks, depending on the location, affluence, economy, and geography of various settlements. It is possible to develop quantitative tools to take many of these factors into account. To facilitate development and application of integrated urban carbon cycle models and to extrapolate local studies to regional, national, and continental scales, useful additional data include:

- common land cover classifications appropriate for characterizing a variety of human settlements across North America,
- emissions inventories at small spatial scales such as individual neighborhoods and municipalities,
- expansion of the national carbon inventory and flux measurement networks to include land cover types within human settlements,
- comparative studies of processes and drivers of development in varying regions and nations, and
- interdisciplinary studies of land-use change that evaluate socioeconomic as well as biophysical drivers of carbon sources and sinks.

In general, there has been a focus in carbon cycle science on measuring carbon stocks and fluxes in natural ecosystems, and consequently highly managed and human-dominated systems such as settlements have been underrepresented in many regional and national inventories. To assess the full carbon balance of settlements ranging from rural developments to large cities, a wide range of measurement techniques and scientific, economic, and social science disciplines are required to understand the dynamics of urban expansion, transportation, economic development, and biological sources and sinks. An advantage to an interdisciplinary focus on the study of human settlements from a carbon cycle perspective is that human activities and biological impacts in and surrounding settled areas encompass many aspects of perturbations to atmospheric CO_2, including a large proportion of national CO_2 emissions and changes in carbon sinks resulting from land-use change.

Coastal Oceans

Lead Authors: Francisco P. Chavez, MBARI; Taro Takahashi, Columbia Univ.

Contributing Authors: Wei-Jun Cai, Univ. Ga.; Gernot Friederich, MBARI; Burke Hales, Oreg. State Univ.; Rik Wanninkhof, NOAA; Richard A. Feely, NOAA

KEY FINDINGS

- The combustion of fossil fuels has increased carbon dioxide in the atmosphere, and the oceans have annually absorbed an equivalent of 20-30% of the carbon dioxide in fossil-fuel emissions. The present annual uptake by the oceans of approximately 1.8 billion tons of carbon (26% of global fossil-fuel emissions in 2003) is well constrained, has slightly acidified the oceans and may ultimately affect ocean ecosystems in unpredictable ways.

- The carbon budgets of ocean margins (coastal regions) are not as well-characterized due to lack of observations coupled with complexity and highly localized geographic variability. Existing data are insufficient, for example, to estimate the amount of carbon derived from human activity stored in the coastal regions of North America or to predict future scenarios.

- New air-sea carbon flux observations reveal that on average, waters within about 100 km (60 miles) of the shores surrounding North America are neither a source nor a sink of carbon dioxide to the atmosphere. A small net source of carbon dioxide to the atmosphere of 19 million tons of carbon per year (with large uncertainty) is estimated mostly from waters around the Gulf of Mexico and the Caribbean Sea. This is equivalent to about 1% of the global ocean uptake.

- With the exception of one or two time-series sites, almost nothing is known about historical trends in air-sea fluxes and the source-sink behavior of North America's coastal oceans.

- The Great Lakes and estuarine systems of North America may be net sources of carbon dioxide where terrestrially-derived organic material is decomposing, while reservoir systems may be storing carbon through sediment transport and burial.

- Options for sequestering carbon in the ocean include iron fertilization in sunlit surface waters and injection of carbon dioxide in subsurface coastal waters. However, sequestration capacity and potential adverse effects on marine environments need to be investigated.

- Highly variable air-sea carbon dioxide fluxes in coastal areas may introduce errors in North American carbon dioxide fluxes calculated by atmospheric inversion methods. Reducing these errors and the uncertainties regarding the variability of carbon cycling in coastal oceans will require observation systems utilizing fixed and mobile platforms, novel instrumentation to measure critical stocks and fluxes, and coordinated national and international research programs. Experimental studies involving coastal carbon cycling should be encouraged.

15.1 INVENTORIES (STOCKS AND FLUXES, QUANTIFICATION)

Climate-driven changes in ocean circulation, chemical properties or biological rates could result in strong feedbacks to the atmosphere.

The uptake of this human-caused CO_2 by the oceans is, on average, turning them more acidic with negative and potentially catastrophic effects on some biota.

This chapter first introduces the role the oceans play in modulating atmospheric carbon dioxide (CO_2), then quantifies air-sea CO_2 fluxes in coastal waters[1] surrounding North America and considers how the underlying processes affect the air-sea fluxes. Stocks of living organisms in marine environments are small relative to those on land, but turnover rates are very high. In addition, aquatic stocks are not well characterized because of their spatial and temporal variability, the complexity of carbon compound transformations, and limited data on these processes. The oceans act as a huge reservoir for inorganic carbon, containing about 50 times as much CO_2 as the atmosphere. The ocean's biological pump converts CO_2 to organic particulate carbon by photosynthesis, transports the organic carbon from the surface by sinking, and therefore plays a critical role in removing atmospheric CO_2 in combination with physical and chemical processes (Gruber and Sarmiento, 2002; Sarmiento and Gruber, 2006). Atmospheric concentration of CO_2 would be much higher in the absence of current ocean processes implying that climate-driven changes in ocean circulation, chemical properties or biological rates could result in strong feedbacks to the atmosphere.

The release of CO_2 into the atmosphere by the combustion of fossil fuels has increased pre-industrial concentrations from around 280 ppm to present day levels of nearly 380 ppm in 2005. This increase in atmospheric concentrations is driving CO_2 into the ocean with the present net air-sea CO_2 flux from the atmosphere into the ocean well constrained to about 1800 million metric tons of carbon (Mt C, See Box 15.1)**** per year (or 1.8 billion tons of carbon [Gt C]**** per year) (Figure 15.1 and Table 15.1) (Chapter 2 for a description of how ocean carbon fluxes relate to the global carbon cycle). The uptake of this human-caused CO_2 by the oceans is, on average, turning them more acidic with negative and potentially catastrophic effects on some biota (Kleypas *et al.*, 2006). The atmosphere is well mixed and nearly homogenous so the large spatial variability in air-sea CO_2 fluxes shown in Figure 15.1 is driven by a combination of physical, chemical, and biological processes in the ocean. The flux over the coastal margins has neither been well characterized (Liu *et al.*, 2000) nor integrated into global calculations because there are large variations over small spatial and temporal scales, and observations have been limited. The need for higher spatial

Table 15.1 Climatological mean distribution of the net air-sea CO_2 flux (in Gt C per year) over the global ocean regions (excluding coastal areas) in reference year 1995. The fluxes are based on about 1.75 million partial pressure measurements for CO_2 in surface ocean waters, excluding the measurements made in the equatorial Pacific (10°N- 10°S) during El Niño periods (see Takahashi et al., 2002). The NCAR/NCEP 42-year mean wind speeds and the (wind speed)^2 dependence for air-sea gas transfer rate are used (Wanninkhof, 1992). Plus signs indicate that the ocean is a source for atmospheric CO_2, and negative signs indicate that ocean is a sink. The ocean uptake has also been estimated on the basis of the following methods: temporal changes in atmospheric oxygen and CO_2 concentrations (Keeling and Garcia, 2002; Bender et al., 2005), $^{13}C/^{12}C$ ratios in sea and air (Battle et al., 2000; Quay et al., 2003), ocean CO_2 inventories (Sabine et al., 2004), and coupled carbon cycle and ocean general circulation models (Sarmiento et al., 2000; Gruber and Sarmiento, 2002). The consensus is that the oceans take up 1.3 to 2.3 Gt C per year.

Latitude bands	Pacific	Atlantic	Indian	Southern Ocean	Global
N of 50°N	+0.01	−0.31			−0.30
14°N-50°N	−0.49	−0.25	+0.05		−0.69
14°N-14°S	+0.65	+0.13	+0.13		+0.91
14°S-50°S	−0.39	−0.21	−0.52		−1.12
S of 50°S				−0.30	−0.30
Total flux	−0.23	−0.64	−0.34	−0.30	−1.50
% of flux	15	42	23	20	100
Area (10⁶ km²)	152.0	74.6	53.0	41.1	320.7
% of area	47	23	17	13	100

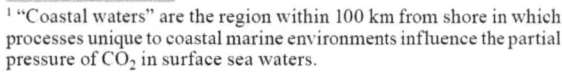

[1] "Coastal waters" are the region within 100 km from shore in which processes unique to coastal marine environments influence the partial pressure of CO_2 in surface sea waters.

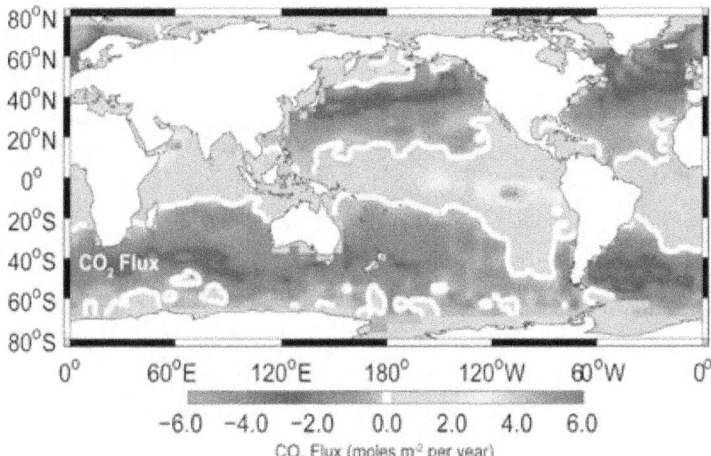

Figure 15.1 Global distribution of sea-air CO_2 flux. The source areas (cyan-green-yellow-orange) are primarily in the tropics with a few high latitiude areas where deep mixing occurs in winter. The sink areas (blue-magenta) are located in mid to high latitiudes. The white line represents zero flux. Updated from Takahashi et al. (2002).

a new analysis of about a half million observations of air-sea flux of CO_2 in coastal waters surrounding the North American continent.

15.1.1 Global Coastal Ocean Carbon Fluxes

The carbon cycle in coastal oceans involves a series of processes, including runoff from terrestrial environments, upwelling and mixing of high CO_2 water from below, photosynthesis at the sea surface, sinking of organic particles, respiration, production and consumption of dissolved organic carbon, and air-sea CO_2 fluxes (Figure 15.2). Although fluxes in the coastal oceans are large relative to surface area (Muller-Karger et al., 2005), there is disagreement as to whether these regions are a net sink or a net source of CO_2 to the atmosphere (Tsunogai et al., 1999; Cai and Dai, 2004; Thomas et al., 2004). Great uncertainties remain in coastal carbon fluxes, which are complex and dynamic, varying rapidly over short distances and at high frequencies. Only recently have new technologies allowed for the measurement of these rapidly changing fluxes (Friederich et al., 1995 and 2002; Hales and Takahashi, 2004).

resolution to resolve the coastal variability has hampered modeling efforts. In the following sections we review existing information on the coastal ocean carbon cycle and its relationship to the global ocean, and we present the results of

Carbon is transported from land to sea mostly by rivers in four components: CO_2 dissolved in water, organic carbon dissolved in water, particulate inorganic carbon (e.g., calcium carbonate [$CaCO_3$]), and particulate organic carbon. The global rate of river input has been estimated to be 1000 Mt C[***] per year, about 38% of it as dissolved CO_2 (or 384 Mt C per year), 25% as dissolved organic matter, 21% as organic particles, and 17% as $CaCO_3$ particles (Gattuso et al., 1998). Estimates for the riverine dissolved CO_2 flux vary from 385 to 429 Mt C per year (Sarmiento and Sundquist, 1992). The Mississippi River, the seventh-largest in freshwater discharge in the world, delivers about 13 Mt C[***] per year as dissolved CO_2 (Cai, 2003). Organic matter in continental

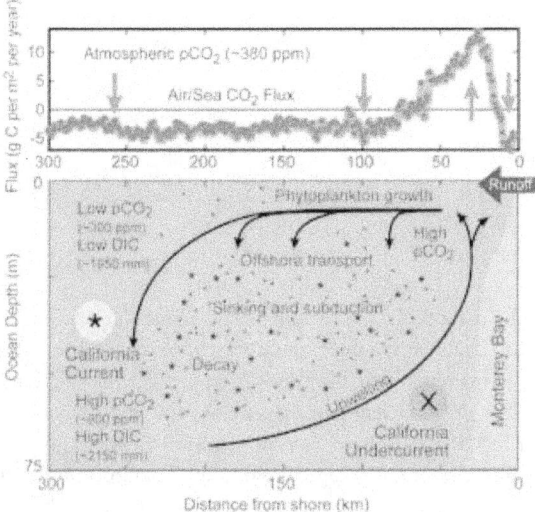

Figure 15.2 Mean air-sea CO_2 flux as calculated from shipboard measurements on a line perpendicular to the central California coast (top panel). Flux within Monterey Bay (~0-20 km offshore) is into the ocean, flux across the active upwelling region (~20-75 km offshore) is from the ocean, and flux in the California Current (75-300 km) is on average into the ocean. These fluxes result from the processes shown in the bottom panel. California Undercurrent water, which has a high CO_2 partial pressure, upwells near shore, and is advected offshore into the California Current and into Monterey Bay. Phytoplankton growing in the upwelled water use CO_2 as a carbon source, and CO_2 is drawn to low levels in those areas. Phytoplankton carbon eventually sinks or is subducted below the euphotic zone, where it decays, elevating the CO_2 levels of subsurface waters. Where the level of surface CO_2 is higher than the level of atmospheric CO_2, diffusion drives CO_2 into the atmosphere. Conversely, where the level of surface CO_2 is lower than that of atmospheric CO_2, diffusion drives CO_2 into the ocean. The net air-sea flux on this spatial scale is near zero. DIC = concentration of inorganic carbon (i.e., all CO_2 species) dissolved in seawater. Updated from Pennington et al. (in press).

Table 15.2 Variability of CO₂ distributions and fluxes in U.S. coastal waters from regional surveys and moored measurements (from Doney *et al.*, 2004).

Location	Surface seawater pCO₂ (μatm)	Instantaneous CO₂ flux (mol/ per m² per year)	Annual average (mol per m² per year)	Sampling method	Reference
New Jersey Coast	211–658	–17 to +12	–0.65	Regional survey	Boehme *et al.* (1998)
Cape Hatteras, North Carolina	ND	–1.0 to +1.2	ND	Moored measurements	DeGrandpre *et al.* (1997)
Middle Atlantic Bight, inner shelf	150–620	ND	–0.9	Regional survey	DeGrandpre *et al.* (2002)
Middle Atlantic Bight, middle shelf	220–480	ND	–1.6	Regional survey	DeGrandpre *et al.* (2002)
Middle Atlantic Bight, outer shelf	300–430	ND	–0.7	Regional survey	DeGrandpre *et al.* (2002)
Florida Bay, Florida	325–725	ND	ND	Regional survey	Millero *et al.* (2001)
Southern California Coastal Fronts	130–580	ND	ND	Regional survey	Simpson (1985)
Coastal Calif. (M-1; Monterey Bay)	245–550	–8 to +50	1997–98: –1.0 1998–99: +1.1	Moored measurements	Friederich *et al.* (2002)
Oregon Coast	250–640	ND	ND	Regional survey	van Geen *et al.* (2000)
Bering Sea Shelf in spring (April–June)	130–400	–8 to –12	–8	Regional survey	Codispoti *et al.* (1986)
South Atlantic Bight	300–1200	ND	2.5	Regional survey	Cai *et al.* (2003)
Miss. River Plume (summer)	80–800	ND	ND	Regional survey	Cai *et al.* (2003)
Bering Sea (Aug–Sep.)	192–400	ND	ND	Regional survey	Park *et al.* (1974)

ND indicates that no data are available.

To convert from "mol" to "grams," multiply the numerical "mol" value by 12.

shelf sediments exhibits only weak isotope and chemical signatures of terrestrial origin, suggesting that riverine organic matter is reprocessed in coastal environments on a time scale of 20 to 130 years (Hedges *et al.*, 1997; Benner and Opsahl, 2001). Of the organic carbon, about 30% is accumulating in estuaries, marshes, and deltas, and a large portion (20% to 60%) of the remaining 70% is readily and rapidly oxidized in coastal waters (Smith and Hollibaugh, 1993). Only about 10% is estimated to be contributed by human activities, such as agriculture and forest clearing (Gattuso *et al.*, 1998), and the rest is a part of the natural carbon cycle.

One of the major differences between coastal and open ocean systems is the activity of the biological pump. In coastal environments, the pump operates much more efficiently, leading to rapid reduction of surface CO₂ and thus complicating the accurate quantification of air-sea CO₂ fluxes. For example, Ducklow and McCallister (2004) constructed a carbon balance for the coastal oceans using the framework of the ocean carbon cycle of Gruber and Sarmiento (2002) and estimated a net CO₂ removal by primary productivity of 1200 Mt C per year and a large CO₂ sink of 900 Mt C per year for the atmosphere. In contrast, Smith and Hollibaugh (1993) estimated a biological pump of about 200 Mt C per year and concluded that the coastal oceans are a weak CO₂ sink of 100 Mt C per year, about one-ninth of the estimate by Ducklow and McCallister (2004). Since the estimated air-sea CO₂ flux depends on quantities that are not well constrained, the mass balance provides widely varying results. For this reason, in this chapter, the net air-sea flux over coastal waters is estimated on the basis of direct measurements of the air-sea difference of partial pressure of CO₂ (pCO₂).

15.1.2 North American Coastal Carbon

Two important types of North American coastal ocean environments can be identified: (1) river-dominated coastal

A)

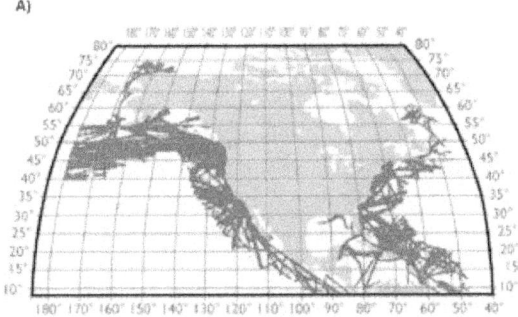

B)

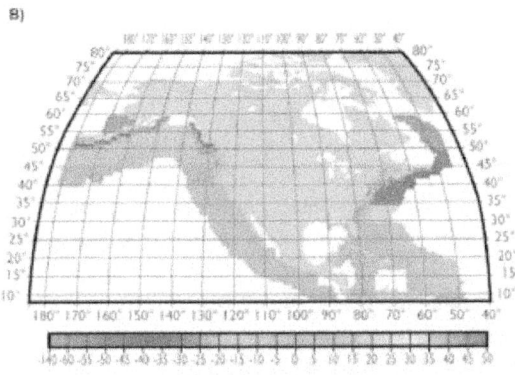

Air-Sea CO$_2$ Flux (g C m^{-2} per year)

Figure 15.3 (A) Distribution of coastal surface water CO$_2$ partial pressure measurements made between 1979 and 2004. (B) The distribution of the annual mean air-sea net CO$_2$ flux over 1° × 1° pixel areas (N-S 100 km, E-W 80 km) around North America. The flux (g C per m2 per year) represents the climatological mean over the 25-year period. The magenta-blue colors indicate that the ocean water is a sink for atmospheric CO$_2$, and the green-yellow-orange colors indicate that the sea is a CO$_2$ source. The data were obtained by the authors and collaborators of this chapter and are archived at the Lamont-Doherty Earth Observatory (www.ldeo.columbia.edu/res/pi/CO$_2$).

margins with large inputs of fresh water, organic matter, and nutrients from land (*e.g.*, Mid- and South-Atlantic Bights) (Cai *et al.*, 2003) and (2) coastal upwelling zones (*e.g.*, the California-Oregon-Washington coasts, along the eastern boundary of the Pacific) where physical processes bring cool, high-nutrient, and high-CO$_2$ waters to the surface. In both environments, the biological uptake of CO$_2$ plays an important role in determining whether an area becomes a sink or a source for the atmosphere.

High biological productivity fueled by nutrients added to coastal waters can lead to seawater becoming a CO$_2$ sink during the summer growing season, as observed in the Bering Sea Shelf (Codispoti and Friederich, 1986) and the northwest waters off Oregon and Washington (van Geen *et al.*, 2000; Hales *et al.*, 2005). Similar CO$_2$ draw-downs may occur in the coastal waters of the Gulf of Alaska and in the Gulf of Mexico near the Mississippi River outflow. Coastal

upwelling results in a very high concentration of CO$_2$ for the surface water (as high as 1000 µatm), and, hence, the surface water becomes a strong CO$_2$ source. This is followed by rapid biological uptake of CO$_2$, which causes the water to become a strong CO$_2$ sink (Friederich *et al.*, 2002; Hales *et al.*, 2005).

A review of North American coastal carbon fluxes has been carried out by Doney *et al.* (2004) (Table 15.2). The information reviewed was very limited in space (only 13 locations) and time, leading Doney *et al.* to conclude that it was unrealistic to reliably estimate an annual flux for North American coastal waters. Measurement programs have increased recently, and we have used the newly available data to calculate annual North American coastal air-sea fluxes for the first time.

15.1.3 Synthesis of Available North American Air-Sea Coastal CO$_2$ Fluxes

A large data set consisting of 550,000 measurements of the pCO$_2$ in surface waters has been assembled and analyzed (Figure 15.3; see Appendix G for details). Partial pressure of CO$_2$ is measured in a carrier gas equilibrated with seawater and, as such, it is a measure of the outflux/influx tendency of CO$_2$ from the atmosphere. Carbon dioxide reacts with seawater and 99.5% of the total amount of CO$_2$ dissolved in sea-

> The open ocean Pacific waters south of 30°N are, on the annual average, a CO$_2$ source to the atmosphere, whereas the area north of 40°N is a sink.

water is in the form of bicarbonate (HCO$_3^-$) and carbonate ions (CO$_3^=$), which do not exchange with the overlying atmosphere. Only CO$_2$ molecules, which constitute about 0.5% of the total dissolved CO$_2$, exchange with the atmosphere. This is expressed as pCO$_2$, which is affected by physical and biological processes; pCO$_2$ increases as seawater warms and decreases when photosynthesis is stimulated. The data were obtained by the authors and collaborators, quality-controlled, and assembled in a uniform electronic format for analysis (available at www.ldeo.columbia.edu/res/pi/CO2). Observations in each 1° × 1° pixel area were compiled into a single year and were analyzed for time-space variability. Seasonal and interannual variations were not well characterized except in a few locations (Friederich *et al.*, 2002). The annual mean air-sea pCO$_2$ difference (ΔpCO$_2$) was computed for 5°-wide zones along the North American continent and was plotted as a function of latitude for four regions (Figure 15.4): North Atlantic, Gulf of Mexico/Caribbean, North Pacific, and Bering/Chukchi Seas. Figure 15.4A shows the fluxes in the first nearshore band, and Figure 15.4B shows the fluxes for a band that is several hundred kilometers from shore. The average fluxes for them and for the intermediate bands are given in Table 15.3. The flux and area data are listed in Table

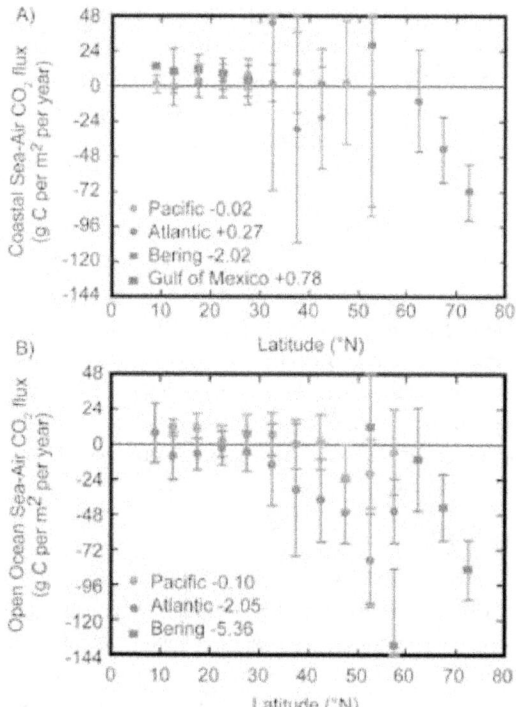

Figure 15.4 Estimated air-sea CO_2 fluxes (g C per per m^2 per year) from 550,000 seawater CO_2 partial pressure (pCO_2) ob-servations made from 1979 to 2004 in ocean waters surrounding the North American continent. (A) Waters within one degree (about 80 km) of the coast and (B) open ocean waters between 300 and 900 km from the shore (see Figure 15.3B). The annual mean air-sea pCO_2 difference (Δ pCO_2) values were calculated from the weekly mean atmospheric CO_2 concentrations in the GLOBALVIEW-CO_2 database (2004) over the same pixel area in the same week and year as the seawater pCO_2 was measured. The monthly net air-sea CO_2 flux was computed from the mean monthly wind speeds in the National Centers for Environmental Prediction/National Center for Atmospheric Research (NCEP/ NCAR) database in the (wind speed)^2 formulation for the air-sea gas transfer rate by Wanninkhof (1992). The ± uncertainties represent one standard deviation.

15.4. A full complement of seasonal observations are lack-ing in the Arctic Sea, including Hudson Bay, the northern Labrador Sea, and the Gulf of St. Lawrence; the northern Bering Sea; the Gulf of Alaska; the Gulf of California; and the Gulf of Mexico and the Caribbean Sea.

In contrast to the Pacific coast, the latitude where Atlantic coastal waters become a CO_2 sink is located further north.

The offshore patterns follow the same general trend found in the global open ocean data set shown in Figure 15.1. On an annual basis the lower latitudes tend to be a source of CO_2 to the atmosphere, whereas the higher latitudes tend to be sinks (Figures 15.3B and 15.4B). The major difference in the coastal waters is that the latitude where CO_2 starts to enter the ocean is further north than it

is in the open ocean, particularly in the Atlantic. A more detailed region-by-region description follows.

15.1.4 Pacific Ocean

Observations made in waters along the Pacific coast of North America illustrate how widely coastal waters vary in space and time, in this case driven by upwelling and relaxation (Friederich *et al.*, 2002). Figure 15.5A shows a summertime quasi-synoptic distribution of temperature, salinity, and pCO_2 in surface waters based on measurements made in July through September 2005. The effects of the Columbia River plume emanating from ~46°N are clearly seen (colder temperature, low salinity, and low pCO_2), as are coastal upwelling effects off Cape Mendocino (~40°N) (colder, high salinity, and very high pCO_2). These coastal features are confined to within 300 km from the coast. The 1997-2005 time-series data for surface water pCO_2 observed off Monterey Bay (Figure 15.5B) show the large, rapidly fluctuating air-sea CO_2 fluxes during the summer upwelling season in each year, as well as the low-pCO_2 periods during the 1997-1998 and 2002-2003 El Niño events. In spite of the large seasonal variability, ranging from 200 to 750 µatm, the annual mean air-sea pCO_2 difference and the net CO_2 flux over the waters off Monterey Bay areas (~37°N) are close to zero (Pennington *et al.*, in press). The seasonal amplitude decreases away from the shore and in the open ocean bands, where the air-sea CO_2 flux changes seasonally in response to seawater temperature (out of the ocean in summer and into the ocean in winter).

The open ocean Pacific waters south of 30°N are, on the annual average, a CO_2 source to the atmosphere, whereas the area north of 40°N is a sink, and the zone between 30° and 40°N is neutral (Takahashi *et al.*, 2002). Coastal waters in the 40°N through 45°N zone (northern California-Oregon coasts) are even a stronger CO_2 sink, associated with nutrient input and stratification by fresh water from the Columbia River (Hales *et al.*, 2005). On the other hand, coastal pCO_2 values in the 15°N through 40°N zones have pCO_2 values similar to open ocean values and to the atmosphere. In the zones 15°N through 40°N, the annual mean values for the net air-sea CO_2 flux are nearly zero, consistent with the finding by Pennington *et al.* (in press).

15.1.5 Atlantic Ocean

With the exception of the 5°N-10°N zone, the open ocean areas are an annual net sink for atmospheric CO_2 with stronger sinks at high latitudes, especially north of 35°N (Figure 15.3B). In contrast, the nearshore waters are a CO_2 source between 15°N and 45°N. Accordingly, in contrast to the Pacific coast, the latitude where Atlantic coastal waters become a CO_2 sink is located further north. In the areas north of 45°N, the open ocean waters are a strong CO_2 sink, due primarily to the cold Labrador Sea waters.

Table 15.3 Climatological mean annual air-sea CO_2 flux (g C per m^2 per year) over the oceans surrounding North America. Negative values indicate that the ocean is a CO_2 sink for the atmosphere. N is the number of seawater pCO_2 measurements. The ± uncertainty is given by one standard deviation of measurements used for analysis and represents primarily the seasonal variability.

Ocean regions	Coastal boxes [a]		First offshore [a]		Second offshore [a]		Third offshore [a]		Open ocean [a]	
	Flux	N	Flux	N	Flux	N	Flux	N	Flux	N
North Atlantic	3.2±142	80,417	−1.4±94	65,148	−7.3±57	35,499	−10.4± 76.4	15,771	−26±83	37,667
North Pacific	−0.2±105	164,838	−6.0±81	69,856	−4.3±66	32,045	−5.3±60	16,174	−1.2±56	84,376
G. Mexico Caribbean	9.4±24	75,496	8.4±23	61,180	11.5±17.0	8,410	13±20	1,646		
Bering/ Chukchi	28.0±110	892	−28±128	868	−44±104	3,399	−53±110	1,465	−63±130	1,848

[a] The pCO_2 data are binned into 1° latitude x 1° longitude box areas. The boxes that include shorelines are named "coastal boxes," and the 1° x 1° boxes located on the ocean side of these "coastal boxes" are called "first offshore" boxes. The next two rows of ocean side boxes are called respectively the "second offshore" and the "third offshore" boxes.

In the coastal zone very high pCO_2 values (up to 2600 µatm) are observed occasionally in areas within 10 km offshore of the barrier islands (see small red dots off the coasts of Georgia and the Carolinas in Figure 15.3B). These waters which have salinities around 20 and high total CO_2 concentrations appear to represent outflow of estuarine/marsh waters rich in carbon (Cai *et al.*, 2003). The large contribution of fresh water that is rich in organic matter relative to the Pacific contributes to this small coastal Atlantic source. Offshore fluxes are in phase with the seasonal cycle of warming and cooling; fluxes are out of the ocean in summer and fall and are the inverse in winter and spring.

15.1.6 Bering and Chukchi Seas

Although measurements in these high-latitude waters are limited, the relevant data for the Bering Sea (south of 65°N) and Chukchi Sea (north of 65°N) are plotted as a function of the latitude in Figure 15.4. The values for the areas north of 55°N are for the summer months only; CO_2 observations are not available during winter seasons. Although data scatter widely, the coastal and open ocean waters are a strong CO_2 sink during the summer months due to photosynthetic draw-down of CO_2. The data in the 70°-75°N zone are from the shallow shelf areas in the Chukchi Sea. These waters are a very strong CO_2 sink (air-sea pCO_2 differences ranging from −80 to -180 µatm) with little changes between the coastal and open ocean areas. The air-sea CO_2 flux during winter months is not known but the summer fluxes are shown in Figure 15.4 for comparison. Bates (2006) estimated a mean-annual air-to-sea CO_2 flux[2] of 39 Mt C[***] per year over the

Chukchi shelf using data from spring and summer of 2002 that suggested that remnant winter waters were as strong a CO_2 sink as summer waters (with air-sea pCO_2 differences of -60 to -160 µatm).

15.1.7 Gulf of Mexico and Caribbean Sea

Although observations are limited, available data suggest that these waters are a strong CO_2 source (Figure 15.4 and Table 15.3). A subsurface anoxic zone has been formed in the Texas-Louisiana coast as a result of the increased addition of anthropogenic nutrients and organic carbon by the Mississippi River (*e.g.*, Lohrenz *et al.*, 1999). The carbon-nutrient cycle in the northern Gulf of Mexico is also being investigated (*e.g.*, Cai, 2003), and the studies suggest that at times those waters are locally a strong CO_2 sink due to high biological production.

15.2 SYNTHESIS

An analysis of half a million measurements of air-sea flux of CO_2 shows that the nearshore (< 100 km) coastal waters surrounding North America are a net CO_2 source for the atmosphere on an annual average of about 19±22 Mt C per year[3] (Table 15.4). Most of the flux (14±9 Mt C per year)[3] occurs in the Gulf of Mexico

[2] The flux was estimated on the basis of measurements made only during the spring and summer months of 2002 at several stations located in a limited area of the Chukchi Sea. The uncertainty of ± 7 Mt C given in the original paper represents one standard deviation of

measured pCO_2, but does not include uncertainties in the sea-air gas transfer coefficient estimated on the basis of wind speeds and those from limited time-space coverage.
[3] The specified uncertainty is ± one standard deviation around the mean.

Table 15.4 Areas (km²) and mean annual air-sea CO₂ flux (Mt C per year) over four ocean regions surrounding North America. Since the observations in the areas north of 60°N in the Chukchi Sea were made only during the summer months, the fluxes from that area are not included. The ± uncertainty is given by one standard deviation of measurements used for analysis and represents primarily the seasonal variability.

Ocean areas (km²)					Mean air-sea CO_2 flux (10^{12} grams or Mt C per year)				
Coastal boxes	First offshore	Second offshore	Third offshore	Open ocean	Coast box	First offshore	Second offshore	Third offshore	Open ocean
North Atlantic coast (8° N to 45°N)									
625,577	651,906	581,652	572,969	3,388,500	2.7±9.5	-0.5±9.3	-4.0±4.9	-6.5±6.3	-41.5±28.1
North Pacific coast (8°N to 55°N)									
1,211,555	855,626	874,766	646,396	7,007,817	2.1±17.1	-7.0±14.1	-4.8±12.5	-3.7±5.3	-53.8±60.7
Gulf of Mexico and Caribbean Sea (8°N to 30°N)									
1,519,335	1,247,413	935,947	1,008,633		13.6±8.9	10.9±7.5	6.8±5.00	6.6±5.0	
Bering and Chukchi Seas (50°N to 70°N)									
481,872	311,243	261,974	117,704	227,609	0.8±3.1	-6.2±9.5	-5.3±7.5	-3.7±3.0	-9.8±3.7
Total ocean areas surrounding North America									
3,838,339	3,066,188	2,654,339	2,300,702	10,623,926	19.1±21.8	-2.8±20.7	-7.4±16.2	-7.3±10.1	-105.2±67.0

and Caribbean Sea. The open oceans are a net CO_2 sink on an annual average (Table 15.4; Takahashi *et al.*, 2002). The reported uncertainties reflect the time-space variability but do not reflect uncertainties due to lack of observations in some portions of the Arctic Sea, Bering Sea, Gulf of Alaska, Gulf of Mexico, or Caribbean Sea. Observations in these areas will be needed to improve estimates. If the estimate of 39 Mt C[***] per year sink for the Chukchi Sea (Bates, 2006) is included, the North American coastal waters might be a small CO_2 sink. These results are consistent with recent global estimates that suggest that nearshore areas receiving terrestrial organic carbon input are sources of CO_2 to the atmosphere and that marginal seas are sinks (Borges, 2005; Borges *et al.*, in press). Hence, the net contribution from North American ocean margins is small and difficult to distinguish from zero. It is not clear how much of the open ocean sink results from photosynthesis driven by nutrients of coastal origin.

15.3 TRENDS AND DRIVERS

The sea-to-air CO_2 flux from the coastal zone is small (about 1%) compared with the global ocean uptake flux, which is about 1800 Mt C per year (or 1.8 Gt C per year), and hence does not influence the global air-sea CO_2 budget. However, coastal waters undergo large variations in air-sea CO_2 flux on daily to seasonal time scales and on small spatial scales (Figure 15.5). Fluxes can change on the order of 250 g C per m² per year or 0.7 g C per m² per day on a day to day basis (Figure 15.5). These large fluctuations can significantly

modulate atmospheric CO_2 concentrations over the adjacent continent and need to be considered when using the distribution of CO_2 in calculations of continental fluxes.

Freshwater bodies have not been treated in this analysis except to note the large surface pCO_2 resulting from estuaries along the east coast. The Great Lakes and rivers also represent net sources of CO_2 as, in the same manner as the estuaries, organic material from the terrestrial environment is oxidized so that respiration exceeds photosynthesis. Interestingly, the effect of fresh water is opposite along the coast of the Pacific northwest, where increased stratification and iron inputs enhance photosynthetic activity (Ware and Thomson, 2005), resulting in a large sink for atmospheric CO_2 (Figure 15.3). A similar process may be at work at the mouth of the Amazon (Körtzinger, 2003). This emphasizes once again the important role of biological processes in controlling the air-sea fluxes of CO_2.

The air-sea fluxes and the underlying carbon cycle processes that determine them (Figure 15.2) vary seasonally, interannually, and on longer time scales. The eastern Pacific, including the United States' west coast, is subject to changes associated with large-scale climate oscillations such as El Niño (Chavez *et al.*, 1999; Feely *et al.*, 2002; Feely *et al.*, 2006) and the Pacific Decadal Oscillation (PDO) (Chavez *et al.*, 2003; Hare and Mantua, 2000; Takahashi *et al.*, 2003). These climate patterns, and others, like the North Atlantic Oscillation (NAO), alter the oceanic CO_2 sink/source conditions directly through seawater temperature changes as well as ecosystem variations that occur via complex physical-

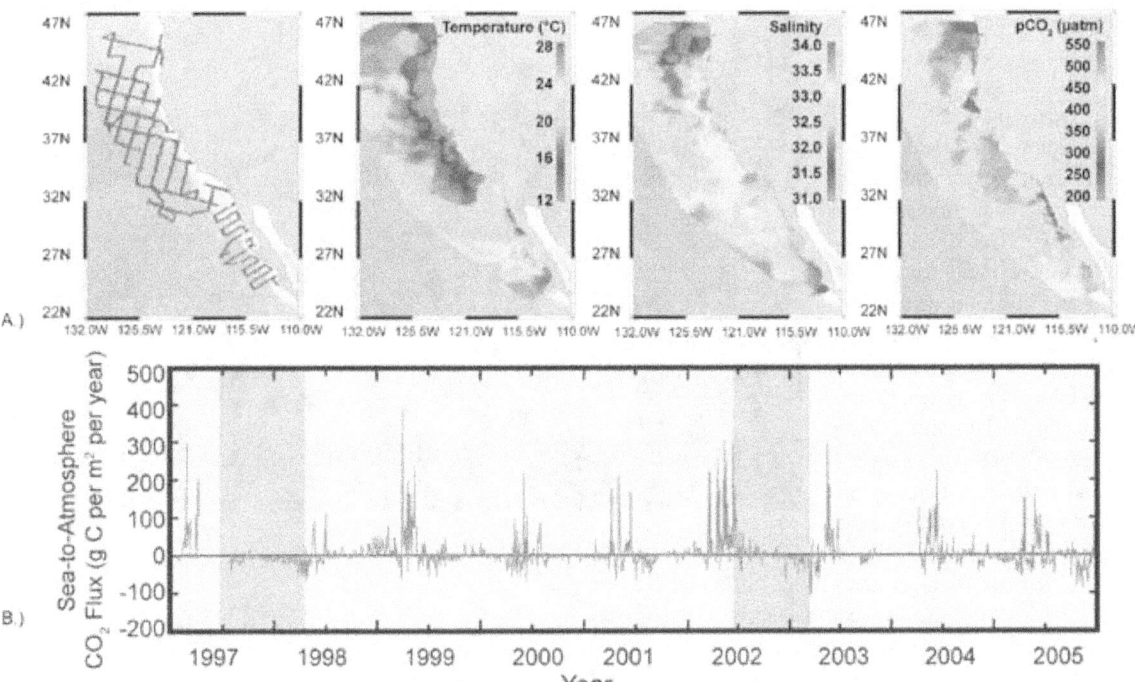

Figure 15.5 Time-space variability of coastal waters off the west coast of North America. (A) Quasi-synoptic distribution of the temperature, salinity, and pCO2 in surface waters during July-September 2005. The Columbia River plume (~46°N) and the upwelling of deep waters off Cape Mendocino (~40°N) are clearly seen. (B) 1997-2005 time-series data for air-sea CO_2 flux from a mooring off Monterey Bay, California. Seawater is a CO_2 source for the atmosphere during the summer upwelling events, but biological uptake reduces levels very rapidly. These rapid fluctuations can affect atmospheric CO_2 levels. For example, if CO_2 from the sea is mixed into a static column, a 500 m thick planetary boundary layer over the course of one day, atmospheric CO_2 concentration would change by 2.5 µatm. If the column of air is mixed vertically through the troposphere to 500 mbar, a change of about 0.5 µatm would occur. The effects would be diluted as the column of air mixes laterally. However, this demonstrates that the large fluctuations of air-sea CO_2 flux observed over coastal waters could affect the concentration of CO_2 significantly enough to affect estimates of air-land flux based on the inversion of atmospheric CO_2 data. Air-sea CO_2 flux was low during the 1997-1998 and 2002-2003 El Niño periods. The shaded areas indicate the 1997-1998 and 2002-2003 El Niño episodes. The greatest El Niño anomalies occur in the winter which is the period of lowest air-sea fluxes.

biological interactions (Hare and Mantua, 2000; Chavez *et al.*, 2003; Patra *et al.*, 2005). For example, during El Niño, upwelling of high CO_2 waters is dramatically reduced along central California (Figure 15.5) so that flux out of the ocean is reduced. At the same time, photosynthetic uptake of CO_2 is also reduced (Chavez *et al.* 2002), reducing ocean uptake. The net effect of climate variability on air-sea fluxes therefore remains uncertain and depends on the time-space integral of the processes.

15.4 OPTIONS FOR MANAGEMENT

Two options for carbon sequestration have been proposed: (1) injection of CO_2 in deep subsurface waters (Brewer, 2003) and (2) ocean iron fertilization (Martin, 1990). The first might be applicable in waters surrounding North America, although potential biological side effects are unresolved. The largest potential for iron fertilization resides in the high nutrient waters of the equatorial Pacific, subarctic Pacific, and

Southern Ocean. Offshore waters of coastal upwelling systems have also been considered to be iron limited. However, efficiency and capacity of sequestration remain unresolved (Bakker *et al.*, 2001; Boyd *et al.*, 2000; Coale *et al.*, 2004; Gervais *et al.*, 2002) as do environmental perturbations that could be induced by fertilization (Chisholm *et al.*, 2001).

15.5 RESEARCH AND DEVELOPMENT NEEDS *VIS-À-VIS* OPTIONS

Waters with highly variable air-sea CO_2 fluxes are located primarily within 100 km of the coast (Figure 15.5). With the exception of a few areas, the available observations are grossly inadequate to resolve the high-frequency, small-spatial-scale variations. These high intensity air-sea CO_2 flux events may introduce errors in continental CO_2 fluxes calculated by atmospheric inversion methods. Achieving a comprehensive understanding of the carbon cycle in waters surrounding the North American continent will

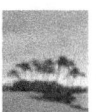

require development of advanced technologies and sustained and inter-disciplinary research efforts. Both of these seem to be on the horizon with (1) the advent of ocean observatories that include novel fixed and mobile platforms together with developing instrumentation to measure critical stocks and fluxes and (2) national and international research programs that include the Integrated Ocean Observing System (IOOS) and Ocean Carbon and Climate Change (OC³). A more comprehensive understanding will require the development of a robust observing program that incorporates time series observations of air-sea and sinking-particulate carbon fluxes in the coastal and open ocean. Our present estimates suggest that the carbon that reaches the bottom over continental margins may be responsible for upwards of 40%*** of the carbon reaching the ocean seafloor (Muller-Karger *et al.*, 2005). Given the importance of aquatic systems to atmospheric CO_2 concentrations, these developing efforts must be strongly encouraged. Ocean carbon sequestration studies should also be continued.

Historical Overview of the Development of United States, Canadian, and Mexican Ecosystem Sources and Sinks for Atmospheric Carbon

Coordinating Lead Author: Stephen Pacala, Princeton Univ.

Lead Authors: Richard A Birdsey, USDA Forest Service; Scott D. Bridgham, Univ. Oreg.; Richard T. Conant, Colo. State Univ.; Kenneth Davis, The Pa. State Univ.; Burke Hales, Oreg. State Univ.; Richard A. Houghton, Woods Hole Research Center; Jennifer C. Jenkins, Univ. Vt.; Mark Johnston, Saskatchewan Research Council; Gregg Marland, ORNL and Mid Sweden Univ. (Östersund); Keith Paustian, Colo. State Univ.;

Contributing Authors: John Casperson, Univ. Toronto; Robert Socolow, Princeton Univ.; Richard S. J. Tol, Hamburg Univ.

Although the lands of the New World were inhabited before the arrival of Europeans, the changes since arrival have been enormous, especially during the last two centuries. Peak United States emissions from land-use change occurred late in the nineteenth century, and the last few decades have experienced a carbon sink (Houghton *et al.*, 1999; Hurtt *et al.*, 2002). In Canada, peak emissions occurred nearly a century later than in the United States, and current data show that land-use change causes a net carbon sink (Environment Canada, 2005). In Mexico, the emissions of carbon continue to increase from net deforestation. All three countries may be in different stages of the same development pattern (Figure 3.2)

The largest changes in land use and the largest emissions of carbon came from the expansion of croplands. In addition to the carbon lost from trees, soils lose 25-30% of their initial carbon content (to a depth of 1 m) when cultivated. In the United States, croplands increased from about 0.25 million hectares (ha) in 1700 to 236 million ha in 1990 (Houghton *et al.*, 1999; Houghton and Hackler, 2000). The most rapid expansion (and the largest emissions) occurred between 1800 and 1900, and since 1920, there has been little net change in cropland area. Pastures expanded nearly as much, from 0.01 million to 231 million ha, most of the increase

taking place between 1850 and 1950. As most pastures were derived from grasslands, the associated changes in carbon stocks were modest.

The total area of forests and woodlands in the United States declined as a result of agricultural expansion by

160 million ha (38%), but this net change obscures the dynamics of forest loss and recovery, especially in the eastern part of the United States. After 1920, forest areas increased by 14 million ha nationwide as farmlands continued to be abandoned in the northeast, southeast, and north central regions. Nevertheless, another 4 million ha of forest were lost in other regions, and the net recovery of 10 million ha offset only 6% of the net loss (Houghton and Hackler, 2000).

Between 1938 and 2002, the total area of forestland in the conterminous United States decreased slightly, by 3 million ha (Smith *et al.*, 2004). This small change is the net result of much larger shifts among land-use classes (Birdsey and Lewis, 2003). Gains of forestland, primarily from cropland and pasture, were about 50 million ha for this period. Losses of forestland to cropland, pasture, and developed use were about 53 million ha for the same period. Gains of forestland were primarily in the Eastern United States, whereas losses to cropland and pasture were predominantly in the South, and losses to developed use were spread around all regions of the United States.

In the United States, harvest of industrial wood (timber) generally followed the periods of major agricultural clearing in each region. In the last few decades, total volume harvested increased until a recent leveling took place (Smith *et al.*, 2004). The volume harvested in the Pacific Coast and Rocky Mountain regions has declined sharply, whereas harvest in the South increased and in the North, stayed level. Fuel wood harvest peaked between 1860 and 1880, after which fossil fuels became the dominant type of fuel (Houghton and Hackler, 2000).

The arrival of Europeans reduced the area annually burned, but a federal program of fire protection was not established until early in the twentieth century. Fire exclusion had begun earlier in California and in parts of the central, mountain, and Pacific regions. However, neither the extent nor the timing of early fire exclusion is well known. After about 1920, the Cooperative Fire Protection Program gradually reduced the areas annually burned by wildfires (Houghton *et al.*, 1999, 2000). The reduction in wildfires led to an increase in carbon storage in forests. How long this "recovery" will last is unclear. There is some evidence that fires are becoming more widespread again, especially in Canada and the western United States. Fire exclusion and suppression are also thought to have led to woody encroachment, especially in the southwestern and western United States. The extent and rate of this process is poorly documented, however, and estimates of a carbon sink are very uncertain. Gains in carbon above-ground may be offset by losses below-ground in some systems, and the spread of exotic annual grasses into semiarid deserts and shrublands may be converting the recent sink to a source (Bradley *et al.*, in preparation).

The consequence of this land-use history is that United States' forests, at present, are recovering from agricultural abandonment, fire suppression, and reduced logging (in some regions), and as a result, are accumulating carbon (Birdsey and Heath, 1995; Houghton *et al.*, 1999; Caspersen *et al.*, 2000; Pacala *et al.*, 2001). The magnitude of the sink is uncertain, and whether any of it has been enhanced by environmental change (CO_2 fertilization, nitrogen deposition, and changes in climate) is unclear. Understanding the mechanisms responsible for the current sink is important for predicting its future behavior (Hurtt *et al.*, 2002).

In the mid-1980s, Mexico lost approximately 668,000 ha of closed forests annually, about 75% of them tropical forests (Masera *et al.*, 1997). Most deforestation was for pastures. Another 136,000 ha of forest suffered major perturbations, and the net flux of carbon from deforestation, logging, fires, degradation, and the establishment of plantations was 52.3 million tons of carbon per year, about 40% of the country's estimated annual emissions of carbon. A later study found the deforestation rate for tropical Mexico to be about 12% higher (1.9% per year) (Cairns *et al.*, 2000).

Eddy-Covariance Measurements Now Confirm Estimates of Carbon Sinks from Forest Inventories

Coordinating Lead Author: Stephen Pacala, Princeton Univ.

Lead Authors: Richard A Birdsey, USDA Forest Service; Scott D. Bridgham, Univ. Oreg.; Richard T. Conant, Colo. State Univ.; Kenneth Davis, The Pa. State Univ.; Burke Hales, Oreg. State Univ.; Richard A. Houghton, Woods Hole Research Center; Jennifer C. Jenkins, Univ. Vt.; Mark Johnston, Saskatchewan Research Council; Gregg Marland, ORNL and Mid Sweden Univ. (Östersund); Keith Paustian, Colo. State Univ.;

Contributing Authors: John Casperson, Univ. Toronto; Robert Socolow, Princeton Univ.; Richard S. J. Tol, Hamburg Univ.

Long-term, tower-based, eddy-covariance measurements (*e.g.*, Wofsy *et al.*, 1993) represent an independent approach to measuring ecosystem-atmosphere carbon dioxide (CO_2) exchange. The method describes fluxes over areas of approximately 1 km² (Horst and Weil, 1994), measures hour-by-hour ecosystem carbon fluxes, and can be integrated over time scales of years. A network of more than 200 sites now exists globally (Baldocchi *et al.*, 2001); more than 50 of these are in North America. None of these sites existed in 1990, so these represent a relatively new source of information about the terrestrial carbon cycle. An increasing number of these measurement sites include concurrent carbon inventory measurements.

Where eddy-covariance and inventory measurements are concurrent, the rates of accumulation or loss of biomass are often consistent to within several tens of g C per m² per year for a one-year sample (10 g C per year is 5% of a typical net sink of two metric tons of carbon per hectare per year for an Eastern deciduous successional forest). Published intercomparisons in

North America exist for western coniferous forests (Law *et al.*, 2001), agricultural sites (Verma *et al.*, 2005), and eastern deciduous forests (Barford *et al.*, 2001; Cook *et al.*, 2004; Curtis *et al.*, 2002; Ehmann *et*

Table B.1 Carbon budget for Harvard Forest from forest inventory and eddy-covariance flux measurements, 1993-2001. Source: Barford *et al.* (2001), Table 1. Numbers in parentheses give the ranges of the 95% confidence intervals. Following the sign convention in Barford *et al.* (2001), positive values represent uptake from the atmosphere (*i.e.*, a sink) and negative values a release (*i.e.*, a source).

Component	Change in carbon stock or flux (Mg C per ha per year)[a]	Totals

Change in live biomass
A. Above-ground
  1. Growth
  2. Mortality
B. Below-ground (estimated)
  1. Growth
  2. Mortality
Subtotal
	
	1.4 (±0.2)
	−0.6 (±0.6)
	
	0.3
	−0.1
	
		1.0 (±0.2)

Change in dead wood
A. Mortality
  1. Above-ground
  2. Below-ground
B. Respiration
Subtotal
	
	0.6 (±0.6)
	0.1
	−0.3 (±0.3)
	
		0.4 (±0.3)

Change in soil carbon (net)		0.2 (±0.1)
Sum of carbon budget figures		1.6 (±0.4)
Sum of eddy-covariance flux measurements		2.0 (±0.4)

[a] 1 Mg C per ha per year = 100g C per m² per year.

al., 2002; Gough *et al.*, in review). Multiyear studies at two sites (Barford *et al.*, 2001; Gough *et al.*, in review) show that 5- to 10-year averages converge toward inventory measurements. Table B.1 from Barford *et al.* (2001) shows the results of nearly a decade of concurrent measurements in an eastern deciduous forest.

This concurrence between eddy-covariance flux measurements and ecosystem carbon inventories is relevant because it provides independent validation of the inventory measurements used to estimate long-term trends in carbon stocks. The eddy-covariance data are also valuable because the assembly of global eddy-covariance data provides independent support for net storage of carbon by many terrestrial ecosystems and the substantial year-to-year variability in this net sink. The existence of the eddy-covariance data also makes the sites suitable for co-locating mechanistic studies of interannual and shorter, time-scale processes governing the terrestrial carbon cycle. Chronosequences show trends consistent with inventory assessments of forest growth, and comparisons across space and plant functional types are beginning to show broad consistency. These results show a consistency across a mixture of observational methods with complementary characteristics, which should facilitate the development of an increasingly complete understanding of continental carbon dynamics (Canadell *et al.*, 2000).

APPENDIX C

Industry and Waste Management - Supplemental Material

Lead Author: John Nyboer, Simon Fraser Univ.

Contributing Authors: Mark Jaccard, Simon Fraser Univ.; Ernest Worrell, LBNL

This appendix presents diagrams of the carbon flows in Canada, the United States, and Mexico, respectively (Figures C.1 through C.3). The numerical data in these figures are shown in thousands of metric tons of carbon, which can be converted into thousands of metric tons of carbon dioxide (CO_2) equivalents by multiplying the carbon values by 44/12 (*i.e.*, the ratio of CO_2 mass to carbon mass). The combined carbon flows for all three nations are presented in Figure 8.2 in Chapter 8 of this report.

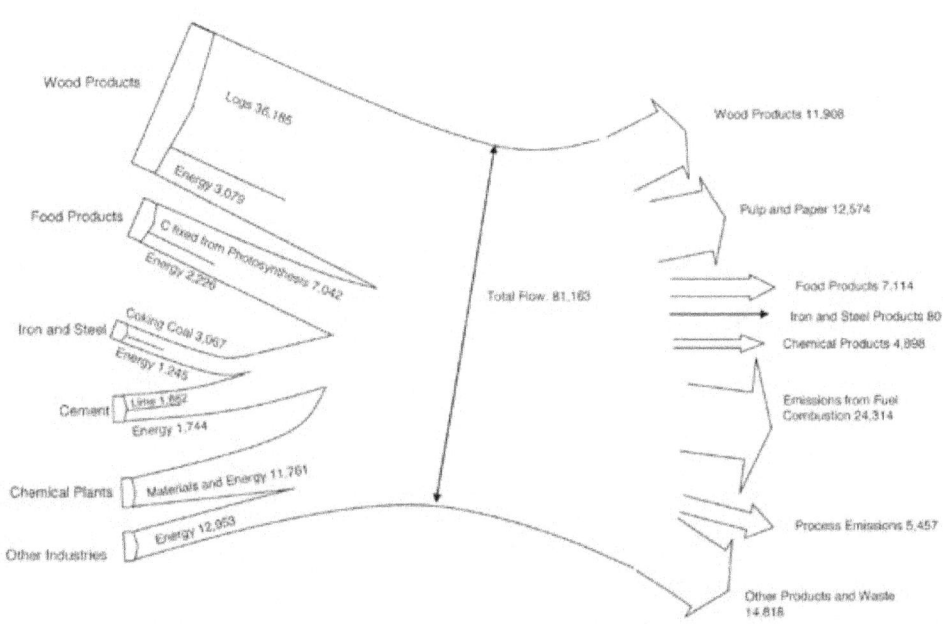

Figure C.1 Carbon flows, Canada. *Source:* Energy data from Statistics Canada Industrial Consumption of Energy survey, conversion coefficients and process emissions from Environment Canada, Canada GHG Inventory (2002). Production data from Statistics Canada, CANSIM Table 002-0010, Tables 303-0010, -0014 to -0021, -0024, -0060, Pub. Cat. Nos.: 21-020, 26-002, 45-002, Canadian Pulp and Paper Association on forestry products.

US Carbon Flows (All Values in Kilotonnes of C)

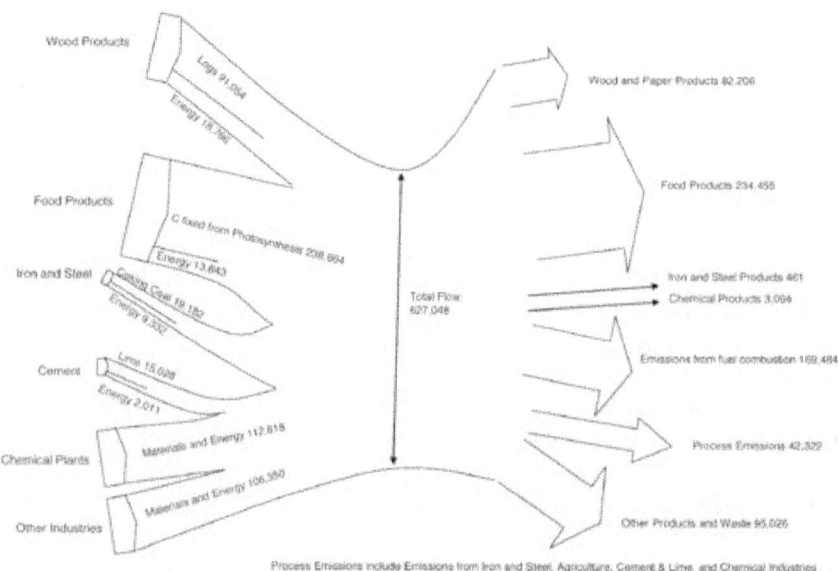

Figure C.2 Carbon flows, United States. *Source:* Energy data from IEA Oil Information (2004), IEA Coal Information (2005), IEA Natural Gas Information (2004). Process emissions: EPA, U.S. Emissions Inventory. Production of forestry products: USDA Database; FO-2471000 and -2472010, U.S. Timber Production, Trade, Consumption, and Price Statistics 1965-2005, Production of organic products (e.g., food): USDA PS&D Official Statistical Results, Steel: International Iron and Steel institute, World steel in figures (2003), Minerals production: USGS mineral publications.

Mexico Carbon Flows (All Values in Kilotonnes of C)

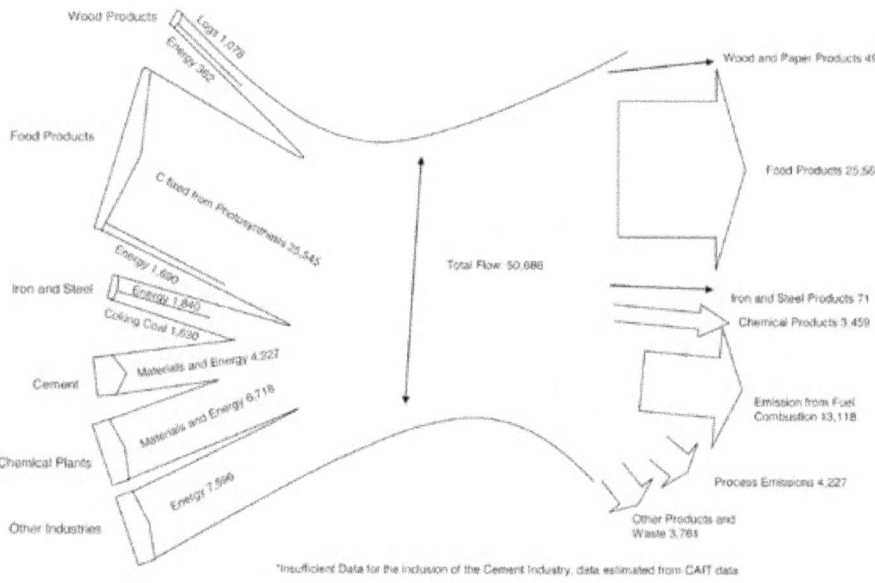

Figure C.3 Carbon flows, Mexico. *Source:* Energy data from IEA Oil Information (2004), IEA Coal Information (2005), IEA Natural Gas Information (2004). Process emissions: EPA, U.S. Emissions Inventory. Production of forestry products: USDA Database; FO-2471000, -2472010, -2482000, -2483040, -6342000, -6342040. Production of organic products (e.g., food): USDA PS&D Official Statistical Results. Steel: International Iron and Steel institute, World steel in figures (2003).

APPENDIX D

Ecosystem Carbon Fluxes

Lead Authors: Richard A. Birdsey, USDA Forest Service; Jennifer C. Jenkins, Univ. Vt.; Mark Johnston, Saskatchewan Research Council; Elisabeth Huber-Sannwald, Instituto Potosino de Investigación Científica y Tecnológica

Contributing Authors: Brian Amiro, Univ. Manitoba; Ben de Jong, ECOSUR; Jorge D. Etchevers Barra, Colegio de Postgraduado; Nancy French, Altarum Inst.; Felipe Garcia-Oliva, UNAM; Mark Harmon, Oreg. State Univ.; Linda S. Heath, USDA Forest Service; Victor J. Jaramillo, UNAM; Kurt Johnsen, USDA Forest Service; Beverly E. Law, Oreg. State Univ.; Erika Marin-Spiotta, Univ. Calif. Berkeley; Omar Masera, UNAM; Ronald Neilson, USDA Forest Service; Yude Pan, USDA Forest Service; Kurt Pregitzer, Mich. Tech. Univ.

The recent history of disturbance largely determines whether a forest system will be a net source or sink of carbon. For example, net ecosystem productivity (NEP, see Table D.1 for a list of definitions and acronyms used in this appendix) is being measured across a range of forest types in Canada using the eddy covariance technique. In mature forests, values range from -19.6 tons of carbon per hectare (t C per ha) per year in a white pine plantation in southern Ontario (Arain and Restrepo-Coupe, 2005) to -3.2 t C per ha per year in a jack pine forest (Amiro *et al.*, 2005; Griffis *et al.*, 2003). In recently disturbed forests, NEP ranges from +58.0 t C per ha per year in a harvested Douglas-fir forest (Humphreys *et al.*, 2005) to +5.7 t C per ha per year in a seven year old harvested jack pine forest (Amiro *et al.*, 2005). In general, forest stands recovering from disturbance are sources of carbon until uptake from growth becomes greater than losses due to respiration, usually within 10 years (Amiro *et al.*, 2005).

Table D.1 Ecosystem Productivity Terms and Definitions. (Terms and definitions apply to Appendices D and E of this report.)

Term	Acronym	Definition
Net Primary Production	NPP	Net uptake of carbon by plants in excess of respiratory loss
Heterotrophic Respiration	R_h	Respiratory loss by above- and below-ground heterotrophs (herbivores, decomposers, *etc.*)
Net Ecosystem Production	NEP	Net carbon accumulation within the ecosystem after all gains and losses are accounted for, typically measured using ground-based techniques. By convention, positive values of NEP represent accumulaitons of carbon by the ecosystem, and negative values represent carbon loss.
Net Ecosystem Exchange	NEE	The net flux of carbon between the land and the atmosphere, typically measured using eddy covariance techniques. Note: NEE and NEP are equivalent terms but are not always identical because of measurement and scaling issues, and the sign conventions are reversed. Positive values of NEE (net ecosystem exchange with the atmoshpere) usually refer to carbon released to the atmosphere (*i.e.*, a source), and negative values refer to carbon uptake (*i.e.*, a sink).

Sources: Randerson *et al.* (2002); Chapin *et al.* (2006).

Table D.2 **Comparison of net ecosystem exchange (NEE) for different types and ages of temperate forests. Negative NEE means the forest is a sink for atmospheric CO_2. Eighty-one site years of data are from multiple published papers from each of the AmeriFlux network sites, and a network synthesis paper (Law et al., 2002). NEE was averaged by site, then the mean was determined by forest type and age class. SD is standard deviation among sites in the forest type and age class.**

NEE (t Carbon per ha per year)			
	Regenerating Clearcut (1 to 3 years after disturbance) (1 site, 5 site-years)	Young forest (8 to 20 years old) (4 sites, 16 site-years)	Mature forest (>20 years old) (13 sites, 60 site-years)
Evergreen Coniferous Forests	−1.7 to +12.7 mean = 7.1, (SD 4.7) (1 site, 5 site-years)	−0.6 to −5.9 mean = −3.1, (SD 2.6) (4 sites, 16 site-years)	−0.6 to −4.5 mean = −2.5, (SD 1.4) (6 sites, 20 site-years)
Mixed Evergreen and Deciduous Forests	NA	NA	−0.3 to −2.1 mean = −1.0, (SD 0.6) (1 site, 6 site-years)
Deciduous Broadleaf Forests	NA	NA	−0.6 to −5.8 mean = −2.7, (SD 1.8) (6 sites, 34 site-years)

In the United States, extensive land-based measurements of forest/atmosphere carbon exchange reveal patterns and causes of sink or source strength (Table D.2). Results show that net ecosystem exchange (NEE) of carbon in temperate forests ranges from a source of +12.7 t C per ha per year to a sink of -5.9 t C per ha per year. Forests identified as sources are primarily forests in the earliest stages of regeneration (up to about eight years) following stand-replacing disturbances such as wildfire and logging (Law et al., 2002). Mature temperate deciduous broadleaf forests and mature evergreen coniferous forests were an average sink of -2.7 and -2.5 t C per ha per year, respectively (12 sites, 54 site-years of data). Values ranged from a source of +0.3 for a mixed deciduous and evergreen forest to a sink of -5.8 for an aggrading deciduous forest, averaged over multiple years. Young temperate evergreen coniferous forests (8 to 20 years) ranged from a sink of -0.6 to -5.9 t C per ha per year (mean -3.1). These forests are still rapidly growing and have not reached the capacity for carbon uptake.

Mature forests can have substantial stocks of sequestered carbon. Disturbances that damage or replace forests can result in the land being a net source of carbon dioxide (CO_2) for a few years in mild climates to 10-20 years in harsh climates while the forests are recovering (Law et al., 2002; Clark et al., 2004). Thus, the range of observed annual NEE of CO_2 ranges from a source of about +13 t C per ha per year in a clearcut forest to a net sink of -6 t C per ha in mature temperate forests.

For Mexican forests, estimates of net ecosystem carbon exchange are unavailable, but estimates from other tropical forests may indicate rates for similar systems in Mexico. In Puerto Rico, aboveground NPP in tropical forests range from -9.2 to -11.0 t C per ha per year (Lugo et al., 1999). Below-ground NPP measurements exist for only one site with -19.5 t C per ha per year (Lugo et al., 1999). In Hawaii, above-ground and below-ground NPP of native forests dominated by *Metreosideros polymorpha* vary depending on substrate age and precipitation regime. Above-ground NPP ranges between -4.0 to -14.0 t C per ha per year, while below-ground NPP ranges between -5.2 and -9.0 t C per ha per year (Giardina et al., 2004). Soil carbon emissions along the substrate age gradient range from +2.2 to +3.3 t C per ha per year, and along the precipitation gradient from +4.0 to +9.7 t C per ha per year (Osher et al., 2003). NEP estimates are not available for these tropical forests, so their net impact on atmospheric carbon stocks cannot be calculated.

APPENDIX E

Principles of Forest Management for Enhancing Carbon Sequestration

Lead Authors: Richard A. Birdsey, USDA Forest Service; Jennifer C. Jenkins, Univ. Vt.; Mark Johnston, Saskatchewan Research Council; Elisabeth Huber-Sannwald, Instituto Potosino de Investigación Científica y Tecnológica

Contributing Authors: Brian Amiro, Univ. Manitoba; Ben de Jong, ECOSUR; Jorge D. Etchevers Barra, Colegio de Postgraduado; Nancy French, Altarum Inst.; Felipe Garcia-Oliva, UNAM; Mark Harmon, Oreg. State Univ.; Linda S. Heath, USDA Forest Service; Victor J. Jaramillo, UNAM; Kurt Johnsen, USDA Forest Service; Beverly E. Law, Oreg. State Univ.; Erika Marin-Spiotta, Univ. Calif. Berkeley; Omar Masera, UNAM; Ronald Neilson, USDA Forest Service; Yude Pan, USDA Forest Service; Kurt Pregitzer, Mich. Tech. Univ.

The net rate of carbon accumulation has been generally understood (Woodwell and Whittaker, 1968) as the difference between gross primary production (gains) and respiration (losses), although this neglects important processes such as leaching of dissolved organic compounds (DOCs), emission of methane (CH_4), fire, harvests, or erosion that may contribute substantially to carbon loss and gain in forest ecosystems (Schulze *et al.*, 1999; Harmon, 2001; Chapin *et al.*, 2006). The net ecosystem carbon balance (NECB) in forests is, therefore, defined as net ecosystem production, or NEP, plus the non-physiological horizontal and vertical transfers into and out of the forest stand.

With respect to the impacts of forest management on the overall carbon balance, some general principles apply (Harmon, 2001; Harmon and Marks, 2002; Pregitzer *et al.*, 2004). First, forest management can impact carbon pool sizes via:
- changing production rates (since NEP = net primary production [NPP] – heterotrophic respiration [R_h]);
- changing decomposition flows (R_h) (*e.g.*, Fitzsimmons *et al.*, 2004);
- changing the amount of material transferred between pools; or
- changing the period between disturbances/ management activities.

The instantaneous balance between production, decomposition, and horizontal or vertical transfers into and out of a forest stand determines whether the forest is a net source or a net sink. Given that these terms all change as forests age, the disturbance return interval is a key driver of stand- and landscape-level carbon dynamics. R_h tends to be enhanced directly after disturbance, so as residue and other organic carbon pools decompose, a forest is often a net source immediately after disturbances such as management activity. NPP tends to increase as forests age, although in older forests it may decline (Ryan, 1997). Eventually, as stands age, NPP and R_h become similar in magnitude, although few managed stands are allowed to reach this age. The longer the average time interval between disturbances, the more carbon is stored. The nature of the disturbance is also important; the less severe the disturbance (*e.g.*, less fire removal), the more carbon is stored.

Several less general principles can be applied to specific carbon pools, fluxes, or situations:
- Management activities that move live carbon to dead pools (such as coarse woody debris [CWD] or soil carbon) over short periods of time will often dramatically enhance decomposition (R_h), although considerable carbon can be stored in decomposing pools (Harmon and Marks, 2002). Regimes seeking to reduce the decomposition-related flows from residue following harvest may enhance overall sink capacity of these forests if these materials are used for energy generation or placed into forest products that last longer than the residue.
- Despite the importance of decomposition rates to the overall stand-level forest carbon balance, management of CWD pools is mostly impacted by recruitment of new CWD rather than by changing decomposition rates (Janisch and Harmon, 2002; Pregitzer and Euskirchen, 2004). Decreasing the interval between harvests can significantly decrease the store in this pool.
- Live coarse root biomass accounts for approximately

20-25% of aboveground forest biomass (Jenkins *et al.*, 2003), and there is additional biomass in fine roots. Following harvest, this pool of live root biomass is transferred to the dead biomass pool, which can form a significant carbon store. Note that roots of various size classes and existing under varying environmental conditions decompose at different rates.

- Some carbon can be sequestered in wood products from harvested wood, though, due to manufacturing losses, only about 60% of the carbon harvested is stored in products (Harmon, 1996). Clearly, longer-lived products will sequester carbon for longer periods of time.

- According to international convention, the replacement of fossil fuel by biomass fuel can be counted as an emissions offset if the wood is produced from sustainably managed forests (Schoene and Netto, 2005)

Little published research has been aimed at quantifying the impacts of specific forest management activities on carbon storage, but examples of specific management activities can be given.

- Practices aimed at increasing NPP: fertilization; genetically improved trees that grow faster (Peterson *et al.*, 1999); any management activity that enhances growth rate without causing a concomitant increase in decomposition (Stanturf *et al.*, 2003; Stainback and Alavalapati, 2005).

- Practices aimed at reducing R_h (*i.e.,* minimizing the time forests are a source to the atmosphere following disturbance): low impact harvesting (that does not promote soil respiration); utilization of logging residues (biomass energy and fuels); incorporation of logging residue into soil during site prep (but note that this could also speed up decomposition); thinning to capture mortality; fertilization.

Since NECB changes with time as forests age, if a landscape is composed of stands with different ages, then carbon gains in one stand can be offset by losses from another stand. The net result of these stand-level changes determines overall landscape-level carbon stores. Note that disturbance-induced R_h losses are typically larger than annual gains, such that a landscape where forest area is increasing might still be neutral with respect to carbon stocks overall. Thus, at the landscape level, practices designed to enhance carbon sequestration must, on balance, replace lower-carbon-density systems with higher-carbon-density systems. Examples of these practices include: reducing fire losses; emphasizing very long-lived forest products; increasing the interval between disturbances; or reducing decomposability of dead material.

Wetlands - Supplemental Material

Lead Author: Scott D. Bridgham, Univ. Oreg.

Contributing Authors: J. Patrick Megonigal, Smithsonian Environmental Research Center; Jason K. Keller, Smithsonian Environmental Research Center; Norman B. Bliss, SAIC, USGS Center for Earth Resources, Observation and Science; Carl Trettin, USDA Forest Service

F.1 INVENTORIES

F.1.1 Current Wetland Area and Rates of Loss

The ability to estimate soil carbon pools and fluxes in North American wetlands is constrained by the national inventories (or lack thereof) for Canada, the United States, and Mexico (Davidson *et al.*, 1999). The National Wetland Inventory (NWI) program of the United States has repeatedly sampled several thousand wetland sites using aerial photographs and more limited field verification. The data are summarized in a series of reports detailing changes in wetland area in the conterminous United States for the periods of the mid-1950s to mid-1970s (Frayer *et al.*, 1983), mid-1970s to mid-1980s (Dahl and Johnson, 1991), and 1986 to 1997 (Dahl, 2000). We used these relatively high-quality data sets extensively for estimating wetland area and loss rates in the conterminous United States, including mud flats. However, the usefulness of the NWI inventory reports for carbon budgeting is limited by the level of classification used to define wetland categories within the Cowardin *et al.* (1979) wetland classification system. At the level used in the national status and trend reports, vegetated freshwater wetlands are classified by dominant physiognomic vegetation type, and it is impossible to make the important distinction between wetlands with deep organic soils (*i.e.,* peatlands) and wetlands with mineral soils. The data are not at an adequate spatial resolution to combine with U.S. Department of Agriculture (USDA) Natural Resources Conservation Service (NRCS) soil maps to discriminate between the two types of wetlands (T. Dahl, personal comm.). Because of these data limitations, we used the NRCS soil inventory of peatlands (*i.e.,* Histosols and Histels, or peatlands with and without permafrost, respectively) to estimate original peatland area (Bridgham *et al.*, 2000) and combined these data with regional estimates of loss (Armentano and Menges, 1986) to estimate current peatland area in the conterminous United States. We calculated the current area of freshwater mineral-soil (FWMS) wetlands in the conterminous United

States by subtracting peatland area from total wetland area (Dahl, 2000). This approach was limited by the Armentano and Menges peatland area data being current only up to the early 1980s, although large losses of peatlands since then are unlikely due to the institution of wetland protection laws.

We used a similar approach for Alaskan peatlands: peatland area was determined by the NRCS soil inventory (N. Bliss, query of the NRCS State Soil Geographic [STATSGO] database, February 2006) and overall wetland inventory was determined by standard NWI methods (Hall *et al.*, 1994). However, our peatland estimate of 132,000 km^2 (Table F.1) is 22% of the often cited value by Kivinen and Pakarinen (1981) of 596,000 km^2.

Kivinen and Pakarinen also used NRCS soils data (Rieger *et al.*, 1979) for their peatland estimates, but they defined a peatland as having a minimum organic layer thickness of 30 cm, whereas the current United States and Canadian soil taxonomies require a 40-cm thickness. The original 1979 Alaska soil inventory has been reclassified with current United States soil taxonomy (J. Moore, Alaska State Soil Scientist, personal comm.). Using the reclassified soil inventory, Alaska has 417,000 km^2 of wetlands with a histic modifier (*i.e.,* a surface organic layer between 20 and 60 cm thick) that are not Histosols or Histels, indicating significant carbon accumulation in the surface horizons of FWMS wetlands. Thus, we conclude that Kivinen and Pakarinen's Alaska peatland area estimate is higher because many Alaskan wetlands have a thin organic horizon that is not deep enough to qualify as a peatland under current soil taxonomy. Our smaller peatland area significantly lowers our estimate of carbon pools and fluxes in Alaskan peatlands compared to earlier studies (see *Carbon Pools* below).

The area of salt marsh in the conterminous United States, Canada, and Alaska were taken from Mendels-

Table F.1 Current and historical area of wetlands in North America and the world (×10³ km²). Historical refers to approximately 1800, unless otherwise specified.

	Permafrost peatlands	Non-permafrost peatlands	Mineral-soil freshwater	Salt marsh	Mangrove	Mudflat	Total
Canada							
Current	422[a]	714[a]	159[b]	0.4[c]	0	6[d]	1301
Historical	424[e]	726[f]	359[g]	1.3[b]	0	7[h]	1517
Alaska							
Current	89[i]	43[i]	556[j]	1.4[c]	0	7[k]	696
Historical	89	43	556	1.4	0	7	696
Conterminous United States							
Current	0	93[l]	312[m]	20[c]	3[c]	2[n]	431
Historical	0	111[l]	762[o]	22[n]	4[n]	3[n]	901
Mexico							
Current	0	10[p]	21[p]	0	5[c]	ND[q]	36
Historical	0	45[p]		0	8[h]	ND	53
North America							
Current	511	861	1,047	22	8	15	2,463
Historical	513	894[r]	1,706[r]	25	12	17	3,167
Global							
Current	3,443[s]		2,315[t]	22[u]	181[v]	ND	5,961
Historical	4,000[w]		5,000[x]	29[y]	278[y]	ND	9,307

[a] Tarnocai et al. (2005).

[b] National Wetlands Working Group (1988).

[c] Brackish and salt marsh areas from Mendelssohn and McKee (2000); freshwater tidal wetlands for the conterminous United States only from Odum et al. (1984) and Field et al. (1991).

[d] Estimated from the area of Canadian salt marshes and the ratio of mudflat to salt marsh area reported by Hanson and Calkins (1996).

[e] Accounting for losses due to permafrost melting in western Canada (Vitt et al., 1994). This is an underestimate, as similar, but undocumented, losses have probably also occurred in eastern Canada and Alaska.

[f] 9000 km² lost to reservoir flooding (Rubec, 1996), 250 km² to forestry drainage (Rubec, 1996), 124 km² to peat harvesting for horticulture (Cleary et al., 2005), and 16 km² to oil sands mining (Turetsky et al., 2002). See note e for permafrost melting estimate.

[g] Rubec (1996).

[h] Estimated loss rate for the Americas from Valiela et al. (2001) for approximately 1980 to 1990.

[i] Historical area from NRCS soil inventory (Bridgham et al., 2000), except Alaska inventory updated by N. Bliss from a February 2006 query of the STATSGO database. Less than 1% wetland losses have occurred in Alaska (Dahl, 1990).

[j] Total freshwater wetland area from Hall et al. (1994) minus peatland area.

[k] Hall et al. (1994).

[l] Historical area from Bridgham et al. (2000) minus losses in Armentano and Menges (1986).

[m] Overall freshwater wetland area from Dahl (2000) minus peatland area.

[n] Dahl (2000). Historical area estimates are only from the 1950s.

[o] Total historical wetland area from Dahl (1990) minus historical peatland area minus historical estuarine area.

[p] Spiers (1999) and Davidson (1999).

[q] ND indicates that no data are available.

[r] Assuming that historical proportion of peatlands to total wetlands in Mexico was the same as today.

[s] Bridgham et al. (2000) for the United States, Tarnocai et al. (2005) for Canada, Joosten, and Clarke (2002) for the rest of world. Recent range in literature 2,974,000–3,985,000 km² (Matthews and Fung, 1987; Aselmann and Crutzen, 1989; Maltby and Immirzi, 1993; Bridgham et al., 2000; Joosten and Clarke, 2002).

[t] Average of 2,289,000 km² from Matthews and Fung (1987) and 2,341,000 km² Aselmann and Crutzen (1989).

[u] Chmura et al. (2003). Underestimated because no inventories were available for the continents Asia, South America, and Australia which are mangrove-dominated but also support salt marsh.

[v] Spalding (1997).

[w] Range from 3,880 to 4,086 in Maltby and Immirzi (1993).

[x] Approximately 50% loss from Moser et al. (1996).

[y] Assumed a 25% loss rate outside North America for tidal marshes; a loss rate of 35% was used for mangroves (Valiela et al., 2001).

sohn and McKee (2000). Because these estimates include brackish tidal marshes, they cannot be compared directly to the area of Canadian salt marshes. Compilations of freshwater, tidal wetland area are difficult to find, but there is approximately 1,640 km² on the east coast of the United States (Odum *et al.*, 1984) and 470 km² on the United States' Gulf Coast (Field *et al.*, 1991). Although some freshwater tidal wetlands are forested, this total was added to the tidal marsh area for the conterminous United States. Mangrove area was taken from Mendelssohn and McKee (2000), and is similar to an estimate by Lugo and Snedaker (1974).

The original area of tidal wetlands in the conterminous United States was based on the NWI (Dahl, 2000), which we considered to be the most defensible estimate available. However, "original" here only refers to the 1950s, when the first national wetland inventory was conducted in the conterminous United States to provide a historic baseline area. It is almost certain that the actual loss of tidal wetland area in the conterminous United States over a longer time frame was larger than the 7.7% figure used in our calculation. Valiela *et al.* (2001) estimated a loss of 31% of mangrove area in the United States from 1958 to 1982, but acknowledged a high level of uncertainty in this figure. We assumed that the original area of Alaskan tidal wetlands was similar to the current area because there has been relatively little development pressure in Alaska. To estimate loss of global tidal wetlands, we arbitrarily used a figure of 35% loss for tidal marshes outside of the United States and Mexico.

A regular national inventory of Canada's wetlands has not been undertaken, although wetland area has been mapped by ecoregion (National Wetlands Working Group, 1988). Extensive recent effort has gone into mapping Canadian peatlands (Tarnocai, 1998; Tarnocai *et al.*, 2005). We calculated the current area of mineral-soil wetlands as the difference between total wetland area and peatland area in National Wetland Working Group (1988). The original area of FWMS wetland area was obtained from Rubec (1996). Canadian salt marsh estimates were taken from a compilation by Mendelssohn and McKee (2000). The compilation does not include brackish or freshwater tidal marshes, and we were unable to locate other estimates of Canadian brackish marsh area. The original area of salt marshes was estimated from the National Wetland Working Group (1988), but it is highly uncertain. There are no reliable country-wide estimates of mud flat area for Canada, but a highly uncertain extrapolation from a limited number of regional estimates was possible based upon the ratio of mudflat to salt marsh area reported by Hanson and Calkins (1996).

No national wetland inventories have been done for Mexico. Current freshwater wetland estimates for Mexico were taken from Davidson *et al.* (1999) and Spiers (1999), who used inventories of discrete wetland regions performed by a variety of organizations. Thus, freshwater wetland area estimates for Mexico are highly unreliable and are possibly a large underestimate. For mangrove area in Mexico, we used the estimates compiled by Mendelssohn and McKee (2000), which are similar to estimates reported in Davidson *et al.* (1999) and Spalding *et al.* (1997). We could find no estimates of tidal marsh or mud flat area for Mexico. Since most vegetated Mexican tidal wetlands are dominated by mangroves (Olmsted, 1993; Mendelssohn and McKee, 2000), the omission of Mexican tidal marshes should not significantly affect our carbon budget. However, there may be large areas of mud flat that would significantly increase our estimate of carbon pools and sequestration in this country. We used the Valiela *et al.* (2001) estimate of 38% for mangrove loss in the Americas, which roughly covers the period 1980 to 1990. This is less than the rough worldwide estimate of 50% wetland loss since the 1880s that is often cited (see Zedler and Kercher, 2005) and is probably conservative. A global loss rate of 35% was used for mangrove area globally based on the analysis of Valiela *et al.* (2001).

F.2 CARBON POOLS

F.2.1 Freshwater Mineral-Soil (Gleysol) Carbon Pools

Gleysol is a soil classification used by the Food and Agriculture Organization (FAO) and many countries that denotes mineral soils formed under waterlogged conditions (FAO-UNESCO, 1974). Tarnocai (1998) reported a soil carbon density of 200 Mg C per hectare (ha) for Canadian Gleysols to 1-m depth. Batjes (1996) determined soil carbon content globally from the Soil Map of the World (FAO, 1991) and a large database of soil pedons. He estimated an average value for soil carbon density of 199 Mg C per ha (CV[1] = 212%, n = 14 pedons) for Gleysols of the world to 2-m depth; to 1-m depth, he reported a soil carbon density of 131 Mg C per ha (CV = 109%, n =142 pedons).

Gleysols are not part of the United States' soil taxonomy scheme, and mineral soils with attributes reflecting waterlogged conditions are distributed among numerous soil groups. We queried the NRCS State Soil Geographic (STATSGO) soils database for soil carbon density in "wet" mineral soils of the conterminous United States (all soils that had a surface texture described as peat, muck, or mucky peat, or appeared on the 1993 list of hydric soils, which were not classified as Histosols) (N. Bliss, query of NRCS STATSGO database, December 2005). We used the average soil carbon densities of 162 Mg C per ha from this query for FWMS wetlands in the conterminous United States and Mexico.

---

[1] CV is the "coefficient of variation," or 100 times the standard deviation divided by the mean.

Table F.2 Soil carbon pools (Gt) and fluxes (Mt per year) of wetlands in North America and the world. "Sequestration in current wetlands" refers to carbon sequestration in extant wetlands; "oxidation in former wetlands" refers to emissions from wetlands that have been converted to non-wetland uses or conversion among wetland types due to human influence; "historical loss in sequestration capacity" refers to the loss in the carbon sequestration function of wetlands that have been converted to non-wetland uses; "change in flux from wetland conversions" is the sum of the two previous fluxes. Positive flux numbers indicate a net flux into the atmosphere, whereas negative numbers indicate a net flux into the ecosystem.

	Permafrost peatlands	Non-permafrost peatlands	Mineral-soil freshwater	Salt marsh	Mangrove	Mudflat	Total
Canada							
Pool Size in Current Wetlands	47.4[a]	102.9[b]	4.6[a]	0.0[c]	0.0[c]	0.1[d]	155.0
Sequestration in Current Wetlands	-5.5[e]	-13.6[e]	-2.7[f]	-0.1	0.0[c]	-1.2[d]	-23.0
Oxidation in Former Wetlands	0.2[g]		0.0[h]	0.0[i]	0.0	0.0	0.2
Historical Loss in Sequestration Capacity	0.0[e]	0.2[e]	3.4[f]	0.2	0.0	0.3	4.2
Change in Flux From Wetland Conversions	0.4		3.4	0.2	0.0	0.3	4.3
Alaska							
Pool Size in Current Wetlands	9.3[i]	6.2[i]	26.0[k]	0.0	0.0	0.1	41.7
Sequestration in Current Wetlands	-1.2[e]	-0.8[e]	-9.4[f]	-0.3	0.0	-1.6	-13.3
Oxidation in Former Wetlands	0.0	0.0	0.0	0.0	0.0	0.0	0.0
Historical Loss in Sequestration Capacity	0.0	0.0	0.0	0.0	0.0	0.0	0.0
Change in Flux From Wetland Conversions	0.0	0.0	0.0	0.0	0.0	0.0	0.0
Conterminous United States							
Pool Size in Current Wetlands	0	14.0[l]	5.1[k]	0.4	0.1	0.0	19.6
Sequestration in Current Wetlands	0	-6.6[m]	-5.3[f]	-4.4	-0.5	-0.5	-17.3
Oxidation in Former Wetlands	0	18.0[n]	0.0[h]	0.0	0.0	0.0	18.0
Historical Loss in Sequestration Capacity	0	1.2[m]	7.6[f]	0.4	0.0	0.1	9.4
Change in Flux from Wetland Conversions	0	19.2	7.6	0.4	0.0	0.1	27.4
Mexico							
Pool Size in Current Wetlands	0	1.5[f]	0.3[k]	0.0	0.1	ND	1.9
Sequestration in Current Wetlands	0	-1.6[o]	-0.4[f]	0.0	-1.6	ND	-3.6
Oxidation in Former Wetlands	0	ND	ND	0.0	0.0	0.0	ND
Historical Loss in Sequestration Capacity	0	ND	ND	0.0	1.0	ND	ND
Change in Flux from Wetland Conversions	0	ND	ND	0.0	1.0	ND	ND
North America							
Pool Size in Current Wetlands	56.7	124.6	36.0	0.4	0.2	0.3	218.2
Sequestration in Current Wetlands	-6.6	-22.6	-17.7	-4.8	-2.1	-3.3	-57.2
Oxidation in Former Wetlands	18.2		0.0	0.0	0.0	0.0	18.2
Historical Loss in Sequestration Capacity	0	1.4	11.0	0.5	1.0	0.5	14.5
Change in Flux from Wetland Conversions	19.6		11.0	0.5	1.0	0.5	32.7
Global							
Pool Size in Current Wetlands	462[p]		46[q]	0.4[r]	4.9[r]	ND	513
Sequestration in Current Wetlands	-55[s]		-39[f]	-4.6[r]	-38.0[r]	ND	-137
Oxidation in Former Wetlands	205[t]		ND	0	0	0	205
Historical Loss in Sequestration Capacity	16[t]		45[f]	0.7[u]	20[v]	ND	82
Change in Flux From Wetland Conversions	221[t]		> 45	0.7	20	ND	287

[a] Tarnocai (1998); mineral soil to 1-m depth.

[b] Tarnocai et al. (2005).

[c] Rates and pools calculated from Chmura et al. (2003) using country-specific data (sedimentation accumulation rates in Mg C per ha per year: Mexican mangroves = 3.3, conterminous United States mangroves = 1.8, conterminous United States tidal marshes = 2.2, tidal marshes in Canada and Alaska = 2.1); areas from Table 13F.1.

[d] Assumed the same carbon density and accumulation rates as the adjacent vegetated wetland ecosystem (mangrove data for Mexico and salt marsh data elsewhere).

[e] Assumed carbon accumulation rate of 0.13 Mg C per ha per year for permafrost peatlands and 0.19 Mg C per ha per year for non-permafrost peatlands. Reported range of long-term apparent accumulation rates from 0.05-0.35 (Ovenden, 1990; Maltby and Immirzi, 1993; Trumbore and Harden, 1997; Vitt et al., 2000; Turunen et al., 2004).

[f] Rate calculated as the geometric mean sediment accumulation rate of 2.2 Mg sediment per ha per year (range 0-80) from Johnston (1991) and Craft and Casey (2000) times 7.7 % C (CV = 109) (Batjes 1996).

[g] Sum of 0.24 Mt C per year from horticulture removal of peat (Cleary et al., 2005) and 0.10 Mt C per year from increased peat sequestration due to permafrost melting (Turetsky et al., 2002).

[h] Assumed that the net oxidation of 8.6% of the soil carbon pool (Euliss et al., 2006) over 50 years after conversion to non-wetland use.

[i] Assumed that conversion of tidal systems is caused by fill and results in burial and preservation of SOM, Sedimentary Organic Matter, rather than oxidation.

[j] Soil carbon densities of 1,441 Mg C per ha for Histosols and 1,048 Mg C per ha for Histels (Tarnocai et al., 2005).

[k] Soil carbon density of 162 Mg C per ha for the conterminous United States and Mexico and 468 Mg C per ha for Alaska based upon NRCS STATSGO database and soil pedon information.

[l] Assumed soil carbon density of 1,500 Mg C per ha.

[m] Webb and Webb (1988).

[n] Estimated loss rate as of early 1980s (Armentano and Menges, 1986). Overall, wetlands losses in the United States have declined dramatically since then (Dahl, 2000) and probably even more so for Histosols, so this number may still be representative.

[o] Using peat accumulation rate of 1.6 Mg C per ha (range 1.0–2.25) (Maltby and Immirzi, 1993).

[p] From Maltby and Immirzi (1993). Range of 234 to 679 GtC (Gorham, 1991; Maltby and Immirzi, 1993; Eswaran et al., 1995; Batjes, 1996; Lappalainen, 1996; Joosten and Clarke, 2002).

[q] Soil carbon density of 199 Mg C per ha (Batjes, 1996).

[r] Chmura et al. (2003).

[s] Joosten and Clarke (2002) reported range of -40 to -70 Mt C per year. Using the peatland estimate in Table F.1 and a C accumulation rate of 0.19 Mg C per ha per year, we calculate a global flux of -65 Mt C per year in peatlands.

[t] Current oxidative flux is the difference between the change in flux and the historical loss in sequestration capacity from this table. The change in flux is from Maltby and Immirzi (1993) (reported range 176 to 266 Mt C per year) and the historical loss in sequestration capacity is from this table for North America, from Armentano and Menges (1986) for other northern peatlands, and from Maltby and Immirzi (1993) for tropical peatlands.

[u] Assumed that global rates approximate the North America rate because most salt marshes inventoried are in North America.

[v] Assumed 25% loss globally since the late 1800s.

ND indicates that no data are available.

Some caution is necessary regarding the use of Gleysol or "wet" mineral soil carbon densities because apparently they include large areas of seasonally wet soils that are not considered wetlands by the more conservative definition of wetlands used by the United States and many other countries and organizations. For example, Eswaran et al. (1995) estimated that global wet mineral-soil area was 8,808,000 km², which is substantially higher than the commonly accepted mineral-soil wetland area estimated by Matthews and Fung (1987) of 2,289,000 km² and Aselmann and Crutzen (1989) of 2,341,000 km², even accounting for substantial global wetland loss. In our query of the NRCS STATSGO database for the United States, we found 1,258,000 km² of wet soils in the conterminous United States versus our estimate of 312,000 km² of FWMS wetlands, currently, and 762,000 km², historically (Table F.1). We assume that including these wet-but-not-wetland soils will decrease the estimated soil carbon density, but to what degree we do not know. However, just considering the differences in area will give large differences in the soil carbon pool. For example, Eswaran et al. (1995) estimated that wet mineral soils globally contain 108 Gt C to 1-m depth, whereas our estimate is 46 Gt C to 2-m depth (Table F.2).

For Alaska, many soil investigations have been conducted since the STATSGO soil data was coded. We updated STATSGO by calculating soil carbon densities from data obtained from the NRCS on 479 pedons collected in Alaska, and then we used this data for both FWMS wetlands and peatlands. For some of the Histosols, missing bulk densities were calculated using averages of measured bulk densities for the closest matching class in the USDA Soil Taxonomy (NRCS, 1999). A matching procedure was developed for relating sets of pedons to sets of STATSGO components. If there were multiple components for each map unit in STATSGO, the percentage of the component was used to scale area and carbon data. We compared matching sets of pedons to sets of components at the four top levels of the United States' soil taxonomy: Orders, Suborders, Great Groups, and Subgroups. For example, the soil carbon for all pedons having the same soil order were averaged, and the carbon content was applied to all of the soil components of the same order (*e.g.,* Histosol pedons are used to characterize Histosol components). At the Order level, all components were matched with pedon data. At the suborder level, pedon data were not available to match approximately 20,000 km² (compared to the nearly 1,500,000-km² area of soil in the state), but the soil characteristics were more closely associated with the appropriate land areas than at the Order level. At the Great Group and Subgroup levels, pedon data were unavailable for much larger areas, even though the quality of the data when available became better. For this study, we used the Suborder-level matching. The resulting soil carbon density for Alaskan FWMS wetlands was 469 Mg C per ha, reflecting large areas of wetlands with a histic epipedon as noted above.

F.2.2 Peatland Soil Carbon Pools

The carbon pool of permafrost and non-permafrost peatlands in Canada had been previously estimated by Tarnocai *et al.* (2005) based upon an extensive database. Good soil-carbon density data are unavailable for peatlands in the United States, as the NRCS soil pedon information typically only goes to a maximum depth of between 1.5 to 2 m, and many peatlands are deeper than this. Therefore, we used the carbon density estimates of Tarnocai *et al.* (2005) of 1,441 Mg C per ha for Histosols and 1,048 Mg C per ha for Histels to estimate the soil carbon pool in Alaskan peatlands.

The importance of our using a smaller area of Alaskan peatlands becomes obvious here. Using the larger area from Kivinen and Pakarinen (1981), Halsey *et al.* (2000) estimated that Alaskan peatlands have a soil carbon pool of 71.5 Gt, almost 5-fold higher than our estimate. However, some of the difference in soil carbon between the two estimates can be accounted for by the 26 Gt C that we calculated resides in Alaskan FWMS wetlands (Table F.2).

The peatlands of the conterminous United States are different in texture, and probably depth, from those in Canada and Alaska, so it is probably inappropriate to use the soil carbon densities for Canadian peatlands for those in the conterminous United States. For example, we compared the relative percentage of the Histosol suborders (excluding the

small area of Folists, as they are predominantly upland soils) for Canada (Tarnocai, 1998), Alaska (updated STATSGO data, J. Moore, personal comm.), and the conterminous United States (NRCS, 1999). The relative percentage of Fibrists, Hemists, and Saprists, respectively, in Canada are 37%, 62%, and 1%, in Alaska are 53%, 27%, and 20%, and in the conterminous United States are 1%, 19%, and 80%. Using the STATSGO database (N. Bliss, query of NRCS STATSGO database, December 2005), the average soil carbon density for Histosols in the conterminous United States is 1,089 Mg C per ha, but this is an underestimate as many peatlands were not sampled to their maximum depth. Armentano and Menges (1986) reported average carbon density of conterminous United States' peatlands to 1-m depth of 1,147 to 1,125 Mg C per ha. Malterer (1996) gave soil carbon densities of conterminous United States' peatlands of 2,902 Mg C per ha for Fibrist, 1,874 Mg C per ha for Hemists, and 2,740 Mg C per ha for Saprists, but it is unclear how he derived these estimates. Batjes (1996) and Eswaran *et al.* (1995) gave average soil carbon densities to 1-m depth for global peatlands of 776 and 2,235 Mg C per ha, respectively. We chose to use an average carbon density of 1,500 Mg C per ha, which is in the middle of the reported range, for peatlands in the conterminous United States and Mexico.

F.2.3 Estuarine Soil Carbon Pools

Tidal wetland soil carbon density was based on a country-specific analysis of data reported in an extensive compilation by Chmura *et al.* (2003). There were more observations for the United States (n = 75) than Canada (n = 34) or Mexico (n = 4), and consequently there were more observations of marshes than mangroves. The Canadian salt marsh estimate was used for Alaskan salt marshes and mud flats. In the conterminous United States and Mexico, country-specific marsh or mangrove estimates were used for mudflats. Although Chmura *et al.* (2003) reported some significant correlations between soil carbon density and mean annual temperature, scatter plots suggest the relationships are weak or driven by a few sites. Thus, we did not separate the data by region or latitude and used mean values for scaling. Chmura *et al.* (2003) assumed a 50-cm-deep profile for the soil carbon pool, which may be an underestimate.

F.2.4 Plant Carbon Pools

While extensive data on plant biomass in individual wetlands have been published, no systematic inventory of wetland plant biomass has been undertaken in North America. Nationally, the forest carbon biomass pool (including above-ground and below-ground biomass) has been estimated to be 54.9 Mg C per ha (Birdsey, 1992), which we used for forested wetlands in the United States and Canada. This approach assumes that wetland forests do not have substantially different biomass carbon densities from upland forests. There is one

Table F.3 Plant carbon pools (Gt) and fluxes (Mt per year) of wetlands in North America and the world. Positive flux numbers indicate a net flux into the atmosphere, whereas negative numbers indicate a net flux into the ecosystem.

	Permafrost peatlands	Non-permafrost peatlands	Mineral-soil freshwater	Salt marsh	Mangrove	Total
Canada						
Pool Size in Current Wetlands	1.4[a]		0.3[b]	0.0[c]	0.0	1.7
Sequestration in Current Wetlands	0.0	ND		0.0	0.0	0.0
Alaska						
Pool Size in Current Wetlands	0.4[a]		1.1[d]	0.0	0.0	1.5
Sequestration in Current Wetlands	0.0	0.0	0.0	0.0	0.0	0.0
Conterminous United States						
Pool Size in Current Wetlands	0.0	1.5[d]		0.0	0.0	1.5
Sequestration in Current Wetlands	0.0	-10.3[e]		0.0	0.0	-10.3
Mexico						
Pool Size in Current Wetlands	0.0	0.0[b]	0.0[b]	0.0	0.1	0.1
Sequestration in Current Wetlands	0.0	ND	ND	0.0	ND	0.0
North America						
Pool Size in Current Wetlands	4.8			0.0	0.1	4.9
Sequestration in Current Wetlands	0.0	-10.3		0.0	ND	-10.3
Global						
Pool Size in Current Wetlands	6.9[b]		4.6[b]	0.0[f]	4.0[g]	15.5
Sequestration in Current Wetlands	0.0	ND	ND	0.0	ND	ND

[a] Biomass for non-forested peatlands from Vitt *et al.* (2000), assuming 50% of biomass is below-ground. Forest biomass density from Birdsey (1992) and forested area from Tarnocai *et al.* (2005) for Canada and from Hall *et al.* (1994) for Alaska.

[b] Assumed 2000 g C per m^2 in above-ground and below-ground plant biomass (Gorham, 1991).

[c] Biomass data from Mitsch and Gosselink (1993).

[d] Biomass for non-forested wetlands from Gorham (1991). Forest biomass density from Birdsey (1992), and forested area from Hall *et al.* (1994) for Alaska and Dahl (2000) for the conterminous United States.

[e] 50 g C per m^2 per yr sequestration from forest growth from a southeastern United States regional assessment of wetland forest growth (Brown *et al.*, 2001).

[f] Assumed that global pools approximate those from North America because most salt marshes inventoried are in North America.

[g] Twilley *et al.* (1992).

ND indicates that no data are available.

regional assessment of forested wetlands in the southeastern United States, which comprise approximately 35% of the total forested wetland area in the conterminous United States. We utilized the southeastern United States regional inventory to evaluate this assumption; above-ground tree biomass averaged 125.2 m^3 per ha for softwood stands and 116.1 m^3 per ha for hardwood stands. Using an average wood density and carbon content, the carbon density for these forests would be 33 Mg C per ha for softwood stands and 42 Mg C per ha for hardwood stands. However, these estimates do not include understory vegetation, below-ground biomass, or dead trees, which account for 49% of the total forest biomass (Birdsey, 1992). Using that factor to make an adjustment for total forest biomass, the range would be 49 to 66 Mg C per ha for the softwood and hardwood stands, respectively. Accordingly, the assumption of using 54.9 Mg C per ha seems reasonable for a national-level estimate.

The area of forested wetlands in Canada came from Tarnocai *et al.* (2005), for Alaska from Hall *et al.* (1994), and for the conterminous United States from Dahl (2000).

Since Tarnocai *et al.* (2005) divided Canadian peatland area into bog and fen, we used above-ground biomass for each community type from Vitt *et al.* (2000), and assumed that 50% of biomass is below-ground. We used the average bog and fen plant biomass from Vitt *et al.* (2000) for Alaskan peatlands. For other wetland areas, we used an average value of 20.0 Mg C per ha for non-forested wetland biomass carbon density (Gorham, 1991).

Tidal marsh root and shoot biomass data were estimated from a compilation in Table 8-7 in Mitsch and Gosselink (1993). There was no clear latitudinal or regional pattern in biomass, so we used mean values for each. Mangrove biomass has been shown to vary with latitude, so we used the empirical relationship from Twilley *et al.* (1992) for this relationship. We made a simple estimate using a single latitude that visually bisected the distribution of mangroves either in the United States (26.9o) or Mexico (23.5o). Total biomass was estimated using a root-to-shoot ratio of 0.82 and a carbon-mass-to-biomass ratio of 0.45, both from Twilley *et al.* (1992).

Plant biomass carbon data are presented in Table F.3.

F.3 CARBON FLUXES

F.3.1 Peatland Soil Carbon Accumulation Rates

Most studies report the long-term apparent rate of carbon accumulation (LORCA) in peatlands based upon basal peat dates, but this assumes a linear accumulation rate through time. However, due to the slow decay of the accumulated peat, the true rate of carbon accumulation will always be less than the LORCA (Clymo *et al.*, 1998), so most reported rates are inherently biased upwards. Tolonen and Turunen (1996) found that the true rate of peat accumulation was about 67% of the LORCA.

For estimates of soil carbon sequestration in conterminous United States' peatlands, we used the LORCA data from 82 sites and 215 cores throughout eastern North America (Webb and Webb III, 1988). They reported a median accumulation rate of 0.066 cm per year (mean = 0.092, sd = 0.085). We converted this value into a carbon accumulation rate of -0.71 Mg C per ha per year by assuming 58% C (see NRCS Soil Survey Laboratory Information Manual,

available on-line at http://soils.usda.gov/survey/nscd/lim/), a bulk density of 0.28 g per cm³, and an organic matter content of 69%. (Positive carbon fluxes indicate net fluxes to the atmosphere, whereas negative carbon fluxes indicate net fluxes into an ecosystem.) The bulk density and organic matter content were the area-weighted and depth-weighted average from all Histosol soil map units greater than 202.5 ha (n = 3,884) in the conterminous United States from the National Soil Information System (NASIS) data base provided by S. Campbell (USDA NRCS, Portland, Oreg.). For comparison, Armentano and Menges (1986) used soil carbon accumulation rates that ranged from -0.48 Mg C per ha per year in northern conterminous United States peatlands to -2.25 Mg C per ha per year in Florida peatlands.

Peatlands accumulate lesser amounts of soil carbon at higher latitudes, with especially low accumulation rates in permafrost peatlands (Ovenden, 1990; Robinson and Moore, 1999). The rates used in this report reflect this gradient, going from -0.13 to -0.19 to -0.71 Mg C per ha per year in permafrost peatlands, non-permafrost Canadian and Alaskan peatlands, and peatlands in the conterminous United States and Mexico, respectively (Table F.2).

F.3.2 Freshwater Mineral-Soil Wetland Carbon Accumulation Rates

Many studies have estimated sediment deposition rates in FWMS wetlands, with a geometric mean rate of 2.2 Mg sediment per ha per year (n = 26, arithmetic mean = 16.3, range 0 to 80.0) in a compilation by Johnston (1991), along with those reported more recently in Craft and Casey (2000). As can be seen by the difference between the geometric and arithmetic means, this dataset is log-normally distributed with several large outliers. Assuming 7.7% carbon for FWMS wetlands (Batjes, 1996), this gives a geometric mean accumulation rate of 0.17 Mg C per ha per year. Johnston (1991) and Craft and Casey (2000) reported more studies with only vertical

sediment accumulation rates, with a geometric mean of 0.23 cm per year (n = 34, arithmetic mean = 0.63 cm per year, range -0.6 to 2.6). If we assume a bulk density of 1.00 g per cm³ for FWMS wetlands (Batjes, 1996; Smith *et al.*, 2001), this converts into an unrealistically large accumulation rate of 1.85 Mg C per ha per year.

We suggest that caution is necessary in interpretation of these data for a number of reasons. There is large variability in sedimentation rates among studies, and even within a site, sedimentation rates are highly variable depending on the local deposition

environment (Johnston *et al.*, 2001). Researchers may have preferentially chosen wetlands with high sedimentation rates to study this process, providing a bias towards greater carbon sequestration. Rates of erosion and resultant deposition have substantially decreased during the last century in the conterminous United States (Craft and Casey, 2000; Trimble and Crosson, 2000). More fundamentally, it is important to distinguish between autochthonous carbon (derived from on-site plant production) and allochthonous carbon (imported from outside the wetland) in soil carbon storage. The soil carbon stored in peatlands is of autochthonous origin and represents sequestration of atmospheric carbon dioxide at the landscape scale. In contrast, a unknown portion of the soil carbon that is stored in FWMS wetlands is of allochthonous origin. However, conterminous United States' soils average between 0.9 and 1.3% soil carbon, which is much less than the average carbon content of FWMS wetlands (7.7%) (Batjes, 1996), suggesting a substantial autochthonous input to FWMS wetlands.

At a landscape scale, redistribution of sediments from uplands to wetlands represents net carbon sequestration only to the extent that the soil carbon is replaced in the terrestrial source area and/or decomposition rates are substantially lower in the receiving wetland (Stallard, 1998; Harden *et al.*, 1999). Agricultural lands are a major source of erosion (Meade *et al.*, 1990, as cited in Stallard, 1998), but it appears that, after large initial losses, soil carbon is relatively stable (Stallard, 1998; Smith *et al.*, 2001) or even increases (Harden *et al.*, 1999) under modern agricultural techniques. It is also generally assumed that sediment carbon deposited in anaerobic environments, such as occur in many wetlands, is relatively recalcitrant (Stallard, 1998; Smith *et al.*, 2001). For example, in a variety of Minnesota wetland soils, carbon

mineralization was approximately six times slower anaerobically than aerobically (Bridgham *et al.*, 1998). However, time since initial deposition and organic quality of sediments appears to be an important constraint on its relative reactivity. Kristensen *et al.* (1995) found that relatively fresh, labile organic matter had similar decomposition rates aerobically and anaerobically, whereas "aged," recalcitrant organic matter decomposed ten times slower anaerobically. Gunnison *et al.* (1983) found that freshly flooded soils had twice as rapid carbon mineralization rates as sediments. In newly constructed reservoirs, sediments maintained these rapid mineralization rates even 6-10 years after initial flooding. Overall, these latter two studies suggest that there may be substantial carbon mineralization in freshly deposited allochthonous sediments in wetlands, but we feel that the data are not adequate to account for this effect quantitatively.

We use a landscape-level sediment sequestration rate of 0.17 Mg C per ha per year for FWMS wetlands in North America, while acknowledging the low level of confidence in this estimate. Johnston (1991) and Craft and Casey (2000) only gave sedimentation rates in FWMS wetlands in the conterminous United States. Since most FWMS wetlands in Canada are in more developed and agricultural regions, we felt that it was reasonable to use the sedimentation estimates from these studies. However, most Alaskan FWMS wetlands are relatively pristine, with little anthropogenic sediment input, but as described above, most have an extensive histic epipedon, so at least historically, they have actively accumulated soil carbon. Given that our soil carbon accumulation rate for Alaskan peatlands is 0.19 Mg C per ha per year, our sediment sequestration rate of 0.17 Mg C per ha per year for Alaskan FWMS wetlands does not seem unreasonable.

Table F.4 Methane fluxes (Mt per year) from wetlands in North America and the world.

	Permafrost peatlands	Non-perma-frost peatlands	Mineral-soil freshwater	Salt marsh	Mangrove	Mudflat	Total
Canada							
CH$_4$ Flux in Current Wetlands	1.1[a]	2.1[b]	1.2	0.0	0.0	0.0[c]	4.4
Historical change in CH$_4$ Flux	0.0	0.3	-1.5	0.0	0.0	0.0	-1.2
Alaska							
CH$_4$ Flux in Current Wetlands	0.2	0.1	1.4	0.0	0.0	0.0	1.7
Historical change in CH$_4$ Flux	0.0	0.0	0.0	0.0	0.0	0.0	0.0
Conterminous United States							
CH$_4$ Flux in Current Wetlands	0.0	0.7	2.4	0.0	0.0	0.0	3.1
Historical change in CH$_4$ Flux	0.0	-0.1	-3.4	0.0	0.0	0.0	-3.5
Mexico							
CH$_4$ Flux in Current Wetlands	0.0	0.1	0.2	0.0	0.0	ND	0.2
Historical change in CH$_4$ Flux	0.0	-0.1		0.0	0.0	ND	-0.1
North America							
CH$_4$ Flux in Current Wetlands	1.3	3.0	5.1	0.0	0.0	0.0	9.4
Historical change in CH$_4$ Flux	0.0	-4.9		0.0	0.0	0.0	-4.9
Global							
CH$_4$ Flux in Current Wetlands	14.1[d]	22.5[d]	68.0[d]	0.0[e]	0.2	ND	105[f]
Historical change in CH$_4$ Flux	-3.6[g]		-79[g]	0.0[e]	-0.1	ND	-83

[a] Used CH$_4$ flux of 2.5 g per m[2] per yr (range 0 to 130, likely mean 2 to 3) (Moore and Roulet, 1995) for Canadian peatlands and all Alaskan freshwater wetlands. Used CH$_4$ flux of 7.6 g per m[2] per yr for Canadian freshwater mineral-soil wetlands and all United States and Mexican freshwater wetlands and 1.3 g per m[2] per yr for estuarine wetlands—from synthesis of published CH$_4$ fluxes for the United States (see Table F.5).

[b] Includes a 17-fold increase in CH$_4$ flux (Kelly et al., 1997) in the 9000 km[2] of reservoirs that have been formed on peatlands (Rubec, 1996) and an estimated CH$_4$ flux of 15 g per m[2] per yr (Moore et al., 1998) from 2,630 km[2] of melted permafrost peatlands (Vitt et al., 1994).

[c] Assumed trace gas fluxes from unvegetated estuarine wetlands (i.e., mudflats) was the same as adjacent wetlands.

[d] Bartlett and Harriss (1993).

[e] Assumed that global rates approximate the North America rate because most salt marsh area is in North America.

[f] Ehhalt et al. (2001), range of 92 to 237 Mt per yr.

[g] Using rates from Bartlett and Harriss (1993) and historical loss of area in Table 1.

ND indicates that no data are available.

F.3.3 Estuarine Carbon Accumulation Rates

Carbon accumulation in tidal wetlands was assumed to be entirely in the soil pool. This should provide a reasonable estimate because marshes are primarily herbaceous, and mangrove biomass should be in steady state unless the site was converted to another use. An important difference between soil carbon sequestration in tidal and non-tidal systems is that tidal sequestration occurs primarily through burial driven by sea level rise. For this reason, carbon accumulation rates can be estimated well with data on changes in soil surface elevation and carbon density. Rates of soil carbon accumulation were calculated from Chmura *et al.* (2003) as described above for the soil carbon pool (rates in Mg C per ha per year are 3.3 for Mexican mangroves; 1.8 and 2.2 for mangroves and tidal marshes, respectively, in the conterminous United States; 2.1 for tidal marshes in Canada and Alaska). These estimates are based on a variety of methods, such as ^{210}Pb dating and soil elevation tables, which integrate vertical soil accumulation rates over periods of time ranging from 1–100 years. The soil carbon sequestered in estuarine wetland sediments is likely to be a mixture of both allochthonous and autochthonous sources. However, without better information, we assumed that *in situ* rates of soil carbon sequestration in estuarine wetlands is representative of the true landscape-level rate.

F.3.4 Extractive Uses of Peat

Use of peat for energy production is, and always has been, negligible in North America, as opposed to other parts of the world (WEC, 2001). However, Canada produces a greater volume of horticultural and agricultural peat than any other country in the world (WEC, 2001). Currently, 124 km^2 of Canadian peatlands have been under extraction now or in the past (Cleary *et al.*, 2005). A life-cycle analysis by these authors estimated that as of 1990 Canada emitted 0.2 Mt per year of CO_2-C equivalents through peat extraction. The United States' production of horticultural peat is about 19% of Canada's (Joosten and Clarke, 2002), which assuming a

similar life-cycle as for Canada, suggests that the United States produces 0.05 Mt of CO_2-C equivalents through peat extraction.

F.3.5 Methane Fluxes

Moore and Roulet (1995) reported a range of methane fluxes from 0 to 130 g CH_4 per m^2 per year from 120 peatland sites in Canada, with the majority <10 g CH_4 per m^2 per year. They estimated a low average flux rate of 2 to 3 g CH_4 per m^2 per year, which equaled an emission of 2–3 Mt CH_4 per year from Canadian peatlands. We used an estimate of 2.5 g CH_4 per m^2 per year for Canadian peatlands and Alaskan freshwater wetlands (Table F.4).

To our knowledge, the last synthesis of field measurements of methane emissions from wetlands was done by Bartlett and Harriss (1993). We supplemented their analysis with all other published field studies (using chamber or eddy covariance techniques) we could find that reported annual or average daily methane fluxes in the conterminous United States (Table F.5). We excluded a few studies that used cores or estimated diffusive fluxes.

In cases where multiple years from the same site were presented, we took the average of those years. Similarly, when multiple sites of the same type were presented in the same paper, we took the average. Studies were separated into freshwater and estuarine systems.

In cases where papers presented both an annual flux and a mean daily flux, we calculated a conversion factor (annual flux/average daily flux) to quantify the relationship between those two numbers (Table F.5). When we looked at all studies (n = 30), this conversion factor was 0.36, suggesting that there is a 360-day emission season. There was surprisingly little variation in this ratio, and it was similar in freshwater (0.36) and estuarine (0.34) wetlands. In contrast, previous syntheses used a 150-day emission season for temperate wetlands (Matthews and Fung, 1987; Bartlett and Harriss, 1993). While substantial winter methane emissions have been found in some studies, it is likely that flux data from most studies have a non-normal distribution with occasional periods of high flux rates that are better captured with annual measurements.

Using the conversion factors for freshwater and estuarine wetlands, we estimated average annual fluxes from the average daily fluxes. The data were highly log-normally distributed, so we used geometric means. For freshwater wetlands, the geometric

mean estimated annual flux rate was 7.1 g CH_4 per m² per year (n = 74, 1 SE = 0.8, arithmetic mean = 38.6), which is very similar to the geometric mean measured rate of 8.1 g CH_4 per m² per year (n = 32, arithmetic mean = 32.1). For estuarine wetlands, the geometric mean estimated annual flux rate was 1.3 g CH_4 per m² per year (n = 25, 1 SE = 0.2, arithmetic mean = 9.8), which is smaller than the geometric mean measured rate of 5.0 g CH_4 per m² per year (n = 13, arithmetic mean = 16.9).

Finally, we combined both approaches. In cases where a paper presented an annual value, we used that number. In cases where only an average daily number was presented, we used that value corrected with the appropriate conversion factor. For conterminous United States wetlands, FWMS Canadian wetlands, and Mexican wetlands, we used a geometric mean flux of 7.6 g CH_4 per m² per year, and for estuarine wetlands, we used a geometric mean flux of 1.3 g CH_4 per m² per year.

F.3.6 Plant Carbon Fluxes

For ecosystems at approximately steady state, plant biomass should be reasonably constant on average because plant production is roughly balanced by mortality and subsequent decomposition. We assumed insignificant plant biomass accumulation in freshwater and estuarine marshes because they are dominated by herbaceous plants that do not accumulate carbon in wood. Sequestration in plants in relatively undisturbed forested wetlands in Alaska and many parts of Canada is probably small, although there may be substantial logging of Canadian forested wetlands for which we do not have data. Similarly, no data was available to evaluate the effect of harvesting of woody biomass in Mexican mangroves on carbon fluxes.

Tree biomass carbon sequestration averages -1.40 Mg C per ha per year in United States' forests across all forest types (Birdsey, 1992). Using the tree growth estimates from the southeastern United States regional assessment of wetland forests (Brown *et al.,* 2001) yields an even lower estimate of sequestration in above-ground tree biomass (approx. -0.50 Mg C per ha per year). We used this lower value and area estimates from Dahl (2000) to estimate that forested wetlands in the conterminous United States currently sequester -10.3 Mt C per year.

Table F.5 Methane fluxes measured in the conterminous United States. The conversion factor is the ratio of the daily average flux to the measured annual flux × 10³. The calculated annual flux was determined based upon the average conversion factor for freshwater (**FW**) and saltwater wetlands (**SW**). The measured annual flux was used if that was available; otherwise, the calculated annual flux was used.

Habitat	State	Method[a]	Salt/Fresh	Daily Average Flux (mg CH₄ per m² per day)	Measured Annual Flux (g CH₄ per m² per year)	Conversion Factor	Estimated Annual Flux (g CH₄ per m² per year)	Used Annual Flux (g CH₄ per m² per year)	Reference
Fens	CO	C	FW		40.7			40.7	Chimner and Cooper (2003)
Wet Alpine Meadow	CO	C	FW	0.1			0.0	0.0	Neff et al. (1994)
Lake - Average	CO	C	FW	25.4			9.2	9.2	Smith and Lewis (1992)
Wetland - Average	CO	C	FW	28.3			10.3	10.3	Smith and Lewis (1992)
Nuphar Bed	CO	C	FW	202.1			73.6	73.6	Smith and Lewis (1992)
Tundra - Carex Meadow	CO	C	FW	2.8			1.0	1.0	West et al. (1999)
Tundra - Acomastylis Meadow	CO	C	FW	-0.5			-0.2	-0.2	West et al. (1999)
Tundra - Kobresia Meadow	CO	C	FW	-0.8			-0.3	-0.3	West et al. (1999)
Moist Grassy	CO	C	FW	6.1	1.9	0.32	2.2	1.9	Wickland et al. (1999)
Moist Mossy	CO	C	FW	1.5	0.5	0.33	0.5	0.5	Wickland et al. (1999)
Wetland	CO	C	FW		41.7			41.7	Wickland et al. (1999)
Hardwood Hammock	FL	C	FW	0.0			0.0	0.0	Bartlett et al. (1989)
Dwarf Cypress / Sawgrass	FL	C	FW	7.5			2.7	2.7	Bartlett et al. (1989)
Spikerush	FL	C	FW	29.4			10.7	10.7	Bartlett et al. (1989)
Sawgrass < 1m	FL	C	FW	38.8			14.1	14.1	Bartlett et al. (1989)
Sawgrass/Spikerush/Periphyton	FL	C	FW	45.1			16.4	16.4	Bartlett et al. (1989)
Swamp Forest	FL	C	FW	68.9			25.1	25.1	Bartlett et al. (1989)
Sawgrass > 1m	FL	C	FW	71.9			26.2	26.2	Bartlett et al. (1989)
Sawgrass	FL	C	FW	107.0			38.9	38.9	Burke et al. (1988)
Pond Open Water	FL	C	FW	624.0			227.1	227.1	Burke et al. (1988)
Everglades - Cladium	FL	C	FW	45.4			16.5	16.5	Chanton et al. (1993)
Everglades - Typha	FL	C	FW	142.9			52.0	52.0	Chanton et al. (1993)
Wet Prairie (Marl)	FL	C	FW	87.0			31.6	31.6	Happell et al. (1993)
Wet Prairie (Marl)	FL	C	FW	27.4			10.0	10.0	Happell et al. (1993)
Marsh (Marl)	FL	C	FW	30.0			10.9	10.9	Happell et al. (1993)
Marsh (Marl)	FL	C	FW	49.6			18.0	18.0	Happell et al. (1993)
Marsh (Peat)	FL	C	FW	45.4			16.5	16.5	Happell et al. (1993)
Marsh (Peat)	FL	C	FW	13.0			4.7	4.7	Happell et al. (1993)
Marsh (Peat)	FL	C	FW	163.6			59.6	59.6	Happell et al. (1993)
Marsh (Peat)	FL	C	FW	20.4			7.4	7.4	Happell et al. (1993)
Wet Prairie / Sawgrass	FL	C	FW	61.0			22.2	22.2	Harriss et al. (1988)
Wetland Forest	FL	C	FW	59.0			21.5	21.5	Harriss et al. (1988)
Cypress Swamp - Flowing Water	FL	C	FW	67.0			24.4	24.4	Harriss and Sebacher (1981)
Open Water Swamp	FL	C	FW	480.0			174.7	174.7	Schipper and Reddy (1994)
Waterlily Slough	FL	C	FW	91.0			33.1	33.1	Schipper and Reddy (1994)

[a] C = chamber, T = tower, eddy covariance, E = ebulition measured separately.

Habitat	State	Method[a]	Salt/ Fresh	Daily Average Flux (mg CH$_4$ per m^2 per day)	Measured Annual Flux (g CH$_4$ per m^2 per year)	Conversion Factor	Estimated Annual Flux (g CH$_4$ per m^2 per year)	Used Annual Flux (g CH$_4$ per m^2 per year)	Reference
Cypress Swamp - Deep Water	GA	C	FW	92.3			33.6	33.6	Harriss and Sebacher (1981)
Bottomland Hardwoods/ Swamps	GA	C	FW		23.0			23.0	Pulliam (1993)
Swamp Forest	LA	C	FW	146.0			53.1	53.1	Alford et al. (1997)
Freshwater Marsh	LA	C	FW	251.0			91.4	91.4	Alford et al. (1997)
Fresh	LA	C	FW	587.0	213.0	0.36	213.6	213.0	DeLaune et al. (1983)
Fresh	LA	C	FW	49.0	18.7	0.38	17.8	18.7	DeLaune et al. (1983)
Sphagnum Bog	MD	C	FW	-1.1			-0.4	-0.4	Yavitt et al. (1990)
Bog	MI	C	FW	193.0			70.2	70.2	Shannon and White (1994)
Bog	MI	C	FW	28.0			10.2	10.2	Shannon and White (1994)
Beaver Meadow	MN	C	FW		2.3			2.3	Bridgham et al. (1995)
Open Bogs	MN	C	FW		0.0			0.0	Bridgham et al. (1995)
Bog (Forested Hummock)	MN	C	FW	10.0	3.5	0.35	3.6	3.5	Dise (1993)
Bog (Forested Hollow)	MN	C	FW	38.0	13.8	0.36	13.8	13.8	Dise (1993)
Fen Lagg	MN	C	FW	35.0	12.6	0.36	12.7	12.6	Dise (1993)
Bog (Open Bog)	MN	C	FW	118.0	43.1	0.37	42.9	43.1	Dise (1993)
Fen (Open Poor Fen)	MN	C	FW	180.0	65.7	0.37	65.5	65.7	Dise (1993)
Poor Fen	MN	C	FW	242.0			88.1	88.1	Dise and Verry (2001)
Sedge Meadow	MN	C	FW		11.7			11.7	Naiman et al. (1991)
Submergent	MN	C	FW		14.4			14.4	Naiman et al. (1991)
Deep Water	MN	C	FW		0.5			0.5	Naiman et al. (1991)
Poor Fen	MN	T	FW		14.6			14.6	Shurpali and Verma (1998)
Submerged Tidal	NC	C, E	FW	144.8			52.7	52.7	Kelly et al. (1995)
Banks Tidal	NC	C, E	FW	20.1			7.3	7.3	Kelly et al. (1995)
Tidal Marsh	NC	C	FW	3.0	1.0	0.34	1.1	1.0	Megonigal and Schlesinger (2002)
Tidal Marsh	NC	C	FW	3.5	2.3	0.65	1.3	2.3	Megonigal and Schlesinger (2002)
Prairie Marsh	NE	T	FW		64.0			64.0	Kim et al. (1999)
Poor Fen	NH	C	FW	503.3	110.6	0.22	183.2	110.6	Carroll and Crill (1997)
Poor Fen	NH	C	FW		69.3			69.3	Frolking and Crill (1994)
Forested Peatland	NY	C	FW	0.6	0.2	0.37	0.2	0.2	Coles and Yavitt (2004)
Pools Forested Swamp	NY	C	FW	224.6	69.0	0.31	81.7	69.0	Miller et al. (1999)
Typha Marsh - Mineral Soils	NY	C	FW	344.4			125.3	125.3	Yavitt (1997)
Typha Marsh - Peat Soils	NY	C	FW	65.1			23.7	23.7	Yavitt (1997)
Typha Marsh - All Soils	NY	C	FW	204.8			74.5	74.5	Yavitt (1997)
Cypress Swamp - Floodplain	SC	C	FW	9.9			3.6	3.6	Harriss and Sebacher (1981)

Habitat	State	Method[a]	Salt/Fresh	Daily Average Flux (mg CH₄ per m² per day)	Measured Annual Flux (g CH₄ per m² per year)	Conversion Factor	Estimated Annual Flux (g CH₄ per m² per year)	Used Annual Flux (g CH₄ per m² per year)	Reference
Swamp	VA	C	FW	470.3			171.2	171.2	Chanton et al. (1992)
Maple/Gum Forested Swamp	VA	C	FW		0.5			0.5	Harriss et al. (1982)
Emergent Tidal Freshwater Marsh	VA	C	FW		96.2			96.2	Neubauer et al. (2000)
Oak Swamp (Bank Site)	VA	C	FW	117.0	43.7	0.37	42.6	43.7	Wilson et al. (1989)
Emergent Macrophytes (Peltandra)	VA	C	FW	155.0			56.4	56.4	Wilson et al. (1989)
Emergent Macrophytes (Smartweed)	VA	C	FW	83.0			30.2	30.2	Wilson et al. (1989)
Ash Tree Swamp	VA	C	FW	152.0			55.3	55.3	Wilson et al. (1989)
Bog	WA	C	FW	73.0			26.6	26.6	Lansdown et al. (1992)
Lowland Shrub and Forested Wetland	WI	T	FW		12.4			12.4	Werner et al. (2003)
Sphagnum/Eriophorum (Poor Fen)	WV	C	FW	6.6			2.4	2.4	Yavitt et al. (1990)
Sphagnum/Shrub (Fen)	WV	C	FW	0.1			0.0	0.0	Yavitt et al. (1990)
Polytrichum/Shrub (Fen)	WV	C	FW	-0.1			0.0	0.0	Yavitt et al. (1990)
Sphagnum/Forest	WV	C	FW	9.6[b]			3.5	3.5	Yavitt et al. (1990)
Sedge Meadow	WV	C	FW	1.5			0.5	0.5	Yavitt et al. (1990)
Beaver Pond	WV	C	FW	250.0			91.0	91.0	Yavitt et al. (1990)
Low Gradient Headwater Stream	WV	C	FW	300.0			109.2	109.2	Yavitt et al. (1990)
Sphagnum/Eriophorum	WV	C	FW	52.1	19.0	0.37	18.9	19.0	Yavitt et al. (1993)
Polytrichum	WV	C	FW	41.1	15.0	0.37	15.0	15.0	Yavitt et al. (1993)
Sphagnum/Shrub	WV	C	FW	4.4	1.6	0.37	1.6	1.6	Yavitt et al. (1993)
Salt Marsh	DE	C	SW	0.5			0.2	0.2	Bartlett et al. (1985)
Red Mangroves	FL	C	SW	4.2			1.4	1.4	Bartlett et al. (1989)
Dwarf Red Mangrove	FL	C	SW	81.9			27.9	27.9	Bartlett et al. (1989)
High Marsh	FL	C	SW	3.9			1.3	1.3	Bartlett et al. (1985)
Salt Marsh	FL	C	SW	0.6			0.2	0.2	Bartlett et al. (1985)
Salt Water Mangroves	FL	C	SW	4.0			1.4	1.4	Harriss et al. (1988)
Salt Marsh	GA	C	SW	13.4			4.6	4.6	Bartlett et al. (1985)
Short Spartina Marsh - High Marsh	GA	C	SW	145.2	53.1	0.37	49.5	53.1	King and Wiebe (1978
Mid Marsh	GA	C	SW	15.8	5.8	0.37	5.4	5.8	King and Wiebe (1978)
Tall Spartina Marsh - Low Marsh	GA	C	SW	1.2	0.4	0.34	0.4	0.4	King and Wiebe (1978)
Intermediate Marsh	LA	C	SW	912[b]					Alford et al. (1997)
Salt Marsh	LA	C	SW	15.7	5.7	0.36	5.4	5.7	DeLaune et al. (1983)
Brackish	LA	C	SW	267.0	97.0		91.1	97.0	DeLaune et al. (1983)
Salt Marsh	LA	C	SW	4.8	1.7	0.35	1.6	1.7	DeLaune et al. (1983)
Brackish	LA	C	SW	17.0	6.4	0.38	5.8	6.4	DeLaune et al. (1983)

[b] Outlier that was removed from further analysis.

Habitat	State	Method[a]	Salt/ Fresh	Daily Average Flux (mg CH_4 per m^2 per day)	Measured Annual Flux (g CH_4 per m^2 per year)	Conversion Factor	Estimated Annual Flux (g CH_4 per m^2 per year)	Used Annual Flux (g CH_4 per m^2 per year)	Reference
Cypress Swamp - Floodplain	SC	C	SW	1.5			0.5	0.5	Bartlett *et al.* (1985)
Salt Marsh	SC	C	SW	0.4			0.1	0.1	Bartlett *et al.* (1985)
Salt Marsh	VA	C	SW	3.0	1.3	0.43	1.0	1.3	Bartlett *et al.* (1985)
Salt Marsh	VA	C	SW	5.0	1.2	0.24	1.7	1.2	Bartlett *et al.* (1985)
Salt Meadow	VA	C	SW	2.0	0.4	0.22	0.7	0.4	Bartlett *et al.* (1985)
Salt Marsh	VA	C	SW	-0.8			-0.3	-0.3	Bartlett *et al.* (1985)
Salt Marsh	VA	C	SW	1.5			0.5	0.5	Bartlett *et al.* (1985)
Salt Meadow	VA	C	SW	-1.9			-0.6	-0.6	Bartlett *et al.* (1985)
Tidal Salt Marsh	VA	C	SW	16.0	5.6	0.35	5.5	5.6	Bartlett *et al.* (1987)
Tidal Brackish Marsh	VA	C	SW	64.6	22.4	0.35	22.0	22.4	Bartlett *et al.* (1987)
Tidal Brackish/Fresh Marsh	VA	C	SW	53.5	18.2	0.34	18.2	18.2	Bartlett *et al.* (1987)
Freshwater									
n					32	18	74	88	
Arithmetic Mean					32.1	0.36	38.6	36.0	
Arithmetic Standard Error					7.9	0.02	6.0	5.0	
Geometric Mean					8.1		7.1	7.6	
Geometric Standard Error					2.1		0.82	2.2	
Saltwater									
n					13	12	25	25	
Arithmetic Mean					16.9	0.34	9.8	10.3	
Arithmetic Standard Error					7.8	0.02	4.1	4.4	
Geometric Mean					5.0		1.3	1.3	
Geometric Standard Error					2.0		0.2	3.3	

APPENDIX G

New pCO_2 Database for Coastal Ocean Waters Surrounding North America

Lead Authors: Francisco P. Chavez, MBARI; Taro Takahashi, Columbia Univ.

Contributing Authors: Wei-Jun Cai, Univ. Ga.; Gernot Friederich, MBARI; Burke Hales, Oreg. State Univ.; Rik Wanninkhof, NOAA; Richard A. Feely, NOAA

A database for the partial pressure of carbon dioxide (pCO_2), temperature, and salinity in surface waters within about 1,000 km from the shore of the North American continent has been assembled. About 550,000 seawater pCO_2 observations were made from 1979 to 2004 by the authors and collaborators of Chapter 15. The pCO_2 data have been obtained by a method using an infrared gas analyzer or gas-chromatograph for the determination of CO_2 concentrations in a carrier gas equilibrated with seawater at a known temperature and total pressure. The precision of pCO_2 measurements has been estimated to be about $\pm 0.7\%$ on average. The quality-controlled data are archived at http://www.ldeo.columbia.edu/res/pi/CO2.

The zonal distribution of the surface water pCO_2, sea surface temperature (SST), and salinity data shows that the greatest variability is confined within 300 km from the shores of both the Atlantic and Pacific. Observations made in various years were combined into a single year and were averaged into $1° \times 1°$ pixels (approximately N-S 100 km by E-W 80 km) for the analysis. Accordingly, the results represent a climatological mean condition over the past 25 years. Finer resolutions (10×10 km) may be desirable for some areas close to shore because of outflow of estuarine and river waters and upwelling. However, for this study, which is aimed at a broad picture of waters surrounding the continent, the fine scale measurements have been incorporated into the $1° \times 1°$ pixels. In addition, data with salinities of less than 16.0 are considered to be inland waters and have been excluded from the analysis.

Climatological monthly and annual mean values for pCO_2 in each zone were computed first. Then, the air-sea pCO_2 difference, which represents the thermodynamic driving potential for air-sea CO_2 gas transfer, was estimated using the atmospheric CO_2 concentration data. Finally, the net air-sea CO_2 flux was computed using transfer coefficients estimated on the basis of climatological mean monthly wind speeds using the (wind speed)^2 formulation of Wanninkhof (1992). The transfer coefficient depends on the state of turbulence above and below the air-sea interface and is commonly parameterized as a function of wind speeds (corrected to 10 m above the sea surface). However, selection of wind data is problematic because wind speeds vary with the time scale (hourly, diurnal, or seasonal). For example, fluxes calculated for the South Atlantic Bight from 6-h mean wind speeds in the NCEP/NCAR version 2 file ($1° \times 1°$ mean) were lower than those estimated using the monthly mean. This discrepancy suggests that ships used commonly for coastal carbon studies tend to be small and, hence, are rarely at sea under high wind conditions, so observations are biased toward lower winds. Taking into account that the observations have been made infrequently over multiple years, the gas transfer coefficients estimated from climatological mean monthly wind speeds may be more representative. The Schmidt number is computed using measured SST and climatological mean salinity (DaSilva *et al.*, 1994). The flux values in a given month are then averaged to yield a climatological mean flux (and standard deviation) for each month. This procedure assumes implicitly that the seawater pCO_2 changes at much slower rates in space and time than the wind speed and that the seawater pCO_2 does not correlate with the wind speed.

GLOSSARY AND ACRONYMS

GLOSSARY

afforestation
the process of establishing trees on land that has lacked forest cover for a very long period of time or has never been forested

anthropogenic
human-induced

apparent consumption
the amount or quantity expressed by the following formula: production + imports – exports +/– changes in stocks

biomass
the mass of living organic matter (plant and animal) in an ecosystem; biomass also refers to organic matter (living and dead) available on a renewable basis for use as a fuel; biomass includes trees and plants (both terrestrial and aquatic), agricultural crops and wastes, wood and wood wastes, forest and mill residues, animal wastes, livestock operation residues, and some municipal and industrial wastes

carbon sequestration
the process of increasing the carbon content of a carbon reservoir other than the atmosphere; often used narrowly to refer to increasing the carbon content of carbon pools in the biosphere and distinguished from physical or chemical collection of carbon followed by injection into geologic reservoirs, which is generally referred to as "carbon capture and storage"

carbon cycle
the term used to describe the flow of carbon (in various forms such as carbon dioxide [CO_2], organic matter, and carbonates) through the atmosphere, ocean, terrestrial biosphere, and lithosphere

carbon equivalent
the amount of carbon in the form of CO_2 that would produce the same effect on the radiative balance of the Earth's climate system; applicable in this report to greenhouse gases such as methane (CH_4)

carbon intensity
the relative amount of carbon emitted per unit of energy or fuels consumed

coastal waters
the region within 100 km from shore in which processes unique to coastal marine environments influence the partial pressure of CO_2 in surface sea waters

CO_2 equivalent
the amount of CO_2 that would produce the same effect on the radiative balance of the Earth's climate system as another greenhouse gas, such as CH_4

CO_2 fertilization
the phenomenon in which plant growth increases (and agricultural crop yields increase) due to the increased rates of photosynthesis of plant species in response to elevated concentrations of CO_2 in the atmosphere

decarbonization
reduction in the use of carbon-based energy sources as a proportion of total energy supplies or increased use of carbon-based fuels with lower values of carbon content per unit of energy content

deforestation
the process of removing or clearing trees from forested land

dry climates
climates where the ratio of mean annual precipitation to potential evapotranspiration is less than 1.0

ecosystem
a community (*i.e.,* an assemblage of populations of plants, animals, fungi, and microorganisms that live in an environment and interact with one another, forming, together, a distinctive living system with its own composition, structure, environmental relations, development, and function) and its environment treated together as a functional system of complementary relationships and transfer and circulation of energy and matter

energy intensity
the relative amount or ratio of the consumption of energy to the resulting amount of output, service, or activity (*i.e.,* expressed as energy per unit of output)

feebates
systems of progressive vehicle taxes on purchases of less efficient new vehicles and subsidies for more efficient new vehicles

fossil fuels

fuels such as coal, petroleum, and natural gas derived from the chemical and physical transformation (fossilization) of the remains of plants and animals that lived during the Carboniferous Period 360–286 million years ago

global warming potential

a factor describing the radiative forcing impact (*e.g.,* warming of the atmosphere) of one unit mass of a given greenhouse gas relative to the warming caused by a similar mass of CO_2; CH_4, for example, has a GWP of 23

greenhouse gases

gases including water vapor, CO_2, CH_4, nitrous oxide, and halocarbons that trap infrared heat, warming the air near the surface and in the lower levels of the atmosphere

leakage

The part of emissions reductions in Annex B countries that may be offset by an increase of the emission in the non-constrained countries above their baseline levels. This can occur through (1) relocation of energy-intensive production in non-constrained regions; (2) increased consumption of fossil fuels in these regions through decline in the international price of oil and gas triggered by lower demand for these energies; and (3) changes in incomes (and thus in energy demand) because of better terms of trade. "Leakage" also refers to the situation in which a carbon sequestration activity (*e.g.,* tree planting) on one piece of land inadvertently, directly or indirectly, triggers an activity, which in whole or part, counteracts the carbon effects of the initial activity

mitigation

a human intervention to reduce the sources of or to enhance the sinks of greenhouse gases

net ecosystem exchange

the net flux of carbon between the land and the atmosphere, typically measured using eddy covariance techniques; note: NEE and NEP are equivalent terms but are not always identical because of measurement and scaling issues, and the sign conventions are reversed; positive values of NEE (net ecosystem exchange with the atmosphere) usually refer to carbon released to the atmosphere (*i.e.,* a source), and negative values refer to carbon uptake (*i.e.,* a sink)

net ecosystem production

the net carbon accumulation within the ecosystem after all gains and losses are accounted for, typically measured using ground-based techniques; by convention, positive values of NEP represent accumulations of carbon by the ecosystem, and negative values represent carbon loss

net primary production

the net uptake of carbon by plants in excess of respiratory loss

North America

the combined land area of Canada, the United States of America, and Mexico and their coastal waters

North American Carbon Program

a multidisciplinary research program, supported by a number of different U.S. federal agencies through a variety of intramural and extramural funding mechanisms and award instruments, to obtain scientific understanding of North America's carbon sources and sinks and of changes in carbon stocks needed to meet societal concerns and to provide tools for decision makers

ocean acidification

the phenomenon in which the pH of the oceans becomes more acidic due to increased levels of CO_2 in the atmosphere which, in turn, increase the amount of dissolved CO_2 in sea water

option

a choice among a set of possible measures or alternatives

peatlands

areas characterized as having an organic layer thickness of at least 40 cm.

permafrost

soils or rocks that remain below 0°C for at least two consecutive years

pool/reservoir

any natural region or zone, or any artificial holding area, containing an accumulation of carbon or carbon-bearing compounds or having the potential to accumulate such substances

reforestation

the process of establishing a new forest by planting or seeding trees in an area where trees have previously been removed

sink

in general, any process, activity, or mechanism which removes a greenhouse gas or a precursor of a greenhouse gas or aerosol from the atmosphere; in this report, a sink is any regime or pool in which the amount of carbon is increasing (*i.e.,* is being accumulated or stored)

source

in general, any process, activity, or mechanism which releases a greenhouse gas or a precursor of a greenhouse gas or aerosol into the atmosphere; in this report, a source is any regime or pool in which the amount of carbon is decreasing (*i.e.,* is being released or emitted)

stocks

the amount or quantity contained in the inventory of a pool or reservoir

temperate zones

regions of the earth's surface located above 30° latitude and below 66.5° latitude

trend

a systematic change over time

tropical zones

regions located between the earth's equator and 30° latitude (this area includes subtropical regions)

uncertainty

a term used to describe the range of possible values around a best estimate, sometimes expressed in terms of probability or likelihood (see Preface, this report)

wet climates

climates where the ratio of mean annual precipitation to potential evapotranspiration is greater than 1.0

wetlands

areas that are inundated or saturated by surface water or groundwater at a frequency and duration sufficient to support—and that, under normal circumstances, do support—a prevalence of vegetation typically adapted for life in saturated soil conditions, including swamps, marshes, bogs, and similar areas

ACRONYMS AND ABBREVIATIONS

μatm	microatmosphere (a measure of pressure)
ACEEE	American Council for an Energy-Efficient Economy
CAFE	Corporate Average Fuel Economy
CAIT	Climate Analysis Indicators Tool
CAST	Council for Agricultural Science and Technology
CBO	U.S. Congressional Budget Office
CCSP	U.S. Climate Change Science Program
CCTP	Climate Change Technology Program
CDIAC	Carbon Dioxide Information Analysis Center
CEC	California Energy Commission
CH_4	methane
CIEEDAC	Canadian Industrial Energy End-Use Data and Analysis Centre
CO	carbon monoxide
CO_2	carbon dioxide
CO_3	carbonate
COP	Conference of Parties
DOC	dissolved organic carbon
DOE	U.S. Department of Energy
DOT	U.S. Department of Transportation
EIA	Energy Information Administration
EPA	U.S. Environmental Protection Agency
ESCOs	energy services companies
FAO	Food and Agriculture Organization
FWMS	freshwater mineral-soil
g	gram
GAO	U.S. Government Accountability Office
GDP	gross domestic product
GHG	greenhouse gas
Gt C	gigatons of carbon (billions of metric tons; *i.e.,* petagrams)
GWP	global warming potential
ha	hectare
HCO_3	bicarbonate
ICLEI	International Council for Local Environmental Initiatives (now known as International Governments for Local Sustainability)
IOOS	Integrated Ocean Observing System
IPCC	Intergovernmental Panel on Climate Change

IWG	Interlaboratory Working Group	**UNFCCC**	United Nations Framework Convention on Climate Change
kg	kilogram		
km	kilometer	**USDA**	U.S. Department of Agriculture
L	liter	**VOCs**	volatile organic compounds
LEED	Leadership in Energy and Environment Design	**WBCSD**	World Business Council for Sustainable Development
m	meter		
MAP	mean annual precipitation		
Mt C	megatons of carbon (millions of metric tons; *i.e.*, teragrams)		
N$_2$O	nitrous oxide (also, dinitrogen oxide)		
NACP	North American Carbon Program		
NAO	North Atlantic oscillation		
NAS	U.S. National Academy of Sciences		
NASA	National Aeronautics and Space Administration		
NATS	North American Transportation Statistics		
NCAR	National Center for Atmospheric Research		
NCEP	National Centers for Environmental Prediction; National Commission on Energy Policy		
NEE	net ecosystem exchange		
NEP	net ecosystem productivity		
NGO	non-governmental organization		
NO$_2$	nitrogen dioxide		
NOAA	National Oceanic and Atmospheric Administration		
NO$_x$	oxides of nitrogen		
NPP	net primary productivity		
NRC	National Research Council		
NRCS	Natural Resources Conservation Service		
NSF	National Science Foundation		
NWI	National Wetland Inventory		
OCCC	Ocean Carbon and Climate Change		
pCO$_2$	partial pressure of carbon dioxide in units of microatmospheres or ppm		
PDO	Pacific decadal oscillation		
PET	potential evapotranspiration		
PJ	petajoules		
ppm	parts per million by volume		
PPP	purchasing power parity		
RGGI	Regional Greenhouse Gas Initiative		
SAP	Synthesis and Assessment Product		
SBSTA	Subsidiary Body for Scientific and Technological Advice		
SOCCR	State of the Carbon Cycle Report		
μatm	microatmospheres or 10^{-6} atmospheres		

EXECUTIVE SUMMARY REFERENCES

IPCC, 2000: *Land Use, Land-use Change and Forestry. A Special Report of the Intergovernmental Panel on Climate Change* [Watson, R.T., I.R. Noble, B. Bolin, N.H. Ravindranath, D.J. Verado, and D.J. Dokken (eds.)]. Cambridge University Press, Cambridge, United Kingdom and New York, NY, USA, 388 pp.

IPCC, 2001: *Climate Change 2001: The Scientific Basis.* Contribution of Working Group I to the Third Assessment Report of the Intergovernmental Panel on Climate Change [Houghton, J.T., Y. Ding, D.J. Griggs, M. Noguer, P.J. van der Linden, X. Dai, K. Maskell, and C.A. Johnson (eds.)]. Cambridge University Press, Cambridge, United Kingdom and New York, NY, USA, 881 pp.

IPCC, 2007: Summary for Policymakers. In: *Climate Change 2007: The Physical Science Basis. Contribution of Working Group I to the Fourth Assessment Report of the Intergovernmental Panel on Climate Change* [Solomon, S., D. Qin, M. Manning, Z. Chen, M. Marquis, K.B. Averyt, M.Tignor and H.L. Miller (eds.)]. Cambridge University Press, Cambridge, United Kingdom and New York, NY, USA.

Marland, G., T.A. Boden, and R.J. Andres, 2006: Global, regional, and national fossil fuel CO_2 emissions. In: *Trends: A Compendium of Data on Global Change.* Carbon Dioxide Information Analysis Center, Oak Ridge National Laboratory, U.S. Department of Energy, Oak Ridge, TN. Available at http://cdiac.esd.ornl.gov/trends/trends.htm

CHAPTER I REFERENCES

Caldeira, K. and M.E. Wickett, 2003: Anthropogenic carbon and ocean pH. *Nature,* **425(6956)**, 365-366.

Caldeira, K., M.G. Morgan, D. Baldocchi, P.G. Brewer, C.-T.A. Chen, G.-J. Nabuurs, N. Nakicenovic, and G.P. Robertson, 2004: A portfolio of carbon management options. In: *The Global Carbon Cycle: Integrating Humans, Climate, and the Natural World* [Field, C.B. and M.R. Raupach (eds.)]. Island Press, Washington, DC, pp. 103-129.

Canadell, J.G., D. Pataki, R. Gifford, R.A. Houghton, Y. Lou, M.R. Raupach, P. Smith, and W. Steffen, 2007: Saturation of the terrestrial carbon sink. In: *Terrestrial Ecosystems in a Changing World,* [Canadell, J.G., D. Pataki, and L. Pitelka (eds.)]. The IGBP Series. Springer-Verlag, Berlin Heidelberg, pp. 59-78.

Cash, D. and W. Clark, 2001: *From Science to Policy: Assessing the Assessment Process.* Faculty Research Working Paper 01-045, Kennedy School of Government, Harvard University, Cambridge, MA. Available at http://ksgnotes1.harvard.edu/Research/wpaper.nsf/RWP/RWP01-045

Cash, D., W. Clark, F. Alcock, N. Dickson, N. Eckley, D. Guston, J. Jäger, and R. Mitchell, 2003: Knowledge systems for sustainable development. *Proceedings of the National Academy of Sciences,* **100(14)**, 8086-8091.

Casperson, J.P., S.W. Pacala, J.C. Jenkins, G.C. Hurtt, P.R. Moorcraft, and R.A. Birdsey, 2000: Contributions of land-use history to carbon accumulation in U.S. Forests. *Science,* **290(5494)**, 1148-1151.

CCSP (U.S. Climate Change Science Program), 2003: *Strategic Plan for the U.S. Climate Change Science Program.* A Report by the Climate Change Science Program and the Subcommittee on Global Change Research, Climate Change Science Program Office, Washington, DC, 211 pp.

DOE (U.S. Department of Energy), 1997: *Technology Opportunities to Reduce Greenhouse Gas Emissions.* U.S. Department of Energy, Washington, DC, 95 pp.

Dilling, L., S.C. Doney, J. Edmonds, K.R. Gurney, R.C. Harriss, D. Schimel, B. Stephens, and G. Stokes, 2003: The role of carbon cycle observations and knowledge in carbon management. *Annual Review of Environment and Resources,* **28**, 521-558.

Ehhalt, D., M. Prather, F. Dentener, E. Dlugokencky, E. Holland, I. Isaksen, J. Katima, V. Kirchhoff, P. Matson, P. Midgley, and M. Wang, 2001: Atmospheric chemistry and greenhouse gases. In: *Climate Change 2001: The Scientific Basis.* Contribution of Working Group I to the Third Assessment Report of the Intergovernmental Panel on Climate Change [Houghton, J.T., Y. Ding, D.J. Griggs, M. Noguer, P.J. van der Linden, X. Dai, K. Maskell, and C.A. Johnson (eds.)]. Cambridge University Press, Cambridge, United Kingdom, pp. 239–287.

EIA (Energy Information Administration), 2005: *Historical Data Overview.* U.S. Department of Energy. Available at http://www.eia.doe.gov/overview_hd.html; http://cdiac.ornl.gov/ftp/trends/emis/meth-reg.htm

Fan, S., M. Gloor, J. Mahlman, S. Pacala, J. Sarmiento, T. Takahashi, and P. Tans, 1998: A large terrestrial carbon sink in North America implied by atmospheric and oceanic carbon dioxide data and models. *Science,* **282(5388)**, 442-446.

Foley, J.A. and N. Ramankutty, 2004: A primer on the terrestrial carbon cycle: hat we don't know but should. In: *The Global Carbon Cycle: Integrating Humans, Climate and the Natural World.* [Field, C.B. and M.R. Raupauch, (eds.)] Island Press, Washington DC, pp. 279-294.

Friedlingstein, P., P. Cox, R. Betts, L. Bopp, W. von Bloh, V. Brovkin, P. Cadule, S. Doney, M. Eby, I. Fung, G. Bala, J. John, C. Jones, F. Joos, T. Kato, M. Kawamiya, W. Knorr,

K. Lindsay, H.D. Matthews, T. Raddatz, P. Rayner, C. Reick, E. Roeckner, K.-G. Schnitzler, R. Schnur, K. Strassmann, A.J. Weaver, C. Yoshikawa, and N. Zeng, 2006: Climate-carbon cycle feedback analysis: results from the C⁴MIP model inter-comparison. *Journal of Climate,* **19(14)**, 3337-3353.

Goodale, C.L., M.J. Apps, R.A. Birdsey, C.B. Field, L.S. Heith, R.A. Houghton, J.C. Jenkins, G.H. Kholmaier, W. Kurz, S. Liu, G.-J. Nabuurs, S. Nilsson, and A.Z. Shvidenko, 2002: Forest carbon sinks in the Northern Hemisphere. *Ecological Applica-tions,* **12(3)**, 891-899.

Greenblatt, J.B. and J.L. Sarmiento, 2004: Variability and climate feedback mechanisms in ocean uptake of CO_2. In: *The Global Carbon Cycle: Integrating Humans, Climate, and the Natural World* [Field, C.B. and M.R. Raupach (eds.)]. Island Press, Washington, DC, pp. 257-275.

Gurney, K.R., R.M. Law, A.S. Denning, P.J. Rayner, D. Baker, P. Bousquet, L. Bruhwiler, Y.-H. Chen, P. Ciais, S. Fan, I.Y. Fung, M. Gloor, M. Heimann, K. Higuchi, J. John, T. Maki, S. Maksyutov, K. Masarie, P. Peylin, M. Prather, B.C. Pak, J. Randerson, J. Sarmiento, S. Taguchi, T. Takahashi, and C.-W. Yue, 2002: Towards robust regional estimates of CO_2 sourc-es and sinks using atmospheric transport models. *Nature,* **415(6872)**, 626-630.

Hoffert, M.I., K. Caldeira, G. Benford, D.R. Criswell, C. Green, H. Herzog, A.K. Jain, H.S. Kheshgi, K.S. Lackner, J.S. Lewis, H.D. Lightfoot, W. Manheimer, J.C. Mankins, M.E. Mauel, L.J. Perkins, M.E. Schlesinger, T. Volk, and T.M.L. Wigley, 2002: Advanced technology paths to global climate stability: energy for a greenhouse planet. *Science,* **298(5595)**, 981-987.

Houghton, R.A., J.L. Hackler, and K.T. Lawrence, 1999: The U.S. carbon budget: contributions from land-use change. *Sci-ence,* **285(5427)**, 574-578.

Houghton, R.A., 2002: Magnitude, distribution and causes of ter-restrial carbon sinks and some implications for policy. *Climate Policy,* **2**, 71-88.

Houghton, R.A., 2003: The contemporary carbon cycle. In: *Trea-tise on Geochemistry, Volume 8 Biogeochemistry* [Schlesinger, W.H. (ed.)]. Elsevier Ltd, New York, pp. 473-513.

IPCC, 2001: *Climate Change 2001: Synthesis Report.* A Contri-bution of Working Groups I, II, and III to the Third Assessment Report of the Intergovernmental Panel on Climate Change [Watson, R.T., and the Core Writing Team (eds.)]. Cambridge University Press, Cambridge, United Kingdom and New York, NY, 398 pp.

Jones, C.D., P.M. Cox, and C. Huntingford, 2006: Climate-carbon cycle feedbacks under stabilization: uncertainty and observa-tional constraints. *Tellus B,* **58(5)**, 603-613.

Joos, F. and I.C. Prentice, 2004: A paleo-perspective on changes in atmospheric CO_2 and climate. In: *The Global Carbon Cycle: Integrating Humans, Climate, and the Natural World* [Field, C.B. and M.R. Raupach (eds.)]. Island Press, Washington, DC, pp. 165-186.

Keeling, C.D. and T.P. Whorf, 2005: Atmospheric CO_2 records from sites in the SIO air sampling network. In: *Trends: A Com-pendium of Data on Global Change.* Carbon Dioxide Informa-tion Analysis Center, Oak Ridge National Laboratory, U.S. De-partment of Energy, Oak Ridge, TN. Available at http://cdiac. esd.ornl.gov/trends/trends.htm

Kirschbaum, M.U.F. and A.L. Cowie, 2004: Giving credit where credit is due: a practical method to distinguish between human and natural factors in carbon accounting. *Climatic Change,* **67(2-3)**, 417-436.

Liu, K.K., K. Iseki, and S.-Y. Chao, 2000: Continental margin car-bon fluxes. In: *The Changing Ocean Carbon Cycle* [Hansen, R., H.W. Ducklow, and J.G. Field (eds.)]. Cambridge Univer-sity Press, Cambridge, United Kingdom, pp. 187–239.

Marland, G., T.A. Boden, and R.J. Andres, 2003: Global, region-al, and national CO_2 emissions. In: *Trends: A Compendium of Data on Global Change.* Carbon Dioxide Information Analy-sis Center, Oak Ridge National Laboratory, U.S. Department of Energy, Oak Ridge, TN, USA. Available at http://cdiac.esd. ornl.gov/trends/trends.htm

Marland, G., T.A. Boden, and R.J. Andres, 2006: Global, region-al, and national CO_2 emissions. In: *Trends: A Compendium of Data on Global Change.* Carbon Dioxide Information Analy-sis Center, Oak Ridge National Laboratory, U.S. Department of Energy, Oak Ridge, TN, USA. Available at http://cdiac.esd. ornl.gov/trends/trends.htm

Nowak, R.S., D.S. Ellsworth, and S.D. Smith, 2004: Functional responses of plants to elevated atmospheric CO_2- do photosyn-thetic and productivity data from FACE experiments support early predictions? *New Phytologist,* **162(2)**, 253-280.

Orr, J.C., V.J. Fabry, O. Aumont, L. Bopp, S.C. Doney, R.A. Feely, A. Gnanadesikan, N. Gruber, A. Ishida, F. Joos, R.M. Key, K. Lindsay, E. Maier-Reimer, R. Matear, P. Monfray, A. Mouchet, R.G. Najjar, G.K. Plattner, K.B. Rodgers, C.L. Sa-bine, J.L. Sarmiento, R. Schlitzer, R.D. Slater, I.J. Totterdell, M.F. Weirig, Y. Yamanaka, and A. Yool, 2005: Anthropogenic ocean acidification over the twenty-first century and its impact on calcifying organisms, *Nature,* **437(7059)**, 681-686.

Pacala, S.W., G.C. Hurtt, D. Baker, P. Peylin, R.A. Houghton, R.A. Birdsey, L. Heath, E.T. Sundquist, R.F. Stallard, P. Ci-ais, P. Moorcroft, J.P. Caspersen, E. Shevliakova, B. Moore, G. Kohlmaier, E. Holland, M. Gloor, M.E. Harmon, S.M. Fan, J.L. Sarmiento, C.L. Goodale, D. Schimel, and C.B. Field, 2001: Consistent land- and atmosphere-based U.S. carbon sink estimates. *Science,* **292(5525)**, 2316-2320.

Parson, E.A., 2003: *Protecting the Ozone Layer: Science and Strategy.* Oxford University Press, Oxford, United Kingdom, 400 pp.

Prentice, I.C., G.D. Farquhar, M.J.R. Fasham, M.L. Goulden, M. Heimann, V.J. Jaramillo, H.S. Kheshgi, C. Le Quéré, R.J. Scholes, and D.W.R. Wallace, 2001: The carbon cycle and atmospheric carbon dioxide. In: *Climate Change 2001: The Scientific Basis.* Contribution of Working Group I to the Third Assessment Report of the Intergovernmental Panel on Climate Change [J. T. Houghton, Y. Ding, D. J. Griggs, M. Noguer, P. J. van der Linden, X. Dai, K. Maskell, and C. A. Johnson (eds.)].

Cambridge University Press, Cambridge, United Kingdom and New York, pp. 183-237.

Prinn, R.G., 2004: Non-CO₂ greenhouse gases. In: *The Global Carbon Cycle: Integrating Humans, Climate, and the Natural World* [Field, C.B. and M.R. Raupach (eds.)]. Island Press, Washington, DC, pp. 205-216.

Running, S.W., R.R. Nemani, F.A. Heinsch, M. Zhao, M. Reeves, and H. Hashimoto, 2004: A continuous satellite-derived measure of global terrestrial primary production. *BioScience*, **54(6)**, 547-560.

Sabine, C.L., M. Heiman, P. Artaxo, D.C.E. Bakker, C.-T.A. Chen, C.B. Field, N. Gruber, C. LeQuéré, R.G. Prinn, J.E. Richey, P. Romero-Lankao, J.A. Sathaye, and R. Valentini, 2004: Current status and past trends of the carbon cycle. In: *The Global Carbon Cycle: Integrating Humans, Climate, and the Natural World* [Field, C.B. and M.R. Raupach (eds.)]. Island Press, Washington, DC, pp. 17-44.

Schaefer, K., A.S. Denning, N. Suits, J. Kaduk, I. Baker, S. Los, and L. Prihodko, 2002: Effect of climate on interannual variability of terrestrial CO₂ fluxes. *Global Biogeochemical Cycles*, **16**, 1102, doi:10.1029/2002GB001928.

Schimel, D.S., J. Melillo, H. Tian, A.D. McGuire, D. Kicklighter, T. Kittel, N. Rosenbloom, S. Running, P. Thornton, D. Ojima, W. Parton, R. Kelly, M. Sykes, R. Neilson, and B. Rizzo, 2000: Contribution of increasing CO₂ and climate to carbon storage by ecosystems in the United States. *Science*, **287(5460)**, 2004-2006.

Sundquist, E.T. and K. Visser, 2003: The geological history of the carbon cycle. In: *Treatise on Geochemistry, Volume 8 Biogeochemistry* [Schlesinger, W.H. (ed.)]. Elsevier Ltd, New York, pp. 425-472.

Turner, D.P., G.J. Koerper, M.E. Harmon, and J.J. Lee, 1995: A carbon budget for forests of the conterminous United States. *Ecological Applications*, **5(2)**, 421-436.

CHAPTER 2 REFERENCES

Andres, R.J., D.J. Fielding, G. Marland, T.A. Boden, N. Kumar, and A.T. Kearney, 1999: Carbon dioxide emissions from fossil-fuel use, 1751-1950. *Tellus B*, **51(4)**, 759-765.

Archer, D., H. Kheshgi, and E. Maier-Reimer, 1998: Dynamics of fossil fuel CO₂ neutralization by marine CaCO₃. *Global Biogeochemical Cycles*, **12(2)**, 259-276.

Bacastow, R. and C.D. Keeling, 1973: Atmospheric carbon dioxide and radiocarbon in the natural carbon cycle. II. Changes from A.D. 1700 to 2070 as deduced from a geochemical reservoir. In: *Carbon and the Biosphere* [Woodwell, G.M. and E.V. Pecan (eds.)]. U.S. Department of Commerce, Springfield, VA, pp. 86-135.

Bachelet, D., R.P. Neilson, T. Hickler, R.J. Drapek, J.M. Lenihan, M.T. Sykes, B. Smith, S. Sitch, and K. Thonicke, 2003: Simulating past and future dynamics of natural ecosystems in the United States. *Global Biogeochemical Cycles*, **17(2)**, 1045, doi:10.1029/2001GB001508.

Baker, D.F., R.M. Law, K.R. Gurney, P. Rayner, P. Peylin, A.S. Denning, P. Bousquet, L. Bruhwiler, Y.H. Chen, P. Ciais, I.Y. Fung, M. Heimann, J. John, T. Maki, S. Maksyutov, K. Masarie, M. Prather, B. Pak, S. Taguchi, and Z. Zhu, 2006: TransCom 3 inversion intercomparison: impact of transport model errors on the interannual variability of regional CO₂ fluxes, 1988-2003. *Global Biogeochemical Cycles*, **20**, GB1002, doi:10.1029/2004GB002439.

Baldocchi, D. and R. Valentini, 2004: Geographic and temporal variation of carbon exchange by ecosystems and their sensitivity to environmental perturbations. In: *The Global Carbon Cycle: Integrating Humans, Climate, and the Natural World* [Field, C.B. and M.R. Raupach (eds.)]. Island Press, Washington, DC, pp. 295-316.

Birdsey, R.A. and L.S. Heath, 1995: Carbon changes in U.S. forests. In: *Productivity of America's Forests and Climate Change* [Joyce, L.A. (ed.)]. General Technical Report RM-GTR-271, U.S. Department of Agriculture, Forest Service, Rocky Mountain Forest and Range Experiment Station, Fort Collins, CO, pp. 56-70.

Bousquet, P., P. Peylin, P. Ciais, C.L. Quéré, P. Friedlingstein, and P.P. Tans, 2000: Regional changes in carbon dioxide fluxes of land and oceans since 1980. *Science*, **290(5495)**, 1342-1346.

Broecker, W.S., T.H. Peng, and T. Takahashi, 1980: A strategy for the use of bomb-produced radiocarbon as a tracer for the transport of fossil fuel CO₂ into the deep-sea source regions. *Earth and Planetary Science Letters*, **49(2)**, 463-468.

Caldeira, K. and M.E. Wickett, 2003: Anthropogenic carbon and ocean pH. *Nature,* **425(6956)**, 365-366.

Chen, J.M., W. Ju, J. Cihlar, D. Price, J. Liu, W. Chen, J. Pan, A. Black, and A. Barr, 2003: Spatial distribution of carbon sources and sinks in Canada's forests. *Tellus B*, **55(2)**, 622-641.

Cramer, W., A. Bondeau, F.I. Woodward, I.C. Prentice, R.A. Betts, V. Brovkin, P.M. Cox, V.A. Fisher, J.A. Foley, A.D. Friend, and C. Kucharik, 2001: Global response of terrestrial ecosystem structure and function to CO₂ and climate change: results from six dynamic global vegetation models. *Global Change Biology*, **7(4)**, 357-373.

Cramer, W., D.W. Kicklighter, A. Bondeau, B. Moore III, G. Churkina, B. Nemry, A. Ruimy, A.L. Schloss, J. Kaduk, and participants of the Potsdam NPP Model Intercomparison, 1999: Comparing global models of terrestrial net primary productivity (NPP): overview and key results. *Global Change Biology*, **5(Suppl. 1)**, 1-15.

DeFries, R.S., C.B. Field, I. Fung, J. Collatz, and L. Bounoua, 1999: Combining satellite data and biogeochemical models to estimate global effects of human-induced land cover change on carbon emissions and primary productivity. *Global Biogeochemical Cycles*, **13(3)**, 803-815.

DOE EIA (U.S. Department of Energy, Energy Information Administration), 2006. Available at http://www.eia.doe.gov/environment.html

Dukes, J., 2003: Burning buried sunshine: human consumption of ancient solar energy. *Climatic Change*, **61(1-2)**, 31-44.

Enting, I.G., 2002: *Inverse Problems in Atmospheric Constituent Transport.* Cambridge University Press, Cambridge, Untied Kingdom, 392 pp.

Falkowski, P.G., M.E. Katz, A.J. Milligan, K. Fennel, B.S. Cramer, M.P. Aubry, R.A. Berner, M.J. Novacek, and W.M. Zapol, 2005: The rise of oxygen over the past 205 million years and the evolution of large placental mammals. *Science,* **309(5744),** 2202-2204.

Feely, R.A., C.L. Sabine, K. Lee, W. Berelson, J. Kleypas, V.J. Fabry, and F.J. Millero, 2004: Impact of anthropogenic CO_2 on the $CaCO_3$ system in the oceans. *Science,* **305(5682),** 362-366.

Friedli, H., H. Lötscher, H. Oeschger, U. Siegenthaler, and B. Stauffer, 1986: Ice core record of $^{13}C/^{12}C$ ratio of atmospheric CO_2 in the past two centuries. *Nature,* **324(6094),** 237-238.

Giampietro, M., S. Ulgiati, and D. Pimentel, 1997: Feasibility of large-scale biofuel production: does an enlargement of scale change the picture? *Bioscience,* **47(9),** 587-600.

Gloor, M., N. Gruber, J. Sarmiento, C.L. Sabine, R.A. Feely, and C. Rodenbeck, 2003: A first estimate of present and preindustrial air-sea CO_2 flux patterns based on ocean interior carbon measurements and models. *Geophysical Research Letters,* **30(1),** 1010, doi:10.1029/2002GL015594.

Goodale, C.L., M.J. Apps, R.A. Birdsey, C.B. Field, L.S. Heath, R.A. Houghton, J.C. Jenkins, G.H. Kohlmaier, W. Kurz, S.R. Liu, G.J. Nabuurs, S. Nilsson, and A.Z. Shvidenko, 2002: Forest carbon sinks in the Northern Hemisphere. *Ecological Applications,* **12(3),** 891-899.

Graham, P.J., 2003: Potential for climate change mitigation through afforestation: an economic analysis of fossil fuel substitution and carbon sequestration benefits. *Agroforestry Systems,* **59(1),** 85-95.

Greenblatt, J.B. and J.L. Sarmiento, 2004: Variability and climate feedback mechanisms in ocean uptake of CO_2. In: *The Global Carbon Cycle: Integrating Humans, Climate, and the Natural World* [Field, C.B. and M.R. Raupach (eds.)]. Island Press, Washington, DC, pp. 257-275.

Gruber, N., P. Friedlingstein, C.B. Field, R. Valentini, M. Heimann, J.E. Richey, P. Romero-Lankao, E.-D. Schulze, and C.-T.A. Chen, 2004: The vulnerability of the carbon cycle in the 21st century: an assessment of carbon-climate-human interactions. In: *The Global Carbon Cycle: Integrating Humans, Climate, and the Natural World* [Field, C.B. and M.R. Raupach (eds.)]. Island Press, Washington, DC, pp. 45-76.

Gurney, K.R., R.M. Law, A.S. Denning, P.J. Rayner, D. Baker, P. Bousquet, L. Bruhwiler, Y.H. Chen, P. Ciais, S.M. Fan, I.Y. Fung, M. Gloor, M. Heimann, K. Higuchi, J. John, E. Kowalczyk, T. Maki, S. Maksyutov, P. Peylin, M. Prather, B.C. Pak, J. Sarmiento, S. Taguchi, T. Takahashi, and C.W. Yuen, 2003: TransCom 3 CO_2 inversion intercomparison: 1. annual mean control results and sensitivity to transport and prior flux information. *Tellus B,* **55(2),** 555-579.

Gurney, K.R., R.M. Law, A.S. Denning, P.J. Rayner, B.C. Pak, D. Baker, P. Bousquet, L. Bruhwiler, Y.H. Chen, P. Ciais, I.Y.

Fung, M. Heimann, J. John, T. Maki, S. Maksyutov, P. Peylin, M. Prather, and S. Taguchi, 2004: Transcom 3 inversion intercomparison: model mean results for the estimation of seasonal carbon sources and sinks. *Global Biogeochemical Cycles,* **18,** GB1010, doi:10.1029/2003GB002111.

Hansen, J., L. Nazarenko, R. Ruedy, M. Sato, J. Willis, A. Del Genio, D. Koch, A. Lacis, K. Lo, S. Menon, T. Novakov, J. Perlwitz, G. Russell, G.A. Schmidt, and N. Tausnev, 2005: Earth's energy imbalance: confirmation and implications. *Science,* **308(5727),** 1431-1435.

Heath, J., E. Ayres, M. Possell, R.D. Bardgett, H.I.J. Black, H. Grant, P. Ineson, and G. Kerstiens, 2005: Rising atmospheric CO_2 reduces sequestration of root-derived soil carbon. *Science,* **309(5741),** 1711-1713.

Hegerl, G.C., F.W. Zwiers, P. Braconnot, N.P. Gillett, Y. Luo, J.A.M. Orsini, N. Nicholls, J.E. Penner, and P.A. Stott, 2007: Understanding and Attributing Climate Change. In: *Climate Change 2007: The Physical Science Basis.* Contribution of Working Group I to Fourth Assessment Report of the Intergovernmental Panel on Climate Change [Solomon, S., Qin, D., Manning, M., Chen, Z., Marquis, M., Averyt, K. B., Tignor, M., and Miller, H. L. (eds.)]. Cambridge University Press, Cambridge, United Kingdom and New York, pp. 665-745.

Hoffert, M.I., K. Caldeira, G. Benford, D.R. Criswell, C. Green, H. Herzog, A.K. Jain, H.S. Kheshgi, K.S. Lackner, J.S. Lewis, H.D. Lightfoot, W. Manheimer, J.C. Mankins, M.E. Mauel, L.J. Perkins, M.E. Schlesinger, T. Volk, and T.M.L. Wigley, 2002: Advanced technology paths to global climate stability: energy for a greenhouse planet. *Science,* **298(5595),** 981-987.

Houghton, R.A. 1999: The annual net flux of carbon to the atmosphere from changes in land use 1850-1990. *Tellus B,* **51(2),** 298-313.

Houghton, R.A., J.L. Hackler, and K.T. Lawrence, 1999: The U.S. carbon budget: contributions from land-use change. *Science,* **285(5427),** 574-578.

Jacobson, A.R., S.E. Mikaloff-Fletcher, N. Gruber, J.L. Sarmiento, M. Gloor, and TransCom Modelers, 2007: A joint atmosphere-ocean inversion for surface fluxes of carbon dioxide. 1. Methods and global-scale fluxes. *Global Biogeochemical Cycles,* **21,** doi:10.1029/2005GB002556.

Joos, F. and I.C. Prentice, 2004: A paleo-perspective on changes in atmospheric CO_2 and climate. In: *The Global Carbon Cycle: Integrating Humans, Climate, and the Natural World* [Field, C.B. and M.R. Raupach (eds.)]. Island Press, Washington, DC, pp. 165-186.

Keeling, C.D., R.B. Bacastow, A.E. Bainbridge, C.A. Ekdahl, P.R. Guenther, and L.S. Waterman, 1976: Atmospheric carbon dioxide variations at Mauna Loa Observatory, Hawaii. *Tellus,* **28(6),** 538-551.

Keeling, R.F., S.C. Piper, and M. Heimann, 1996: Global and hemispheric CO_2 sinks deduced from changes in atmospheric O_2 concentration. *Nature,* **381(6579),** 218-221.

Keeling, R.F. and B.B. Stephens, 2001: Antarctic sea ice and the control of Pleistocene climate instability. *Paleoceanography*, **16(1)**, 112-131.

Kirschbaum, M.U.F., 2003: To sink or burn? A discussion of the potential contributions of forests to greenhouse gas balances through storing carbon or providing biofuels. *Biomass and Bioenergy*, **24(4-5)**, 297-310.

Kurz, W.A. and M.J. Apps, 1999: A 70-year retrospective analysis of carbon fluxes in the Canadian forest sector. *Ecological Applications*, **9(2)**, 526-547.

Law, R.M., Y.-H. Chen, K.R. Gurney, and TransCom 3 Modellers, 2003: TransCom 3 CO_2 inversion intercomparison: 2. sensitivity of annual mean results to data choices. *Tellus B*, **55(2)**, 580-595.

Le Quéré, C. and N. Metzl, 2004: Natural processes regulating the ocean uptake of CO_2. In: *The Global Carbon Cycle: Integrating Humans, Climate, and the Natural World* [Field, C.B. and M.R. Raupach (eds.)]. Island Press, Washington, DC, pp. 243-256.

Marland, G. and R.M. Rotty, 1984: Carbon dioxide emissions from fossil fuels: a procedure for estimation and results for 1950-1982. *Tellus B*, **36(4)**, 232-261.

Martin, J.H., 1990: Glacial-interglacial CO_2 change: the iron hypothesis. *Paleoceanography*, **5(1)**, 1-13.

Masarie, K.A. and P.P. Tans, 1995: Extension and integration of atmospheric carbon dioxide data into a globally consistent measurement record. *Journal of Geophysical Research (Atmospheres)*, **100(D6)**, 11593-11610.

Matear, R.J. and B.I. McNeil, 2003: Decadal accumulation of anthropogenic CO_2 in the Southern Ocean: a comparison of CFC-age derived estimates to multiple-linear regression estimates. *Global Biogeochemical Cycles*, **17(4)**, 1113, doi:10.1029/2003GB002089.

Matsumoto, K., J.L. Sarmiento, R.M. Key, O. Aumont, J.L. Bullister, K. Caldeira, J.M. Campin, S.C. Doney, H. Drange, J.C. Dutay, M. Follows, Y. Gao, A. Gnanadesikan, N. Gruber, A. Ishida, F. Joos, K. Lindsay, E. Maier-Reimer, J.C. Marshall, R.J. Matear, P. Monfray, A. Mouchet, R. Najjar, G.K. Plattner, R. Schlitzer, R. Slater, P.S. Swathi, I.J. Totterdell, M.F. Weirig, Y. Yamanaka, A. Yool, and J.C. Orr, 2004: Evaluation of ocean carbon cycle models with data-based metrics. *Geophysical Research Letters*, **31**, L07303-07304.

McGillis, W.R., J.B. Edson, J.E. Hare, and C.W. Fairall. 2001: Direct covariance air-sea CO_2 fluxes. *Journal of Geophysical Research*, **106(C8)**:16729-16745.

Newsam, G.N. and I.G. Enting, 1988: Inverse problems in atmospheric constituent studies: I. determination of surface sources under a diffusive transport approximation. *Inverse Problems*, **4(4)**, 1037-1054.

Norby, R.J., E.H. DeLucia, B. Gielen, C. Calfapietra, C.P. Giardina, J.S. King, J. Ledford, H.R. McCarthy, D.J.P. Moore, R. Ceulemans, P. De Angelis, A.C. Finzi, D.F. Karnosky, M.E. Kubiske, M. Lukac, K.S. Pregitzer, G.E. Scarascia-Mugnozza, W.H. Schlesinger, and R. Oren, 2005: Forest response to elevated CO_2 is conserved across a broad range of productivity. *Proceedings of the National Academy of Sciences*, **102(50)**, 18052-18056.

Oren, R., D.S. Ellsworth, K.H. Johnsen, N. Phillips, B.E. Ewers, C. Maier, K.V.R. Schäfer, H. McCarthy, G. Hendrey, S.G. McNulty, and G.G. Katul, 2001: Soil fertility limits carbon sequestration by forest ecosystems in a CO_2-enriched atmosphere. *Nature*, **411(6836)**, 469-472.

Orr, J.C., V.J. Fabry, O. Aumont, L. Bopp, S.C. Doney, R.A. Feely, A. Gnanadesikan, N. Gruber, A. Ishida, F. Joos, R.M. Key, K. Lindsay, E. Maier-Reimer, R. Matear, P. Monfray, A. Mouchet, R.G. Najjar, G.K. Plattner, K.B. Rodgers, C.L. Sabine, J.L. Sarmiento, R. Schlitzer, R.D. Slater, I.J. Totterdell, M.F. Weirig, Y. Yamanaka, and A. Yool, 2005: Anthropogenic ocean acidification over the twenty-first century and its impact on calcifying organisms. *Nature*, **437(7059)**, 681-686.

Pacala, S. and R. Socolow, 2004: Stabilization wedges: solving the climate problem for the next 50 years with current technologies. *Science*, **305(5686)**, 968-972.

Pacala, S.W., G.C. Hurtt, D. Baker, P. Peylin, R.A. Houghton, R.A. Birdsey, L. Heath, E.T. Sundquist, R.F. Stallard, P. Ciais, P. Moorcroft, J.P. Caspersen, E. Shevliakova, B. Moore, G. Kohlmaier, E. Holland, M. Gloor, M.E. Harmon, S.M. Fan, J.L. Sarmiento, C.L. Goodale, D. Schimel, and C.B. Field, 2001: Consistent land- and atmosphere-based U.S. carbon sink estimates. *Science*, **292(5525)**, 2316-2320.

Petit, J.R., J. Jouzel, D. Raynaud, N.I. Barkov, J.-M. Barnola, I. Basile, M. Bender, J. Chappellaz, M. Davis, G. Delaygue, M. Delmotte, V.M. Kotlyakov, M. Legrand, V.Y. Lipenkov, C. Lorius, L. Pépin, C. Ritz, E. Saltzman, and M. Stievenard, 1999: Climate and atmospheric history of the past 420,000 years from the Vostok ice core, Antarctica. *Nature*, **399(6735)**, 429-436.

Peylin, P., P. Bousquet, C. Le Quere, S. Sitch, P. Friedlingstein, G. McKinley, N. Gruber, P. Rayner, and P. Ciais, 2005: Multiple constraints on regional CO_2 flux variations over land and oceans. *Global Biogeochemical Cycles*, **19**, GB1011, doi:10.1029/2003GB002214.

Post, W.M., T.H. Peng, W.R. Emanuel, A.W. King, V.H. Dale, and D.L. Deangelis, 1990: The global carbon cycle. *American Scientist*, **78(4)**, 310-326.

Prentice, I.C., G.D. Farquhar, M.J.R. Fasham, M.L. Goulden, M. Heimann, V.J. Jaramillo, H.S. Kheshgi, C. Le Quéré, R.J. Scholes, and D.W.R. Wallace, 2001: The carbon cycle and atmospheric carbon dioxide. In: *Climate Change 2001: The Scientific Basis*. Contribution of Working Group I to the Third Assessment Report of the Intergovernmental Panel on Climate Change [J. T. Houghton, Y. Ding, D. J. Griggs, M. Noguer, P. J. van der Linden, X. Dai, K. Maskell, and C. A. Johnson (eds.)]. Cambridge University Press, Cambridge, United Kingdom and New York, pp. 183-237.

Rodenbeck, C., S. Houweling, M. Gloor, and M. Heimann, 2003: CO_2 flux history 1982-2001 inferred from atmospheric data using a global inversion of atmospheric transport. *Atmospheric Chemistry and Physics*, **3(6)**, 1919-1964.

Rousteenoja, K., T.R. Carter, K. Jylha, and H. Tuomenvirta, 2003: *Future Climate in World Regions: An Intercomparison of Model-Based Projections for the New IPCC Emissions Scenarios*. Finnish Environment Institute, Helsinki, pp. 83.

Running, S.W., R.R. Nemani, F.A. Heinsch, M.S. Zhao, M. Reeves, and H. Hashimoto, 2004: A continuous satellite-derived measure of global terrestrial primary production. *Bioscience*, **54(6)**, 547-560.

Sabine, C.L., R.A. Feely, N. Gruber, R.M. Key, K. Lee, J.L. Bullister, R. Wanninkhof, C.S. Wong, D.W.R. Wallace, B. Tilbrook, F.J. Millero, T.H. Peng, A. Kozyr, T. Ono, and A.F. Rios, 2004a: The oceanic sink for anthropogenic CO_2. *Science*, **305(5682)**, 367-371.

Sabine, C.L., M. Heiman, P. Artaxo, D.C.E. Bakker, C.-T.A. Chen, C.B. Field, N. Gruber, C. LeQuéré, R.G. Prinn, J.E. Richey, P. Romero-Lankao, J.A. Sathaye, and R. Valentini, 2004b: Current status and past trends of the carbon cycle. In: *The Global Carbon Cycle: Integrating Humans, Climate, and the Natural World* [Field, C.B. and M.R. Raupach (eds.)]. Island Press, Washington, DC, pp. 17-44.

Schimel, D., J. Melillo, H. Tian, A.D. McGuire, D. Kicklighter, T. Kittel, N. Rosenbloom, S. Running, P. Thornton, D. Ojima, W. Parton, R. Kelly, M. Sykes, R. Neilson, and B. Rizzo, 2000: Contribution of increasing CO_2 and climate to carbon storage by ecosystems in the United States. *Science*, **287(5460)**, 2004-2006.

Schimel, D.S., J.I. House, K.A. Hibbard, P. Bousquet, P. Ciais, P. Peylin, B.H. Braswell, M.J. Apps, D. Baker, A. Bondeau, J. Canadell, G. Churkina, W. Cramer, A.S. Denning, C.B. Field, P. Friedlingstein, C. Goodale, M. Heimann, R.A. Houghton, J.M. Melillo, B. Moore, D. Murdiyarso, I. Noble, S.W. Pacala, I.C. Prentice, M.R. Raupach, P.J. Rayner, R.J. Scholes, W.L. Steffen, and C. Wirth, 2001: Recent patterns and mechanisms of carbon exchange by terrestrial ecosystems. *Nature*, **414(6860)**, 169-172.

Shaw, M.R., E.S. Zavaleta, N.R. Chiariello, E.E. Cleland, H.A. Mooney, and C.B. Field, 2002: Grassland responses to global environmental changes suppressed by elevated CO_2. *Science*, **298(5600)**, 1987-1990.

Siegenthaler, U. and H. Oeschger, 1987: Biospheric CO_2 emissions during the past 200 years reconstructed by deconvolution of ice core data. *Tellus B*, **39(1-2)**, 140-154.

Sigman, D.M. and E.A. Boyle, 2000: Glacial/interglacial variations in atmospheric carbon dioxide. *Nature*, **407(6806)**, 859-869.

Takahashi, T., R.A. Feely, R.F. Weiss, R. Wanninkhof, D.W. Chipman, S.C. Sutherland, and T.T. Takahashi, 1997: Global air-sea flux of CO_2: an estimate based on measurements of sea-air pCO_2 difference. *Proceedings of the National Academy of Sciences*, **94(16)**, 8292-8299.

Takahashi, T., S.C. Sutherland, C. Sweeney, A. Poisson, N. Metzl, B. Tilbrook, N. Bates, R. Wanninkhof, R.A. Feely, C. Sabine, J. Olafsson, and Y. Nojiri, 2002: Global sea-air CO_2 flux based on climatological surface ocean pCO_2, and seasonal biological and temperature effects. *Deep-Sea Research II*, **49(9-10_**, 1601-1622.

Tarantola, A., 1987: *Inverse Problem Theory: Methods for Data Fitting and Model Parameter Estimation*. Elsevier, New York, 630 pp.

Thoning, K.W., P.P. Tans, and W.D. Komhyr, 1989: Atmospheric carbon dioxide at Mauna Loa Observatory 2. Analysis of the NOAA GMCC data, 1974-1985. *Journal of Geophysical Research*, **94(D6)**, 8549-8565.

van der Werf, G.R., J.T. Randerson, G.J. Collatz, L. Giglio, P.S. Kasibhatla, A.F. Arellano, S.C. Olsen, and E.S. Kasischk, 2004: Continental-scale partitioning of fire emissions during the 1997 to 2001 El Niño/La Niña period. *Science*, **303(5654)**, 73-74.

Wanninkhof, R. and W. McGillis, 1999: A cubic relationship between air-sea CO_2 exchange and wind speed. *Geophysical Research Letters*, **26(13)**, 1889-1892.

Wofsy, S.C., M.L. Goulden, J.W. Munger, S.-M. Fan, P.S. Bakwin, B.C. Daube, S.L. Bassow, and F.A. Bazzaz, 1993: Net exchange of CO_2 in a mid-latitude temperate forest. *Science*, **260(5112)**, 1314-1317.

CHAPTER 3 REFERENCES

Aldy, J.E., 2005: An environmental kuznets curve analysis of US state level carbon dioxide emissions. *Journal of Environment and Development*, **14(1)**, 58-72.

Ang, B.W. and F.Q. Zhang, 2000: A survey of index decomposition analysis in energy and environmental studies. *Energy*, **25(12)**, 1149-1176.

Bradley, B.A., R.A. Houghton, J.F. Mustard, and S.P. Hamburg, 2006: Invasive grass reduces aboveground carbon stocks in shrublands of the Western US. *Global Change Biology*, **12(10)**, 1815-1822.

Cairns, M.A., P.K. Haggerty, R. Alvarez, B.H.J. De Jong, and I. Olmsted, 2000: Tropical Mexico's recent land-use change: a region's contribution to the global carbon cycle. *Ecological Applications*, **10(5)**, 1426-1441.

Casler, S.D. and A.Z. Rose, 1998: Carbon dioxide emissions in the US economy. *Environmental and Resource Economics*, **11(3-4)**, 349-363.

Caspersen, J.P., S.W. Pacala, J.C. Jenkins, G.C. Hurtt, P.R. Moorcroft, and R.A. Birdsey, 2000: Contributions of land-use history to carbon accumulation in U.S. forests. *Science*, **290(5495)**, 1148-1151.

Cox, P.M., R.A. Betts, C.D. Jones, S.A. Spall, and I.J. Totterdell, 2000: Acceleration of global warming due to carbon-cycle feedbacks in a coupled climate model. *Nature*, **408(6809)**, 184-187.

Davis, W.B., A.H. Sanstad, and J.G. Koomey, 2002: Contributions of weather and fuel mix to recent declines in US energy and carbon intensity. *Energy Economics*, **25(4)**, 375-396.

Defries, R.S., R.A. Houghton, M.C. Hansen, C.B. Field, D. Skole, and J. Townshend, 2002: Carbon emissions from tropical deforestation and regrowth based on satellite observations for the 1980s and 1990s. *Proceedings of the National Academy of Sciences*, **99(22)**, 14256-14261.

EIA (Energy Information Administration), 2005: *Historical Data Overview*. U.S. Department of Energy. Available at http://www.eia.doe.gov/overview_hd.html; http://cdiac.ornl.gov/ftp/trends/emis/meth-reg.htm

Environment Canada, 2005: *Canada's Greenhouse Gas Inventory 1990-2003: Initial Submission*. Greenhouse Gas Division, Environment Canada, Ottawa, Ontario, Canada. Available at http://unfccc.int/national_reports/annex_i_ghg_inventories/national_inventories_submissions/items/2761.php

Environment Canada, 2006: *National Inventory Report 1990-2004: Greenhouse Gas Sources and Sinks in Canada*. Greenhouse Gas Division, Environment Canada, Ottowa, Ontario, Canada. Available at http://www.cc.gc.ca/pdb/ghg/inventory_report/2004_report/toc_e.cfm

Fan, S.-M., M. Gloor, J. Mahlman, S. Pacala, J. Sarmiento, T. Takahashi, and P. Tans, 1998: Atmospheric and oceanic CO_2 data and models imply a large terrestrial carbon sink in North America. *Science*, **282(5388)**, 442-446.

Flannigan, M.D., K.A. Logan, B.D. Amiro, W.R. Skinner, and B.J. Stocks, 2005: Future area burned in Canada. *Climatic Change*, **72(1)**, 1-16.

Gillett, N.P., A.J. Weaver, F.W. Zwiers, and M.D. Flannigan, 2004: Detecting the effect of climate change on Canadian forest fires. *Geophysical Research Letters*, **31**, L18211, doi:10.1029/2004GL020876

Goetz, S.J., A. Bunn, G. Fiske, and R.A. Houghton. 2005: Satellite-observed photosynthetic trends across boreal North America associated with climate and fire disturbance. *Proceedings of the National Academy of Sciences*, **102(38)**, 13521-13525.

Golove, W.H. and L.J. Schipper, 1998: Long-term trends in U.S. manufacturing energy consumption and carbon dioxide emissions. *Energy*, **21(7/8)**, 683-692.

Goodale, C.L., M.J. Apps, R.A. Birdsey, C.B. Field, L.S. Heath, R.A. Houghton, J.C. Jenkins, G.H. Kohlmaier, W. Kurz, S. Liu, G.J. Nabuurs, S. Nilsson, and A.Z. Shvidenko, 2002: Forest carbon sinks in the Northern Hemisphere. *Ecological Applications*, **12(3)**, 891-899.

Greening, L.A., W.B. Davis, L. Schipper, and M. Khrushch, 1997: Comparison of six decomposition methods: application to aggregate energy intensity for manufacturing in 10 OECD countries. *Energy Economics*, **19(3)**, 375-390.

Greening, L.A., W.B. Davis, and L. Schipper, 1998: Decomposition of aggregate carbon intensity for the manufacturing sector: comparison of declining trends from 10 OECD countries for the period 1971-1993. *Energy Economics*, **20(1)**, 43-65.

Greening, L.A., M. Ting, and W.B. Davis, 1999: Decomposition of aggregate carbon intensity for freight: trends from 10 OECD countries for the period 1971-1993. *Energy Economics*, **21(4)**, 331-361.

Greening, L.A., M. Ting, and T.J. Krackler, 2001: Effects of changes in residential end-uses on aggregate carbon intensity: comparison of 10 OECD countries for the period 1970 through 1993. *Energy Economics*, **23(2)**, 153-178.

Greening, L.A., 2004: Effects of human behavior on aggregate carbon intensity of personal transportation: comparison of 10 OECD countries for the period 1970-1993. *Energy Economics*, **26(1)**, 1-30.

Grossman, G.M. and A.B. Krueger, 1995: Economic growth and the environment. *Quarterly Journal of Economics*, **60(2)**, 353-375.

Guo, L.B. and R.M. Gifford, 2002: Soil carbon stocks and land use change: a meta analysis. *Global Change Biology*, **8(4)**, 345-360.

Gurney, K.R., R.M. Law, A.S. Denning, P.J. Rayner, B.C. Pak, D. Baker, P. Bousquet, L. Bruhwiler, Y.H. Chen, P. Ciais, I.Y. Fung, M. Heimann, J. John, T. Maki, S. Maksyutov, P. Peylin, M. Prather, and S. Taguchi, 2004: Transcom 3 inversion intercomparison: model mean results for the estimation of seasonal carbon sources and sinks. *Global Biogeochemical Cycles*, **18**, GB1010, doi:10.1029/2003GB002111.

Houghton, R.A., J.L. Hackler, and K.T. Lawrence, 1999: The U.S. carbon budget: contributions from land-use change. *Science*, **285(5427)**, 574-578.

Houghton, R.A. and J.L. Hackler, 2000: Changes in terrestrial carbon storage in the United States. 1. The roles of agriculture and forestry. *Global Ecology and Biogeography*, **9(12)**, 125-144.

Houghton, R.A., J.L. Hackler, and K.T. Lawrence, 2000: Changes in terrestrial carbon storage in the United States. 2. The role of fire and fire management. *Global Ecology and Biogeography*, **9(2)**, 145-170.

Hungate, B.A., J.S. Dukes, M.R. Shaw, Y. Luo, and C.B. Field, 2003: Nitrogen and climate change. *Science*, **302(5650)**, 1512-1513.

Hurtt, G.C., S.W. Pacala, P.R. Moorcroft, J. Caspersen, E. Shevliakova, R.A. Houghton, and B. Moore III, 2002: Projecting the future of the U.S. carbon sink. *Proceedings of the National Academy of Sciences*, **99(3)**, 1389-1394.

Jackson, R.B., J.L. Banner, E.G. Jobbagy, W.T. Pockman, and D.H. Wall, 2002: Ecosystem carbon loss with woody plant invasion of grasslands. *Nature*, **418(6898)**, 623-626.

Kahn, M.E., 2003: The geography of US pollution intensive trade: evidence from 1958 to 1994. *Regional Science and Urban Economics*, **33(4)**, 383-400.

Körner, C., R. Asshoff, O. Bignucolo, S. Hättenschwiler, S.G. Keel, S. Peláez-Riedl, S. Pepin, R.T.W. Siegwolf, and G. Zotz, 2005: Carbon flux and growth in mature deciduous forest trees exposed to elevated CO_2. *Science*, **309(5739)**, 1360-1362.

Lenzen, M., M. Wier, C. Cohen, H. Hayami, S. Pachauri, and R. Schaeffer, 2006: A comparative multivariate analysis of household energy requirements in Australia, Brazil, Denmark, India and Japan. *Energy*, **31(2-3)**, 181-207.

Lindmark, M., 2004: Patterns of historical CO_2 intensity transitions among high and low income countries. *Explorations in Economic History*, **41(4)**, 426-447.

Luo, Y., D. Hui, and D. Zhang, 2006: Elevated carbon dioxide stimulates net accumulations of carbon and nitrogen in terrestrial ecosystems: a meta-analysis. *Ecology*, **87(1)**, 53-63.

Masera, O.R., M.J. Ordóñez, and R. Dirzo, 1997: Carbon emissions from Mexican forests: current situation and long-term scenarios. *Climatic Change*, **35(3)**, 265-295.

Maddison, A., 2003: *The World Economy: Historical Statistics.* OECD, Paris, 384 pp.

Marland, G., T.A. Boden, and R.J. Andres, 2005: Global, regional, and national CO_2 emissions. In: *Trends: A Compendium of Data on Global Change.* Carbon Dioxide Information Analysis Center, Oak Ridge National Laboratory, U.S. Department of Energy, Oak Ridge, TN. Available at http://cdiac.esd.ornl.gov/trends/trends.htm

Mitchell, B.R., 1998: *International Historical Statistics: The Americas, 1750-1993.* Stockton Press, New York, 4th ed., 830 pp.

Oren, R., D.S. Ellsworth, K.H. Johnsen, N. Phillips, B.E. Ewers, C. Maier, K.V.R. Schäfer, H. McCarthy, G. Hendrey, S.G. McNulty, and G.G. Katul, 2001: Soil fertility limits carbon sequestration by forest ecosystems in a CO_2-enriched atmosphere. *Nature*, **411(6836)**, 469-472.

Pacala, S.W., G.C. Hurtt, D. Baker, P. Peylin, R.A. Houghton, R.A. Birdsey, L. Heath, E.T. Sundquist, R.F. Stallard, P. Ciais, P. Moorcroft, J.P. Caspersen, E. Shevliakova, B. Moore, G. Kohlmaier, E. Holland, M. Gloor, M.E. Harmon, S.M. Fan, J.L. Sarmiento, C.L. Goodale, D. Schimel, and C.B. Field, 2001: Consistent land- and atmosphere-based U.S. carbon sink estimates. *Science*, **292(5525)**, 2316-2320.

Pacala, S. and R. Socolow, 2004: Stabilization wedges: solving the climate problem for the next 50 years with current technologies. *Science*, **305(5686)**, 968-972.

Rothman, D.S., 1998: Environmental Kuznets curves—real progress or passing the buck: a case for consumption-based approaches. *Ecological Economics*, **25(2)**, 177-194.

Schaphoff, S., W. Lucht, D. Gerten, S. Sitch, W. Cramer, and I.C. Prentice, 2006: Terrestrial biosphere carbon storage under alternative climate projections. *Climatic Change*, **74(1-3)**, 97-122.

Selden, T.M. and D. Song, 1994: Environmental quality and development—is there a Kuznets curve for air pollution emissions? *Journal of Environmental Economics and Management*, **27(2)**, 147-162.

Skog, K.E. and G.A. Nicholson, 1998: Carbon cycling through wood products: the role of wood and paper products in carbon sequestration. *Forest Products Journal*, **48(7)**, 75-83. Available at http://www.fpl.fs.fed.us/documnts/pdf1998/skog98a.pdf

Skog, K.E., K. Pingoud, and J.E. Smith, 2004: A method countries can use to estimate changes in carbon stored in harvested wood products and the uncertainty of such estimates. *Environmental Management*, **33 (Supplement 1)**, S65-S73.

Smith, J.E. and L.S. Heath, 2005: Land use change and forestry and related sections. In: *Inventory of U.S. Greenhouse Gas Emissions and Sinks: 1990-2003.* Excerpted, EPA 430-R-05-003, U.S. Environmental Protection Agency. Available at http://yosemite.epa.gov/oar/globalwarming.nsf/content/ResourceCenterPublicationsGHGEmissions.html

Stallard, R.F., 1998: Terrestrial sedimentation and the carbon cycle: coupling weathering and erosion to carbon burial. *Global Biogeochemical Cycles*, **12(2)**, 231-257.

Suri, V. and D. Chapman, 1998: Economic growth, trade and energy: implications for the environmental Kuznets curve. *Ecological Economics*, **25(2)**, 195-208.

Westerling, A., H.G. Hidalgo, D.R. Cayan, and T.W. Swetnam, 2006: Warming and earlier Spring increase western U.S. forest wildfire activity. *Science*, **313(5789)**, 940-943.

World Forest Institute, 2006: *Wood Products Trade: North America* Available at http://wfi.worldforestrycenter.org/trade-2.htm

WRI (World Resources Institute), 2005: *EarthTrends—The Environmental Information Portal.* Available at http://earthtrends.wri.org/

CHAPTER 4 REFERENCES

AF&PA (American Forest & Paper Association and the U.S. Department of Energy) 2006: *Forest Products Industry Technology Roadmap.* Agenda 2020 Technology Alliance, Washington, DC, 54 pp.

Bohm, P. and C. Russell 1986: Comparative analysis of alternative policy instruments. In: *Handbook of Natural Resource and Energy Economics* [Kneese, A. and J. Sweeney (eds.)]. Vol. 2, Elsevier, New York, pp. 395-460.

Casler, S.D. and A. Rose, 1998. Carbon dioxide emissions in the U.S. economy: a structural decomposition analysis. *Environmental and Resource Economics*, **11(3-4)**, 349-363.

CBO (Congressional Budget Office), 2003: *The Economic Costs of Fuel Economy Standards Versus a Gasoline Tax.* Congress of the United States, Washington, DC, 24 pp.

CBO (Congressional Budget Office), 2006: *Evaluating the Role of Prices and R&D in Reducing Carbon Dioxide Emissions.* Congress of the United States, Washington, DC, 19 pp.

Croci, E. (ed.), 2005: *The Handbook of Environmental Voluntary Agreements: Design, Implementation and Evaluation Issues*, Springer, The Netherlands, 391 pp.

Darnall, N. and J. Carmin, 2003: *The Design and Rigor of U.S. Voluntary Environmental Programs: Results from the Survey*, North Carolina State University, Raleigh, 55 pp.

DOE (U.S. Department of Energy), 2006: *Carbon Sequestration R&D Overview*, CO_2 capture and storage costs. Available at http://www.fossil.energy.gov/programs/sequestration/overview.html

DOE/EERE (U.S. Department of Energy, Energy Efficiency and Renewable Energy), 2006: *Building America Puts Residential Building Research to Work.* Washington, DC. Available at http://www.eere.energy.gov/buildings/building_america/

DOE/EIA (Energy Information Administration), 2006: *Annual Energy Review 2006*: U.S. Department of Energy, , Washington, DC, 401 pp.

EIA (Energy Information Administration), 2003a: *International Energy Outlook: 2003*. DOE/EIA-0484(2003), U.S. Department of Energy, Washington, DC, 249 pp.

EIA (Energy Information Administration), 2003b: *Analysis of S.139, the Climate Stewardship Act of 2003*. SR/OIAF/2003-02, U.S. Department of Energy, Washington, DC, 515 pp.

EIA (Energy Information Administration), 2005. *Emissions of Greenhouse Gases in the United States, 2005.*, U.S. Department of Energy, Washington, DC, 106 pp.

Ellerman, D., P. Joskow, R. Schmalansee, J. Montero, and E. Bailey, 2000: *Markets for Clean Air: The U.S. Acid Rain Program*, Cambridge University Press, New York, 384 pp.

Energy Modeling Forum (EMF), 2000: *Costs of GHG Emissions Reduction*, Stanford University, Palo Alto, CA.

EPA (Environmental Protection Agency), 2005: *Greenhouse Gas Mitigation Potential in U.S. Forestry and Agriculture.* EPA-430-R-05-006, U.S. Environmental Protection Agency, Washington, DC, 154 pp.

Feng, Hongli, C.L. Kling, L.A. Kurkalova, and S. Secchi, 2003: Subsidies! The Other Incentive-Based Instrument: The Case of the Conservation Reserve Program. Working Paper 03-WP 345, Center for Agricultural and Rural Development, Iowa State University, Ames, IA, 29 pp.

Greene, D.L., P.D. Patterson, M. Singh, and J. Li, 2005: Feebates, rebates and gas guzzler taxes: a study of incentives for increased fuel economy. Energy Policy, 33(6), 757–776.

Goulder, L.H., 2004: Induced Technological Change and Climate Policy. Pew Center on Global Climate Change, Washington, DC, 38 pp.

Grubb, M., C. Carraro and J. Schellnhuber 2006: Technological change for atmospheric stabilization: introductory overview to the innovation modeling comparison project, The Energy Journal, Special Issue on Endogenous Technological Change and the Economics of Atmospheric Stabilization, pp. 1-16.

Gupta, S., D. A. Tirpak, N. Burger, J. Gupta, N. Höhne, A. I. Boncheva, G. M. Kanoan, C. Kolstad, J. A. Kruger, A. Michaelowa, S. Murase, J. Pershing, T. Saijo, A. Sari, 2007: Policies, instruments and co-operative arrangements. In: Climate Change 2007: Mitigation. Contribution of Working Group III to the Fourth Assessment Report of the Intergovernmental Panel on Climate Change [B. Metz, O.R. Davidson, P.R. Bosch, R. Dave, L.A. Meyer (eds.)]. Cambridge University Press, Cambridge, United Kingdom and New York, NY, USA, pp 745-807.

Harrison, K., 1999: Talking with the donkey: cooperative approaches to environmental protection, Journal of Industrial Ecology, 2(3), 51–72.

Herzog, H, 1999: The economics of CO2 capture. In: Greenhouse Gas Control Technologies [Reimer, P., B. Eliasson, A. Wokaum (eds.)]. Elsevier Science Ltd., Oxford, United Kingdom, pp. 101–106.

Hoffert, M.I., K. Calderia, A.K. Jain, E.F. Haites, L.D.D. Harvey, S.D. Potter, M.E. Schlesinger, S.H. Schneider, R.G. Watson, T.M.L. Wigley, and D.J. Wuebbles, 1998: Energy implications of future stabilization of atmospheric CO2 content. Nature, 395(6705), 881–884.

Hoffert, M.I., K. Caldeira, G. Benford, D.R. Criswell, C. Green, H. Herzog, A.K. Jain, H.S. Kheshgi, K.S. Lackner, J.S. Lewis, H.D. Lightfoot, W. Manheimer, J.C. Mankins, M.E. Mauel, L.J. Perkins, M.E. Schlesinger, T. Volk, and T.M.L. Wigley, 2002: Advanced technology paths to global climate stability: energy for a greenhouse planet. *Science*, **298(5595)**, 981-987.

IEA (International Energy Agency), 2006a: *Key World Energy Statistics, 2006*, IEA, Paris, France, 82 pp.

IEA (International Energy Agency), 2006b: *Energy Technology Perspectives: Scenarios and Strategies to 2050*, IEA, Paris, France, 484 pp.

IPCC, 2000: *Land Use, Land-use Change and Forestry. A Special Report of the Intergovernmental Panel on Climate Change* [Watson, R.T., I.R. Noble, B. Bolin, N.H. Ravindranath, D.J. Verado, and D.J. Dokken (eds.)]. Cambridge University Press, Cambridge, United Kingdom and New York, NY, USA, 388 pp.

IPCC, 2001: *Climate Change 2001: The Scientific Basis.* Contribution of Working Group I to the Third Assessment Report of the Intergovernmental Panel on Climate Change [Houghton, J.T., Y. Ding, D.J. Griggs, M. Noguer, P.J. van der Linden, X. Dai, K. Maskell, and C.A. Johnson (eds.)]. Cambridge University Press, Cambridge, United Kingdom and New York, NY, USA, 881 pp.

IPCC, 2005: *IPCC Special Report on Carbon Dioxide Capture and Storage, Summary for Policymakers*. Approved by the 8th Session of IPCC Working Group III, Montreal, Canada, 53 pp.

Jaccard, M., J. Nyboer, and B. Sadownik, 2002: *The Cost of Climate Policy*. University of British Columbia Press, Vancouver, British Columbia, Canada. 242 pp.

Jaccard, M., J. Nyboer, C. Bataille, and B. Sadownik, 2003a: Modeling the cost of climate policy: distinguishing between alternative cost definitions and long-run cost dynamics. *The Energy Journal*, **24(1)**, 49-73.

Jaccard, M., R. Loulou, A. Kanudia, J. Nyboer, A. Bailie, and M. Labriet, 2003b: Methodological contrasts in costing GHG abatement policies: optimization and simulation modeling of micro-economic effects in Canada. *European Journal of Operations Research*, **145(1)**, 148-164.

Jaccard, M., N. Rivers, C. Bataille, R. Murphy, J. Nyboer and B. Sadownik, 2006: *Burning Our Money to Warm the Planet*. Commentary No. 234, C.D. Howe Institute, Toronto, 31 pp.

Kauppi, P., R. Sedjo, M. Apps, C. Cerri, T. Fujimori, H. Janzen, O. Krankina, W. Makundi, G. Marland, O. Masera, G.-J. Nabuurs, W. Razali, N.H. Ravindranath, 2001. Technological and economic potential of options to enhance, maintain and manage biological carbon reservoirs and geo-engineering. In: *Climate Change 2001 - Mitigation*. Contribution of Working Group III to the Third Assessment Report of Inter-Governmental Panel on Climate Change [Metz, B., O. Davidson, R. Swart, and J. Pan (eds.)]. UNEP-WMO. Cambridge University Press, Cambridge, United Kingdom and New York, NY, USA, pp. 301-343.

Kates, R., and T. Wilbanks, 2003: Making the global local: responding to climate change concerns from the bottom up. *Environment*, **45(3)**, 12-23.

King, A., and M. Lenox, 2000: Industry self-regulation without sanctions: the chemical industry's responsible care program. *Academy of Management Journal*, **43(4)**, 698-716.

Marland, G., B.A. McCarl, and U.A. Schneider, 2001: Soil carbon: policy and economics. *Climatic Change*, **51(1)**, 101–117.

Martin, N., E. Worrell, M. Ruth, L. Price, R.N. Elliott, A.M. Shipley, and J. Thorne, 2001: *Emerging Energy-Efficient Industrial Technologies*. LBNL Report Number 46990, New York State Edition, published by American Council for an Energy-Efficient Economy (ACEEE), Washington DC, 183 pp.

National Academies, 2004: *The Hydrogen Economy: Opportunities, Costs, Barriers, and R&D Needs*. The National Academies Press, Washington, DC, 240 pp.

NRC (National Research Council) and National Academy of Engineering, 2004: *Committee on Alternatives and Strategies for Future Hydrogen Production and Use*. The National Academies. Washington, DC, 148 pp.

NCEP (National Commission on Energy Policy), 2005: *Ending the Energy Stalemate: A Bipartisan Strategy to Meet America's Energy Challenge*. Washington, DC, 148 pp.

OECD (Organization for Economic Co-operation and Development), 2003a: *Technology Innovation, Development and Diffusion*. OECD and IEA Information Paper, COM/ENV/EPOC/IEA/SLT(2003)4, Paris, France, 48 pp.

OECD (Organization for Economic Co-operation and Development), 2003b: *Voluntary Approaches for Environmental Policy: Effectiveness, Efficiency and Usage in Policy Mixes*. OECD, Paris, France, 143 pp.

Pacala, S. and R. Socolow, 2004: Stabilization wedges: solving the climate problem for the next 50 years with current technologies. *Science*, **305(5686)**, 968-972.

Parry, I.W.H., R. Williams, and L.H. Goulder, 1999: When can carbon abatement policies increase welfare? The fundamental role of distorted factor markets. *Journal of Environmental Economics and Management*, **37(1)**, 52–84.

Raupach, M., J.G. Canadell, D.C. Bakker, P. Ciais, M.J. Sans, J.Y. Fank, J.M. Melillo, P. Romero-Lankao, J.A. Sathaye, E.D. Schulze, P. Smith, and J. Tschirley, 2004: Atmospheric stabilization in the context of carbon-climate-human interactions. In: *Toward CO$_2$ Stabilization: Issues, Strategies, and Consequences* [Field, C. and M. Raupach (eds.)]. Island Press, Washington, DC.

Rose, A., and G. Oladosu, 2002: Greenhouse gas reduction in the U.S.: identifying winners and losers in an expanded permit trading system. *Energy Journal*, **23(1)**, 1–18.

Sedjo, R.A., 2001: Forest 'sinks' as a tool for climate-change policymaking: a look at the advantages and challenges. *Resources*, **143**, 21–23.

Sedjo, R.A. and S.K. Swallow, 2002: Voluntary eco-labeling and the price premium. *Land Economics*, **87(2)**, 272–284.

Stern, N., 2006: *Stern Review on the Economics of Climate Change*, Cambridge University Press, Cambridge, United Kingdom. Available at http://www.hm-treasury.gov.uk/independent_reviews/stern_review_economics_climate_change/sternreview_index.cfm

Stern, P.C and Fineberg, H.V. (eds.), 1996: *Understanding Risk: Informing Decisions in a Democratic Society*. National Academy Press, Washington, DC, 249 pp.

Swift, B., 2001: How environmental laws work: an analysis of the utility sector's response to regulation of nitrogen oxides and sulfur dioxide under the Clean Air Act, *Tulane Environmental Law Journal*, **14(2)**, 309-425.

Tietenberg, T., 2000: *Environmental and Natural Resource Economics*. 5th Edition, Addison-Wesley, New York, 630 pp.

Tietenberg, T. and D. Wheeler, 2001: Empowering the community: information strategies for pollution control. In: *Frontiers of Environmental Economics* [Folmer, H., H.L. Gabel, S. Gerking, and A. Rose (eds.)]. Edward Elgar, Cheltenham, United Kingdom, 417 pp.

USGBC (U.S. Green Building Council), 2005: LEED for New Construction-Rating System 2.2 U.S. Greeen Building Council, Washington, DC. 83pp.

Welch, E.W., A. Mazur, and S. Bretschneider, 2000: Voluntary behavior by electric utilities: levels of adoption and contribution of the Climate Challenge Program to the reduction of carbon dioxide, *Journal of Public Policy Analysis and Management*, **19(3)**, 407-426.

Worrell, E., L.K. Price, and C. Galitsky, 2004: *Emerging Energy-efficient Technologies in Industry: Case Studies of Selected Technologies*. Environmental Technologies Division, Lawrence Berkeley Laboratory, University of California at Berkeley, 67 pp.

CHAPTER 5 REFERENCES

Adler, P., R. Barrett, M. Bean, J. Birkoff, C. Ozawa, and E. Rudin, 1999: *Managing Scientific and Technical Information in Environmental Cases: Principles and Practices for Mediators*. U.S. Institute for Environmental Conflict Resolution, Tucson, AZ, 74 pp.

Agrawala, S., K. Broad, and D.H. Guston, 2001: Integrating climate forecasts and societal decision making: challenges to an emergent boundary organization. *Science, Technology and Human Values*, **26(4)**, 454-477.

Apps, M., J. Canadell, M. Heimann, V. Jaramillo, D. Murdiyarso, D. Schimel, and M. Manning, 2003: *Expert Meeting Report: IPCC Meeting on Current Understanding of the Processes Affecting Terrestrial Carbon Stocks and Human Influences Upon Them*. Geneva, Switzerland, July 21-23, 2003. Available at http://www.ipcc.ch/pub/carbon.pdf. 37 pp.

Cash, D.W., 2001: In order to aid in diffusing useful and practical information: agricultural extension and boundary organizations. *Science, Technology and Human Values*, **26**, 431-453.

Cash, D., W. Clark, F. Alcock, N. Dickson, N. Eckley, D. Guston, J. Jäger, and R. Mitchell, 2003: Knowledge systems for sustainable development. *Proceedings of the National Academy of Sciences,* **100(14)**, 8086-8091.

Cash, D.W., J.C. Borck, A.G. Patt, 2006: Countering the loading-dock approach to linking science and decision making. *Science, Technology and Human Values,* **31(4)**, 465-494.

Curry, T., D. Reiner, S. Ansolabehere, and H. Herzog, *How Aware is the Public of Carbon Capture and Storage?* Presented at the Seventh International Conference on Greenhouse Gas Control Technologies, Vancouver, Canada, September 2004. Available at http://sequestration.mit.edu/bibliography/policy.html

Denning, A.S., *et al.,* 2005: *Science Implementation Strategy for the North American Carbon Program.* Report of the NACP Implementation Strategy Group, U.S. Carbon Cycle Interagency Working Group, U.S. Carbon Cycle Science Program, Washington, DC, 68 pp. Available at http://www.nacarbon.org/nacp/documents.html

Dilling, L., 2007: Towards science in support of decision making: characterizing the supply of carbon cycle science. *Environmental Science and Policy,* **10(1),** 48-61.

Dilling, L., S.C. Doney, J. Edmonds, K.R. Gurney, R.C. Harris, D. Schimel, B. Stephens, and G. Stokes, 2003: The role of carbon cycle observations and knowledge in carbon management. *Annual Review of Environment and Resources,* **28**, 521-558.

Douglas, M. and A. Wildavsky, 1984: *Risk and Culture.* University of California Press, Berkeley, CA, 221 pp. (1982 ed.)

Ehrmann, J. and B. Stinson, 1999: Joint fact-finding and the use of technical experts. In: *The Consensus Building Handbook* [Susskind, L., J.T. Larmer, and S. McKearnan (eds.)]. Sage Publications, Thousand Oaks, CA, 1147 pp.

Farrell, A. and J. Jäger (eds.), 2005: *Assessments of Regional and Global Environmental Risks: Designing Processes for the Effective Use of Science in Decision-Making.* Resources for the Future, Washington, DC, 301 pp.

Gibbons, M., C. Limoges, and H. Nowotny, 1994: *The New Production of Knowledge: The Dynamics of Science and Research in Contemporary Societies.* Sage, London, 179 pp.

Environment Canada, 2005: *Project Green: Moving Forward on Climate Change: A Plan for Honoring Our Kyoto Commitment.* Available at http://www.climatechange.gc.ca/english/newsroom/2005/plan05.asp

Guston, D.H., 2001: Boundary organizations in environmental policy and science: An introduction. *Science, Technology, & Human Values,* **26(4)**, 399-408, Special Issue: Boundary Organizations in Environmental Policy and Science (Autumn 2001).

Holling, C.S. (ed.), 1978: *Adaptive Environmental Assessment and Management.* John Wiley, New York, 377 pp.

Holling, C.S., 1995: What barriers? What bridges? In: *Barriers and Bridges to the Renewal of Ecosystems and Institutions* [Gunderson L.H., C.S. Holling, and S.S. Light (eds.)]. Columbia University Press, New York, pp. 3-34.

IPCC, 2000: *Land Use, Land-use Change and Forestry. A Special Report of the Intergovernmental Panel on Climate Change* [Watson, R.T., I.R. Noble, B. Bolin, N.H. Ravindranath, D.J. Verado, and D.J. Dokken (eds.)]. Cambridge University Press, Cambridge, United Kingdom and New York, NY, USA, 388 pp.

IPCC, 2007: *Climate Change 2007: The Physical Science Basis.* Contribution of Working Group I to the Fourth Assessment Report of the Intergovernmental Panel on Climate Change [Solomon, S., D. Qin, M. Manning, Z. Chen, M. Marquis, K.B. Averyt, M. Tignor and H.L. Miller (eds.)]. Cambridge University Press, Cambridge, United Kingdom and New York, NY, USA, 1009 pp.

Lahsen, M., 1999: The detection and attribution of conspiracies: the controversy over chapter 8. In: *Paranoia Within Reason: A Casebook on Conspiracy as Explanation.* [Marcus, G.E. (ed.)]. University of Chicago Press, Chicago, IL, pp. 111-136.

Lahsen, M., 2007: Trust through participation? Problems of knowledge in climate decision making. In: *The Social Construction of Climate Change.* [Pettinger, Mary E. (ed.)]. Ashgate Publishing, pp. 173-196. Available at http://sciencepolicy.colorado.edu/our_science_their_science/pubs/lahsen_2007_pettenger1.pdf

Lahsen, M. and C. A. Nobre, 2007: Challenges of connecting international science and local level sustainability efforts: the case of the Large-Scale Biosphere-Atmosphere Experiment in Amazonia. *Environmental Science & Policy,* **10(1)**, 62-74.

Lemos, M.C. and B.J. Morehouse, 2005: The co-production of science and policy in integrated climate assessments. *Global Environmental Change,* **15(1)**, 57-68.

Martinez, J. and A. Fernandez-Bremauntz (eds.), 2004: Cambio climático: una visión desde Mexico. Secretaría de Medio Ambiente y Recursos Naturales, Instituto Nacional de Ecología, Mexico City, Mexico, 525 pp.

Mitchell, R.B., W.C. Clark, D.W. Cash, and F. Alcock, 2004: Science, scientists, and the policy process: lessons from global environmental assessments for the northwest forest. In: *Forest Futures: Science, Politics and Policy for the Next Century* [Arabas, K. and J. Bowersox (eds.)]. Rowman and Littlefield, pp. 95-111.

Mitchell, R.B., W.C. Clark, D.W. Cash, and N.M. Dickson (eds.), 2006: *Global Environmental Assessments: Information and Influence.* The MIT Press, Cambridge, MA, 344 pp.

Moser, S., 2005: Stakeholder involvement in the first U.S. national assessment of the potential consequences of climate variability and change: an evaluation, finally. In: *Public Participation in Environmental Assessment and Decision Making.* National Research Council, Committee on Human Dimensions of Global Change, NAS/NRC, Washington, DC (in press).

NRC (National Research Council), 1999: *Making Climate Forecasts Matter.* National Academy Press, Washington, DC, 175 pp.

NRC (National Research Council), 2004: *Committee to Review the U.S. Climate Change Science Program Strategic Plan.* Implementing Climate and Global Change Research: A Review of the Final U.S. Climate Change Science Program Strategic Plan, National Academy Press, Washington, DC, 96 pp.

NRC (National Research Council), 2005: *Roundtable on Science and Technology for Sustainability*. Knowledge-Action Systems for Seasonal to Interannual Climate Forecasting: Summary of a Workshop. National Academy Press, Washington, DC, 32 pp.

Parson, E.A., 2003: *Protecting the Ozone Layer: Science and Strategy*. Oxford University Press, Oxford, United Kingdom, 400 pp.

Patt, A., P. Suarez, and C. Gwata, 2005a: Effects of seasonal climate forecasts and participatory workshops among subsistence farmers in Zimbabwe. *Proceedings of the National Academy of Sciences*, **102(35)**, 12623-12628.

Patt, A.G., R. Klein, and A. de la Vega-Leinert, 2005b: Taking the uncertainties in climate change vulnerability assessment seriously. *Comptes Rendus Geosciences*, **337(4)**, 411-424.

Pulwarty, R. S. and K.T. Redmond, 1997: Climate and salmon restoration in the Columbia River Basin: the role and usability of seasonal forecasts. *Bulletin of the American Meteorological Society*, **78(3)**, 381-396.

Richards, K., 2004: A brief overview of carbon sequestration economics and policy. *Environmental Management*, **33(4)**, 545-558.

Sarmiento, J.L. and S.C. Wofsy, 1999: *A U.S. Carbon Cycle Science Plan: A Report of the Carbon and Climate Working Group*. U.S. Global Change Research Program, Washington, DC. Available at http://www.nacarbon.org/nacp/documents.html

Schröter, D., W. Cramer, R. Leemans, I.C. Prentice, M.B. Araújo, N.W. Arnell, A. Bondeau, H. Bugmann, T.R. Carter, C.A. Gracia, A.C. de la Vega-Leinert, M. Erhard, F. Ewert, M. Glendining, J.I. House, S. Kankaanpää, R.J.T. Klein, S. Lavorel, M. Lindner, M.J. Metzger, J. Meyer, T.D. Mitchell, I. Reginster, M. Rounsevell, S. Sabaté, S. Sitch, B. Smith, J. Smith, P. Smith, M.T. Sykes, K. Thonicke, W. Thuiller, G. Tuck, S. Zaehle, B. Zierl, 2005: Ecosystem service supply and vulnerability to global change in Europe. *Science*, **310(5752)**, 1333-1337.

Shackley, S., C. McLachlan, and C. Gough, 2005: The public perception of carbon dioxide capture and storage in the UK: results from focus groups and a survey. *Climate Policy*, **4(4)**, 377-398.

Stokes, D.E., 1997: *Pasteur's Quadrant: Basic Science and Technological Innovation*. Brookings Institution Press, Washington, DC, 180 pp.

U.S. Climate Change Science Program, 2003: *Strategic Plan for the U.S. Climate Change Science Program*. Last accessed February 20, 2006. Available at http://www.climatescience.gov

U.S. Climate Change Science Program, 2006: *Our Changing Planet: The US Climate Change Science Program for Fiscal Year 2006*. A Report by the Climate Change Science Program and the Subcommittee on Global Change Research. Last accessed February 23, 2006, Washington, DC. Available at: http://www.climatescience.gov

U.S. Department of State, 2004: *U.S. Climate Change Policy: The Bush Administration's Actions on Global Climate Change*. Fact sheet released by the White House, Office of the Press Secretary Washington, DC, November 19, 2004. Available at http://www.state.gov/g/oes/rls/fs/2004/38641.htm

Van den Belt, M., 2004: *Mediated Modeling: A Systems Dynamic Approach to Environmental Consensus Building*. Island Press, Washington, DC, 296 pp.

Van House, N.A., 2003: Digital libraries and collaborative knowledge construction. In: *Digital Library Use: Social Practice in Design and Evaluation* [Bishop, A.P., B.P. Buttenfield, and N.A. Van House (eds.)]. MIT Press, pp. 271-295.

Western Economic Diversification Canada, 2006: *Government of Canada and Government of Alberta Announce $16.6 Million Worth of Joint Projects*. Available at http://www.wd.gc.ca/mediacentre/2006/may23-02a_e.asp

Yaniv, I., 2004: Receiving other people's advice: influence and benefit. *Organizational Behavior and Human Decision Processes*, **93(1)**, 1-13.

PART II OVERVIEW REFERENCES

Andres, R.J., D.J. Fielding, G. Marland, T.A. Boden, N. Kumar, and A.T. Kearney, 1999: Carbon dioxide emissions from fossil-fuel use, 1751–1950. *Tellus B*, **51(4)**, 759–765.

Andres, R.J., J.S. Gregg, L.M. Losey, and G. Marland, 2005: *Monthly Resolution Fossil-Fuel-Derived Carbon Dioxide Emissions for the Countries of the North American Carbon Program*. Proceedings of the Seventh International Carbon Dioxide Conference, Broomfield, CO, September, 2005, pp. 157–158.

Blasing, T.J., C.T. Broniak, and G. Marland, 2005a: The annual cycle of fossil-fuel carbon dioxide emissions in the United States. *Tellus B*, **57(2)**, 107–115. Available at http://cdiac.esd.ornl.gov

Blasing, T.J., C. Broniak, and G. Marland, 2005b: State-by-state carbon dioxide emissions from fossil-fuel use in the United States 1960–2000. *Mitigation and Adaptation Strategies for Global Change*, **10(4)**, 659–674.

Environment Canada, 2005: *Canada's Greenhouse Gas Inventory: 1990–2003*. National Inventory Report, April 15, 2005, Greenhouse Gas Division, Environment Canada, 339 pp.

Gregg, J.S., 2005: *Improving the Temporal and Spatial Resolution of Carbon Dioxide Emissions Estimates from Fossil-Fuel Consumption*. A thesis submitted to the graduate faculty of the University of North Dakota, August, 2005, 404 pp. Available at http://cdiac.esd.ornl.gov

Gurney, K.R., Y.H. Chen, T. Maki, S.R. Kawa, A. Andrews, and Z. Zhu, 2005: Sensitivity of atmospheric CO_2 inversion to seasonal and interannual variations in fossil-fuel emissions. *Journal of Geophysical Research*, **110(D10)**, 10308, doi:10.1029/2004JD005373.

IEA (International Energy Agency), 2005: CO_2 *Emissions from Fuel Combustion: 1971–2003*. OECD/IEA, Paris, France, 556 pp.

IPCC, 1997: *Revised 1996 IPCC Guidelines for National Greenhouse Gas Inventories (3 Volumes)*. IPCC Technical Support Unit, Bracknell, United Kingdom, 3 volumes (looseleaf).

Losey, L.M., 2004: *Monthly and Seasonal Estimates of Carbon Dioxide Emissions from Fossil Fuel Consumption in Canada, Mexico, Brazil, The United Kingdom, France, Spain, Italy, and Poland*. A thesis submitted to the graduate faculty of the University of North Dakota, May, 2004, 328 pp. Available at http://cdiac.essd.ornl.gov

Marland, G., T. Boden, and R.J. Andres, 1995: Carbon dioxide emissions from fossil fuel burning: emissions coefficients and the global contribution of eastern European countries. *Időjárás*, **99**, 157–170.

Marland, G., T.A. Boden, and R.J. Andres, 2005: Global, regional, and national CO_2 emissions. In: *Trends: A Compendium of Data on Global Change*. Carbon Dioxide Information Analysis Center, Oak Ridge National Laboratory, U.S. Department of Energy, Oak Ridge, TN. Available at http://cdiac.esd.ornl.gov/trends/trends.htm

Mexico, 2001: *México: Segunda Comunicación Nacional ante la Convención Marco de las Naciones Unidas sobre el Cambio Climático*. Comité Intersecretarial Sobre Cambio Climático, Secretaria de Medio Ambiente y Recursos Naturales (Semarnat), Mexico City, 374 pp.

Neumayer, E., 2004: National carbon dioxide emissions: geography matters. *Area*, **36(1)**, 33–40.

USEPA (U.S. Environmental Protection Agency), 2005: *Inventory of U.S. Greenhouse Gas Emissions and Sinks: 1990–2003*. EPA 430-R-05-003, EPA, Washington, DC.

CHAPTER 6 REFERENCES

AAG (Association of American Geographers), 2003: *Global Change and Local Places: Estimating, Understanding, and Reducing Greenhouse Gases*. Cambridge University Press, Cambridge, United Kingdom, 270 pp.

Caldeira, K, D. Day, W. Fulkerson, M. Hoffert, 2005: *Climate Change Technology Exploratory Research*. Working paper, Climate Policy Center, Washington, DC, 10 pp. Available online at http://www.cpc-inc.org/

EIA (Energy Information Administration), 2004: *Emissions of Greenhouse Gases in the United States, 2004*. U.S. Department of Energy, , Washington, DC, 134 pp.

EIA (Energy Information Administration), 2005: *International Energy Outlook, 2005*. U.S. Department of Energy, , Washington, DC, 186 pp.

EIA (Energy Information Administration), 2006a: *International Energy Outlook, 2006*. U.S. Department of Energy, , Washington, DC, 192 pp.

EIA (Energy Information Administration), 2006b: *Annual Energy Review 2006*: U.S. Department of Energy, , Washington, DC, 401 pp.

Environment Canada, 2003: *Canada's Greenhouse Gas Inventory, 1990-2003*. Available at http://www.ec.gc.ca/pdb/ghg/inventory_report/2003_report/ts_2_e.cfm

Environment Canada, 2005: *The Green Lane: Climate Change: The Greenhouse Gas Emissions Outlook to 2020*. Available at http://www.ec.gc.ca/climate/overview_2020-e.html

GAO (General Accounting Office), 2004: *Climate Change: Analysis of Two Studies of Estimated Costs of Implementing the Kyoto Protocol*. General Accounting Office, Washington, DC, 20 pp.

Government of Canada, 2005: *Project Green: Moving Forward on Climate Change*. Government of Canada, Ottawa, 48 pp

Government of Mexico, 2001: *Second National Communication*. Submitted to UNFCCC by the Secretaria de Medio Ambiente y Recursos Naturales, Mexico City.

Hoffert, M.I., K. Caldeira, G. Benford, D.R. Criswell, C. Green, H. Herzog, A.K. Jain, H.S. Kheshgi, K.S. Lackner, J.S. Lewis, H.D. Lightfoot, W. Manheimer, J.C. Mankins, M.E. Mauel, L.J. Perkins, M.E. Schlesinger, T. Volk, and T.M.L. Wigley, 2002: Advanced technology paths to global climate stability: energy for a greenhouse planet. *Science*, **298(5595)**, 981-987.

Interlaboratory Working Group, 1997: *Scenarios of U.S. Carbon Reductions*. Oak Ridge National Laboratory, Oak Ridge, TN.

Interlaboratory Working Group, 2000: *Scenarios for a Clean Energy Future*. Oak Ridge National Laboratory, Oak Ridge, TN, 371 pp.

IPCC, 2001: *Climate Change, 2001: Mitigation*. Contribution of Working Group III to the Third Assessment Report of the Intergovernmental Panel on Climate Change [B. Metz (ed.)]. Cambridge University Press, Cambridge, United Kingdom, 752 pp.

IPCC, 2006: *Carbon Dioxide Capture and Storage*. IPCC Special Report. [B. Metz (ed.)]. Cambridge University Press, Cambridge, United Kingdom, 431 pp.

Lewis, N., 2005: *Global Energy Perspective*. Paper presented to the U.S. DOE Laboratory Energy and Development Working Group (LERDWG), Washington, DC. Available at http://nsl.caltech.edu/energy.html

NAS (National Academy of Sciences), 1992: *Policy Implications of Greenhouse Warming: Mitigation, Adaptation, and the Science Base*. Washington, DC, 127 pp.

NAS (National Academy of Sciences), 1999: *Our Common Journey: A Transition Toward Sustainability*. National Academy Press, Washington, DC, 363 pp.

National Commission on Energy Policy, 2004: *Ending the Energy Stalemate: A Bipartisan Strategy to Meet America's Energy Challenges*. NCEP, Washington, DC, 148 pp.

National Laboratory Directors, 1997: *Technology Opportunities to Reduce U.S. Greenhouse Gas Emissions*. Prepared for the U.S. Department of Energy, 95 pp.

OTA (Office of Technology Assessment), 1991: *Changing By Degrees: Steps to Reduce Greenhouse Gases*. OTA-0-482, Washington, DC, 354 pp.

Pacala, S. and R. Socolow, 2004: Stabilization wedges: solving the climate problem for the next 50 years with current technologies. *Science*, **305(5686)**, 968-972.

Pew Center on Global Climate Change, 2002: *Climate Change Mitigation in Developing Countries: Brazil, China, Mexico, South Africa and Turkey.* Report prepared by Chandler W., R. Schaffer, Z. Dadi, P.R. Shukla, F. Tudela, O. Davidson, and S. Alpan-Atamar, Washington, DC, 64 pp.

Tremblay, A., 2004: *Greenhouse Gas Emissions - Fluxes and Processes: hydroelectric Reservoirs and Natural Environments.* Springer, New York, 732 pp.

U.S. Climate Change Technology Program, 2005: *Strategic Plan: Draft for Public Comment.* Available at http://www.climatetechnology.gov/stratplan/draft/index.htm

U.S. DOE (U.S. Department of Energy), 2004: National energy policy/overview/Canada. In: *Energy Trends.* Available at http://energytrends.pnl.gov/Canada/ca004.htm

Wilbanks, T., 1992: Energy policy responses to concerns about global climate change. In: *Global Climate Change: Implications, Challenges and Mitigation Measures* [Majumdar, S.K., L.S. Kalkstein, B. Yamal, E.W. Miller, and L.M. Rosenfeld (eds.)]. Pennsylvania Academy of Sciences, Easton, PA, pp. 452-470.

CHAPTER 7 REFERENCES

CBO (Congressional Budget Office), 2003: *The Economic Costs of Fuel Economy Standards Versus a Gasoline Tax.* Congress of the United States, Washington, DC, December, 24 pp.

CIA (Central Intelligence Agency), 2005: *The World Factbook.* Washington, DC. Available at http://www.cia.gov/cia/publications/factbook

Davis, S.C. and S.W. Diegel, 2004: *Transportation Energy Data Book: Edition 24.* ORNL-6973, Oak Ridge National Laboratory, Oak Ridge, TN.

Environment Canada, 2005a: *Canada's Greenhouse Gas Inventory: 1990-2003.* National Inventory Report, Ottawa, Ontario, Canada, 339 pp.

Environment Canada, 2005b: *Moving Forward on Climate Change: A Plan for Honouring our Kyoto Commitment.* Ottawa, Canada. Available at http://www.climatechange.gc.ca

Fulton, L. and G. Eads, 2004: *IEA/SMP Model Documentation and Reference Case Projection.* World Business Council for Sustainable Development. Available at http://www.wbcsd.ch/web/publications/mobility/smp-model-document.pdf, July.

GAO (U.S. General Accounting Office), 2003: *Climate Change, Selected Nations' Reports on Greenhouse Gas Emissions Varied in Their Adherence to Standards.* GAO-04-98, Washington, DC, December. Available at http://www.gao.gov/cgi-bin/getrpt?GAO-04-98.

General Motors Corporation, Argonne National Laboratory, ExxonMobil, and Shell, 2001: *Well-to-Wheel Energy Use and Greenhouse Gas Emissions of Advanced Fuel/Vehicle Systems: North American Analysis.* Vol. 2, Argonne National Laboratory, Argonne, IL.

Greene, D.L. and A. Schafer, 2003: *Reducing Greenhouse Gas Emissions from U.S. Transportation.* Pew Center on Global Climate Change, Arlington, VA, May, 68 pp.

Greene, D.L., P.D. Patterson, M. Singh, and J. Li, 2005: Feebates, rebates and gas-guzzler taxes: a study of incentives for increased fuel economy. *Energy Policy*, **33**(6), 757-776.

Harrington, W. and V. McConnell, 2003: *Motor Vehicles and the Environment.* RFF Report, Resources for the Future, Washington, DC, April, 92 pp.

IEA (International Energy Agency), 2005a: *World Energy Outlook 2005: Middle East and North Africa Insights.* OECD, Paris, France, 629 pp.

IEA (International Energy Agency), 2005b: *Prospects for Hydrogen and Fuel Cells.* OECD, Paris, France, 253 pp.

IEA (International Energy Agency), 2004: *Biofuels for Transport.* OECD, Paris, France, 210 pp.

IEA (International Energy Agency), 2001: *Saving Oil and Reducing CO_2 Emissions in Transport.* OECD, Paris, France, 194 pp.

ILWG (Interlaboratory Working Group), 2000: *Scenarios for a Clean Energy Future.* Prepared by Lawrence Berkeley National Laboratory (LBNL-44029) and Oak Ridge National Laboratory (ORNL/CON-476) for the U.S. Department of Energy, 371 pp.

INE (Instituto Nacional de Ecología), 2003: *Energía. Sector Transporte 2000-2001*, Inventario Nacional de Emisiones de Gases de Efecto Invernadero, INGEI/2000/ENC, Mexico D.F. Available at http://www.ine.gob.mx/dgicurg/cclimatico/inventario.html

Marland, G., T. Boden, and R.J. Andres, 2005: *Global CO_2 Emissions from Fossil Fuel Burning, Cement Manufacture and Gas Flaring, 1751-2002.* Available at http://cdiac.esd.ornl.gov/ftp/ndp030/global.1751_2002.ems, November 8.

Moomaw, W.R. and J.R. Moreira, 2001: Technological and economic potential of greenhouse gas emissions reduction (Chapter 3). In: *Climate Change 2001: Mitigation* [Metz, Davidson, Swart, and Pan (eds.)]. Cambridge University Press, Cambridge, United Kingdom, pp. 167-277.

Nakićenović, N., A. Grübler, and A. McDonald, 1998: *Global Energy Perspectives.* Cambridge University Press, Cambridge, United Kingdom, 299 pp.

NAS (National Academy of Sciences), 2004: *The Hydrogen Economy: Opportunities, costs, barriers and research and development needs.* National Academies Press, Washington, DC, 240 pp.

NAS (National Academy of Sciences), 2002: *Effectiveness and Impact of Corporate Average Fuel Economy (CAFE) Standards.* National Academies Press, Washington, DC, 166 pp.

NATS (North American Transportation Statistics), 2005: *Various Tables.* A joint project of the U.S. Bureau of Transportation Statistics, Statistics Canada and Instituto Nacional de Estadistica Gegrafica Informatica (INEGI), Mexico. Available at http://nats.sct.gob.mx/lib/series

NCEP (National Commission on Energy Policy), 2004: *Ending the Energy Stalemate, A Bipartisan Strategy to Meet America's Energy Challenges*, Chapter 3, Washington, D.C.: National

Commission on Energy Policy, 148 pp. Available at http://www.energycommission.org

NRCan (Natural Resources Canada), 2006: *Comprehensive Energy Use Database Tables*. Transportation sector, table 1: Secondary energy use by source, table 8: GHG emissions by transportation mode. Available at http://oee.nrcan.gc.ca/corporate/statistics/neud/dpa/trends_tran_ca.cfm

NRTEE (National Round Table on the Environment and the Economy), 2005: *Economic Instruments for Long-term Reductions in Energy-based Carbon Emissions*. Renouf Publishing Co., Ltd., Ottawa, Ontario, Canada, 132 pp.

Patterson, P., D. Greene, E. Steiner, S. Plotkin, M. Singh, A. Vyas, M. Mintz, D. Santini, S. Folga, J. Moore, P. Reilly-Roe, K. Cliffe, R. Talbot, P. Khannna, and V. Stanciulescu, 2003: *Joint DOE/NRCan Study of North American Transportation Energy Futures*. Energy Efficiency and Renewable Energy, U.S. Department of Energy, Washington, DC, May. Available at http://www.eere.energy.gov/ba/pdfs/final_2050_pres.pdf

Perlack, R.D., L.L. Wright, A.F. Turhollow, R.L. Graham, B.J. Stokes, and D.C. Erbach, 2005: *Biomass as Feedstock for a Bioenergy and Bioproducts Industry: The Technical Feasibility of a Billion-Ton Annual Supply*. DOE/GO-102995-2135, U.S. Department of Energy, Washington, DC.

Rodríguez, H.M., 2005: *Perspectivas del Uso de los Hidrocarburos a Nivel México*. Presentation, Subsecretario de Hidrocarburos, Mexico City, Mexico, slides online.

SENER (Secretaria de Energia), 2005: *Sistema de Información Energetica, información estadistica*. Available at http://sie.energia.gob.mx/sie, under *Consumo final de energia en el sector transporte*.

Turrentine, T. and K. Kurani, 2004: *Automotive Fuel Economy in the Purchase and Use Decisions of Households*. ITS-RR-04-31, Institute for Transportation Studies, University of California at Davis, Davis, CA, 35 pp.

U.S. DOE (U.S. Department of Energy), 2005: *Hydrogen, Fuel Cells and Infrastructure Technologies Program: Multi-Year Research, Development and Demonstration Plan*. DOE/GO-102003-1741, Energy Efficiency and Renewable Energy. Available at http://www.eere.energy.gov

U.S. DOE/EIA (U.S. Department of Energy, Energy Information Administration), 2005a: *Annual Energy Review 2004*. DOE/EIA-0384(2004), Washington, DC, August. Available at http://www.eia.doe.gov

U.S. DOE/EIA (U.S. Department of Energy, Energy Information Administration), 2005b: *International Energy Outlook 2005*. DOE/EIA-0484(2005), Washington, DC, 186 pp.

U.S. DOE/EIA (U.S. Department of Energy, Energy Information Administration), 2003: *Analysis of S.139, the Climate Stewardship Act of 2003*. SR/OIAF/2003-02, Washington, DC, 515 pp.

U.S. DOT (U.S. Department of Transportation), 1998: *Transportation and Global Climate Change: A Review and Analysis of the Literature*. Federal Highway Administration, Washington, DC.

U.S. EPA (U.S. Environmental Protection Agency), 2005: *Inventory of U.S. Greenhouse Gas Emissions and Sinks: 1990-2003*. EPA 430-R-05-003, Environmental Protection Agency, Office of Atmospheric Programs, Washington, DC.

Wang, M.Q., 2005: Argonne Expert Addresses Energy and Environmental Impacts of Fuel Ethanol. *TransForum*, **5(2)**, 3-4.

Weiss, M.A., J.B. Heywood, E.M. Drake, A. Schafer, and F.F. AuYeung, 2000: *On the Road in 2020: A life-cycle analysis of new automobile technologies*. Energy Laboratory Report #MIT EL 00-003, Energy Laboratory, Massachusetts Institute of Technology, Cambridge, MA.

Williams, R.H., 2005: *CO_2 Capture and Storage Strategies for Coal and Biomass to Reduce GHG Emissions for Synfuels*. Princeton Environmental Institute, Princeton University, Princeton, NJ, March. Presentation, 154 pp.

WBCSD (World Business Council for Sustainable Development), 2004: *Mobility 2030*. The Sustainable Mobility Project, Geneva, Switzerland. Available at http://www.wbcsd.org

CHAPTER 8 REFERENCES

Barlaz, M.A. and R.K. Ham, 1990: *The Use of Mass Balances for Calculation of the Methane Potential of Fresh and Anaerobically Decomposed Refuse*. Proceedings from the GRCDA 13th Annual International Landfill Gas Symposium, March 27-29, 1990, Silver Spring, MD, GRCDA—The Association of Solid Waste Management Professionals, 1990, 235 pp.

Barlaz, M., 1994: *Measurement of the Methane Potential of the Paper, Yard Waste, and Food Waste Components of Municipal Solid Waste*. Unpublished paper, Department of Civil Engineering, North Carolina State University.

Bogner, J. and K. Spokas, 1993: Landfill CH_4: rates, fates, and role in the global carbon cycle. *Chemosphere*, **26(1-4)**, 369-386.

Böhringer, C., 1998: The synthesis of bottom-up and top-down in energy policy modeling. *Energy Economics*, **20(3)**, 233-48.

California Environmental Protection Agency, 2003: *Environmental Technologies and Service Opportunities in the Baja California Peninsula*. International Affairs Unit. 95 pp.

CIEEDAC (Canadian Industrial Energy End-Use Data and Analysis Centre), 2005: *Development of Energy Intensity Indicators for Canadian Industry: 1990-2004*. Simon Fraser University, Vancouver, Canada, 166 pp.

DOE, (U.S. Department of Energy), 2006: Accessed on March 27, 2006, *Carbon Sequestration: Research and Development Overview*. Available at http://www.fossil.energy.gov/programs/sequestration/overview.html

Edenhofer, O., C. Carraro, J. Kohler, and M. Grubb, 2006: Endogenous technological change and the economics of atmospheric stabilisation. *The Energy Journal*, special issue, 284 pp.

Edmonds, J., J. Roop, and M. Scott, 2000: *Technology and the Economics of Climate Change Policy*. Prepared for the Pew Center on Climate Change by Battelle National Laboratories, 35 pp.

Energy Information Administration, 2005: *International Energy Outlook, 2005*, 186 pp.

EPA (Environmental Protection Agency), 2003a: *International Analysis of Methane and Nitrous Oxide Abatement Opportunities: Report to Energy Modeling Forum Working Group 21*. Environmental Proection Agency, Washington DC, 9 pp.

EPA (Environmental Protection Agency), 2003b: *Municipal Solid Waste in the United States: 2001 Facts and Figures*. Environmental Protection Agency, Office of Solid Waste and Emergency Response, Washington DC, 170 pp.

EPA (Environmental Protection Agency), 2005. *Inventory of U.S. Greenhouse Gas Emissions and Sinks, 1990-2003*. Environmental Protection Agency, Office of Atmospheric Programs, Washington DC.

EPIC (Environment and Plastics Industry Council), 2002: *Opportunities for Reducing Greenhouse Gas Emissions through Residential Waste Management*. Environment and Plastics Industry Council, 34 pp.

Grubb, M., J.A. Edmonds, P. ten Brink, and M. Morrison, 1993: The cost of limiting fossil fuel CO_2 emissions: a survey and analysis. *Annual Review of Energy and the Environment*, **18**, 397-478.

Grubb, M., I. Kohler, and D. Anderson, 2002: Induced technical change in energy and environmental modeling: analytical approaches and policy implications. *Annual Review of Energy and the Environment*, **27**, 271-308.

Frei, C., P. Haldi, and G. Sarlos, 2003: Dynamic formulation of a top-down and bottom-up merging energy policy model. *Energy Policy*, **31(10)**, 1017-1031.

Hershkowitz, A., 1997: *Too Good to Throw Away: Recycling's Proven Record*. National Resources Defense Council, New York, February 1997, 86 pp.

Herzog, H., 1999: The economics of CO_2 capture. In: *Greenhouse Gas Control Technologies* [Reimer P., B. Eliasson, and A. Wokaum (eds.)]. Elsevier Science Ltd., Oxford, pp. 101-106.

Humphreys, K. and M. Mahasenan, 2002: *Toward A Sustainable Cement Industry - Substudy 8: Climate Change*. World Business Council for Sustainable Development (WBCSD), Geneva, Switzerland.

IEA (International Energy Agency), 2004: *Oil crisis and climate challenges:30 Years of Energy Use in IEA Countries*. International Energy Agency, Paris, France. 211 pp.

IEA (International Energy Agency), 2006: *Energy Technology Perspectives 2006: Scenarios and Strategies to 2050*. International Energy Agency, Paris, France, 484 pp.

IPCC, 2001: *Climate Change, 2001: Mitigation*. Contribution of Working Group III to the Third Assessment Report of the Intergovernmental Panel on Climate Change [B. Metz (ed.)]. Cambridge University Press, Cambridge, United Kingdom, 752 pp

Jaccard, M., J. Nyboer, and B. Sadownik, 2002: *The Cost of Climate Policy*. University of British Columbia Press, Vancouver, British Columbia, Canada, 242 pp.

Jaccard, M., J. Nyboer, C. Bataille, and B. Sadownik, 2003a: Modeling the cost of climate policy: distinguishing between alternative cost definitions and long-run cost dynamics. *The Energy Journal*, **24(1)**, 49-73.

Jaccard, M., R. Loulou, A. Kanudia, J. Nyboer, A. Bailie, and M. Labriet, 2003b: Methodological contrasts in costing GHG abatement policies: optimization and simulation modeling of micro-economic effects in Canada. *European Journal of Operations Research*, **145(1)**, 148-164.

Jacobsen, H., 1998: Integrating the bottom-up and top-down approach to energy-economy modeling: the case of Denmark. *Energy Economics*, **20(4)**, 443-461.

Jaffe, A., R. Newell, and R. Stavins, 2002: Environmental policy and technological change. *Environmental and Resource Economics*, **22(1-2)**, 41-69.

Keith, D.W., and M. Ha-Duong, 2003: *CO_2 Capture From the Air: Technology Assessment and Implications for Climate Policy*. Proceedings of the 6th Greenhouse Gas Control Conference, Kyoto, Japan [J. Gale and Y. Kaya (eds.)]. Permagon Press, Oxford, United Kingdom, pp. 187-197.

Kim, Y. and E. Worrell, 2002: International comparison of CO_2 emissions trends in the iron and steel industry. *Energy Policy*, **30(10)**, 827-838.

Koopmans, C.C. and D.W. te Velde, 2001: Bridging the energy efficiency gap: using bottom-up information in a top-down energy demand model. *Energy Economics*, **23(1)**, 57-75.

Löschel, A., 2002: Technological change in economic models of environmental policy: a survey. *Ecological Economics*, **43(2-3)**, 105-126.

Martin, N., E. Worrell, M. Ruth, L. Price, R.N. Elliott, A.M. Shipley, and J. Thorne, 2001: *Emerging Energy-Efficient Industrial Technologies: New York State Edition*. LBNL Report Number 46990, American Council for an Energy-Efficient Economy (ACEEE), 195 pp.

Matysek, A., M. Ford, G. Jakeman, A. Gurney, K. Low, and B.S. Fisher, 2006: *Technology: It's role in economic development and climate change*. ABARE Research Report 06.6, Canberra, Australia.

McFarland, J., J. Reilly, and H. Herzog, 2004: Representing energy technologies in top-down economic models using bottom-up information. *Energy Economics*, **26(4)**, 685-707.

McKitrick, R., 1996: *The Economic Consequences of Taxing Carbon Emissions in Canada*. Department of Economics, University of British Columbia.

Mohareb, A.K., M. Warith, and R.M. Narbaitz, 2004: Strategies for the municipal solid waste sector to assist Canada in meeting its Kyoto Protocol commitments. *Environmental Review*, **12(2)**, 71-95.

Morris, S., G. Goldstein, and V. Fthenakis, 2002: NEMS and MARKAL-MACRO models for energy-environmental-economic analysis: a comparison of the electricity and carbon reduction projections. *Environmental Modeling and Assessment*, **7(3)**, 207-216.

Newell, R., A. Jaffe, and R. Stavins, 1999: The induced innovation hypothesis and energy-saving technological change. *Quarterly Journal of Economics*, **14(2)**, 941-975.

Sands, R., 2002: Dynamics of carbon abatement in the second generation model. *Energy Economics*, **26(4)**, 721-738.

Schäfer, A. and H. Jacoby, 2005: Technology detail in a multi-sector CGE model: transport under climate policy. *Energy Economics*, **27(1)**, 1-24.

Statistics Canada, 2004: *Human Activity and the Environment*. Statistics Canada, Cat no.16-201-XIE. Ottawa, Canada, 100 pp.

Sutherland, R., 2000: "No cost" efforts to reduce carbon emissions in the U.S.: an economic perspective. *Energy Journal*, **21(3)**, 89-112.

Weyant, J., H. Jacoby, J. Edmonds, and R. Richels (eds.), 1999: The costs of the Kyoto Protocol - a multi-model evaluation. *The Energy Journal*, special issue. 398 pp.

WRI (World Resources Institute), 2005: *Climate Analysis Indicators Tool (CAIT)*, Version 3.0, Washington, DC. Available at http://cait.wri.org

Worrell, E., L.K. Price, and C. Galitsky, 2004: *Emerging Energy-Efficient Technologies in Industry: Case Studies of Selected Technologies*. Environmental Technologies Division, Lawrence Berkeley Laboratory, University of California at Berkeley.

CHAPTER 9 REFERENCES

CEC (California Energy Commission), 2005: *California's Water Energy Relationship*. Staff Final Report, California Energy Commission, Sacramento, CA, 174 pp.

Chaudhari, M., L. Frantzis, T.E. Hoff, 2004: *PV Grid Connected Market Potential in 2010 under a Cost Breakthrough Scenario*. 1174373. Navigant Consulting Inc. 93 pp.

CONAFOVI (Comisión Nacional de Fomento a la Vivienda), 2001: *Programa Sectoral de Vivienda 2001-2006*.

DeCanio, S., 1993: Barriers within firms to energy-efficient investments. *Energy Policy*, **21(9)**, 906-914.

DeCanio, S., 1994: Why do profitable energy-saving investiment projects languish? *Journal of General Management*, **20(1)**, 62-71.

DOE/EERE (U.S. Department of Energy, Energy Efficiency and Renewable Energy), 2005: *2005 Buildings Energy Data Book*. Office of Energy Efficiency and Renewable Energy, Washington, DC.

DOE/EERE (U.S. Department of Energy, Energy Efficiency and Renewable Energy), 2006: *Building America Puts Residential Building Research to Work*. Washington, DC. Available at http://www.eere.energy.gov/buildings/building_america/

DOE/EIA (U.S. Departmenet of Energy and Energy Information Administration), 2003: *Carbon Coefficients Used in Emissions of Greenhouse Gases in the United States*. Washington, DC. Available at http://www.eia.doe.gov/oiaf/1605/ggrpt/pdf/tab6.1.pdf

DOE/EIA (U.S. Department of Energy and Energy Information Administration), 2005: *Annual Energy Outlook 2005*. Energy Information Administration, EIA-0383(2005), Washington, DC, 136 pp.

IEA (International Energy Agency), 2005: *CO$_2$ Emissions from Fuel Combustion: 1971-2003*. International Energy Agency, OECD/IEA, Paris, France.

INEGI (Instituto Nacional de Estadística Geografía e Informática), 2005: *Censo general de población y vivienda 2005*. Mexico, D.F., 2005.

Interlaboratory Working Group, 2000: *Scenarios for a Clean Energy Future*. Oak Ridge National Laboratory, Oak Ridge, TN, 371 pp.

Kinzey, B.R., S. Kim, J.D. Ryan, 2002: *The Federal Buildings Research and Development Program: A Sharp Tool for Climate Policy*. ACEEE Buildings Summer Study 2002, Pacific Grove. 11 vols.

NRCan (Natural Resources Canada), 2005a: *Residential Sector Secondary Energy Use and GHG Emissions by End Use 2005*. Ottawa, Canada.

NRCan (Natural Resources Canada), 2005b: *Office of Energy Efficiency National Energy Use Database 2005*. Ottawa, Canada. Available at http://oee.nrcan.gc.ca/corporate/statistics/neud/dpa/data_e/database_e.cfm

Public Technology Inc. and U.S. Green Building Council, 1996: *Sustainable Building Technical Manual*. Washington D.C.,

Rabl, A., and J.V. Spadaro, 2007: *Environmental Impacts and Costs of Energy*, Chapter 4 in: *Handbook of Energy Efficiency and Renewable Energy*, [F. Krieth and Y. Goswami (eds)], CRC Press, New York.

SENER México, 2004: *Balance Nacional de Energía 2003*. Subsecretaría de Paneación Energética y Desarrollo Tecnológico. Secretaría de Energía, México D.F., 215 pp.

SENER México, 2005: *Secretaria de Energia—Sistema de Información Energética*. México D.F. Available at http://sie.energia.gob.mx/sie/bdiController

USGBC (U.S. Green Building Council), 2005: *LEED for New Construction—Rating System 2.2*. U.S. Green Building Council, LEED (NC) 2.2, Washington, DC, 416 pp.

Wiel, S., and J.E. McMahon, 2005: *Energy-Efficiency Labels and Standards: A Guidebook for Appliances, Equipment, and Lighting, 2nd Edition*. Collaborative Labeling and Standards Program, Washington, DC, 218 pp.

PART III OVERVIEW REFERENCES

Bradley, B.A., R.A. Houghton, J.F. Mustard, and S.P. Hamburg, 2006: Invasive grass reduces aboveground carbon stocks in shrublands of the Western US. *Global Change Biology*, **12(10)**, 1815-1822.

Caspersen, J.P., S.W. Pacala, J.C. Jenkins, G.C. Hurtt, P.R. Moorcroft, and R.A. Birdsey, 2000: Contributions of land-use history to carbon accumulation in United States forests. *Science*, **290(5494)**, 1148-1151.

Davidson, E.A., and I.A. Janssens, 2006: Temperature sensitivity of soil carbon decomposition and feedbacks to climate change. *Nature*, **440(7081)**, 165-173.

215

Friedlingstein, P., P. Cox, R. Betts, L. Bopp, W. von Bloh, V. Brovkin, P. Cadule, S. Doney, M. Eby, I. Fung, G. Bala, J. John, C. Jones, F. Joos, T. Kato, M. Kawamiya, W. Knorr, K. Lindsay, H.D. Matthews, T. Raddatz, P. Rayner, C. Reick, E. Roeckner, K.-G. Schnitzler, R. Schnur, K. Strassmann, A.J. Weaver, C. Yoshikawa, and N. Zeng, 2006: Climate-carbon cycle feedback analysis: results from the C4MIP model inter-comparison. *Journal of Climate*, **19(14)**, 3337-3353.

Houghton, R.A., J.L. Hackler, and K.T. Lawrence, 1999: The U.S. carbon budget: contributions from land-use change. *Science*, **285(5427)**, 574-578.

Hurtt, G.C., S.W. Pacala, P.R. Moorcroft, J. Caspersen, E. Shevliakova, R.A. Houghton, and B. Moore III, 2002: Projecting the future of the U.S. carbon sink. *Proceedings of the National Academy of Sciences*, **99(3)**, 1389-1394.

Jones, C., C. McConnell, K. Coleman, P. Cox, P. Falloon, D. Jenkinson, and D. Powlson, 2005: Global climate change and soil carbon stocks; predictions from two contrasting models for the turnover of organic carbon in soil. *Global Change Biology*, **11(1)**, 154-166.

Lal, R., 2001: Fate of eroded soil carbon: emission or sequestration. In: *Soil Carbon Sequestration and the Greenhouse Effect* [R. Lal (ed.)]. Soil Science Society of America Special Publication, vol. 57; Madison, Wisconsin, pp. 173-181.

Osterkamp, T.E., and V.E. Romanovsky, 1999: Evidence for warming and thawing of discontinuous permafrost in Alaska. *Permafrost and Periglacial Processes*, **10(1)**, 17-37.

Osterkamp, T.E., L. Viereck, Y. Shur, M.T. Jorgenson, C. Racine, A. Doyle, and R.D. Boone, 2000: Observations of thermokarst and its impact on boreal forests in Alaska, United States. *Arctic, Antarctic and Alpine Research*, **32(3)**, 303-315.

Potter, C., P. Tan, V. Kumar, C. Kucharik, S. Klooster, V. Genovese, W. Cohen, and S. Healey, 2005. Recent history of large-scale ecosystem disturbances in North America derived from the AVHRR satellite record. *Ecosystems*, **8(7)**, 808.

Potter, C., S. Klooster, P. Tan, M. Steinbach, V. Kumar, and V. Genovese, 2003: Variability in terrestrial carbon sinks over two decades: Part 1—North America. *Earth Interactions*, **7**, Paper 12.

Prentice, I.C., G.D. Farquhar, M.J.R. Fasham, M.L. Goulden, M. Heimann, V.J. Jaramillo, H.S. Kheshgi, C. Le Quéré, R.J. Scholes, and D.W.R. Wallace, 2001: The carbon cycle and atmospheric carbon dioxide. In: *Climate Change 2001: The Scientific Basis.* Contribution of Working Group I to the Third Assessment Report of the Intergovernmental Panel on Climate Change [J. T. Houghton, Y. Ding, D. J. Griggs, M. Noguer, P. J. van der Linden, X. Dai, K. Maskell, and C. A. Johnson (eds.)]. Cambridge University Press, Cambridge, United Kingdom and New York, pp. 183-237.

Schimel, D., J. Melillo, H. Tian, A.D. McGuire, D. Kicklighter, T. Kittel, N. Rosenbloom, S. Running, P. Thornton, D. Ojima, W. Parton, R. Kelly, M. Sykes, R. Neilson, and B. Rizzo, 2000: Contribution of increasing CO_2 and climate to carbon storage by ecosystems in the United States. *Science*, **287(5460)**, 2004-2006.

Smith, L.C., Y. Sheng, G.M. MacDonald, L.D. Hinzman, 2005a: Disappearing Arctic Lakes. *Science*, **308(5727)**, 1429.

Smith, S.L., M.M. Burgess, and F.M. Nixon, 2001: Response of active layer and permafrost temperatures to warming during 1998 in the Mackenzie Delta, Northwest Territories and at Canadian Forces Station Alert and Baker Lake, Nunavut. *Geological Survey of Canada Current Research*, 2001-E5, 8 pp.

Smith, S.V., R.O. Sleezer, W.H. Renwick, and R.W. Buddemeier, 2005b: Fates of eroded soil organic carbon: Mississippi Basin case study. *Ecological Applications*, **15(6)**, 1929-1940.

Stallard, R.F., 1998: Terrestrial sedimentation and the carbon cycle: coupling weathering and erosion to carbon burial. *Global Biogeochemical Cycles*, **12(2)**, 231-257.

Wienert, A., 2006: From forestland to house lot: carbon stock changes and greenhouse gas emissions from exurban land development in central New Hampshire. Master's Thesis, Brown University, Providence, Rhode Island. 29 pp.

CHAPTER 10 REFERENCES

Ågren, G.I. and E. Bosatta, 2002: Reconciling differences in predictions of temperature response of soil organic matter. *Soil Biology and Biochemistry*, **34(1)**, 129–132.

Agriculture and Agri-Food Canada, 1999: The health of our air: toward sustainable agriculture in Canada. In: *Publication 1981/E* [Janzen, H.H., R.L. Desjardins, J.M.R. Asselin, and B. Grace (eds.)]. Agriculture and Agri-Foods Canada, Ottawa, Ontario, Canada, 40 pp.

Antle, J.M., S. Capalbo, S. Mooney, E.T. Elliott, and K. Paustian, 2001: Economic analysis of agricultural soil carbon sequestration: an integrated assessment approach. *Journal of Agricultural and Resource Economics*, **26(2)**, 344–367.

Antle, J.M., S.M. Capalbo, S. Mooney, D.K. Elliott, and K.H. Paustian, 2003: Spatial heterogeneity, contract design, and the efficiency of carbon sequestration policies for agriculture. *Journal of Environmental Economics and Management*, **46(2)**, 231–250.

Bellamy, P.H., P.J. Loveland, R.I. Bradley, R.M. Lark, and G.J.D. Kirk, 2005: Carbon losses from all soils across England and Wales 1978–2003. *Nature*, **437(7056)**, 245–248.

Boadi, D., C. Benchaar, J. Chiquette, and D. Masse, 2004: Mitigation strategies to reduce enteric methane emissions from dairy cows: Update review. *Canadian Journal of Animal Science*, **84(3)**, 319–335.

Boehm, M., B. Junkins, R. Desjardins, S.N. Kulshreshtha, and W. Lindwall, 2004: Sink potential of Canadian agricultural soils. *Climatic Change*, **65(3)**, 297–314.

Bradley, B.A., R.A. Houghton, J.F. Mustard, and S.P. Hamburg, 2006: Invasive grass reduces aboveground carbon stocks in shrublands of the Western US. *Global Change Biology*, **12(10)**, 1815–1822.

Buyanovsky, G.A. and G.H. Wagner, 1998: Carbon cycling in cultivated land and its global significance. *Global Change Biology*, **4(2)**, 131–141.

CAST (Council for Agricultural Science and Technology), 2004: *Climate Change and Greenhouse Gas Mitigation: Challenges and Opportunities for Agriculture*. [Paustian, K., B.A. Babcock, J. Hatvield, C.L. Kling, R. Lal, B.A. McCarl, S. McLaughlin, A.R. Mosier, W.M. Post, C.W. Rice, G.P. Robertson, N.J. Rosenberg, C. Rosenzweig, W.H. Schlesinger, and D. Zilberman (Task Force Members)]. CAST, Ames, IA, 120 pp.

CISCC (Comité Intersecretarial Sobre Cambio Climático)., 2001: *Second National Communication of Mexico to the UN Framework Convention on Climate Change*. Secretaría de Medio Ambiente y Recursos Naturales: Instituto Nacional de Ecología, México, D.F. México 374 pp. Available at http://unfccc.int/resource/docs/natc/mexnc2.pdf

Conant, R.T., S.J. Del Grosso, W.J. Parton, and K. Paustian, 2005: Nitrogen pools and fluxes in grassland soils sequestering carbon. *Nutrient Cycling in Agroecosystems*, **71(3)**, 239–248.

Conant, R.T. and K. Paustian, 2002: Potential soil carbon sequestration in overgrazed grassland ecosystems. *Global Biogeochemical Cycles*, **16(4)**, 1143.

Conant, R.T., K. Paustian, and E.T. Elliott, 2001: Grassland management and conversion into grassland: Effects on soil carbon. *Ecological Applications*, **11(2)**, 343–355.

Cox, P.M., R.A. Betts, C.D. Jones, S.A. Spall, and I.J. Totterdell, 2000: Acceleration of global warming due to carbon-cycle feedbacks in a coupled climate model. *Nature*, **408(6809)**, 184–187.

Davidson, E.A. and I.L. Ackerman, 1993: Change in soil carbon inventories following cultivation of previously untilled soils. *Biogeochemistry*, **20(3)**, 161–193.

Enquete Commission, 1995: *Protecting our Green Earth: How to Manage Global Warming Through Environmentally Sound Farming and Preservation of the World's Forests*. Economica Verlag, Bonn, Germany, 683 pp.

EPA (U.S. Environmental Protection Agency), 2000: *Options for Reducing Methane Emissions Internationally*. USEPA #430-R-90-006, EPA, Washington, DC.

EPA (U.S. Environmental Protection Agency), 2005: *Greenhouse Gas Mitigation Potential in U.S. Forestry and Agriculture*. USEPA #430-R-05-006, EPA Office of Atmospheric Programs, Washington, DC, 154 pp. Available at http://www.epa.gov/sequestration/greenhouse_gas.html

EPA (U.S. Environmental Protection Agency), 2006: *Inventory of U.S. Greenhouse Gas Emissions and Sinks: 1990–2004*. USEPA #430-R-06-002, EPA, Washington, DC, 459 pp. Available at http://epa.gov/climatechange/emissions/usgginv_archive.html

Eve, M.D., M. Sperow, K. Paustian, and R.F. Follett, 2002: National-scale estimation of changes in soil carbon stocks on agricultural lands. *Environmental Pollution*, **116(3)**, 431–438.

Follett, R.F., J.M. Kimble, and R. Lal, 2001a: *The Potential of U.S. Grazing Lands to Sequester Carbon and Mitigate the Greenhouse Effect*. CRC Press, Chelsea, MI, 442 pp.

Follett, R.F., E.G. Pruessner, S. Samson-Liebig, J.M. Kimble, and S. Waltman, 2001b: Carbon sequestration under the Conservation Reserve Program in the historical grassland soils of the United States of America. In: *Soil Management for Enhancing Carbon Sequestration* [Lal, R. and K. McSweeney (eds.)]. Soil Science Society of America, Madison, WI, pp. 1–14.

Friedl, M.A., A.H. Strahler, X. Zhang, and J. Hodges, 2002: The MODIS land cover product: multi-attribute mapping of global vegetation and land cover properties from time series MODIS data. *Proceedings of the International Geoscience and Remote Sensing Symposium*, **4**, 3199–3201.

Frye, W.W., 1984: Energy requirements in no-tillage. In: *No Tillage Agricultural Principles And Practices* [Phillips, R.E. and S.H. Phillips (eds.)]. Van Nostrand Reinhold, New York, pp. 127–151.

Giardina, C.P. and M.G. Ryan, 2000: Evidence that decomposition rates of organic carbon in mineral soil do not vary with temperature. *Nature*, **404(6780)**, 858–861.

Gregorich, E.G., P. Rochette, A.J. VandenBygaart, and D.A. Angers, 2005: Greenhouse gas contributions of agricultural soils and potential mitigation practices in Eastern Canada. *Soil & Tillage Research*, **83(1)**, 53–72.

Helgason, B. L., H.H. Janzen, M.H. Chantigny, C.F. Drury, B.H. Ellert, E.G. Gregorich, R.L. Lemke, E. Pattey, P. Rochette, and C. Wagner-Riddle, 2005: Toward improved coefficients for predicting direct N_2O emissions from soil in Canadian agroecosystems. *Nutrient Cycling in Agroecosystems*, **72(1)**, 87–99.

Houghton, R.A. and C.L. Goodale, 2004: Effects of land-use change on the carbon balance of terrestrial ecosystems. In: *Ecosystems and Land Use Change* [DeFries, R.S., G. P. Asner, and R.A. Houghton (eds.)]. Geophysical Monograph Series 153, American Geophysical Union, Washington DC, pp 85–98.

Houghton, R. A., J.L. Hackler, and K.T. Lawrence, 1999: The U.S. carbon budget: contributions from land-use change. *Science*, **285(5427)**, 574–578.

IPCC, 2001: *Third Assessment Report*. Cambridge University Press, Cambridge, United Kingdom, 87 pp.

ISRIC (International Soil Reference and Information Centre) 2002: *FAO Soil Database*. CD ROM, Rome, Italy.

Jackson, R.B., J.L. Banner, E.G. Jobbagy, W.T. Pockman, and D.H. Wall, 2002: Ecosystem carbon loss with woody plant invasion of grasslands. *Nature*, **418(6898)**, 623–626.

Jenkinson, D.S., D.E. Adams, and A. Wild, 1991: Model estimates of CO_2 emissions from soil in response to global warming. *Nature*, **351(6324)**, 304–306.

Johnson, K.A. and D.E. Johnson, 1995: Methane emissions from cattle. *Journal of Animal Science*, **73(8)**, 2483–2492.

Kätterer, T., M. Reichstein, O. Andren, and A. Lomander, 1998: Temperature dependence of organic matter decomposition: a critical review using literature data analyzed with different models. *Biology and Fertility of Soils*, **27(3)**, 258–262.

Keppler, F., J.T.G. Hamilton, M. Brass, and T. Rockmann, 2006: Methane emissions from terrestrial plants under aerobic conditions. *Nature*, **439(7073)**, 187–191.

Knorr, W., I.C. Prentice, J.I. House, and E.A. Holland, 2005: Long-term sensitivity of soil carbon turnover to warming. *Nature*, **433(7023)**, 298–301.

Kulshreshtha, S.N., B. Junkins, and R. Desjardins, 2000: Prioritizing greenhouse gas emission mitigation measures for agriculture. *Agricultural Systems*, **66(3)**, 145–166.

Lal, R., 2002: Why carbon sequestration in agricultural soils? In: *Agricultural Practices and Policies for Carbon Sequestration in Soil* [Kimble, J., R. Lal, and R.F. Follett (eds.)]. CRC Press, Boca Raton, FL, pp. 21–30.

Lal, R., 2004: Carbon emission from farm operations. *Environment International*, **30(7)**, 981–990.

Lal, R., R.F. Follett, and J.M. Kimble, 2003: Achieving soil carbon sequestration in the United States: a challenge to policy makers. *Soil Science*, **168**, 827–845.

Lal, R., J.M. Kimble, R.F. Follett, and C.V. Cole, 1998: *The Potential of U.S. Cropland to Sequester Carbon and Mitigate the Greenhouse Effect*. Ann Arbor Press, Chelsea, MI, 128 pp.

Lewandrowski, J., M. Peters, C. Jones, R. House, M. Sperow, M.D. Eve, and K. Paustian, 2004: *Economics of Sequestering Carbon in the U.S. Agricultural Sector*. Technical Bulletin No. TB 1909, Economic Research Service, Washington, DC.

Long, S.P., E.A. Ainsworth, A.D.B. Leakey, J. Nösberger and D.R. Ort, 2006: Food for thought: lower-than-expected crop yield stimulation with rising CO_2 concentrations. *Science*, **312(5782)**, 1918–1921.

Lynch, D.H., R.D.H. Cohen, A. Fredeen, G. Patterson, and R.C. Martin, 2005: Management of Canadian prairie region grazed grasslands: Soil C sequestration, livestock productivity and profitability. *Canadian Journal of Soil Science*, **85(2)**, 183–192.

Matin, A., P. Collas, D. Blain, C. Ha, C. Liang, L. MacDonald, S. McKibbon, C. Palmer, and R. Kerry, 2004: *Canada's Greenhouse Gas Inventory: 1990–2002*. Greenhouse Gas Division, Environment Canada, 13 pp.

Ministry of the Environment, 2006: National Inventory Report. In: *Trends in GHG Sources and Sinks in Canada, 1990-2004*. Greenhouse Gas Division, Environment Canada, 16 pp..

McCarl, B.A. and E.K. Schneider, 2001: The Cost of Greenhouse Gas Mitigation in U.S. Agriculture and Forestry. *Science*, **294(5551)**, 2481–2482.

Mosier, A., C. Kroeze, C. Nevison, O. Oenema, S. Seitzinger, and O. van Cleemput, 1998a: Closing the global N_2O budget: nitrous oxide emissions through the agricultural nitrogen cycle - OECD/IPCC/IEA phase II development of IPCC guidelines for national greenhouse gas inventory methodology. *Nutrient Cycling in Agroecosystems*, **52(2-3)**, 225–248.

Mosier, A.R., J.M. Duxbury, J.R. Freney, O. Heinemeyer, K. Minami, and D.E. Johnson, 1998b: Mitigating agricultural emissions of methane. *Climatic Change*, **40(1)**, 39–80.

Murray, B.C., B.A. McCarl, and H.C. Lee, 2004: Estimating leakage from forest carbon sequestration programs. *Land Economics*, **80(1)**, 109–124.

Nabuurs, G.-J., N.H. Ravindranath, K. Paustian, A. Freibauer, B. Hohenstein, W. Makundi, H. Aalde, A.Y. Abdelgadir, S.A.K. Anwar, J. Barton, K. Bickel, S. Bin-Musa, D. Blain, R. Boer, K. Byrne, C.C. Cerri, L. Ciccarese, D.-C. Choque, E. Duchemin, L. Dja, J. Ford-Robertson, W. Galinski, J.C. Germon, H. Ginzo, M. Gytarsky, L. Heath, D. Loustau, T. Mandouri, J. Mindas, K. Pingoud, J. Raison, V. Savchenko, D. Schone, R. Sievanen, K. Skog, K.A. Smith, D. Xu, M. Bakker, M. Bernoux, J. Bhatti, R.T. Conant, M.E. Harmon, Y. Hirakawa, T. Iehara, M. Ishizuka, E.G. Jobbagy, J. Laine, M. van der Merwe, I.K. Murthy, D. Nowak, S.M. Ogle, P. Sudha, R.J. Scholes, and X. Zhang, 2004: LUCF-sector good practice guidance. In: *IPCC Good Practice Guidance for LULUCF* [Penman, J., M. Gytarsky, T. Hirishi, T. Krug, and D. Kruger (eds.)]. Institute for Global Environmental Strategies, Hayama, Japan.

NAS (National Academy of Sciences), 2001: *Climate Change Science: An Analysis of Some Key Questions*. NAS, Committee on the Science of Climate Change, National Research Council, Washington, DC. 29 pp.

Nowak, R.S., D.S. Ellsworth, and S.D. Smith, 2004: Functional responses of plants to elevated atmospheric CO_2- do photosynthetic and productivity data from FACE experiments support early predictions? *New Phytologist*, **162(2)**, 253–280.

NRCS (Natural Resources Conservation Service), 2005: *Anaerobic Digestion Practice Standards*. U.S. Department of Agriculture, Washington, DC.

Ogle, S.M., F.J. Breidt, M.D. Eve, and K. Paustian, 2003: Uncertainty in estimating land use and management impacts on soil organic carbon storage for U.S. agricultural lands between 1982 and 1997. *Global Change Biology*, **9(11)**, 1521–1542.

Ogle, S.M., R.T. Conant, and K. Paustian, 2004: Deriving grassland management factors for a carbon accounting method developed by the Intergovernmental Panel on Climate Change. *Environmental Management*, **33(4)**, 474–484.

Pacala, S.W., G.C. Hurtt, D. Baker, P. Peylin, R.A. Houghton, R.A. Birdsey, L. Heath, E.T. Sundquist, R.F. Stallard, P. Ciais, P. Moorcroft, J.P. Caspersen, E. Shevliakova, B. Moore, G. Kohlmaier, E. Holland, M. Gloor, M.E. Harmon, S.M. Fan, J.L. Sarmiento, C.L. Goodale, D. Schimel, and C.B. Field, 2001: Consistent land- and atmosphere-based U.S. carbon sink estimates. *Science*, **292(5525)**, 2316-2320.

Paustian, K., O. Andren, H.H. Janzen, R. Lal, P. Smith, G. Tian, H. Tiessen, M. Van Noordwijk, and P.L. Woomer, 1997: Agricultural soils as a sink to mitigate CO_2 emissions. *Soil Use and Management*, **13(s4)**, 230–244.

Paustian, K., J.M. Antle, J. Sheehan, and E.A. Paul, 2006: *Agriculture's Role in Greenhouse Gas Mitigation*. Pew Center on Global Climate Change, Washington, DC, 76 pp.

Paustian, K., C.V. Cole, D. Sauerbeck, and N. Sampson, 1998: CO_2 mitigation by agriculture: an overview. *Climatic Change*, **40(1)**, 135–162.

Peoples, M.B., E.W. Boyer, K.W.T. Goulding, P. Heffer, V.A. Ochwoh, B. Vanlauwe, S. Wood, K. Yagi, and O. van Cleemput, 2004: Pathways of nitrogen loss and their impacts on human health and the environment. In: *Agriculture and the Nitrogen Cycle* [Mosier, A.R., J.K. Syers, and J.R. Freney (eds.)]. Island Press, Washington, DC, pp. 53–69.

Pimentel, D., P. Hepperly, J. Hanson, D. Douds, and R. Seidel, 2005: Environmental, energetic, and economic comparisons of organic and conventional farming systems. *Bioscience*, **55(7)**, 573–582.

Post, W.M. and K.C. Kwon, 2000: Soil carbon sequestration and land-use change: processes and potential. *Global Change Biology*, **6(3)**, 317–327.

Raymond, P.A. and J.J. Cole, 2003: Increase in the export of alkalinity from North America's largest river. *Science*, **301**, 88–91.

Reilly, J.M. and K.O. Fuglie, 1998: Future yield growth in field crops: what evidence exists? *Soil Tillage Research*, **47**, 275–290.

Robertson, G.P. and P.R. Grace, 2004: Greenhouse gas fluxes in tropical and temperate agriculture: the need for a full-cost accounting of global warming potentials. *Environment, Development and Sustainability*, **6**, 51–63.

Robertson, G.P., E.A. Paul, and R.R. Harwood, 2000: Greenhouse gases in intensive agriculture: contributions of individual gases to the radiative forcing of the atmosphere. *Science*, **289(5486)**, 1922–1925.

Sheinbaum, C. and O. Masera, 2000: Mitigating carbon emissions while advancing national development priorities: the case of Mexico. *Climatic Change*, **47(3)**, 259–282.

Six, J., S.M. Ogle, F.J. Briedt, R.T. Conant, A.R. Mosier, and K. Paustian, 2004: The potential to mitigate global warming with no-tillage management is only realized when practiced in the long term. *Global Change Biology*, **10(2)**, 155–160.

Smith, K.A. and F. Conen, 2004: Impacts of land management on fluxes of trace greenhouse gases. *Soil Use and Management*, **20**, 255–263.

Smith, P., K.W. Goulding, K.A. Smith, D.S. Powlson, J.U. Smith, P. Falloon, and K. Coleman, 2001: Enhancing the carbon sink in European agricultural soils: including trace gas flux estimates of carbon mitigation potential. *Nutrient Cycling in Agroecosystems*, **60(1-3)**, 237–252.

Smith, S.V., R.O. Sleezer, W.H. Renwick, and R.W. Buddemeier, 2005: Fates of eroded soil organic carbon: Mississippi basin case study. *Ecological Applications*, **15(6)**, 1929–1940.

Sobool, D. and S. Kulshreshtha, 2005: *Greenhouse Gas Emissions from Agriculture and Agri-Food Systems in Canada*. Department of Agricultural Economics, University of Saskatchewan, Saskatoon, Saskatchewan, Canada, 156 pp.

Sombroek, W.G., F.O. Nachtergaele, and A. Hebel, 1993: Amounts, dynamics and sequestering of carbon in tropical and subtropical soils. *Ambio*, **22(7)**, 417–426.

Sperow, M., M.D. Eve, and K. Paustian, 2003: Potential soil C sequestration on U.S. agricultural soils. *Climatic Change*, **57(3)**, 319–339.

Van Auken, O.W., 2000: Shrub invasions of North American semiarid grasslands. *Annual Review of Ecology and Systematics*, **31**, 197–205.

VandenBygaart, A.J., E.G. Gregorich, and D.A. Angers, 2003: Influence of agricultural management on soil organic carbon: A compendium and assessment of Canadian studies. *Canadian Journal of Soil Science*, **83(4)**, 363–380.

West, T.O. and G. Marland, 2003: Net carbon flux from agriculture: Carbon emissions, carbon sequestration, crop yield, and land-use change. *Biogeochemistry*, **63(1)**, 73–83.

West, T.O., G. Marland, A.W. King, W.M. Post, A.K. Jain, and K. Andrasko, 2004: Carbon management response curves: estimates of temporal soil carbon dynamics. *Environmental Management*, **33(4)**, 507–518.

Yoo, K., R. Amundson, A.M. Heimsath, and W.E. Dietrich, 2005: Erosion of upland hillslope soil organic carbon: coupling field measurements with a sediment transport model. *Global Biogeochemical Cycles*, **19(3)**, GB3003, doi:10.1029/2004GB002271.

CHAPTER 11 REFERENCES

Aber, J., R.P. Neilson, S. McNulty, J.M. Lenihan, D. Bachelet, and R.J. Drapek, 2001: Forest processes and global change: predicting the effects of individual and multiple stressors. *BioScience* **51(9)**, 735-751.

Albrecht, A. and S.T. Kandji, 2003: Carbon sequestration in tropical agroforestry systems. *Agriculture, Ecosystems and Environments*, **99(1-3)**, 15-27.

Amiro, B.D., J.B. Todd, B.M. Wotton, K.A. Logan, M.D. Flannigan, B.J. Stocks, J.A. Mason, D.L. Martell and K.G. Hirsch, 2001: Direct carbon emissions from Canadian forest fires, 1959-1999. *Canadian Journal of Forest Research*, **31(3)**, 512-525.

Apps, M.J., W.A. Kurz, S.J. Beukema, and J.S. Bhatti, 1999: Carbon budget of the Canadian forest product sector. *Environmental Science & Policy*, **2(1)**, 25-41.

Bachelet, D., R.P. Neilson, J.M. Lenihan, and R.J. Drapek, 2001: Climate change effects on vegetation distribution and carbon budget in the United States. *Ecosystems*, **4(3)**, 164-185.

Baldocchi, D.D. and J.S. Amthor, 2001: Canopy photosynthesis: history, measurements and models. In: *Terrestrial Global Productivity* [Roy, J., B. Saugier, and H. Mooney (eds.)]. Academic Press, San Diego, CA, pp. 9-31.

Barford, C.C., S.C. Wofsy, M.L. Goulden, J.W. Munger, E.H. Pyle, S.P. Urbanski, L. Hutyra, S.R. Saleska, D. Fitzjarrald, and K. Moore, 2001: Factors controlling long- and short-term sequestration of atmospheric CO_2 in a mid-latitude forest. *Science*, **294(5547)**, 1688-1691.

Bechtold, W.A. and P.L. Patterson (eds.), 2005: *The Enhanced Forest Inventory and Analysis Program - National Sampling Design and Estimation Procedures*. General Technical Report SRS-80, U.S. Department of Agriculture, Forest Service, Southern Research Station, Asheville, NC, 85 pp.

Birdsey, R.A. and L.S. Heath, 1995: Carbon changes in U.S. forests. In: *Productivity Of America's Forests* and Climate Change [Joyce, L.A. (ed.)]. General Technical Report RM-GTR-271, U.S. Department of Agriculture, Forest Service, Rocky Mountain Forest and Range Experiment Station, Fort Collins, CO, pp. 56-70.

Birdsey, R.A., R. Alig, and D. Adams, 2000: Mitigation activities in the forest sector to reduce emissions and enhance sinks of greenhouse gases. In: *The Impact of Climate Change on America's Forests: A Technical Document Supporting the 2000 USDA Forest Service RPA Assessment* [Joyce, L.A. and R.A. Birdsey (eds.)]. RMRS-GTR-59, U.S. Department of Agriculture, Forest Service, Rocky Mountain Research Station, Fort Collins, CO, pp. 112-131.

Birdsey, R.A., 2004: Data gaps for monitoring forest carbon in the United States: an inventory perspective. In: *Environmental Management* [Mickler, R.A. (ed.)]. **33(Suppl. 1)**, S1-S8.

Birdsey, R.A. and G.M. Lewis, 2003: Current and historical trends in use, management, and disturbance of U.S. forestlands. In: *The Potential of U.S. Forest Soils to Sequester Carbon and Mitigate the Greenhouse Effect* [Kimble, J.M., L.S. Heath, and R.A. Birdsey (eds.)]. CRC Press LLC, New York, pp. 15-33.

Birdsey, R., K. Pregitzer, and A. Lucier, 2006: Forest carbon management in the United States, 1600-2100. *Journal of Environmental Quality,* **35(4)**, 1461-1469.

Caldeira, K., M.G. Morgan, D. Baldocchi, P.G. Brewer, C.-T.A. Chen, G.-J. Nabuurs, N. Nakicenovic, and G.P. Robertson, 2004: A portfolio of carbon management options. In: *The Global Carbon Cycle: Integrating Humans, Climate, and the Natural World* [Field, C.B. and M.R. Raupach (eds.)]. Island Press, Washington, DC, pp. 103-129.

Canadian Forest Service, 2005: *State of the Forest Report, 2004-2005.* Canadian Forest Service, Natural Resources Canada, Ottawa, Ontario, Canada. Available at http://www.nrcan-rncan.gc.ca/cfs-scf/national/what-quoi/sof/latest_e.html

Caspersen, J.P., S.W. Pacala, J.C. Jenkins, G.C. Hurtt, P.R. Moorcraft, and R.A. Birdsey, 2000: Contributions of land-use history to carbon accumulation in U.S. forests. *Science,* **290(5494)**, 1148-1151.

Chen, J.M., W. Ju, J. Cihlar, D. Price, J. Liu, W. Chen, J. Pan, A. Black, and A. Barr, 2003: Spatial distribution of carbon sources and sinks in Canada's forests. *Tellus B,* **55(2)**, 622-641.

Dale, V.H., L.A. Joyce, S. McNulty, R.P. Neilson, M.P. Ayres, M.D. Flannigan, P.J. Hanson, L.C. Irland, A.E. Lugo, C.J. Peterson, D. Simberloff, F.J. Swanson, B.J. Stocks, and B. Wotton, 2001: Climate change and forest disturbances. *Bioscience,* **51(9)**, 723-734.

De Jong, B.H.J., S. Ochoa-Gaona, M.A. Castillo-Santiago, N. Ramirez-Marcial, and M.A. Cairns, 2000: Carbon fluxes and patterns of land-use/land-cover change in the Selva Lacandona, Mexico. *Ambio,* **29(8)**, 504-511.

Environment Canada, 2006: *National Inventory Report, 1990-2004: Greenhouse Gas Sources and Sinks in Canada,* Environment Canada, Ottawa, Ontario. Available at http://unfccc.int/national_reports/annex_i_ghg_inventories/national_inventories_submissions/items/3734.php

EPA (U.S. Environmental Protection Agency), 2005: *Greenhouse Gas Mitigation Potential in U.S. Forestry and Agriculture.* EPA, Washington, DC, 154 pp.

Finzi, A.C., D.J.P. Moore, E.H. DeLucia, J. Lichter, K.S. Hofmockel, R.P. Jackson, H.-S. Kim, R. Matamala, H.R. McCarthy, R. Oren, J.S. Pippen, and W.H. Schlesinger, 2006: Progressive nitrogen limitation of ecosystem processes under elevated CO_2 in a warm-temperate forest. *Ecology,* **87(1)**, 15-25.

Flannigan, M.D., K.A. Logan, B.D. Amiro, W.R. Skinner, and B.J. Stocks, 2005: Future area burned in Canada. *Climatic Change,* **72(1-2)**, 1-16.

Foley, J.A. and N. Ramankutty, 2004: A primer on the terrestrial carbon cycle: what we don't know but should. In: *The Global Carbon Cycle: Integrating Humans, Climate and the Natural World.* [Field, C.B. and M.R. Raupauch, (eds.)] Island Press, Washington DC, pp. 279-294.

FAO (Food and Agriculture Organization), 2001: *Global Forest Resources Assessment 2000. Main Report.* FAO Forestry Paper 140, Rome, Italy, 481 pp.

Goetz, S.J., A.G. Bunn, G.J. Fiske, and R.A. Houghton, 2005: Satellite-observed photosynthetic trends across boreal North America associated with climate and fire disturbance. *Proceedings of the National Academy of Sciences,* **102(38)**, 13521-13525.

Goodale, C.L., M.J. Apps, R.A. Birdsey, C.B. Field, L.S. Heath, R.A. Houghton, J.C. Jenkins, G.H. Kohlmaier, W. Kurz, S. Liu, G.-J. Nabuurs, S. Nilsson, and A.Z. Shvidenko, 2002: Forest carbon sinks in the Northern Hemisphere. *Ecological Applications,* **12(3)**, 891-899.

Haynes, R.W. (ed.), 2003: *An Analysis of the Timber Situation in the United States: 1952-2050.* General Technical Report PNW-GTR-560, U.S. Department of Agriculture, Forest Service, Portland, OR, 254 pp.

Heath, L.S. and J.E. Smith, 2004: Criterion 5, indicator 26: total forest ecosystem biomass and carbon pool, and if appropriate, by forest type, age class and successional change. In: *Data Report: A Supplement to the National Report on Sustainable Forests—2003* [Darr, D.R. (coord.)]. FS-766A, U.S. Department of Agriculture, Washington, DC, 14 pp. Available at http://www.fs.fed.us/research/sustain/contents.htm

Heath, L.S. and J. E. Smith, 2000: An assessment of uncertainty in forest carbon budget projections. *Environmental Science & Policy,* **3(2-3)**, 73-82.

Hogg, E.H. and P.Y. Bernier, 2005: Climate change impacts on drought-prone forests in western Canada. *Forestry Chronicle,* **81(5)**, 675-682.

Houghton, R.A., J.L. Hackler, and K.T. Lawrence, 1999: The U.S. carbon budget: contributions from land-use change, *Science,* **285(5427)**, 574-578.

Johnson, D.W., R.B. Thomas, K.L. Griffen, and D.T. Tissue, J.T. Ball, B.R. Strain, and R.F. Walker, 1998: Effects of carbon dioxide and nitrogen on growth and nitrogen uptake in ponderosa and loblolly pine. *Journal of Environmental Quality*, **27**, 414-425.

Johnston, M. and T. Williamson, 2005: Climate change implications for stand yields and soil expectation values: a northern Saskatchewan case study. *Forestry Chronicle*, **81**, 683-690.

Joyce, L., J. Baer, S. McNulty, V. Dale, A. Hansen, L. Irland, R. Neilson, and K. Skog, 2001: Potential consequences of climate variability and change for the forests of the United States. In: *Climate Change Impacts in the United States*. Report for the U.S Global Change Research Program. Cambridge University Press, Cambridge, United Kingdom, pp. 489-521.

Karnosky, D.F., D.R. Zak, K.S. Pregitzer, C.S. Awmack, J.G. Bockheim, R.E. Dickson, G.R. Hendrey, G.E. Host, J.S. King, B.J. Kopper, E.L. Kruger, M.E. Kubiske, R.L. Lindroth, W.J. Mattson, E.P. Mcdonald, A. Noormets, E. Oksanen, W.F.J. Parsons, K.E. Percy, G.K. Podila, D.E. Riemenschneider, P. Sharma, R. Thakur, A. Sôber, J. Sôber, W.S. Jones, S. Anttonen, E. Vapaavuori, B. Mankovska, W. Heilman, J.G. Isebrands, 2003: Tropospheric ozone moderates responses of temperate hardwood forests to elevated CO_2: a synthesis of molecular to ecosystem results from the Aspen FACE project. *Functional Ecology*, **17(3)**, 289-304.

Körner, C., 2000: Biosphere responses to CO_2 enrichment. *Ecological Applications*, **10(6)**, 1590-1619.

Körner, C., R. Asshoff, O. Bignucolo, S. Hättenschwiler, S.G. Keel, S. Peláez-Riedl, S. Pepin, R.T.W. Siegwolf, and G. Zotz, 2005: Carbon flux and growth in mature deciduous forest trees exposed to elevated CO_2. *Science*, **309(5739)**, 1360-1362.

Kurz, W.A., M.J. Apps, T.M. Webb, and P.J. McNamee, 1992: *The Carbon Budget of the Canadian Forest Sector: Phase 1*. Information Report NOR-X-326, Forestry Canada, Northern Forestry Centre, Edmonton, Alberta, Canada, 93pp.

Kurz, W.A., S. Beukema, and M.J. Apps, 1998: Carbon budget implications of the transition from natural to managed disturbance regimes in forest landscapes. *Mitigation and Adaptation Strategies for Global Change*, **2(4)**, 405-421.

Kurz, W.A. and M.J. Apps, 1999: A 70-year retrospective analysis of carbon fluxes in the Canadian forest sector. *Ecological Applications*, **9(2)**, 526-547.

Kurz, W., M. Apps, E. Banfield, and G. Stinson, 2002: Forest carbon accounting at the operational scale. *The Forestry Chronicle*, **78**, 672-679.

Leenhouts, B., 1998: Assessment of biomass burning in the conterminous United States. *Conservation Ecology*, **2(1)**, 1.

Lewandrowski, J., M. Sperow, M. Peters, M. Eve, C. Jones, K. Paustian, and R. House, 2004: *Economics of Sequestering Carbon in the U.S. Agricultural Sector*. Technical Bulletin 1909, U.S. Department of Agriculture, Economic Research Service, Washington, DC, 61 pp.

Lichter, J., S.H. Barron, C.E. Bevacqua, A.C. Finzi, K.F. Irving, E.A. Stemmler, and W.H. Schlesinger, 2005: Soil carbon sequestration and turnover in a pine forest after six years of atmospheric CO_2 enrichment. *Ecology*, **86(7)**, 1835-1847.

Lippke, B., J. Wilson, J. Perez-Garcia, J. Bowyer, and J. Miel, 2004: CORRIM: Life cycle environmental performance of renewable building materials. *Forest Products Journal*, **54(6)**, 8-19.

Loya, W.M., K.S. Pregitzer, N.J. Karberg, J.S. King, and C.P. Giardina, 2003: Reduction of soil carbon formation by tropospheric ozone under increased carbon dioxide levels. *Nature*, **425(6959)**, 705-707.

Masera, O., M.J. Ordóñez, and R. Dirzo, 1997: Carbon emissions from Mexican forests: the current situation and long-term scenarios. *Climatic Change*, **35(3)**, 265-295.

Masera, O., A. Delia Cerón, and A. Ordóñez, 2001: Forestry mitigation options for Mexico: finding synergies between national sustainable development priorities and global concerns. *Mitigation and Adaptation Strategies for Global Change*, **6(3-4)**, 291-312.

McNulty, S.G., 2002: Hurricane impacts on U.S. forest carbon sequestration. *Environmental Pollution*, **116(Supplement 1)**, S17-S24.

Nair, P.K.R. and V.D. Nair, 2003: Carbon storage in North American agroforestry systems. In: *The Potential of U.S. Forest Soils to Sequester Carbon and Mitigate the Greenhouse Gas Effect*. [Kimble, J., L.S. Heath, R.A. Birdsey, and R. Lal (eds.)]. CRC Press, Boca Raton, FL, pp. 333-346.

NRCan (Natural Resources Canada), 2005: *The State of Canada's Forests*. Canadian Forest Service, NRC, Ottawa, Ontario, Canada. Available at http://www.nrcan-rncan.gc.ca/cfs-scf/national/what-quoi/sof/latest_e.html

Neilson, R.P., I.C. Prentice, B. Smith, T.G.F. Kittel, and D. Viner, 1998: Simulated changes in vegetation distribution under global warming. In: *The Regional Impacts of Climate Change: An Assessment of Vulnerability* [Watson, R.T., M.C. Zinyowera, R.H. Moss, and D.J. Dokken (eds.)]. Cambridge University Press, Cambridge, United Kingdom, pp. 439-456.

Nelson, K.C. and B.H.J. de Jong, 2003: Making global initiatives local realities: carbon mitigation projects in Chiapas, Mexico. *Global Environmental Change*, **13(1)**, 19-30.

NFDP (National Forestry Database Program), 2005: *Compendium of Canadian Forestry Statistics*. National Forestry Database Program, Canadian Council of Forest Ministers, Ottawa, Ontario, Canada. Available at http://nfdp.ccfm.org/compendium/index_e.php

Norby, R.J., E.H. DeLucia, B. Gielen, C. Calfapietra, C.P. Giardina, J.S. King, J. Ledford, H.R. McCarthy, D.J.P. Moore, R. Ceulemans, P. De Angelis, A.C. Finzi, D.F. Karnosky, M.E. Kubiske, M. Lukac, K.S. Pregitzer, G.E. Scarascia-Mugnozza, W.H. Schlesinger, and R. Oren, 2005: Forest response to elevated CO_2 is conserved across a broad range of productivity. *Proceedings of the National Academy of Sciences*, **102(50)**, 18052-18056.

Nowak, R.S., D.S. Ellsworth, and S.D. Smith, 2004: Functional responses of plants to elevated atmospheric CO_2- do photosynthetic and productivity data from FACE experiments support early predictions? *New Phytologist*, **162(2)**, 253-280.

Ollinger, S.V., J.D. Aber, P.B. Reich, and R.J. Freuder, 2002: Interactive effects of nitrogen deposition, tropospheric ozone, elevated CO_2 land use history on the carbon dynamics of northern hardwood forests. *Global Change Biology*, **8(6)**, 545-562.

Oren, R., D.S. Ellsworth, K.H. Johnsen, N. Phillips, B.E. Ewers, C. Maier, K.V.R. Schäfer, H. McCarthy, G. Hendrey, S.G. McNulty, and G.G. Katul, 2001: Soil fertility limits carbon sequestration by forest ecosystems in a CO_2-enriched atmosphere. *Nature*, **411(6836)**, 469-472.

Osher, L.J., P.A. Matson, and R. Amundson, 2003: Effect of land use change on soil carbon in Hawaii. *Biogeochemistry*, **65(2)**, 213-232.

Pacala, S.W., G.C. Hurtt, D. Baker, P. Peylin, R.A. Houghton, R.A. Birdsey, L. Heath, E.T. Sundquist, R.F. Stallard, P. Ciais, P. Moorcroft, J.P. Caspersen, E. Shevliakova, B. Moore, G. Kohlmaier, E. Holland, M. Gloor, M.E. Harmon, S.M. Fan, J.L. Sarmiento, C.L. Goodale, D. Schimel, and C.B. Field, 2001: Consistent land- and atmosphere-based U.S. carbon sink estimates. *Science*, **292(5525)**, 2316-2320.

Palacio-Prieto, J.L., G. Bocco, A. Velázquez, J.-F. Mas, F. Takaki-Takaki, A. Victoria, L. Luna-González, G. Gómez-Rodríguez, J. López-Garcia, M.P. Muñoz, I. Trejo-Vázquez, A.P. Higuera, J. Prado-Molina, A. Rodriguez-Aguilar, R. Mayorga-Saucedo, and F.G. Medrano, 2000: La condición actual de los recursos forestales en México: resultados del Inventario Forestal Nacional 2000. *Investigaciones Geográficas, Boletín del Instituto de Geografía, UNAM*, **43**, 183-203.

Pan, Y., J. Melillo, A.D. McGuire, D. Kicklighter, L.F. Pitelka, K.A. Hibbard, L.L. Pierce, S.W. Running, D.S. Ojima, W.J. Parton, and D.S. Schimel, 1998: Modeled responses of terrestrial ecosystems to elevated atmospheric CO_2: a comparison of simulations by the biogeochemistry models of the Vegetation/Ecosystem Modeling and Analysis Project (VEMAP). *Oecologia*, **114(4)**, 389-404.

Parisien, M.-A., V. Kafka, N. Flynn, K.G. Hirsch, J.B. Todd, and M.D. Flannigan, 2005: *Fire Behavior Potential in Central Saskatchewan Under Predicted Climate Change*. PARC Summary Document 05-01, PARC (Prairie Adaptation Research Collaborative), Regina, Saskatchewan, Canada, 12 pp.

Potter, C., S.A. Klooster, R. Myneni, V. Genovese, P. Tan, and V. Kumar, 2003: Continental scale comparisons of terrestrial carbon sinks estimated from satellite data and ecosystem modeling 1982-98. *Global and Planetary Change*, **39(3-4)**, 201-213.

Price, D.T., D.H. Halliwell, M.J. Apps, W.A. Kurz, and S.R. Curry, 1997: Comprehensive assessment of carbon stocks and fluxes in a Boreal-Cordilleran forest management unit. *Canadian Journal of Forest Research*, **27(12)**, 2005-2016.

Price, D.T., D.W. McKenney, P. Papadopol, T. Logan, and M.F. Hutchinson, 2004: *High Resolution Future Scenario Climate Data for North America*. Proceedings of the American Meteorological Society 26th Conference on Agricultural and Forest Meteorology, Vancouver, British Columbia, Canada, 23-26 August 2004, 13 pp.

Price, D.T., C.H. Peng, M.J. Apps, and D.H. Halliwell, 1999: Simulating effects of climate change on boreal ecosystem carbon pools in central Canada. *Journal of Biogeography*, **26(6)**, 1237-1248.

Proctor, P., L.S. Heath, P.C. Van Deusen, J.H. Gove, and J.E. Smith, 2005: COLE: a web-based tool for interfacing with forest inventory data. In: *Proceedings of the Fourth Annual Forest Inventory and Analysis Symposium* [McRoberts, R.E., G.A. Reams, P.C. Van Deusen, W.H. McWilliams, C.J. Cieszewski (eds.)]. GTR-NC-252, U.S. Department of Agriculture, Forest Service. St. Paul, MN, 258 pp.

Running, S.W., R.R. Nemani, F.A. Heinsch, M.S. Zhao, M. Reeves, and H. Hashimoto, 2004: A continuous satellite-derived measure of global terrestrial primary production. *Bioscience*, **54(6)**, 547-560.

Schimel, D., J. Melillo, H. Tian, A.D. McGuire, D. Kicklighter, T. Kittel, N. Rosenbloom, S. Running, P. Thornton, D. Ojima, W. Parton, R. Kelly, M. Sykes, R. Neilson, and B. Rizzo, 2000: Contribution of increasing CO_2 and climate to carbon storage by ecosystems in the United States. *Science*, **287(5460)**, 2004-2006.

Schoene, D. and M. Netto, 2005: The Kyoto Protocol: what does it mean for forests and forestry? *Unasylva*, **222(56)**, 3-11.

Secretaría de Medio Ambiente, Recursos Naturales y Pesca (SEMARNAP), 1996: *Programa Forestal y de Suelo 1995-2000*. Poder ejecutivo Federal, SEMARNAP, México City, 79 pp.

Skog, K.E. and G.A. Nicholson, 1998: Carbon cycling through wood products: the role of wood and paper products in carbon sequestration. *Forest Products Journal*, **48(7)**, 75-83. Available at http://www.fpl.fs.fed.us/documnts/pdf1998/skog98a.pdf

Smith, W.B., P.D. Miles, J.S. Vissage, and S.A. Pugh, 2004: *Forest Resources of the United States, 2002*. General Technical Report NC-241, U.S. Department of Agriculture, Forest Service, North Central Research Station, St. Paul, MN, 137 pp.

Smith, J.E. and L.S. Heath, 2005: Land use change and forestry and related sections. In: *Inventory of U.S. Greenhouse Gas Emissions and Sinks: 1990-2003*. Excerpted, EPA 430-R-05-003, U.S. Environmental Protection Agency. Available at http://yosemite.epa.gov/oar/globalwarming.nsf/content/ResourceCenterPublicationsGHGEmissions.html

Smith, J.E. and L.S. Heath, 2000: Considerations for interpreting probabilistic estimates of uncertainty of forest carbon. In: *The Impact of Climate Change on America's Forests* [Joyce, L.A. and R. Birdsey (eds.)]. General Technical Report RMRS-GTR-59, U.S. Department of Agriculture, Forest Service, pp. 102-111.

Smith, J.E., L.S. Heath, K.E. Skog, and R.A. Birdsey, 2006: *Methods for Calculating Forest Ecosystem and Harvested Carbon with Standard Estimates for Forest Types of the United States*. General Technical Report NE-343 U.S. Department of Agriculture, Forest Service, Newtown Square, PA. 216 pp.

Soto-Pinto, L., G. Jimenez-Ferrer, A.V. Guillen, B. de Jong Bergsma, and E. Esquivel-Bazan, 2001: Experiencia agroforestral para la captura de carbono en comunidades indigenas de Mexico. *Revista Forestal Iberoamericana*, **1**, 44-50.

Stavins, R.N. and K.R. Richards, 2005: *The Cost of U.S. Forest-Based Carbon Sequestration.* The Pew Center on Global Climate Change, Arlington, VA, 40 pp. Available at http://www.pewclimate.org

Torres, R.J.M., 2004: *Estudio de tendencias y perspectivas del sector forestal en América Latina al año 2020.* Informe Nacional México, FAO. Available at http://www.fao.org/documents/show_cdr.asp?url_file=/docrep/006/j2215s/j2215s11.htm

Totten, M., 1999: *Getting it Right: Emerging Markets for Storing Carbon in Forests.* World Resources Institute, Washington, DC, 49 pp.

Turner, D.P., W.D. Ritts, W.B. Cohen, S.T. Gower, S.W. Running, M. Zhao, M.H. Costa, A. Kirschbaum, J. Ham, S. Saleska, and D.E. Ahl, 2006: Evaluation of MODIS NPP and GPP products across multiple biomes. *Remote Sensing of Environment*, **102(3-4)**, 282-292.

U.S. Climate Change Science Program, 2003: *Strategic Plan for the Climate Change Science Program.* Washington, DC. Available at http://www.climatescience.gov/Library/stratplan2003/default.htm

Van Tuyl, S., B.E. Law, D.P. Turner, and A.I. Gitelman, 2005: Variability in net primary production and carbon storage across Oregon forests - an assessment integrating data from forest inventories, intensive sites, and remote sensing. *Forest Ecology and Management*, **209(3)**, 273-291.

VEMAP Members (Melillo, J. M., J. Borchers, J. Chaney, H. Fisher, S. Fox, A. Haxeltine, A, Janetos, D.W. Kicklighter, T.G.F. Kittel, A.D. McGuire, R. McKeown, R. Neilson, R. Nemani, D.S. Ojima, T. Painter, Y. Pan, W.J. Parton, L. Pierce, L. Pitelka, C. Prentice, B. Rizzo, N.A. Rosenbloom, S. Running, D.S. Schimel, S. Sitch, .S. T.; Smith, I. Woodward, (VEMAP Members) 1995: Vegetation/ecosystem modeling and analysis project (VEMAP): comparing biogeography and biogeochemistry models in a continental-scale study of terrestrial ecosystem responses to climate change and CO_2 doubling. *Global Biogeochemical Cycles*, **9(4)**, 407-437.

Volney, J.A.W. and K. Hirsch, 2005: Disturbing forest disturbances. *Forestry Chronicle*, **81(5)**, 662-668.

Weber, M.G. and M.D. Flannigan, 1997: Canadian boreal forest ecosystem structure and function in a changing climate: impact on fire regimes. *Environmental Reviews*, **5(3-4)**, 145-166.

Winrock International, 2005: *Ecosystem Services.* Date accessed unknown. Available at http://www.winrock.org/what/projects.cfm?BU=9086

CHAPTER 12 REFERENCES

AMAP (Arctic Monitoring and Assessment Programme), 2004: *AMAP Assessment 2002: Persistent Organic Pollutants in the Arctic.* AMAP, Oslo, Norway, 310 pp.

Bailey, R. and C.T. Cushwa, 1981: *Ecoregions of North America.* 1:12 million scale map, U.S. Forest Service and U.S. Fish and Wildlife Service.

Bockheim, J.G. and C. Tarnocai, 1998: Recognition of cryoturbation for classifying permafrost-affected soils. *Geoderma*, **81(3-4)**, 281-293.

Brown, J., O.J. Ferrians, Jr., J.A. Heginbottom, and E.S. Melnikov, 1997: *Circum-Arctic Map of Permafrost and Ground Ice Conditions.* 1:10 million scale map, International Permafrost Association.

Burgess, M.M. and C. Tarnocai, 1997: Peatlands in the discontinuous permafrost zone along the Norman Wells pipeline, Canada. In: *Proceedings of the International Symposium on Physics, Chemistry, and Ecology of Seasonally Frozen Soils* Fairbanks, Alaska, June 10-12, 1997 [Iskandr, I.K., E.A. Wright, J.K. Radke, B.S. Sharratt, P.H. Groenevelt, and L.D. Hinzman (eds.)]. Special Report 97-10, U.S. Army Cold Regions Research and Engineering Laboratory, Hanover, NH, pp. 417-424.

Christensen, T., 1991: Arctic and sub-Arctic soil emissions: possible implications for global climate change. *Polar Record*, **27**, 205-210.

Cryosol Working Group, 2001: *Northern and Mid Latitudes Soil Database, Version 1.* National Soil Database, Research Branch, Agriculture and Agri-Food Canada, Ottawa, Ontario, Canada.

Davidson, E.A. and I.A. Janssens, 2006: Temperature sensitivity of soil carbon decomposition and feedbacks to climate change. *Nature*, **440(7081)**, 165-173.

Driscoll, C.T., J. Holsapple, C.L. Schofield, and R. Munson, 1998: The chemistry and transport of mercury in a small wetland in the Adirondack region of New York, USA. *Biogeochemistry*, **40(2/3)**, 137-146.

Ecoregions Working Group, 1989: *Ecoclimatic Regions of Canada, First Approximation.* Ecoregions Working Group of the Canada Committee on Ecological Land Classification, Ecological Land Classification Series, No. 23, Sustainable Development Branch, Canadian Wildlife Service, Conservation and Protection, Environment Canada, Ottawa, Ontario, Canada, 119 pp.

Gorham, E., 1988: Canada's peatlands: their importance for the global carbon cycle and possible effect of "greenhouse" climate warming. *Transactions of the Royal Society of Canada*, *Series V*, **3**, 21-23.

Hengeveld, H.G., 2000: Projections for Canada's climate future: a discussion of recent simulations with the Canadian Global Climate Model. In: *Climate Change Digest.* CCD 00-01, Special Edition, 27 pp. Last accessed April 6, 2005, Meteorological Service of Canada, Environment Canada, Downsview, Ontario, Canada. Available at http://www.msc.ec.gc.ca/saib/climate/docs/ccd_00-01.pdf

Jorgenson, M.T., C.H. Racine, J.C. Walters, and T.E. Osterkamp, 2001: Permafrost degradation and ecological changes associated with a warming climate in central Alaska. *Climatic Change*, **48(4)**, 551-579.

Jorgenson, M.Y. and J. Brown. 2005: Classification of the Alaskan Beaufort Sea Coast and estimation of carbon and sediment inputs from coastal erosion. *Geo-marine Letters.* **25(2-3)**, 69-80.

Kettles, I.M. and C. Tarnocai, 1999: Development of a model for estimating the sensitivity of Canadian peatlands to climate warming. *Géographie physique et Quaternaire*, **53(3)**, 323-338.

Kokelj, S.V. and C.R. Burn, 2005: Geochemistry of the active layer and near-surface permafrost, Mackenzie delta region, Northwest Territories, Canada. *Canadian Journal of Earth Sciences*, **42(1)**, 37-48.

Kuhry, G.P., 1994: The role of fire in the development of Sphagnum-dominated peatlands in the western boreal Canada. *Journal of Ecology*, **82(4)**, 899-910.

Lacelle, B., C. Tarnocai, S. Waltman, J. Kimble, N. Bliss, B. Worstell, F. Orozco-Chavez, and B. Jakobsen, 2000: *North American Soil Organic Carbon Map.* 1:10 million scale map, Agriculture and Agri-Food Canada, USDA, USGS, INEGI and Institute of Geography, University of Copenhagen.

Liblik, L.K., T.R. Moore, J.L. Bubier, and S.D. Robinson, 1997: Methane emissions from wetlands in the zone of discontinuous permafrost: Fort Simpson, Northwest Territories, Canada. *Global Biogeochemical Cycles*, **11(4)**, 485-494.

Mackay, J.R., 1980: The origin of hummocks, western Arctic coast, Canada. *Canadian Journal of Earth Sciences*, **13(7)**, 889-897.

Melillo, J.M., T.V. Callaghan, F.I. Woodward, E. Salati, and S.K. Sinha, 1990: Climate change - effects on ecosystems (Chapter 10). In: *Climate Change: The IPCC Scientific Assessment* [Houghton, J.T., G.J. Jenkins, and J.J. Ephraums (eds.)]. Cambridge University Press, New York, pp. 283-310.

Michaelson, G.J., C.L. Ping, and J.M. Kimble, 1996: Carbon storage and distribution in tundra soils of Arctic Alaska, U.S.A. *Arctic and Alpine Research*, **28(4)**, 414-424.

Moore, T.R., 1997: Dissolved organic carbon: sources, sinks, and fluxes and role in the soil carbon cycle (Chapter 19). In: *Soil Processes and the Carbon Cycle* [Lal, R., J.M. Kimble, R.F. Follett, and B.A. Stewart (eds.)]. *Advances in Soil Science*, CRC Press, Boca Raton, FL, pp. 281-292.

Moore, T.R. and N.T. Roulet, 1995: Methane emissions from Canadian peatlands (Chapter 12). In: *Soils and Global Change* [Lal, R., J. Kimble, E. Levine, and B.A. Stewart (eds.)]. CRC Lewis Publishers, Boca Raton, FL, pp. 153-164.

National Wetlands Working Group, 1988: *Wetlands of Canada.* Ecological Land Classification Series No. 24, Sustainable Development Branch, Environment Canada, Ottawa, Ontario, and Polyscience Publications, Inc., Montreal, Quebec. 452 pp.

Oechel, W. and G.L. Vourlitis, 1994: The effect of climate change on land-atmosphere feedbacks in arctic tundra regions. *Trends in Ecology and Evolution*, **9(9)**, 324-329.

Peterson, R.A. and W.B. Krantz, 2003: A mechanism for differential frost heave and its implications for patterned-ground formation. *Journal of Glaciology*, **49(164)**, 69-80.

Ping, C.L., T. Jorgenson, J. Brown, L.D. Guo, and Y. Shur, 2006: Coastal erosion across northern Alaska and community action. *Arctic Forum 2006*, Arctic Research Consortium of the U.S. (ARCUS), Fairbanks, AK, p 53.

Reader, R.J. and J.M. Stewart, 1972: The relationship between net primary production and accumulation for a peatland in southeastern Manitoba. *Ecology*, **53(6)**, 1024-1037.

Ritchie, J.C., 1987: *Postglacial Vegetation of Canada.* Cambridge University Press, New York, 178 pp.

Robinson, S.D. and T.R. Moore, 1999: Carbon and peat accumulation over the past 1200 years in a landscape with discontinuous permafrost, northwestern Canada. *Global Biogeochemical Cycles*, **13(2)**, 591-601.

Robinson, S.D. and T.R. Moore, 2000: The influence of permafrost and fire upon carbon accumulation in High Boreal peatlands, Northwest Territories, Canada. *Arctic, Antarctic, and Alpine Research*, **32(2)**, 155-166.

Robinson, S.D., M.R. Turetsky, I.M. Kettles, and R.K. Wieder, 2003: Permafrost and peatland carbon sink capacity with increasing latitude. In: *Permafrost* [Phillips, M., S.M. Springman, and L.U. Arenson (eds.)]. Proceedings of the 8th International Conference on Permafrost, 21-25 July 2003, Zurich, Switzerland, **2**, 965-970.

Soil Carbon Database Working Group, 1993: *Soil Carbon for Canadian Soils.* Digital database, Centre for Land and Biological Resources Research, Research Branch, Agriculture and Agri-Food Canada, Ottawa, Ontario, Canada, 137 pp.

Suchanek, T.H., P.J. Richerson, J.R. Flanders, D.C. Nelson, L.H. Mullen, L.L. Brester, and J.C. Becker, 2000: Monitoring interannual variability reveals sources of mercury contamination in Clear Lake, CA. *Environmental Monitoring and Assessment*, **64(1)**, 299-310.

Tarnocai, C., 1998: The amount of organic carbon in various soil orders and ecological provinces in Canada. In: *Soil Processes and the Carbon Cycle* [Lal, R., J.M. Kimble, R.L.F. Follett, and B.A. Stewart (eds.)]. *Advances in Soil Science*, CRC Press, New York, 81-92.

Tarnocai, C., 1999: The effect of climate warming on the carbon balance of Cryosols in Canada. In: *Cryosols and Cryogenic Environments* [Tarnocai, C., R. King, and S. Smith (eds.)]. Special issue of *Permafrost and Periglacial Processes*, **10(3)**, 251-263.

Tarnocai, C., 2000: Carbon pools in soils of the Arctic, Subarctic and Boreal regions of Canada. In: *Global Climate Change and Cold Regions Ecosystems* [Lal, R., J.M. Kimble, and B.A. Stewart (eds.)]. *Advances in Soil Science*, Lewis Publishers, Boca Raton, FL, pp. 91-103.

Tarnocai, C., 2006: The effect of climate change on carbon in Canadian peatlands. *Global and Planetary Change*, **53(3)**, 222-232.

Tarnocai, C., I.M. Kettles, and B. Lacelle, 2005: *Peatlands of Canada Database.* Digital database, Agriculture and Agri-Food Canada, Research Branch, Ottawa, Ontario, Canada.

Trumbore, S.E. and J.W. Harden, 1997: Accumulation and turnover of carbon in organic and mineral soils of the BOREAS (Boreal Ecosystem-Atmosphere Study) northern study area. *Journal of Geophysical Research*, **102(D24)**, 28817-28830.

Turetsky, M.R., B.D. Amiro, E. Bosch, and J.S. Bhatti, 2004: Historical burn area in western Canadian peatlands and its relationship to fire weather indices. *Global Biogeochemical Cycles*, **18**, GB4014, doi:10.1029/2004GB002222.

Vandenberghe, J., 1992: Cryoturbations: a sediment structural analysis. *Permafrost and Periglacial Processes*, **4**, 121-135.

Van Everdingen, R. (ed.), 1998 revised May 2005: *Multi-language Glossary of Permafrost and Related Ground-Ice Terms*. National Snow and Ice Data Center/World Data Center for Glaciology, Boulder, CO, 90 pp., Available at http://nsidc.org/fgdc/glossary

Van Vliet-Lanoë, B., 1991: Differential frost heave, load casting and convection: converging mechanisms; a discussion of the origin of cryoturbations. *Permafrost and Periglacial Processes*, **2**, 123-139.

Vitt, D.H., L.A. Halsey, I.E. Bauer, and C. Campbell, 2000: Spatial and temporal trends in carbon storage of peatlands of continental western Canada through the Holocene. *Canadian Journal of Earth Sciences*, **37(5)**, 683-693.

Walker, D.A., V.E. Romanovsky, W.B. Krantz, C.L. Ping, R.A. Peterson, M.K. Raynolds, H.E. Epstein, J.G. Jia, and D.C. Wirth, 2002: *Biocomplexity of Frost Boil Ecosystem on the Arctic Slope, Alaska*. ARCUS 14th Annual Meeting and Arctic Forum 2002, Arlington, VA. Available at http://siempre.arcus.org/4DACTION/wi_pos_displayAbstract/5/391

Zoltai, S.C., C. Tarnocai, and W.W. Pettapiece, 1978: *Age of Cryoturbated Organic Material in Earth Hummocks From the Canadian Arctic*. Proceedings of the Third International Conference on Permafrost, Edmonton, Alberta, Canada, pp. 325-331.

Zoltai, S.C., S. Taylor, J.K. Jeglum, G.F. Mills, and J.D. Johnson, 1988: Wetlands of Boreal Canada. In: *Wetlands of Canada*. Ecological Land Classification Series, No. 24, National Wetlands Working Group, Sustainable Development Branch, Environment Canada, Ottawa, Canada, and Polyscience Publications, Montreal, Quebec, Canada, pp. 97-154.

CHAPTER 13 REFERENCES

Armentano, T.B. and E.S. Menges, 1986: Patterns of change in the carbon balance of organic soil-wetlands of the temperate zone. *Journal of Ecology*, **74(3)**, 755–774.

Aselmann, I. and P.J. Crutzen, 1989: Global distribution of natural freshwater wetlands and rice paddies, their net primary productivity, seasonality and possible methane emissions. *Journal of Atmospheric Chemistry*, **8(4)**, 307–359.

Barker, J.R., G.A. Baumgardner, D.P. Turner, and J.J. Lee, 1996: Carbon dynamics of the conservation and wetland reserve program. *Journal of Soil and Water Conservation*, **51(4)**, 340–346.

Bartlett, K.B. and R.C. Harriss, 1993: Review and assessment of methane emissions from wetlands. *Chemosphere*, **26(1-4)**, 261–320.

Blunier, T., J. Chappellaz, J. Schwander, B. Stauffer, and D. Raynaud, 1995: Variations in atmospheric methane concentration during the Holocene epoch. *Nature*, **374(6517)**, 46–49.

Bourne, J., 2000: Louisiana's vanishing wetlands: going, going... *Science*, **289(5486)**, 1860–1863.

Bridgham, S.D., C.L. Ping, J.L. Richardson, and K. Updegraff, 2000: Soils of northern peatlands: Histosols and Gelisols. In: *Wetland Soils: Genesis, Hydrology, Landscapes, and Classification* [Richardson, J.L. and M.J. Vepraskas (eds.)]. CRC Press, Boca Raton, FL, pp. 343–370.

Bridgham, S.D., K. Updegraff, and J. Pastor, 1998: Carbon, nitrogen, and phosphorus mineralization in northern wetlands. *Ecology*, **79(5)**, 1545–1561.

Cao, M., K. Gregson, and S. Marshall, 1998: Global methane emission from wetlands and its sensitivity to climate change. *Atmospheric Environment*, **32(19)**, 3293–3299.

Chappellaz, J., T. Bluniert, D. Raynaud, J.M. Barnola, J. Schwander, and B. Stauffert, 1993: Synchronous changes in atmospheric CH_4 and Greenland climate between 40 and 8 kyr B.P. *Nature*, **366(6454)**, 443–445.

Cleary, J., N.T. Roulet, and T.R. Moore, 2005: Greenhouse gas emissions from Canadian peat extraction, 1990–2000: a life-cycle analysis. *Ambio*, **34(6)**, 456–461.

Cowardin, L.M., V. Carter, F.C. Golet, and E.T. LaRoe, 1979: *Classification of Wetlands and Deepwater Habitats of the United States*. FWS/OBS-79/31, Fish and Wildlife Service, U.S. Department of the Interior, Washington, DC.

Dahl, T.E., 2000: *Status and Trends of Wetlands in the Conterminous United States, 1986 to 1997*. U.S. Department of the Interior, Fish and Wildlife Service, Washington, DC.

Day Jr., J.W., G.P. Shafer, L.D. Britsch, D.J. Reed, S.R. Hawes, and D. Cahoon, 2000: Pattern and process of land loss in the Mississippi Delta: a spatial and temporal analysis of wetland habitat change. *Estuaries*, **23(4)**, 425–438.

Day Jr., J.W., G.P. Shaffer, D.J. Reed, D.R. Cahoon, L.D. Britsch, and S.R. Hawes, 2001: Patterns and processes of wetland loss in coastal Louisiana are complex: a reply to Turner 2001. estimating the indirect effects of hydrologic change on wetland loss: if the earth is curved, then how would we know it? *Estuaries*, **24(4)**, 647–651.

Dugan, P. (ed.), 1993: *Wetlands in Danger—A World Conservation Atlas*. Oxford University Press, New York, 192 pp.

Ehhalt, D., M. Prather, F. Dentener, E. Dlugokencky, E. Holland, I. Isaksen, J. Katima, V. Kirchhoff, P. Matson, P. Midgley, and M. Wang, 2001: Atmospheric chemistry and greenhouse gases. In: *Climate Change 2001: The Scientific Basis*. Contribution of Working Group I to the Third Assessment Report of the Intergovernmental Panel on Climate Change [Houghton, J.T., Y. Ding, D.J. Griggs, M. Noguer, P.J. van der Linden, X. Dai, K. Maskell, and C.A. Johnson (eds.)]. Cambridge University Press, Cambridge, United Kingdom, pp. 239–287.

Euliss, N.H., R.A. Gleason, A. Olness, R.L. McDougal, H.R. Murkin, R.D. Robarts, R.A. Bourbonniere, and B.G. Warner, 2006: North American prairie wetlands are important nonforested land-based carbon storage sites. *Science of the Total Environment*, **361(1-3)**, 179–188.

Finlayson, C.M., N.C. Davidson, A.G. Spiers, and N.J. Stevenson, 1999: Global wetland inventory—current status and future priorities. *Marine Freshwater Research*, **50**, 717–727.

Fletcher, S.E.M., P.P. Tans, L.M. Bruhwiler, J.B. Miller, and M. Heimann, 2004a: CH$_4$ sources estimated from atmospheric observations of CH$_4$ and its ^{13}C/^{12}C isotopic ratios: 1. inverse modeling of source processes. *Global Biogeochemical Cycles*, **18**, doi:10.1029/2004GB002223.

Fletcher, S.E.M., P.P. Tans, L.M. Bruhwiler, J.B. Miller, and M. Heimann, 2004b: CH$_4$ sources estimated from atmospheric observations of CH$_4$ and its ^{13}C/^{12}C isotopic ratios: 2. inverse modeling of CH$_4$ fluxes from geographical regions. *Global Biogeochemical Cycles*, **18**, doi:10.1029/2004GB002224.

Freeman, C., N. Fenner, N.J. Ostle, H. Kang, D.J. Dowrick, B. Reynolds, M.A. Lock, D. Sleep, S. Hughes, and J. Hudson, 2004: Export of dissolved organic carbon from peatlands under elevated carbon dioxide levels. *Nature*, **430(6996)**, 195-198.

Frolking, S., N. Roulet, and J. Fuglestvedt, 2006: How northern peatlands influence the earth's radiative budget: sustained methane emission versus sustained carbon sequestration. *Journal of Geophysical Research-Biogeosciences*, **111**, G01008, doi:10.1029/2005JG000091.

Frolking, S., N.T. Roulet, T.R. Moore, P.M. Lafleur, J.L. Bubier, and P.M. Crill, 2002: Modeling seasonal to annual carbon balance of Mer Bleue Bog, Ontario, Canada. *Global Biogeochemical Cycles*, **16**, doi:10.1029/2001GB001457.

Gauci, V., E. Matthews, N. Dise, B. Walter, D. Koch, G. Granberg, and M. Vile, 2004: Sulfur pollution suppression of the wetland methane source in the 20th and 21st centuries. *Proceedings of the National Academy of Sciences*, **101(34)**, 12583–12587, doi:10.1073/pnas.0404412101.

Gedney, N., P.M. Cox, and C. Huntingford, 2004: Climate feedbacks from methane emissions. *Geophysical Research Letters*, **31**, L20503, doi:20510.21029/22004GL020919.

Gorham, E., 1991: Northern peatlands: Role in the carbon cycle and probable responses to climatic warming. *Ecological Applications*, **1(2)**, 182–195.

Harden, J.W., J.M. Sharpe, W.J. Parton, D.S. Ojima, T.L. Fries, T.G. Huntington, and S.M. Dabney, 1999: Dynamic replacement and loss of soil carbon on eroding cropland. *Global Biogeochemical Cycles*, **13(4)**, 885-901.

Hines, M.E. and K.N. Duddleston, 2001: Carbon flow to acetate and C$_1$ compounds in northern wetlands. *Geophysical Research Letters*, **28(22)**, 4251–4254.

Hoosbeek, M.R., N. van Breemen, F. Berendse, P. Brosvernier, H. Vasander, and B. Wallén, 2001: Limited effect of increased atmospheric CO$_2$ concentration on ombrotrophic bog vegetation. *New Phytologist*, **150(2)**, 459-463.

Hoosbeek, M.R., M. Lukac, D. van Dam, D.L. Godbold, E.J. Velthorst, F.A. Biondi, A. Peressotti, M.F. Cotrufo, P. de Angelis, and G. Scarascia-Mugnozza, 2004: More new carbon in the mineral soil of a poplar plantation under Free Air Carbon Enrichment (POPFACE): Cause of increased priming effect?

Global Biogeochemical Cycles, **18**, GB1040, doi:10.1029/2003GB002127.

Joosten, H. and D. Clarke, 2002: *Wise Use of Mires and Peatlands - Background Principles Including a Framework for Decision-Making*. International Mire Conservation Group and International Peat Society, Saarijärvi, Finland, 253 pp.

Kearney, M.S., A.S. Rogers, J.R.G. Townshend, E. Rizzo, D. Stutzer, J.C. Stevenson, and K. Sundborg, 2002: Landsat imagery shows decline of coastal marshes in Chesapeake and Delaware Bays. *EOS*, **83**, 173.

Kearney, M.S. and J.C. Stevenson, 1991: Island land loss and marsh vertical accretion rate evidence for historical sea-level changes in Chesapeake Bay. *Journal of Coastal Research*, **7**, 403–415.

Keller, J.K., S.D. Bridgham, C.T. Chapin, and C.M. Iversen, 2005: Limited effects of six years of fertilization on carbon mineralization dynamics in a Minnesota fen. *Soil Biology and Biochemistry*, **37(6)**, 1197–1204.

Kim, H.Y., M. Lieffering, S. Miura, K. Kobayashi, and M. Okada, 2001: Growth and nitrogen uptake of CO$_2$-enriched rice under field conditions. *New Phytologist*, **150(2)**, 223-229.

Lichter, J., S.H. Barron, C.E. Bevacqua, A.C. Finzi, K.F. Irving, E.A. Stemmler, and W.H. Schlesinger, 2005: Soil carbon sequestration and turnover in a pine forest after six years of atmospheric CO$_2$ enrichment. *Ecology*, **86(7)**, 1835-1847.

Lynch-Stewart, P., I. Kessel-Taylor, and C. Rubec, 1999: *Wetlands and Government: Policy and Legislation for Wetland Conservation in Canada*. Sustaining Wetlands Issue Paper No. 1999-1, North American Wetlands Conservation Council (Canada), 57 pp.

Maltby, E. and P. Immirzi, 1993: Carbon dynamics in peatlands and other wetland soils, regional and global perspectives. *Chemosphere*, **27(6)**, 999–1023.

Marsh, A.S., D.P. Rasse, B.G. Drake, and J.P. Megonigal, 2005: Effect of elevated CO$_2$ on carbon pools and fluxes in a brackish marsh. *Estuaries*, **28(5)**, 694-704.

Matthews, E. and I. Fung, 1987: Methane emission from natural wetlands: global distribution, area, and environmental characteristics of sources. *Global Biogeochemical Cycles*, **1**, 61–86.

Megonigal, J.P. and W.H. Schlesinger, 1997: Enhanced CH$_4$ emissions from a wetland soil exposed to elevated CO$_2$. *Biogeochemistry*, **37(1)**, 77–88.

Megonigal, J.P., C.D. Vann, and A.A. Wolf, 2005: Flooding constraints on tree (*Taxodium distichum*) and herb growth responses to elevated CO$_2$. *Wetlands*, **25(2)**, 430–438.

Mitra, S., R. Wassmann, and P.L.G. Vlek, 2005: An appraisal of global wetland area and its organic carbon stock. *Current Science*, **88(1)**, 25–35.

Moore, T.R., 1997: Dissolved organic carbon: sources, sinks, and fluxes and role in the soil carbon cycle. In: *Soil Processes and the Carbon Cycle*. [Lal, R., J.M. Kimble, R.F. Follett, and B.A. Stewart (eds.)]. Lewis Publishers, Boca Raton, FL, pp: 281-292.

Moore, T.R. and N.T. Roulet, 1995: Methane emissions from Canadian peatlands. In: *Soils and Global Change* [Lal, R., J. Kimble, E. Levine, and B.A. Stewart (eds.)]. Lewis Publishers, Boca Raton, FL, pp. 153–164.

Moore, T.R., N.T. Roulet, and J.M. Waddington, 1998: Uncertainty in predicting the effect of climatic change on the carbon cycling of Canadian peatlands. *Climatic Change*, **40(2)**, 229–245.

Moser, M., C. Prentice, and S. Frazier, 1996: A global overview of wetland loss and degradation. In: *Proceedings of the 6th Meeting of the Conference of the Contracting Parties*, Brisbane, Australia, Papers, Technical Session B, Vol 10/12B, 19-27 March 1996, Ramsar Convention Bureau, Gland, Switzerland, 21-31.

Najjar, R.G., H.A. Walker, P.J. Anderson, E.J. Barron, R.J. Bord, J.R. Gibson, V.S. Kennedy, C.G. Knight, J.P. Megonigal, R.E. O'Conner, C.D. Polsky, N.P. Psuty, B.A. Richards, L.G. Sorenson, E.M. Steele, and R.S. Swanson, 2000: The potential impacts of climate change on the mid-Atlantic coastal region. *Climate Research*, **14(3)**, 219–233.

NRC (National Research Council), 1995: *Wetlands: Characteristics and Boundaries*. National Academy Press, Washington, DC, 307 pp.

NRC (National Research Council), 2001: *Compensating for Wetland Losses Under the Clean Water Act*. National Academy Press, Washington, DC, 322 pp.

National Wetlands Working Group, 1997: *The Canadian Wetland Classification System, Second Edition* [Warner, B.G. and C.D.A. Rubec (eds.)]. Wetlands Research Centre, University of Waterloo, Waterloo, Ontario, Canada, 68 pp.

OECD (Organization for Economic Co-operation and Development), 1996: *Guidelines for Aid Agencies for Improved Conservation and Sustainable Use of Tropical and Sub-tropical Wetlands*. OECD, Paris, France, 69 pp.

Oechel, W.C., S. Cowles, N. Grulke, S.J. Hastings, B. Lawrence, T. Prudhomme, G. Riechers, B. Strain, D. Tissue, and G. Vourlitis, 1994: Transient nature of CO_2 fertilzation in arctic tundra. *Nature*, **371(6497)**, 500-502.

Petit, J.R., J. Jouzel, D. Raynaud, N.I. Barkov, J.-M. Barnola, I. Basile, M. Bender, J. Chappellaz, M. Davis, G. Delaygue, M. Delmotte, V.M. Kotlyakov, M. Legrand, V.Y. Lipenkov, C. Lorius, L. Pepin, C. Ritz, E. Saltzman, and M. Stievenard, 1999: Climate and atmospheric history of the past 420,000 years from the Vostok ice core, Antarctica. *Nature*, **399(6735)**, 429-436.

Ramaswamy, V., O. Boucher, J. Haigh, D. Hauglustaine, J. Haywood, G. Myhre, T. Nakajima, G.Y. Shi, and S. Solomon, 2001: Radiative forcing of climate change. In: *Climate Change 2001: The Scientific Basis*. Contribution of Working Group I to the Third Assessment Report of the Intergovernmental Panel on Climate Change [Houghton, J.T., Y. Ding, D.J. Griggs, M. Noguer, P.J. van der Linden, X. Dai, K. Maskell, and C.A. Johnson (eds.)]. Cambridge University Press, Cambridge, United Kingdom, pp. 349–416.

Rasse, D.P., G. Peresta, and B.G. Drake, 2005: Seventeen years of elevated CO_2 exposure in a Chesapeake Bay wetland: sustained but contrasting responses of plant growth and CO_2 uptake. *Global Change Biology*, **11(3)**, 369-377.

Roulet, N.T., 2000: Peatlands, carbon storage, greenhouse gases, and the Kyoto Protocol: prospects and significance for Canada. *Wetlands*, **20(4)**, 605–615.

Rubec, C., 1996: The status of peatland resources in Canada. In: *Global Peat Resources* [Lappalainen, E. (ed.)]. International Peat Society and Geological Survey of Finland, Jyskä, Finland, pp. 243–252.

Smith, S.V., W.H. Renwick, R.W. Buddemeier, and C.J. Crossland, 2001: Budgets of soil erosion and deposition for sediments and sedimentary organic carbon across the conterminous United States. *Global Biogeochemical Cycles,* **15(3)**, 697-707.

Stallard, R.F., 1998: Terrestrial sedimentation and the carbon cycle: coupling weathering and erosion to carbon burial. *Global Biogeochemical Cycles*, **12(2)**, 231-257.

Strack, M., J.M. Waddington, and E.S. Tuittila, 2004: Effect of water table drawdown on northern peatland methane dynamics: implications for climate change. *Global Biogeochemical Cycles*, **18**, GB4003, doi:4010.1029/2003GB002209.

Tissue, D.T. and W.C. Oechel, 1987: Response of *Eriophorum vaginatum* to elevated CO_2 and temperature in the Alaskan tussock tundra. *Ecology*, **68(2)**, 401-410.

Turetsky, M.R., B.D. Amiro, E. Bosch, and J.S. Bhatti, 2004: Historical burn area in western Canadian peatlands and its relationship to fire weather indices. *Global Biogeochemical Cycles*, **18**, GB4014, doi:1029/2004GB002222.

Turner, R.E., 1997: Wetland loss in the Northern Gulf of Mexico: multiple working hypotheses. *Estuaries*, **20(1)**, 1–13.

Updegraff, K., S.D. Bridgham, J. Pastor, P. Weishampel, and C. Harth, 2001: Response of CO_2 and CH_4 emissions in peatlands to warming and water-table manipulation. *Ecological Applications*, **11(2)**, 311–326.

Vann, C.D. and J.P. Megonigal, 2003: Elevated CO_2 and water depth regulation of methane emissions: comparison of woody and non-woody wetland plant species. *Biogeochemistry*, **63(2)**, 117–134.

Vile, M.A., S.D. Bridgham, R.K. Wieder, and M. Novák, 2003: Atmospheric sulfur deposition alters pathways of gaseous carbon production in peatlands. *Global Biogeochemical Cycles*, **17(2)**, 1058–1064.

Wang, J.S., J.A. Logan, M.B. McElroy, B.N. Duncan, I.A. Megretskaia, and R.M. Yantosca, 2004: A 3-D model analysis of the slowdown and interannual variability in the methane growth rate from 1988 to 1997. *Global Biogeochemical Cycles*, **18**, GB3011, doi:101029/102003GB002180.

Watson, R.T., I.R. Noble, B. Bolin, N.H. Ravindranath, D.J. Verardo, and D.J. Dokken, 2000: *IPCC Special Report on Land Use, Land-Use Change and Forestry*. Cambridge University Press, Cambridge, United Kingdom, 377 pp.

Whiting, G.J. and J.P. Chanton, 1993: Primary production control of methane emissions from wetlands. *Nature*, **364(6440)**, 794–795.

Wylynko, D. (ed.), 1999: *Prairie Wetlands and Carbon Sequestration: Assessing Sinks Under the Kyoto Protocol*. Institute for Sustainable Development, Ducks Unlimited Canada, and Wetlands International, Winnipeg, Manitoba, Canada.

Zedler, J.B. and S. Kercher, 2005: Wetland resources: status, trends, ecosystem services, and restorability. *Annual Review of Environmental Resources*, **30**, 39-74.

Zhuang, Q., J.M. Melillo, D.W. Kicklighter, R.G. Prin, A.D. McGuire, P.A. Steudler, B. Felzer, and S. Hu, 2004: Methane fluxes between terrestrial ecosystems and the atmosphere at northern high latitudes during the past century: a restrospective analysis with a process-based biogeochemistry model. *Global Biogeochemical Cycles*, **18**, GB 3010, doi:3010.1029/2004GB002239.

CHAPTER 14 REFERENCES

Agarwal, C., G.M. Green, J.M. Grove, T.P. Evans, and C.M. Schweik, 2000: *A Review and Assessment of Land-Use Change Models: Dynamics of Space, Time and Human Choice*. CIPEC Collaborative Report Series No. 1, Center for the Study of Institutions, Populations, and Environmental Change, Indiana University and the USDA Forest Service, 61 pp.

Akbari, H., 2002: Shade trees reduce building energy use and CO_2 emissions from power plants. *Environmental Pollution*, **116(Supplement 1)**, S119-S126.

Akbari, H. and S. Konopacki, 2005: Calculating energy-saving potentials of heat-island reduction strategies. *Energy Policy*, **33(6)**, 721-756.

Akbari, H., D.M. Kurn, S.E. Bretz, and J.W. Hanford, 1997: Peak power and cooling energy savings of shade trees. *Energy and Buildings*, **25(2)**, 139-148.

Akbari, H. and H. Taha, 1992: The impact of trees and white surfaces on residential heating and cooling energy use in four Canadian cities. *Energy*, **17(2)**, 141-149.

Alig, R.J., J.D. Kline, and M. Lichtenstein, 2004: Urbanization on the U.S. landscape: Looking ahead in the 21st century. *Landscape and Urban Planning*, **69(2-3)**, 219-234.

Alig, R.J., A. Plantinga, S. Ahn, and J.D. Kline, 2003: *Land Use Changes Involving Forestry for the United States: 1952 to 1997, With Projections to 2050*. General Technical Report 587, USDA Forest Service, Pacific Northwest Research Station, Portland, OR, 92 pp.

Anderson, W.P., P.S. Kanaroglou, E.J. Miller, 1996: Urban form, energy and the environment: a review of issues, evidence and policy. *Urban Studies*, **33(1)**, 7-35.

Betsill, M.M., 2001: Mitigating climate change in U.S. cities: opportunities and obstacles. *Local Environment*, **6(4)**, 393-406.

CEC (Commission for Environmental Cooperation), 2001: *The North American Mosaic: A State of the Environment Report*. CEC, Montreal, Canada, 100 pp.

CIESIN (Center for International Earth Science Network), Columbia University, International Food Policy Research Institute (IPFRI), the World Bank, Centro Internacional de Agricultura Tropical (CIAT), 2004: *Global Rural-Urban Mapping Project (GRUMP): Urban Extents*. Last accessed 3 Dec 2005. Available at http://sedac.ciesin.columbia.edu/gpw

Cifuentes, L., V.H. Borja-Aburto, N. Gouveia, G. Thurston, and D.L. Davis, 2001: Assessing health benefits of urban air pollution reductions associated with climate change mitigation (2000-2020): Santiago, Sao Paulo, Mexico City, and New York City. *Environmental Health Perspectives*, **109(4)**, 419-425.

Decker, E.H., S. Elliot, F.A. Smith, D.R. Blake, and F.S. Rowland, 2000: Energy and material flow through the urban ecosystem. *Annual Review of Energy and the Environment*, **25**, 685-740.

Easterling, W.E., C. Polsky, D.G. Goodin, M.W. Mayfield, W.A. Muraco, and B. Yarnal, 2003: Changing places and changing emissions: comparing local, state, and United States emissions. In: *Global Change and Local Places: Estimating, Understanding and Reducing Greenhouse Gases* [Association of American Geographers Global Change in Local Places Research Group (eds.)]. Cambridge University Press, Cambridge, United Kingdom, pp. 143-157.

EPA (U.S. Environmental Protection Agency), 2000: *Projecting Land-Use Change: A Summary of Models for Assessing the Effects of Community Growth and Change on Land-Use Patterns*. EPA/600/R-00/098, Washington, DC, 142 pp.

Ewing, R., R. Pendall, and D. Chen, 2003: Measuring sprawl and its transportation impacts. *Transportation Research Record*, **1831**, 175-183.

Folke, C., A. Jansson, J. Larsson, and R. Costanza, 1997: Ecosystem appropriation by cites. *Ambio*, **26**, 167-172.

Golubiewski, N.E., 2006: Urbanization transforms prairie carbon pools: effects of landscaping in Colorado's Front Range. *Ecological Applications*, **16(2)**, 555-51.

Gomez-Ibanez, J.A., 1991: A global view of automobile dependence. *Journal of the American Planning Association*, **57(3)**, 376-379.

Gonzalez, G.A., 2005: Urban sprawl, global warming and the limits of ecological modernisation. *Environmental Politics*, **14(3)**, 344-362.

Gordon, P. and H.W. Richardson, 1989: Gasoline consumption and cities: a reply. *Journal of the American Planning Association*, **55**, 342-346.

Grimm, N.B., J.M. Grove, S.T.A. Pickett, and C.L. Redman, 2000: Integrated approaches to long-term studies of urban ecological systems. *Bioscience*, **50(7)**, 571-584.

Grimmond, C.S.B., T.S. King, F.D. Cropley, D.J. Nowak, and C. Souch, 2002: Local-scale fluxes of carbon dioxide in urban environments: methodological challenges and results from Chicago. *Environmental Pollution*, **116(Supplement 1)**, S243-S254.

Grimmond, C.S.B., J.A. Salmond, T.R. Oke, B. Offerle, and A. Lemonsu, 2004: Flux and turbulence measurements at a densely built-up site in Marseille: heat, mass (water and carbon dioxide), and momentum. *Journal of Geophysical Research-Atmospheres*, **109**, doi:10.1029/2004JD004936.

Huang, Y.J., H. Akbari, H. Taha, and H. Rosenfeld, 1987: The potential of vegetation in reducing summer cooling loads in residential buildings. *Journal of Climate and Applied Meteorology*, **26(9)**, 1103-1116.

Hunt, J.D., D.S. Kriger, and E.J. Miller, 2005: Current operation urban land-use-transport modeling frameworks: a review. *Transport Reviews*, **25(3)**, 329-376.

ICLEI (The International Council for Local Environmental Initiatives), 1993: *Cities for Climate Protection: An International Campaign to Reduce Urban Emissions of Greenhouse Gases*. Last accessed 30 Mar 2006. Available at http://www.iclei.org/index.php?id=1651

ICLEI (The International Council for Local Environmental Initiatives), 2000: *Best Practices for Climate Protection: A Local Government Guide*. ICLEI, Berkeley, CA.

Imhoff, M.L., L. Bounoua, R.S. DeFries, W.T. Lawrence, D. Stutzer, J.T. Compton, and T. Ricketts, 2004: The consequences of urban land transformations on net primary productivity in the United States. *Remote Sensing of the Environment*, **89(4)**, 434-443.

Ironmonger, D.S., C.K. Aitken, and B. Erbas, 1995: Economies of scale in energy use in adult-only households. *Energy Economics*, **17(4)**, 301-310.

Jaccard, M., L. Failing, and T. Berry, 1997: From equipment to infrastructure: community energy management and greenhouse gas emission reduction. *Energy Policy*, **25(13)**, 1065-1074.

Kaye, J.P., I.C. Burke, A.R. Mosier, and J.P. Guerschman, 2004: Methane and nitrous oxide fluxes from urban soils to the atmosphere. *Ecological Applications*, **14(4)**, 975-981.

Kaye, J.P., R.L. McCulley, and I.C. Burke, 2005: Carbon fluxes, nitrogen cycling, and soil microbial communities in adjacent urban, native and agricultural ecosystems. *Global Change Biology*, **11(4)**, 575-587.

Kenworthy, J.R. and P.W.G. Newman, 1990: Cities and transport energy: lessons from a global survey. *Ekistics*, **34**, 258-268.

Koerner, B. and J Klopatek, 2002: Anthropogenic and natural CO_2 emission sources in an arid urban environment. *Environmental Pollution*, **116(Supplement 1)**, S45-S51.

Kousky, C. and S.H. Schneider, 2003: Global climate policy: will cities lead the way? *Climate Policy*, **3(4)**, 359-372.

Lin, J.C., C. Gerbig, S.C. Wofsy, A.E. Andrews, B.C. Daube, B.C. Grainger, B.B. Stephens, P.S. Bakwin, and D.Y. Hollinger, 2004: Measuring fluxes of trace gases at regional scales by Lagrangian observations: application to the CO_2 budget and rectification airborne (COBRA study). *Journal of Geophysical Research-Atmospheres*, **109**, doi:10.1029/2004JD004754.

Liu, J., G.C. Daily, P.R. Ehrich, G.W. Luck, 2003: Effects of household dynamics on resource consumption and biodiversity. *Nature*, **421(6922)**, 530-533.

MacKellar, F.L., W. Lutz, C. Prinz, and A. Goujon, 1995: Population, households, and CO_2 emissions. *Population and Development Review*, **21(4)**, 849-865.

McPherson, E.G., J.R. Simpson, P.F. Peper, S.E. Maco, and Q. Xiao, 2005: Municipal forest benefits and costs in five U.S. cities. *Journal of Forestry*, **103(8)**, 411-416.

Mindali, O., A. Raveh, and I. Saloman, 2004: Urban density and energy consumption: a new look at old statistics. *Transportation Research Record Part A*, **38(2)**, 143-162.

NAHB (National Association of Home Builders), 2005: *Housing Facts, Figures and Trends.*, Washington, DC.

Nemitz, E., K. Hargreaves, A.G. McDonald, J.R. Dorsey, and D. Fowler, 2002: Micrometeorological measurements of the urban heat budget and CO_2 emissions on a city scale. *Environmental Science and Technology*, **36(14)**, 3139-3146.

Newman, P.W.G., 1999: Sustainability and cities: extending the metabolism model. *Landscape and Urban Planning*, **44(4)**, 219-226.

Nowak, D.J. and D.E. Crane, 2002: Carbon storage and sequestration by urban trees in the USA. *Environmental Pollution*, **116(3)**, 381-389.

Nowak, D.J., J.T. Walton, J.F. Dwyer, L.G. Kaya, and S. Myeong, 2005: The increasing influence of urban environments on U.S. forest management. *Journal of Forestry*, **103(8)**, 377-382.

Oke, T.R., 1989: The micrometeorology of the urban forest. *Philosophical Transactions of the Royal Society of London, Series B*, **324(1223)**, 335-349.

Pataki, D.E., R.J. Alig, A.S. Fung, N.E. Golubiewski, C.A. Kennedy, E.G. McPherson, D.J. Nowak, R.V. Pouyat, and P. Romero Lankao, 2006a: Urban ecosystems and the North American carbon cycle. *Global Change Biology* **12(11)**, 2092-2102.

Pataki, D.E., D.R. Bowling, and J.R. Ehleringer, 2003: The seasonal cycle of carbon dioxide and its isotopic composition in an urban atmosphere: anthropogenic and biogenic effects. *Journal of Geophysical Research-Atmospheres*, **108(D23)**, 4735.

Pataki, D.E., D.R. Bowling, J.R. Ehleringer, and J.M. Zobitz, 2006b: High resolution monitoring of urban carbon dioxide sources. *Geophysical Research Letters*, **33**, L03813, doi:10.1029/2005GL024822.

Pickett, S.T.A., M.L. Cadenasso, J.M. Grove, C.H. Nilon, R.V. Pouyat, W.C. Zipperer, and R. Costanza, 2001: Urban ecological systems: linking terrestrial ecological, physical, and socioeconomic components of metropolitan areas. *Annual Review of Ecology and Systematics*, **32**, 127-157.

Pouyat, R., P. Groffman, I. Yesilonis, and L. Hernandez, 2002: Soil carbon pools and fluxes in urban ecosystems. *Environmental Pollution*, **116(Supplement 1)**, S107-S118.

Pouyat, R.V., I. Yesilonis, and D.J. Nowak, 2006: Carbon storage by urban soils in the USA. *Journal of Environmental Quality*, **35(4)**, 1566-1575.

Qian, Y. and R.F. Follet, 2002: Assessing soil carbon sequestration in turfgrass systems using long-term soil testing data. *Agronomy Journal*, **94(4)**, 930-935.

Romero Lankao, P., H. Lopez, A. Rosas, G. Gunther, and Z. Correa, 2004: *Can Cities Reduce Global Warming?* Urban Development and the Carbon Cycle in Latin America. IAI, UAM-X, IHDP, GCP, Mexico City, 92 pp. Available at http//www.globalcarbonproject.org/global/pdf/MeetingAgenda.pdf

Sahely, H.R., S. Dudding, and C.A. Kennedy, 2003: Estimating the urban metabolism of Canadian cities: Greater Toronto Area case study. *Canadian Journal of Civil Engineering*, **30(2)**, 468-483.

Soegaard, H. and L. Moller-Jensen, 2003: Toward a spatial CO_2 budget of metropolitan region based on textural image classification and flux measurements. *Remote Sensing of the Environment*, **87(2-3)**, 283-294.

Taha, H., 1997: Urban climates and heat islands: albedo, evapotranspiration, and anthropogenic heat. *Energy and Buildings*, **25(2)**, 99-103.

United Nations, 2002: *Demographic Yearbook.* Available at http://unstats.un.org/unsd/demographic/products/dyb/default.htm

United Nations, 2004: *World Urbanization Prospects: The 2003 Revision.* E.04.XIII.6, U.N. Dept. of Economic and Social Affairs, Population Division, New York, 323 pp.

United Nations Habitat, 2003: *Global Observatory Database.* Last accessed 10 Nov 2005. Available at http://www.unchs.org/programmes/guo

Warren-Rhodes, K. and A. Koenig, 2001: Ecosystem appropriation by Hong Kong and its implications for sustainable development. *Ecological Economics*, **39(2)**, 347-359.

West, J.J., P. Osnaya, I. Laguna, J. Martinez, and A. Fernandez, 2004: Co-control of urban air pollutants and greenhouse gases in Mexico City. *Environmental Science and Technology*, **38(13)**, 3474-3481.

CHAPTER 15 REFERENCES

Bakker, D.C.E., A.J. Watson, and C.S. Law, 2001: Southern Ocean iron enrichment promotes inorganic carbon drawdown. *Deep-Sea Research II*, **48(11-12)**, 2483-2507.

Bates, N.R., 2006: Fluxes and the continental shelf pump of carbon in the Chukchi Sea adjacent to the Arctic Ocean. *Journal of Geophysical Research*, **111**, C10013, doi:10.1029/2005JC003083.

Battle, M., M.L. Bender, P.P. Tans, J.W.C. White, J.T. Ellis, T. Conway, and R.J. Francey, 2000: Global carbon sinks and their variability inferred from atmospheric O_2 and $\delta^{13}C$. *Science*, **287(5462)**, 2467-2470.

Bender, M.L., D.T. Ho, M.B. Hendricks, R. Mika, M.O. Bazttle, P.P. Tans, T.J. Conway, B. Sturtevant, and N. Cassar, 2005: Atmospheric O_2/N_2 changes, 1993-2002: implications for the partitioning of fossil fuel CO_2 sequestration. *Global Biogeochemical Cycles*, **19**, GB4017, doi:10.1029/2004GB002410.

Benner, R. and S. Opsahl, 2001: Molecular indicators of the sources and transformations of dissolved organic matter in the Mississippi River plume. *Organic Geochemistry*, **32(4)**, 597-611.

Boehme, S.E., C.L. Sabine, and C.E. Reimers, 1998: CO_2 fluxes from a coastal transect: a time-series approach. *Marine Chemistry*, **63(1-2)**, 49-67.

Borges, A.V., 2005: Do we have enough pieces of the jigsaw to integrate CO_2 fluxes in the Coastal Ocean? *Estuaries*, **28(1)**, 3-27.

Borges, A.V., B. Delille, and M. Frankignoulle, 2005: Budgeting sinks and sources of CO_2 in the coastal ocean: diversity of ecosystems counts. *Geophysical Research Letters*, **32(14)**, L14601, doi:10.1029/2005GL023053.

Boyd P.W., A.J. Watson, C.S. Law, E.R. Abraham, T. Trull, R. Murdoch, D.C.E. Bakker, A.R. Bowie, K.O. Buesseler, H. Chang, M. Charette, P. Croot, K. Downing, R. Frew, M. Gall, M. Hadfield, J. Hall, M. Harvey, G. Jameson, J. LaRoche, M. Liddicoat, R. Ling, M.T. Maldonado, R.M. McKay, S. Nodder, S. Pickmere, R. Pridmore, S. Rintoul, K. Safi, P. Sutton, R. Strzepek, K. Tanneberger, S. Turner, A. Waite, and J. Zeldis, 2000: A mesoscale phytoplankton bloom in the polar Southern Ocean stimulated by iron fertilization. *Nature*, **407(6805)**, 695-702.

Brewer, P.G., 2003: Direct injection of carbon dioxide into the oceans. In: *The Carbon Dioxide Dilemma: Promising Technologies and Policies.* National Academies Press, pp. 43-51.

Cai, W.J., 2003: Riverine inorganic carbon flux and rate of biological uptake in the Mississippi River plume. *Geophysical Research Letters*, **30(2)**, 1032.

Cai, W.-J., Z.A. Wang, and Y.C. Wang, 2003: The role of marsh-dominated heterotrophic continental margins in transport of CO_2 between the atmosphere, the land-sea interface and the oceans. *Geophysical Research Letters*, **30(16)**, 1849, doi:10.1029/2003GL017633.

Cai, W.J. and M. Dai, 2004: Comment on enhanced open ocean storage of CO_2 from shelf sea pumping. *Science*, **306(5701)**, 1477c.

Chavez, F.P., P.G. Strutton, G.E. Friederich, R.A. Feely, G.C. Feldman, D.G. Foley, and M.J. McPhaden, 1999: Biological and chemical response of the equatorial Pacific Ocean to 1997-98 El Niño. *Science*, **286(5447)**, 2126-2131.

Chavez, F.P., J.T. Pennington, C.G. Castro, J.P. Ryan, R.M. Michisaki, B. Schlining, P. Walz, K.R. Buck, A. McFayden, and C.A. Collins, 2002: Biological and chemical consequences of the 1997-98 El Niño in central California waters. *Progress in Oceanography*, **54(1-4)**, 205-232.

Chavez, F.P., J. Ryan, S. Lluch-Cota, and M. Ñiquen C., 2003: From anchovies to sardines and back: multidecadal change in the Pacific Ocean. *Science*, **299(5604)**, 217-221.

Chisholm, S.W., P.G. Falkowski, and J. Cullen, 2001: Discrediting ocean fertilization. *Science*, **294(5541)**, 309-310.

Codispoti, L.A. and G.E. Friederich, 1986: Variability in the inorganic carbon system over the southeastern Bering Sea shelf during the spring of 1980 and spring-summer 1981. *Continental Shelf Research*, **5(1)**, 133-160.

Coale, K.H., K.S. Johnson, F.P. Chavez, K.O. Buesseler, R.T. Barber, M.A. Brzezinski, W.P. Cochlan, F.J. Millero, P.G.

Falkowski, J.E. Bauer, R.H. Wanninkhof, R.M. Kudela, M.A. Altabet, B.E. Hales, T. Takahashi, M.R. Landry, R.R. Bidigare, X. Wang, Z. Chase, P.G. Strutton, G.E. Friederich, M.Y. Gorbunov, V. P. Lance, A.K. Hilting, M.R. Hiscock, M. Demarest, W.T. Hiscock, K.F. Sullivan, S.J. Tanner, R.M. Gordon, C.N. Hunter, V.A. Elrod, S.E. Fitzwater, J.L. Jones, S. Tozzi, M. Koblizek, A.E. Roberts, J. Herndon, J. Brewster, N. Ladizinsky, G. Smith, D. Cooper, D. Timothy, S.L. Brown, K.E. Selph, C.C. Sheridan, B.S. Twining, and Z.I. Johnson, 2004: Southern Ocean iron enrichment experiment: carbon cycling in high- and low-Si waters. *Science*, **304(5669)**, 408-414.

DeGrandpre, M.D., T.R. Hammar, D.W.R. Wallace, and C.D. Wirick, 1997: Simultaneous mooring-based measurements of seawater CO_2 and O_2 off Cape Hatteras, North Carolina. *Limnology and Oceanography*, **42(1)**, 21-28.

DeGrandpre, M.D., G.J. Olbu, C.M. Beatty, and T.R. Hammar, 2002: Air-sea CO_2 fluxes on the U.S. Middle Atlantic Bight. *Deep-Sea Research II*, **49(20)**, 4355-4367.

Doney, S.C., R. Anderson, J. Bishop, K. Caldeira, C. Carlson, M.E. Carr, R. Feely, M. Hood, C. Hopkinson, R. Jahnke, D. Karl, J. Kleypas, C. Lee, R. Letelier, C. McClain, C. Sabine, J. Sarmiento, B. Stephens, and R. Weller, 2004: *Ocean Carbon and Climate Change (OCCC): An Implementation Strategy for U.S. Ocean Carbon Cycle Science*. UCAR, Boulder, CO, 108 pp.

Ducklow, H.W. and S.L. McCallister, 2004: The biogeochemistry of carbon dioxide in the coastal oceans. In: *The Sea*, Vol. 13 [Robinson, A.R. and K.H. Brink (eds.)]. John Wiley & Sons, New York, pp. 269-315.

Feely, R.A., J. Boutin, C.E. Coasca, Y. Dandonneau, J. Etcheto, H. Inoue, M. Ishii, C. LeQuere, D.J. Mackey, M. McPhaden, N. Metzl, A. Poisson, and R. Wanninkhof, 2002: Seasonal and interannual variability of CO_2 in the equatorial Pacific. *Deep-Sea Research II*, **49(13-14)**, 2443-2469.

Feely, R.A., T. Takahashi, R. Wanninkhof, M.J. McPhaden, C.E. Cosca, S.C. Sutherland, and M.E. Carr, 2006: Decadal variability of the air-sea CO_2 fluxes in the equatorial Pacific Ocean. *Journal of Geophysical Research,* **111**, C07S03, doi:10.1029/2005jc003129.

Friederich, G.E., P.G. Brewer, R. Herlein, and F.P. Chavez, 1995: Measurement of sea surface partial pressure of CO_2 from a moored buoy. *Deep-Sea Research I*, **42(7)**, 1175-1186.

Friederich, G., P. Walz, M. Burczynski, and F.P. Chavez, 2002: Inorganic carbon in the central California upwelling system during the 1997-1999 El Niño -La Niña Event. *Progress in Oceanography*, **54(1-4)**, 185-204.

Gattuso, J.M., M. Frankignoulle, and R. Wollast, 1998: Carbon and carbonate metabolism in coastal aquatic ecosystem. *Annual Review of Ecology and Systematics*, **29**, 405-434.

Gervais, F., U. Riebesell, and M.Y. Gorbunov, 2002: Changes in primary productivity and chlorophyll: a in response to iron fertilization in the Southern Polar Frontal Zone. *Limnology and Oceanography*, **47(5)**, 1324.

Gruber, N. and J.L. Sarmiento, 2002: Large-scale biogeochemical-physical interactions in elemental cycles. In: *The Sea*, Vol. 12 [Robinson, A.R., J. McCarthy, and B.J. Rothschild (eds.)]. John Wiley & Sons, New York, pp. 337-399.

Hales, B. and T. Takahashi, 2004: High-resolution biogeochemical investigation of the Ross Sea, Antarctica, during the AESOPS (U. S. JGOFS) Program. *Global Biogeochemical Cycles*, **18(3)**, GB3006, doi:10.1029/2003GB002165.

Hales, B., T. Takahashi, and L. Bandstra, 2005: Atmospheric CO_2 uptake by a coastal upwelling system. *Global Biogeochemical Cycles*, **19**, GB1009, doi:10.1029/2004GB002295.

Hare, S.R. and N.J. Mantua, 2000: Empirical evidence for North Pacific regime shifts in 1977 and 1989. *Progress in Oceanography*, **47(1)**, 103-145.

Hedges, J.I., R.G. Keil, and R. Benner, 1997: What happens to terrestrial organic matter in the ocean? *Organic Geochemistry*, **27(5-6)**, 195-212.

Keeling, R.F. and H. Garcia, 2002: The change in oceanic O_2 inventory associated with recent global warming. *Proceedings of the National Academy of Sciences*, **99(12)**, 7848-7853.

Kleypas, J.A., R.A. Feely, V.J. Fabry, C. Langdon, C.L. Sabine, and L.Robbins, 2006: *Impacts of ocean acidification on coral reefs and other marine calcifiers: A guide for future research*. Report of a workshop held 18-20, April, 2005, St. Petersburg, FL Sponsored by NSF, NOAA and USGS, 88 pp. Available at http://www.issue.icar.edu/florida/

Körtzinger, A., 2003: A significant CO_2 sink in the tropical Atlantic Ocean associated with the Amazon river plume. *Geophysical Research Letters*, **30**, 2287, doi:10.1029/2003GL018841.

Liu, K.K., K. Iseki, and S.-Y. Chao, 2000: Continental margin carbon fluxes. In: *The Changing Ocean Carbon Cycle* [Hansen, R., H.W. Ducklow, and J.G. Field (eds.)]. Cambridge University Press, Cambridge, United Kingdom, pp. 187-239.

Lohrenz, S.E., M.J. Dagg, and T.E. Whitledge, 1999: Nutrients, irradiance, and mixing as factors regulating primary production in coastal waters impacted by the Mississippi River plume. *Continental Shelf Research*, **19(9)**, 1113-1141.

Martin, J.H., 1990: Glacial-interglacial CO_2 change: the iron hypothesis. *Paleoceanography*, **5(1)**, 1-13.

Millero, F.J., W.T. Hiscock, F. Huang, M. Roche, and J.-Z. Zhang, 2001: Seasonal variation of the carbonate system in Florida Bay. *Bulletin of Marine Science*, **68(1)**, 101-123.

Muller-Karger, F.E., R. Varela, R. Thunell, R. Luerssen, C. Hu, and J. J. Walsh, 2005. The importance of continental margins in the global carbon cycle. *Geophysical Research Letters*, **32**, L01602, doi:10.1029/2004GL021346.

Park, P.K., L.I. Gordon, and S. Alvarez-Borrego, 1974: The carbon dioxide system of the Bering Sea. In: *Oceanography of the Bering Sea* [Hood, D.W. (ed.)]. Occasional Publication No. 2, Institute of Marine Science, University of Alaska, Fairbanks, AK, 623 pp.

Patra, P.K., S. Maksyutov, M. Ishizawa, T. Nakazawa, T. Takahashi, and J. Ukita, 2005: Interannual and decadal changes in the sea-air CO_2 flux from atmospheric CO_2 inverse modeling.

Global Biogeochemical Cycles, **19**, GB4013, doi:10.1029/2004GB002257.

Pennington, J.T., C.G. Castro, C.A. Collins, W.W. Evans IV, G.E. Friederich, R.P. Michisaki, and F.P. Chavez: *A Carbon Budget for the Northern and Central California Coastal Upwelling System*. Continental Margins Task Team, The Synthesis Book, Chapter 2.2, California Current System, Springer-Verlag, New York (in press), 32 mss. pp.

Quay, P., R. Sommerup, T. Westby, J. Sutsman, and A. McNichol, 2003: Changes in the $^{13}C/^{12}C$ of dissolved inorganic carbon in the ocean as a tracer of anthropogenic CO_2 uptake. *Global Biogeochemical Cycles*, **17(1)**, doi:10.1029/2001GB001817.

Sabine, C.L., R.A. Feely, N. Gruber, R.M. Key, K. Lee, J.L. Bullister, R. Wanninkhof, C.S. Wong, D.W.R. Wallace, B. Tilbrook, F.J. Millero, T.H. Peng, A. Kozyr, T. Ono, and A.F. Rios, 2004a: The oceanic sink for anthropogenic CO_2. *Science*, **305(5682)**, 367-371.

Sarmiento, J.L. and Gruber, N., 2006: *Ocean Biogeochemical Dynamics*, Princeton University Press, Princeton, NJ, pp. 503.

Sarmiento, J.L. and E.T. Sundquist, 1992: Revised budget for the oceanic uptake of anthropogenic carbon dioxide. *Nature*, **356(6370)**, 589-593.

Sarmiento, J.L., P. Monfray, E. Maier-Reimer, O. Aumont, R.J. Murnane, and J.C. Orr, 2000: Sea-air CO_2 fluxes and carbon transport: a comparison of three ocean general circulation models. *Global Biogeochemical Cycles*, **14(4)**, 1267-1282.

Simpson, J.J., 1985: Air-sea exchange of carbon dioxide and oxygen induced by phytoplankton: methods and interpretation. In: *Mapping Strategies in Chemical Oceanography* [Zirino, A. (ed.)]. American Chemical Society, Washington, DC, pp. 409-450.

Smith, S.V. and J.T. Hollibaugh, 1993: Coastal metabolism and the oceanic organic carbon balance. *Review of Geophysics*, **31(1)**, 75-89.

Takahashi, T., S.C. Sutherland, C. Sweeney, A. Poisson, N. Metzl, B. Tillbrook, N. Bates, R. Wanninkhof, R.A. Feely, C. Sabine, J. Olafsson, and Y. Nojiri, 2002: Global sea-air CO_2 flux based on climatological surface ocean pCO_2, and seasonal biological and temperature effects. *Deep-Sea Research II*, **49(9-10)**, 1601-1622.

Takahashi, T., S.C. Sutherland, R.A. Feely, and C. Cosca, 2003: Decadal variation of the surface water pCO_2 in the western and central Equatorial Pacific. *Science*, **302(5646)**, 852-856.

Thomas, H., Y. Bozec, K. Elkalay, and H.J.W. De Baar, 2004: Enhanced open ocean storage of CO_2 from shelf sea pumping. *Science*, **304(5673)**, 1005-1008.

Tsunogai, S., S. Watanabe, and T. Sato, 1999: Is there a "continental shelf pump" for the absorption of atmospheric CO_2? *Tellus B*, **5(3)**, 701-712.

van Geen, A., R.K. Takesue, J. Goddard, T. Takahashi, J.A. Barth, and R.L. Smith, 2000: Carbon and nutrient dynamics during upwelling off Cape Blanco, Oregon. *Deep-Sea Research II*, **49(20)**, 4369-4385.

Wanninkhof, R., 1992: Relationship between wind speed and gas exchange. *Journal of Geophysical Research*, **97(C5)**, 7373-7382.

Ware, D.M. and R.D. Thomson, 2005: Bottom-up ecosystem trophic dynamics determine fish production in the Northeast Pacific. *Science*, **308(5726)**, 1280-1284.

APPENDIX A REFERENCES

Birdsey, R.A. and L.S. Heath, 1995: Carbon changes in U.S. forests. In: *Productivity of America's Forests and Climate Change* [Joyce, L.A. (ed.)]. General Technical Report RM-GTR-271, U.S. Department of Agriculture, Forest Service, Rocky Mountain Forest and Range Experiment Station, Fort Collins, CO, pp. 56-70.

Birdsey, R.A. and G.M. Lewis, 2003: Current and historical trends in use, management, and disturbance of U.S. forestlands. In: *The Potential of U.S. Forest Soils to Sequester Carbon and Mitigate the Greenhouse Effect* [Kimble, J.M., L.S. Heath, and R. A. Birdsey (eds.)]. CRC Press LLC, New York, pp. 15-33.

Bradley, B.A., R.A. Houghton, J.F. Mustard, and S.P. Hamburg, 2006: Invasive grass reduces aboveground carbon stocks in shrublands of the Western US. *Global Change Biology*, **12(10)**, 1815-1822.

Cairns, M.A., P.K. Haggerty, R. Alvarez, B.H.J. De Jong, and I. Olmsted, 2000: Tropical Mexico's recent land-use change: a region's contribution to the global carbon cycle. *Ecological Applications*, **10(5)**, 1426-1441.

Caspersen, J.P., S.W. Pacala, J.C. Jenkins, G.C. Hurtt, P.R. Moorcroft, and R.A. Birdsey, 2000: Contributions of land-use history to carbon accumulation in U.S. forests. *Science*, **290(5494)**, 1148-1151.

Environment Canada, 2005: *Canada's Greenhouse Gas Inventory 1990-2003: Initial Submission*. Greenhouse Gas Division, Environment Canada, Ottawa, Ontario, Canada. Available at http://unfccc.int/national_reports/annex_i_ghg_inventories/national_inventories_submissions/items/2761.php

Houghton, R.A., J.L. Hackler, and K.T. Lawrence, 1999: The U.S. carbon budget: contributions from land-use change. *Science*, **285(5427)**, 574-578.

Houghton, R.A. and J.L. Hackler, 2000: Changes in terrestrial carbon storage in the United States. 1. The roles of agriculture and forestry. *Global Ecology and Biogeography*, **9(2)**, 125-144.

Houghton, R.A., J.L. Hackler, and K.T. Lawrence, 2000: Changes in terrestrial carbon storage in the United States. 2. The role of fire and fire management. *Global Ecology and Biogeography*, **9(2)**, 145-170.

Hurtt, G.C., S.W. Pacala, P.R. Moorcroft, J. Caspersen, E. Shevliakova, R.A. Houghton, and B. Moore III, 2002: Projecting the future of the U.S. carbon sink. *Proceedings of the National Academy of Sciences*, **99(3)**, 1389-1394.

Masera, O.R., M.J. Ordonez, and R. Dirzo, 1997: Carbon emissions from Mexican forests: current situation and long-term scenarios. *Climatic Change*, **35(3)**, 265-295.

Pacala, S.W., G.C. Hurtt, D. Baker, P. Peylin, R.A. Houghton, R.A. Birdsey, L. Heath, E.T. Sundquist, R.F. Stallard, P. Ciais, P. Moorcroft, J.P. Caspersen, E. Shevliakova, B. Moore, G. Kohlmaier, E. Holland, M. Gloor, M.E. Harmon, S.M. Fan, J.L. Sarmiento, C.L. Goodale, D. Schimel, and C.B. Field, 2001: Consistent land- and atmosphere-based U.S. carbon sink estimates. *Science*, **292(5525)**, 2316-2320.

Smith, W.B., P.D. Miles, J.S. Vissage, and S.A. Pugh, 2004: *Forest Resources of the United States, 2002*. General Technical Report NC-241, U.S. Department of Agriculture, Forest Service, St. Paul, MN, 137 pp.

APPENDIX B REFERENCES

Baldocchi, D., E. Falge, L.H. Gu, R. Olson, D. Hollinger, S. Running, P. Anthoni, C. Bernhofer, K. Davis, R. Evans, J. Fuentes, A. Goldstein, G. Katul, B. Law, X.H. Lee, Y. Malhi, T. Meyers, W. Munger, W. Oechel, K.T. Paw U, K. Pilegaard, H.P. Schmid, R. Valentini, S. Verma, T. Vesala, K. Wilson, and S. Wofsy, 2001: FLUXNET: a new tool to study the temporal and spatial variability of ecosystem-scale carbon dioxide, water vapor, and energy flux densities, *Bulletin of the American Meteorological Society*, **82(11)**, 2415-2434.

Barford, C.C., S.C. Wofsy, M.L. Goulden, J.W. Munger, E.H. Pyle, S.P. Urbanski, L. Hutyra, S.R. Saleska, D. Fitzjarrald, and K. Moore, 2001: Factors controlling long- and short-term sequestration of atmospheric CO_2 in a mid-latitude forest. *Science*, **294(5547)**, 1688-1691.

Canadell, J.G., H.A. Mooney, D.D. Baldocchi, J.A. Berry, J.R. Ehleringer, C.B. Field, S.T. Gower, D.Y. Hollinger, J.E. Hunt, R.B. Jackson, S.W. Running, G.R. Shaver, W. Steffen, S.E. Trumbore, R. Valentini, B.Y. Bond, 2000: Carbon metabolism of the terrestrial biosphere: a multitechnique approach for improved understanding. *Ecosystems*, **3(2)**, 115-130.

Cook, B.D., K.J. Davis, W. Wang, A. Desai, B.W. Berger, R.M. Teclaw, J.G. Martin, P.V. Bolstad, P.S. Bakwin, C. Yi, and W. Heilman, 2004: Carbon exchange and venting anomalies in an upland deciduous forest in northern Wisconsin, USA. *Agricultural and Forest Meteorology*, **126(3-4)**, 271-295.

Curtis, P.S., P.J. Hanson, P. Bolstad, C. Barford, J.C. Randolph, H.P. Schmid, and K.B. Wilson, 2002: Biometric and eddy-covariance based estimates of annual carbon storage in five eastern North American deciduous forests. *Agricultural and Forest Meteorology*, **113(1-4)**, 3-19.

Ehman, J.L., H.P. Schmid, C.S.B. Grimmond, J.C. Randolph, P.J. Hanson, C.A. Wayson, and F.D. Cropley, 2002: An initial intercomparison of micrometerological and ecological inventory estimates of carbon exchange in a mid-latitude deciduous forest. *Global Change Biology*, **8(6)**, 575-589.

Gough, C.M., P.S. Curtis, J.G. Vogel, H.P. Schmid, and H.B. Su: Annual carbon storage from 1999 to 2003 in a Northern hardwood forest assessed using eddy-covariance and biometric methods. *Agricultural and Forest Meteorology* (in review).

Horst, T.W. and J.C. Weil, 1994: How far is far enough? The fetch requirements for micrometeorological measurement of surface fluxes. *Journal of Atmospheric and Oceanic Technology*, **11(4)**, 1018-1025.

Law, B.E., P.E. Thornton, J. Irvine, P.M. Anthoni, and S. Van Tuyl, 2001: Carbon storage and fluxes in ponderosa pine forests at different developmental stages. *Global Change Biology*, **7(7)**, 755-777.

Verma, S.B., A. Dobermann, K.G. Cassman, D.T. Walters, J.M. Knops, T.J. Arkebauer, A.E. Suyker, G.G. Burba, B. Amos, H.S. Yang, D. Ginting, K.G. Hubbard, A.A. Gitelson, and E.A. Walter-Shea, 2005: Annual carbon dioxide exchange in irrigated and rainfed maize-based agroecosystems. *Agricultural and Forest Meteorology*, **131(1-2)**, 77-96.

Wofsy, S.C., M.L. Goulden, J.W. Munger, S.-M. Fan, P.S. Bakwin, B.C. Daube, S.L. Bassow, and F.A. Bazzaz, 1993: Net exchange of CO_2 in a mid-latitude forest. *Science*, **260(5112)**, 1314-1317.

APPENDIX D REFERENCES

Amiro, B.D., A.G. Barr, T.A. Black, H. Iwashita, N. Kljun, J.H. McCaughey, K. Morgenstern, S. Murayama, Z. Nesic, A.L. Orchansky, and N. Saigusa, 2005: Carbon, energy and water fluxes at mature and disturbed forest sites, Saskatchewan, Canada. *Agricultural and Forest Meteorology*, **136(3-4)**, 237-251.

Arain, M.A. and N. Restrepo-Coupe, 2005: Net ecosystem production in an eastern white pine plantation in southern Canada. *Agricultural and Forest Meteorology*, **128(3-4)**, 223-241.

Chapin, F.S. III, G.M. Woodwell, J.T. Randerson, G.M. Lovett, E.B. Rastetter, D.D. Baldocchi, D.A. Clark, M.E. Harmon, D.S. Schimel, R. Valentini, C. Wirth, J.D. Aber, J.J. Cole, M.L. Goulden, J.W. Harden, M. Heimann, R.W. Howarth, P.A. Matson, A.D. McGuire, J.M. Melillo, H.A. Mooney, J.C. Neff, R.A. Houghton, M.L. Pace, M.G. Ryan, S.W. Running, O.E. Sala, W.H. Schlesinger, E. D. Schulze, 2006. Reconciling carbon-cycle concepts, terminology, and methodology. *Ecosystems* **9(7)**, 1041-1050.

Clark, K.L., H.L. Gholz, and M.S. Castro, 2004: Carbon dynamics along a chronosequence of slash pine plantations in north Florida. *Ecological Applications*, **14(4)**, 1154-1171.

Giardina, C.P., D. Binkley, M.G. Ryan, J.H. Fownes, and R.S. Senock, 2004: Belowground carbon cycling in a humid tropical forest decreases with fertilization. *Oecologia*, **139(4)**, 545-550.

Griffis, T.J., T.A. Black, K. Morgenstern, A.G. Barr, Z. Nesic, G.B. Drewitt, D. Gaumont-Guay, and J.H. McCaughey, 2003: Ecophysiological controls on the carbon balances of three southern boreal forests. *Agricultural and Forest Meteorology*, **117(1-2)**, 53-71.

Humphreys, E.R., T.A. Black, K. Morgenstern, Z. Li, and Z. Nesic, 2005: Net ecosystem production of a Douglas-fir stand for three years following clearcut harvesting. *Global Change Biology*, **11(3)**, 450-464.

Law, B.E., E. Falge, D.D. Baldocchi, P. Bakwin, P. Berbigier, K. Davis, A.J. Dolman, M. Falk, J.D. Fuentes, A. Goldstein, A. Granier, A. Grelle, D. Hollinger, I.A. Janssens, P. Jarvis,

N.O. Jensen, G. Katul, Y. Mahli, G. Matteucci, R. Monson, W. Munger, W. Oechel, R. Olson, K. Pilegaard, K.T. Paw, H. Thorgeirsson, R. Valentini, S. Verma, T. Vesala, K. Wilson, and S. Wofsy, 2002: Environmental controls over carbon dioxide and water vapor exchange of terrestrial vegetation. *Agricultural and Forest Meteorology,* **113(1-2)**, 97-120.

Lugo, A.E., J.F. Colón, and F.N. Scatena, 1999: The Caribbean. In: *North American Terrestrial Vegetation* [Barbour, M.G. and W.D. Billings (eds.)]. Cambridge University Press, Cambridge, United Kingdom, 530 pp.

Osher, L.J., P.A. Matson, and R. Amundson, 2003: Effect of land use change on soil carbon in Hawaii. *Biogeochemistry,* **65(2)**, 213-232.

Randerson, J.T., F.S. Chapin, III, J.W. Hardin, J.C. Neff, and M.E. Harmon, 2002. Net ecosystem production: a comprehensive measure of net carbon accumulation by ecosystems. *Ecological Applications* **12(4)**: 937-947.

APPENDIX E REFERENCES

Chapin, F.S. III, G.M. Woodwell, J.T. Randerson, G.M. Lovett, E.B. Rastetter, D.D. Baldocchi, D.A.Clark, M.E. Harmon, D.S. Schimel, R. Valentini, C. Wirth, J.D. Aber, J.J. Cole, M.L. Goulden, J.W. Harden, M. Heimann, R.W. Howarth, P.A. Matson, A.D. McGuire, J.M. Melillo, H.A. Mooney, J.C. Neff, R.A. Houghton, M.L. Pace, M.G. Ryan, S.W. Running, O.E. Sala, W.H. Schlesinger, E. D. Schulze, 2006. Reconciling carbon-cycle concepts, terminology, and methodology. *Ecosystems* **9**, 1041-1050.

Fitzsimmons, M.J., D.J. Pennock, and J. Thorpe, 2004: Effects of deforestation on ecosystem carbon densities in central Saskatchewan, Canada. *Forest Ecology and Management,* **188(1-3)**, 349-361.

Harmon, M.E., J.M. Harmon, W.K. Ferrell, and D. Brooks, 1996: Modeling carbon stores in Oregon and Washington forest products: 1900-1992. *Climatic Change,* **33(4)**, 521-550.

Harmon, M., 2001: Carbon sequestration in forests - addressing the scale question. *Journal of Forestry,* **99**, 24-29.

Harmon, M. and P. Marks, 2002: Effects of silvicultural practices on carbon stores in Douglas-fir-western hemlock forests in the Pacific Northwest, USA: results from a simulation model. *Canadian Journal of Forest Research,* **32(5)**, 863-877.

Janisch, J. and M. Harmon, 2002: Successional changes in live and dead wood carbon stores: implications for net ecosystem productivity. *Tree Physiology,* **22(2-3)**, 77-89.

Jenkins, J.C., D.C. Chojnacky, L.S. Heath, and R.A. Birdsey, 2003: National-scale biomass estimators for United States tree species. *Forest Science,* **49(1)**, 12-35.

Peterson, E.B., G.M. Bonnor, G.C. Robinson, and N.M. Peterson, 1999: *Carbon Sequestration Aspects of an Afforestation Program in Canada's Prairie Provinces.* Joint Forest Sector Table/Sinks Table, National Climate Change Process, Ottawa, Ontario, Canada. Available at http://www.nccp.ca/NCCP/national_process/issues/sinks_e.html

Pregitzer, K.S. and E.S. Euskirchen, 2004: Carbon cycling and storage in world forests: biomes patterns related to forest age. *Global Change Biology,* **10(12)**, 2052-2077.

Ryan, M.G., D. Binkley, and J.H. Fownes, 1997: Age-related decline in forest productivity: pattern and process. *Advances in Ecological Research,* **27**, 213-262.

Schulze, E., J. Lloyd, F. Kelliher, C. Wirth, C. Rebmann, B. Luhker, M. Mund, A. Knohl, I. Milyukova, W. Schulze, W. Ziegler, A. Varlagin, A. Sogachev, R. Valentini, S. Dore, S. Grigoriev, O. Kolle, M. Panfyorov, N. Tchebakova, and N. Vygodskaya, 1999: Productivity of forests in the Eurosiberian boreal region and their potential to act as a carbon sink - a synthesis. *Global Change Biology,* **5(6)**, 703-722.

Stainback, G.A. and J.R.R. Alavalapati, 2005: Effects of carbon markets on the optimal management of slash pine (Pinus elliottii) plantations. *Southern Journal of Applied Forestry,* **29(1)**, 27-32.

Stanturf, J.A., R.C. Kellison, F.S. Broerman, and S.B. Jones, 2003: Productivity of southern pine plantations - where we are and how did we get here? *Journal of Forestry,* **101(3)**, 26-31.

Woodwell, G. and R. Whittaker, 1968: Primary production in terrestrial communities. *American Zoologist,* **8**, 19-30.

APPENDIX F REFERENCES

Alford, D.P., R.D. Delaune, and C.W. Lindau, 1997: Methane flux from Mississippi River deltaic plain wetlands. *Biogeochemistry,* **37(3)**, 227–236.

Armentano, T.B. and E.S. Menges, 1986: Patterns of change in the carbon balance of organic soil-wetlands of the temperate zone. *Journal of Ecology,* **74(3)**, 755–774.

Aselmann, I. and P.J. Crutzen, 1989: Global distribution of natural freshwater wetlands and rice paddies, their net primary productivity, seasonality and possible methane emissions. *Journal of Atmospheric Chemistry,* **8(4)**, 307–359.

Bartlett, D.S., K.B. Bartlett, J.M. Hartman, R.C. Harriss, D.I. Sebacher, R. Pelletier-Travis, D.D. Dow, and D.P. Brannon, 1989: Methane emissions from the Florida Everglades: patterns of variability in a regional wetland ecosystem. *Global Biogeochemical Cycles,* **3(4)**, 363–374.

Bartlett, K.B., D.S. Bartlett, R.C. Harriss, and D.I. Sebacher, 1987: Methane emissions along a salt marsh salinity gradient. *Biogeochemistry,* **4(3)**, 183–202.

Bartlett, K.B. and R.C. Harriss, 1993: Review and assessment of methane emissions from wetlands. *Chemosphere,* **26(1-4)**, 261–320.

Bartlett, K.B., R.C. Harriss, and D.I. Sebacher, 1985: Methane flux from coastal salt marshes. *Journal of Geophysical Research,* **90(D3)**, 5710–5720.

Batjes, N.H., 1996: Total carbon and nitrogen in the soils of the world. *European Journal of Soil Science,* **47(2)**, 151–163.

Birdsey, R.A., 1992: *Carbon Storage and Accumulation in United States Forest Ecosystems.* General Technical Report WO-59, USDA Forest Service, Washington, DC.

Bridgham, S.D., C.-L. Ping, J.L. Richardson, and K. Updegraff, 2000: Soils of northern peatlands: Histosols and Gelisols. In: *Wetland Soils: Genesis, Hydrology, Landscapes, and Classification* [Richardson, J.L. and M.J. Vepraskas (eds.)]. CRC Press, Boca Raton, FL, pp. 343–370.

Bridgham, S.D., K. Updegraff, and J. Pastor, 1998: Carbon, nitrogen, and phosphorus mineralization in northern wetlands. *Ecology*, **79(5)**, 1545–1561.

Brown, M.J., G.M. Smith, and J. McCollum, 2001: *Wetland Forest Statistics for the South Atlantic States*. RB-SRS-062, Southern Research Station, U.S. Forest Service, Asheville, NC, 52 pp.

Burke, R.A., T.R. Barber, and W.M. Sackett, 1988: Methane flux and stable hydrogen and carbon isotope composition of sedimentary methane from the Florida Everglades. *Global Biogeochemical Cycles*, **2(4)**, 329–340.

Carroll, P.C. and P.M. Crill, 1997: Carbon balance of a temperate poor fen. *Global Biogeochemical Cycles*, **11(3)**, 349–356.

Chanton, J.P., G.J. Whiting, J.D. Happell, and G. Gerard, 1993: Contrasting rates and diurnal patterns of methane emission from emergent aquatic macrophytes. *Aquatic Botany*, **46(2)**, 111–128.

Chanton, J.P., G.J. Whiting, W.J. Showers, and P.M. Crill, 1992: Methane flux from *Peltandra virginica*: stable isotope tracing and chamber effects. *Global Biogeochemical Cycles*, **6(1)**, 15–31.

Chimner, R.A. and D.J. Cooper, 2003: Carbon dynamics of pristine and hydrologically modified fens in the southern Rocky Mountains. *Canadian Journal of Botany*, **81(5)**, 477–491.

Chmura, G.L., S.C. Anisfeld, D.R. Cahoon, and J.C. Lynch, 2003: Global carbon sequestration in tidal, saline wetland soils. *Global Biogeochemical Cycles*, **17(4)**, 1111, doi:10.1029/2002GB001917.

Cleary, J., N.T. Roulet, and T.R. Moore, 2005: Greenhouse gas emissions from Canadian peat extraction, 1990–2000: a life-cycle analysis. *Ambio*, **34(6)**, 456–461.

Clymo, R.S., J. Turunen, and K. Tolonen, 1998: Carbon accumulation in peatland. *Oikos*, **81(2)**, 368–388.

Coles, J.R.P. and J.B. Yavitt, 2004: Linking belowground carbon allocation to anaerobic CH_4 and CO_2 production in a forested peatland, New York State. *Geomicrobiology Journal*, **21(7)**, 445–454.

Cowardin, L.M., V. Carter, F.C. Golet, and E.T. LaRoe, 1979: *Classification of Wetlands and Deepwater Habitats of the United States*. FWS/OBS-79/31, Fish and Wildlife Service, U.S. Department of the Interior, Washington, DC, 131 pp.

Craft, C.B. and W.P. Casey, 2000: Sediment and nutrient accumulation in floodplain and depressional freshwater wetlands of Georgia, USA. *Wetlands*, **20(2)**, 323–332.

Dahl, T.E., 1990: *Wetland Losses in the United States 1970's to 1980's*. U.S. Department of the Interior, Fish and Wildlife Service, Washington, DC, 13 pp.

Dahl, T.E., 2000: *Status and Trends of Wetlands in the Conterminous United States, 1986 to 1997*. U.S. Department of the Interior, Fish and Wildlife Service, Washington, DC, 82 pp.

Dahl, T.E. and C.E. Johnson, 1991: *Status and Trends of Wetlands in the Conterminous United States, Mid-1970's to Mid-1980's*. U.S. Department of the Interior, Fish and Wildlife Service, Washington, DC, 22 pp.

Davidson, I., R. Vanderkam, and M. Padilla, 1999: Review of wetland inventory information in North America. In: *Global Review of Wetland Resources and Priorities for Wetland Inventory* [Finlayson, C.M. and A.G. Spiers (eds.)]. Supervising Scientist Report 144, Supervising Scientist, Canberra, Australia, 35 pp.

DeLaune, R.D., C.J. Smith, and W.H. Patrick Jr., 1983: Methane release from Gulf coast wetlands. *Tellus*, **35B(1)**, 8–15.

Dise, N., 1993: Methane emissions from Minnesota peatlands: spatial and seasonal variability. *Global Biogeochemical Cycles*, **7(1)**, 123–142.

Dise, N.B. and E.S. Verry, 2001: Suppression of peatland methane emission by cumulative sulfate deposition in simulated acid rain. *Biogeochemistry*, **53(2)**, 143–160.

Ehhalt, D., M. Prather, F. Dentener, E. Dlugokencky, E. Holland, I. Isaksen, J. Katima, V. Kirchhoff, P. Matson, P. Midgley, and M. Wang, 2001: Atmospheric chemistry and greenhouse gases. In: *Climate Change 2001: The Scientific Basis*. Contribution of Working Group I to the Third Assessment Report of the Intergovernmental Panel on Climate Change [Houghton, J.T., Y. Ding, D.J. Griggs, M. Noguer, P.J. van der Linden, X. Dai, K. Maskell, and C.A. Johnson (eds.)]. Cambridge University Press, Cambridge, United Kingdom, pp. 239–287.

Eswaran, H., E. Van Den Berg, and J. Kimble, 1995: Global soil carbon resources. In: *Soils and Global Change* [Lal, R., J. Kimble, E. Levine, and B.A. Stewart (eds.)]. Lewis Publishers, Boca Raton, FL, pp. 27–43.

Euliss, N.H., R.A. Gleason, A. Olness, R.L. McDougal, H.R. Murkin, R.D. Robarts, R.A. Bourbonniere, and B.G. Warner, 2006: North American prairie wetlands are important nonforested land-based carbon storage sites. *Science of the Total Environment*, **361(1-3)**, 179–188.

FAO (Food and Agriculture Organization of the United Nations), 1991: *The Digitized Soil Map of the World*. World Soil Resource Report 67, FAO, Rome, Italy.

FAO-UNESCO (Food and Agriculture Organization, United Nations Educational, Scientific and Cultural Organization), 1974: *Soil Map of the World*. 1:5,000,000 scale map, UNESCO, Paris, France.

Field, D.W., A.J. Reyer, P.V. Genovese, and B.D. Shearer, 1991: *Coastal Wetlands of the United States: An Accounting of a Valuable Natural Resource*. Strategic Assessment Branch, Ocean Assessments Division, Office of Oceanography and Marine Assessment, National Ocean Service, National Oceanic and Atmospheric Administration, Washington, DC, 59 pp.

Frayer, W.E., T.J. Monahan, D.C. Bowden, and F.A. Graybill, 1983: *Status and Trends of Wetlands and Deepwater Habitats in the Conterminous United States, 1950s to 1970s*. Department of Forest and Wood Sciences, Colorado State University, Fort Collins, CO, 31 pp.

Frolking, S. and P. Crill, 1994: Climate controls on temporal variability of methane flux from a poor fen in southeastern New Hampshire: measurement and modeling. *Global Biogeochemical Cycles*, **8(4)**, 385–397.

Gorham, E., 1991: Northern peatlands: role in the carbon cycle and probable responses to climatic warming. *Ecological Applications*, **1(2)**, 182–195.

Gunnison, D., R.L. Chen, and J.M. Brannon, 1983: Relationship of materials in flooded soils and sediments to the water-quality of reservoirs. 1. Oxygen-consumption rates. *Water Research*, **17(11)**, 1609–1617.

Hall, J.V., W.E. Frayer, and B.O. Wilen, 1994: *Status of Alaska Wetlands*. U.S. Fish and Wildlife Service, Anchorage, Alaska, 32 pp.

Halsey, L.A., D.H. Vitt, and L.D. Gignac, 2000: Sphagnum-dominated peatlands in North America since the last glacial maximum: their occurence and extent. *The Bryologist*, **103(2)**, 334–352.

Hanson, A.R. and L. Calkins, 1996: *Wetlands of the Maritime Provinces: Revised Documentation for the Wetlands Inventory*. Technical Report No. 267, Canadian Wildlife Service, Atlantic Region, Sackville, New Brunswick, Canada, 67 pp.

Happell, J.D., J.P. Chanton, G.J. Whiting, and W.J. Showers, 1993: Stable isotopes as tracers of methane dynamics in Everglades marshes with and without active populations of methane oxidizing bacteria. *Journal of Geophysical Research*, **98(D8)**, 14771–14782.

Harden, J.W., J.M. Sharpe, W.J. Parton, D.S. Ojima, T.L. Fries, T.G. Huntington, and S.M. Dabney, 1999:. Dynamic replacement and loss of soil carbon on eroding cropland. *Global Biogeochemical Cycles*, **13(4)**, 885–901.

Harriss, R.C. and D.I. Sebacher, 1981: Methane flux in forested freshwater swamps of the southeastern United States. *Geophysical Research Letters*, **8(9)**, 1002–1004.

Harriss, R.C., D.I. Sebacher, K.B. Bartlett, D.S. Bartlett, and P.M. Crill, 1988: Sources of atmospheric methane in the south Florida environment. *Global Biogeochemical Cycles*, **2(3)**, 231–243.

Harriss, R.C., D.I. Sebacher, and F.P. Day, Jr., 1982: Methane flux in the Great Dismal Swamp. *Nature*, **297(5868)**, 673–674.

Johnston, C.A., 1991: Sediment and nutrient retention by freshwater wetlands: effects on surface water quality. *Critical Reviews in Environmental Control*, **21**, 491–565.

Johnston, C.A., S.D. Bridgham, and J.P. Schubauer-Berigan, 2001: Nutrient dynamics in relation to geomorphology of riverine wetlands. *Soil Science Society of America Journal*, **65(2)**, 557–577.

Joosten, H. and D. Clarke, 2002: *Wise Use of Mires and Peatlands - Background Principles Including a Framework for Decision-Making*. International Mire Conservation Group and International Peat Society, Saarijärvi, Finland, 304 pp.

Kelly, C.A., C.S. Martens, and W. Ussler III, 1995: Methane dynamics across a tidally flooded riverbank margin. *Limnology and Oceanography*, **40(6)**, 1112–1129.

Kelly, C.A., J.W.M. Rudd, R.A. Bodaly, N.T. Roulet, V.L. St. Louis, A. Heyes, T.R. Moore, S. Schiff, R. Aravena, K.J. Scott, B. Dyck, R. Harris, B. Warner, and G. Edwards, 1997: Increase in fluxes of greenhouse gases and methyl mercury following flooding of an experimental reservoir. *Environmental Science & Technology*, **31(5)**, 1334–1344.

Kim, J., S.B. Verma, and D.P. Billesbach, 1999: Seasonal variation in methane emission from a temperate Phragmites-dominated marsh: effect of growth stage and plant-mediated transport. *Global Change Biology*, **5(4)**, 433-440.

King, G.M. and W.J. Wiebe, 1978: Methane release from soils of a Georgia salt marsh. *Geochimica et Cosmochimica Acta*, **42(4)**, 343–348.

Kivinen, E. and P. Pakarinen, 1981: Geographical distribution of peat resources and major peatland complex types in the world. *Annales Academiae Scientiarum Fennicae, Series A*, **3(132)**, 1–28.

Kristensen, E., S.I. Ahmed, and A.H. Devol, 1995: Aerobic and anaerobic decomposition of organic matter in marine sediment: Which is fastest? *Limnology and Oceanography*, **40(9)**, 1430–1437.

Lansdown, J., P. Quay, and S. King, 1992: CH_4 production via CO_2 reduction in a temperate bog: a source of ^{13}C-depleted CH_4. *Geochimica et Comsochimica Acta*, **56(9)**, 3493-3503.

Lappalainen, E., 1996: General review on world peatland and peat resources. In: *Global Peat Resources* [Lappalainen, E. (ed.)]. International Peat Society and Geological Survey of Finland, Jyskä, Finland, pp. 53–56.

Lugo, A.E. and S.C. Snedaker, 1974: The ecology of mangroves. *Annual Review of Ecology and Systematics*, **5**, 39–64.

Maltby, E. and P. Immirzi, 1993: Carbon dynamics in peatlands and other wetland soils, regional and global perspectives. *Chemosphere*, **27(6)**, 999–1023.

Malterer, T.J., 1996: Peat resources of the United States. In: *Global Peat Resources* [Lappalainen, E. (ed.)]. International Peat Society and Geological Survey of Finland, Jyskä, Finland, pp. 253–260.

Matthews, E. and I. Fung, 1987: Methane emission from natural wetlands: global distribution, area, and environmental characteristics of sources. *Global Biogeochemical Cycles*, **1**, 61–86.

Meade, R.H., T.R. Yuzyk, and T.J. Day, 1990: Movement and storage of sediments in rivers of the United States and Canada. In: *Surface Water Hydrology*. The Geology of North America, Vol. 0-1 [Wolman, M.G. and H.C. Riggs (eds.)]. Geological Society of America, Boulder, CO, pp. 255–280.

Megonigal, J.P. and W.H. Schlesinger, 2002: Methane-limited methanotrophy in tidal freshwater swamps. *Global Biogeochemical Cycles*, **16(4)**, 1088, doi:1010.1029/2001GB001594.

Mendelssohn, I.A. and K.L. McKee, 2000: Saltmarshes and mangroves. In: *North American Terrestrial Vegetation* [Barbour, M.G. and W.D. Billings (eds.)]. Cambridge University Press, Cambridge, United Kingdom, pp. 501–536.

Miller, D.N., W.C. Ghiorse, and J.B. Yavitt, 1999: Seasonal patterns and controls on methane and carbon dioxide fluxes in forested swamp pools. *Geomicrobiology Journal*, **16(4)**, 325–331.

Mitsch, W.J. and J.G. Gosselink, 1993: *Wetlands*. Van Nostrand Reinhold, New York, 722 pp.

Moore, T.R. and N.T. Roulet, 1995: Methane emissions from Canadian peatlands. In: *Soils and Global Change* [Lal, R., J. Kimble, E. Levine, and B.A. Stewart (eds.)]. Lewis Publishers, Boca Raton, FL, pp. 153–164.

Moore, T.R., N.T. Roulet, and J.M. Waddington, 1998: Uncertainty in predicting the effect of climatic change on the carbon cycling of Canadian peatlands. *Climatic Change*, **40(2)**, 229–245.

Moser, M., C. Prentice, and S. Frazier, 1996: *A Global Overview of Wetland Loss and Degradation*. Proceeding of the 6th Meeting of the Conference of Contracting Parties, Brisbane, Australia, Papers, Technical Session B, /Vol 10/12B, 19-27 March 1996, Ramsar Convention Bureau, Gland, Switzerland, 21-31.

Naiman, R.J., T. Manning, and C.A. Johnston, 1991: Beaver population fluctuations and tropospheric methane emissions in boreal wetlands. *Biogeochemistry*, **12(1)**, 1–15.

National Wetlands Working Group, 1988: *Wetlands of Canada*. Sustainable Development Branch, Environment Canada, Ottawa, Ontario, and Polyscience Publications Inc, Montreal, Quebec, 452 pp.

Neff, J.C., W.D. Bowman, E.A. Holland, M.C. Fisk, and S.K. Schmidt, 1994: Fluxes of nitrous oxide and methane from nitrogen-amended soils in a Colorado alpine ecosystem. *Biogeochemistry*, **27(1)**, 23–33.

Neubauer, S.C., W.D. Miller, and I.C. Anderson, 2000: Carbon cycling in a tidal freshwater marsh ecosystem: a carbon gas flux study. *Marine Ecology Progress Series*, **199**, 13–30.

NRCS (Natural Resources Conservation Service), 1999: *Soil Taxonomy: A Basic System of Soil Classification for Making and Interpreting Soil Surveys, Second Edition*. NRCS, U.S. Department of Agriculture, Washington, DC, 869 pp.

Odum, W.E., T.J. Smith, III, J.K. Hoover, and C.C. McIvor, 1984: *The Ecology of Tidal Freshwater Marshes of the United States East Coast: A Community Profile*. FWS/OBS-83/17, U.S. Fish and Wildlife Service, Washington, DC, 177 pp.

Olmsted, I., 1993: Wetlands of Mexico. In: *Wetlands of the World* [Whigham, D.F., D. Dykjová, and S. Hejný (eds.)]. Kluwer Academic Publishers, Dordrecht, The Netherlands, pp. 637–677.

Ovenden, L., 1990: Peat accumulation in northern wetlands. *Quaternary Research*, **33(3)**, 377–386.

Pulliam, W.M., 1993: Carbon dioxide and methane exports from a southeastern floodplain swamp. *Ecological Monographs*, **63**, 29–53.

Rieger, S., D.B. Schoephoster, and C.E. Furbush, 1979: *Exploratory Soil Survey of Alaska*. U.S. Department of Agriculture Soil Conservation Service, Anchorage, Alaska, 213 pp.

Robinson, S.D. and T.R. Moore, 1999: Carbon and peat accumulation over the past 1200 years in a landscape with discontinuous permafrost, northwestern Canada. *Global Biogeochemical Cycles*, **13(2)**, 591–602.

Rubec, C., 1996: The status of peatland resources in Canada. In: *Global Peat Resources* [Lappalainen, E. (ed.)]. International Peat Society and Geological Survey of Finland, Jyskä, Finland, pp. 243–252.

Schipper, L.A. and K.R. Reddy, 1994: Methane production and emissions from four reclaimed and pristine wetlands of southeastern United States. *Soil Science Society of America*, **58**, 1270–1275.

Shannon, R.D. and J.R. White, 1994: A three year study of controls on methane emissions from two Michigan peatlands. *Biogeochemistry*, **27(1)**, 35–60.

Shurpali, N.J. and S.B. Verma, 1998: Micrometeorological measurements of methane flux in a Minnesota peatland during two growing seasons. *Biogeochemistry*, **40(1)**, 1–15.

Smith, L.K. and W.M. Lewis Jr., 1992: Seasonality of methane emissions from five lakes and associated wetlands of the Colorado Rockies. *Global Biogeochemical Cycles*, **6(4)**, 323–338.

Smith, S.V., W.H. Renwick, R.W. Buddemeier, and C.J. Crossland, 2001: Budgets of soil erosion and deposition for sediments and sedimentary organic carbon across the conterminous United States. *Global Biogeochemical Cycles*, **15(3)**, 697–707.

Spalding, M., F. Blasco, and C. Field (eds.), 1997: *World Mangrove Atlas*. The International Society for Mangrove Ecosystems, Okinawa, Japan, 178 pp.

Spiers, A.G., 1999: Review of international/continental wetland resources. In: *Global Review of Wetland Resources and Priorities for Wetland Inventory* [Finlayson, C.M. and A.G. Spiers (eds.)]. Supervising Scientist Report 144, Supervising Scientist, Canberra, Australia.

Stallard, R.F., 1998: Terrestrial sedimentation and the carbon cycle: coupling weathering and erosion to carbon burial. *Global Biogeochemical Cycles*, **12(2)**, 231–257.

Tarnocai, C., 1998: The amount of organic carbon in various soil orders and ecological provinces in Canada. In: *Soil Processes and the Carbon Cycle* [Lal, R., J.M. Kimble, R.F. Follett, and B.A. Stewart (eds.)]. CRC Press, Boca Raton, FL, pp. 81–92.

Tarnocai, C., I.M. Kettles, and B. Lacelle, 2005: *Peatlands of Canada*. 1:6,500,000 scale map, Agriculture and Agri-Food Canada, Research Branch, Ottawa, Ontario, Canada.

Tolonen, K. and J. Turunen, 1996: Accumulation rates of carbon in mires in Finland and implications for climate change. *Holocene*, **6**, 171–178.

Trimble, S.W. and P. Crosson, 2000: Land use - US soil erosion rates - Myth and reality. *Science*, **289(5477)**, 248–250.

Trumbore, S.E. and J.W. Harden, 1997: Accumulation and turnover of carbon in organic and mineral soils of the BOREAS northern study area. *Journal of Geophysical Research*, **102(D24)**, 28817–28830.

Turetsky, M.R., R.K. Wieder, L.A. Halsey, and D. Vitt, 2002: Current disturbance and the diminishing peatland carbon sink. *Geophysical Research Letters*, **29**, doi:10.1029/2001GL014000.

Turunen, J., N.T. Roulet, and T.R. Moore, 2004: Nitrogen deposition and increased carbon accumulation in ombrotrophic peatlands in eastern Canada. *Global Biogeochemical Cycles*, **18**, GB3002, doi:3010.1029/2003GB002154.

Twilley, R.R., R.H. Chen, and T. Hargis, 1992: Carbon sinks in mangroves and their implications to carbon budget of tropical coastal ecosystems. *Water, Air and Soil Pollution*, **64(1)**, 265–288.

Valiela, I., J.L. Bowen, and J.K. York, 2001: Mangrove forests: one of the world's threatened major tropical environments. *BioScience*, **51(10)**, 807–815.

Vitt, D.H., L.A. Halsey, I.E. Bauer, and C. Campbell, 2000: Spatial and temporal trends in carbon storage of peatlands of continental western Canada through the Holocene. *Canadian Journal of Earth Sciences*, **37(5)**, 683–693.

Vitt, D.H., L.A. Halsey, and S.C. Zoltai, 1994: The bog landforms of continental western Canada in relation to climate and permafrost patterns. *Arctic and Alpine Research*, **26(1)**, 1–13.

Webb, R.S. and T. Webb III, 1988: Rates of sediment accumulation in pollen cores from small lakes and mires of eastern North America. *Quaternary Research*, **30(3)**, 284–297.

WEC (World Energy Council), 2001: *Survey of Energy Resources*. http://www.worldenergy.org/wec-geis/publications/reports/ser/peat/peat.asp

Werner, C., K. Davis, P. Bakwin, C. Yi, D. Hurst, and L. Lock, 2003: Regional-scale measurements of CH_4 exchange from a tall tower over a mixed temperate/boreal lowland and wetland forest. *Global Change Biology*, **9(9)**, 1251–1261.

West, A.E., P.D. Brooks, M.C. Fisk, L.K. Smith, E.A. Holland, C.H. Jaeger III, S. Babcock, R.S. Lai, and S.K. Schmidt, 1999: Landscape patterns of CH_4 fluxes in an alpine tundra ecosystem. *Biogeochemistry*, **45(3)**, 243–264.

Wickland, K.P., R.G. Striegl, S.K. Schmidt, and M.A. Mast, 1999: Methane flux in subalpine wetland and unsaturated soils in the southern Rocky Mountains. *Global Biogeochemical Cycles*, **13(1)**, 101–113.

Wilson, J.O., P.M. Crill, K.B. Bartlett, D.I. Sebacher, R.C. Harriss, and R.L. Sass, 1989: Seasonal variation of methane emissions from a temperate swamp. *Biogeochemistry*, **8(1)**, 55–71.

Yavitt, J.B., 1997: Methane and carbon dioxide dynamics in *Typha latifolia* (L.) wetlands in central New York state. *Wetlands*, **17(3)**, 394–406.

Yavitt, J.B., G.E. Lang, and A.J. Sexstone, 1990: Methane fluxes in wetland and forest soils, beaver ponds, and low-order streams of a temperate forest ecosystem. *Journal of Geophysical Research*, **95(D13)**, 22463–22474.

Yavitt, J.B., R.K. Wieder, and G.E. Lang, 1993: CO_2 and CH_4 dynamcis of a *Sphagnum*-dominated peatland in West Virginia. *Global Biogeochemical Cycles*, **7(2)**, 259–274.

Zedler, J.B. and S. Kercher, 2005: Wetland resources: status, trends, ecosystem services, and restorability. *Annual Review of Environmental Resources*, **30**, 39–74.

APPENDIX G REFERENCES

DaSilva, A., C. Young, and S. Levitus, 1994: *Atlas of Surface Marine Data 1994, Vol. 1: Algorithms and Procedures*. NOAA Atlas NESDIS 6, U.S. Department of Commerce, Washington, DC. v6, 83pp.

Wanninkhof, R., 1992: Relationship between wind speed and gas exchange. *Journal of Geophysical Research*, **97(C5)**, 7373-7382.

PHOTOGRAPHY CREDITS

Cover/Title Page/Table of Contents

Cover, table of contents, pg 95, Image for Chapter 9 (Aerial view), Grant Goodge, STG Inc., Asheville, N.C..

Executive Summary

Page 1, (CP & L stacks), Grant Goodge, STG. Inc., Asheville, N.C.

Page 7, (Autumnal pond), Dave McGuirk, NOAA, NCDC, Asheville, N.C.

Page 13, (Power plant), Grant Goodge, STG. Inc., Asheville, N.C.

Chapter 1

Page 19, (Nuclear power plant), Deborah Misch, STG Inc., Asheville, N.C.

Chapter 2

Page 21 (Inversion), Grant Goodge, STG. Inc., Asheville, N.C.

Page 24 (Low visibility), Grant Goodge, STG. Inc., Asheville, N.C.

Page 25 (Aerial view), Grant Goodge, STG. Inc., Asheville, N.C.

Page 27, (Creek), Sara Veasey, NOAA, NCDC, Asheville, N.C.

Chapter 3

Page 35, (Aerial view), Grant Goodge, STG. Inc., Asheville, N.C.

Chapter 4

Page 39, (Palm Springs Wind Farm), Grant Goodge, STG. Inc., Asheville, N.C.

Page 46, (Aerial view Fontana Dam), Grant Goodge, STG. Inc., Asheville, N.C.

Chapter 6

Page 67, (Plume smoke), Grant Goodge, STG. Inc., Asheville, N.C.

Page 70, (Coal fired plant), Grant Goodge, STG. Inc., Asheville, N.C.

Chapter 7

Page 83, (Jet plane), Grant Goodge, STG. Inc., Asheville, N.C.

Chapter 8

Page 85, (Paper mill aerial), Grant Goodge, STG. Inc., Asheville, N.C.

Page 89, (Clear cut aerial), Grant Goodge, STG. Inc., Asheville, N.C.

Chapter 9

Page 95, Chapter Heading, Aerial urban view, Grant Goodge, STG. Inc., Asheville, N.C.

Page 95, (San Francisco, Calif.), Grant Goodge, STG. Inc., Asheville, N.C.

Page 96, 99, 102 (Manhatten, NY), Anne Waple, STG. Inc., Asheville, N.C.

Chapter 11

Page 117, (Trees and contrails), Grant Goodge, STG. Inc., Asheville, N.C.

Page 121, (Damaged shrub), Deborah Misch, STG. Inc., Asheville, N.C.

Page 122, (Rockies and yellow flowers), Dave McGuirk, NOAA, NCDC, Asheville, N.C.

Chapter 12

Page 136, (Alaskan Pipeline), Grant Goodge, STG. Inc., Asheville, N.C.

Chapter 13

Page 139, (Cattails, wetlands), Grant Goodge, STG. Inc., Asheville, N.C.

Chapter 14

Page 152,153, (Aerial building and traffic), Grant Goodge, STG. Inc., Asheville, N.C.

Page 154,155, (Urban trees), Sara Veasey, NOAA, NCDC, Asheville, N.C.

Appendix A

Page 167, (Mountain waterfall), Dave McGuirk, NOAA, NCDC, Asheville, N.C.

Appendix F

Page 184, (Wetlands), Grant Goodge, STG. Inc., Asheville, N.C.

Contact Information

Global Change Research Information Office
c/o Climate Change Science Program Office
1717 Pennsylvania Avenue, NW
Suite 250
Washington, DC 20006
202-223-6262 (voice)
202-223-3065 (fax)

The Climate Change Science Program
incorporates the U.S. Global Change Research
Program and the Climate Change Research
Initiative.

To obtain a copy of this document, place
an order at the Global Change Research
Information Office (GCRIO) web site:
http://www.gcrio.org/orders

Climate Change Science Program and the Subcommittee on Global Change Research

William J. Brennan, Chair
Department of Commerce
National Oceanic and Atmospheric Administration
Acting Director, Climate Change Science Program

Jack Kaye, Vice Chair
National Aeronautics and Space Administration

Allen Dearry
Department of Health and Human Services

Jerry Elwood
Department of Energy

Mary Glackin
National Oceanic and Atmospheric Administration

Patricia Gruber
Department of Defense

William Hohenstein
Department of Agriculture

Linda Lawson
Department of Transportation

Mark Myers
U.S. Geological Survey

Jarvis Moyers
National Science Foundation

Patrick Neale
Smithsonian Institution

Jacqueline Schafer
U.S. Agency for International Development

Joel Scheraga
Environmental Protection Agency

Harlan Watson
Department of State

EXECUTIVE OFFICE AND OTHER LIAISONS

Melissa Brandt
Office of Management and Budget

Stephen Eule
Department of Energy
Director, Climate Change Technology Program

Katharine Gebbie
National Institute of Standards & Technology

George Banks
Council on Environmental Quality

Gene Whitney
Office of Science and Technology Policy

www.ingramcontent.com/pod-product-compliance
Lightning Source LLC
Chambersburg PA
CBHW080636180526
45168CB00008B/3185